ARCTIC OCEAN
Franz Josef Land
Svalbard
Barents Sea
Novaya Zemlya
Kara Sea
Laptev Sea
East Siberian Sea
Arctic Circle
SWEDEN
FINLAND
NORWAY
RUSSIA
ASIA
Baltic Sea
EUROPE
Lake Baikal
Sea of Okhotsk
Kamchatka Peninsula
Bering Sea
Kuril Islands
NORTH PACIFIC OCEAN
Caspian Sea
Aral Sea
Black Sea
Sea of Japan
CHINA
JAPAN
Mediterranean Sea
IRAN
PAKISTAN
BANGLADESH
East China Sea
Persian Gulf
Red Sea
MYANMAR
Tropic of Cancer
Gulf of Oman
OMAN
INDIA
Arabian Sea
Bay of Bengal
THAILAND
VIETNAM
Philippine Sea
AFRICA
South China Sea
PHILIPPINES
NIGERIA
CAMBODIA
SRI LANKA
SOMALIA
Equator
Sumatra
Borneo
GABON
INDONESIA
New Guinea
Soloman Islands
TANZANIA
MOZAMBIQUE
ANGOLA
INDIAN OCEAN
MADAGASCAR
NAMIBIA
Tropic of Capricorn
AUSTRALIA
SOUTH AFRICA
Tasman Sea
Great Australian Bight
NEW ZEALAND
Tasmania
Prince Edward Islands
Crozet Islands
Kerguelen Islands
Antarctic Circle
Ross Sea

NATIONAL AUDUBON SOCIETY

Guide to MARINE MAMMALS *of the World*

A Chanticleer Press Edition

NATIONAL AUDUBON SOCIETY

Guide to MARINE

ALFRED A. KNOPF ▪ NEW YORK

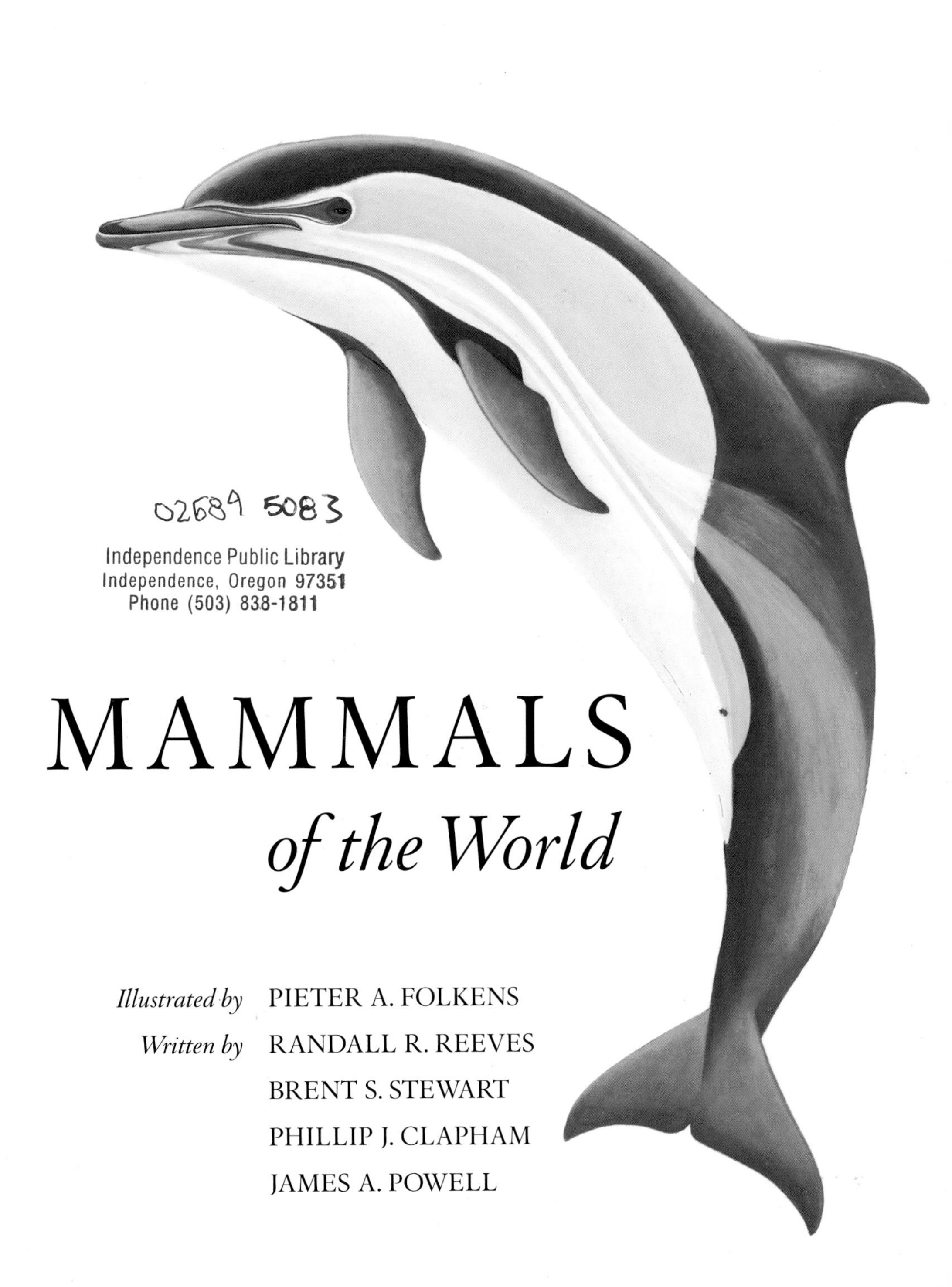

MAMMALS
of the World

Illustrated by PIETER A. FOLKENS
Written by RANDALL R. REEVES
BRENT S. STEWART
PHILLIP J. CLAPHAM
JAMES A. POWELL

This is a Borzoi Book.
Published by Alfred A. Knopf, Inc.

www.randomhouse.com

Prepared and produced by Chanticleer Press, Inc., New York.

Printed and bound by Dai Nippon Printing Co., Ltd., Hong Kong.

First Edition
Published March 2002

Library of Congress Cataloging-in-Publication Data

National Audubon Society guide to marine mammals of the world /
illustrated by Pieter Arend Folkens; written by Randall R.
Reeves . . . [et al.]. — 1st ed.
p. cm.
"A Chanticleer Press edition."
ISBN 0-375-41141-0
1. Marine mammals. 2. Marine mammals—Identification. I.
Title: Guide to marine mammals of the world. II. Folkens,
Pieter A. III. Reeves, Randall R. IV. National Audubon Society.
QL713.2 .N37 2002
599.5—dc21

Illustrations by Pieter A. Folkens: page 1, Killer Whale; page 3, Long-beaked Common Dolphin; page 6, Polar Bear (top), Northern Fur Seals (bottom); page 7, Stejneger's Beaked Whale (top), Amazonian Manatee (bottom).

National Audubon Society

The mission of NATIONAL AUDUBON SOCIETY, *founded in 1905, is to conserve and restore natural ecosystems, focusing on birds, other wildlife, and their habitats for the benefit of humanity and the earth's biological diversity.*

One of the largest, most effective environmental organizations, AUDUBON has nearly 550,000 members, numerous state offices and nature centers, and 500+ chapters in the United States and Latin America, plus a professional staff of scientists, educators, and policy analysts. Through its nationwide sanctuary system AUDUBON manages 160,000 acres of critical wildlife habitat and unique natural areas for birds, wild animals, and rare plant life.

The award-winning *Audubon* magazine, which is sent to all members, carries outstanding articles and color photography on wildlife, nature, environmental issues, and conservation news. AUDUBON also publishes *Audubon Adventures,* a children's newsletter reaching 450,000 students. Through its ecology camps and workshops in Maine, Connecticut, and Wyoming, AUDUBON offers nature education for teachers, families, and children; through *Audubon Expedition Institute* in Belfast, Maine, AUDUBON offers unique, traveling undergraduate and graduate degree programs in Environmental Education.

AUDUBON sponsors books and on-line nature activities, plus travel programs to exotic places like Antarctica, Africa, Baja California, the Galápagos Islands, and Patagonia. For information about how to become an AUDUBON member, to subscribe to *Audubon Adventures,* or to learn more about any of our programs, please contact:

NATIONAL AUDUBON SOCIETY
Membership Dept.
700 Broadway
New York, NY 10003
(800) 274-4201 or (212) 979-3000
www.audubon.org

Contents

Introduction *8*

What Is a Marine Mammal? 10
Range and Habitat 14
Behavior 17
Reproduction 23
Food and Foraging 26
Status and Conservation 29
Watching Marine Mammals 32
Organization of the Guide 34

Polar Bear and Otters *36*

Polar Bear 38
Otters 42

Pinnipeds *49*

Eared Seals 58
Walrus 110
True Seals 114

Cetaceans *180*

Baleen Whales 184
Sperm Whales 238
Beaked Whales 248
River Dolphins 299
Beluga and Narwhal 316
Ocean Dolphins 326
Porpoises 452

Sirenians *474*

Dugong 478
Manatees 482

Appendices *493*

Regional Assemblages 494
Marine Mammal Morphology 499
Illustrated Glossary 500
Photo Credits 518
Index 522
Contributors 526
Acknowledgements 527

Introduction

Humans and marine mammals have a long history of interaction. Written references to this remarkable group of animals go back millennia. In the 2nd century A.D., the Greek poet Oppian recorded his delight at the antics of dolphins, and there are descriptions of whale hunts by the Phoenicians in the Mediterranean Sea as early as 1,000 B.C. While in many places marine mammals were for centuries no more than a passing curiosity in the daily lives of coastal peoples, in others they were an important source of food and other life necessities and thus represented an integral part of the local culture.

Today, as in the past, the relationship between humans and marine mammals is defined by extremes. Commercial and aboriginal hunts continue to exploit whales, dolphins, seals, and others, with more than one species driven to extinction over the past 250 years and the fate of others hanging in the balance. In recent years, however, human interest in marine mammals has increasingly been driven by a desire to observe and appreciate these animals rather than to exploit them. Every year, millions of tourists all over the world join excursions to watch marine mammals, and such ventures have become a burgeoning industry in places as far apart as New England and New Zealand.

Marine mammals are a diverse and fascinating group, and watching them in their natural habitats can bring the patient observer considerable enjoyment. It is relatively easy to observe some species from a

Atlantic Spotted Dolphins, Bahamas

simple coastal lookout without ever boarding a boat. Seals, sea otters, manatees, and even some dolphins and whales are in many areas that are readily accessible to the land-bound observer. Other species, however, are difficult or impossible to find, even by the scientists who study them. For example, there are at least 20 species of beaked whale, collectively constituting more than one-fourth of the recognized species of cetaceans (whales, dolphins, and porpoises). However, most beaked whales live reclusively in deep water far from land, and remarkably, several species of beaked whale have yet to be observed alive.

Similarly, the ease with which different species can be identified in the field varies considerably. While few people with an interest in natural history would fail to identify a Walrus or a Polar Bear, correctly distinguishing among the many species of seals or dolphins can present a challenging task for even a professional observer. This guide is intended to introduce the reader to the appearance and habits of the many species of mammals that live much or all of their lives in marine environments. The descriptions of the species vary in length and detail, according to how much is known about them. Scientists have learned a great deal about many marine mammals, especially those that inhabit environments close to land; however, our knowledge of others is much more limited, and in the case of a few rarely glimpsed species of cetaceans it is virtually nonexistent.

What Is a Marine Mammal?

In the strictest sense, a marine mammal would be defined as any mammal that makes the sea its home for part or all of its life. This category includes cetaceans (whales, dolphins, and porpoises), pinnipeds (seals, sea lions, and the Walrus), sirenians (the Dugong and the manatees), marine mustelids (the Sea Otter and the Marine Otter), and the Polar Bear. Most species included in this book easily meet this criterion. Whales are obviously marine mammals, as are dolphins, porpoises, seals, sea lions, manatees, the Dugong, and the Sea Otter. Many people might be surprised to learn that the Polar Bear is considered a marine mammal, yet this animal spends a great deal of its time in marine habitats, albeit the "water" concerned exists in frozen form as sea ice.

A few species, however, are included in this book despite the fact that their habitat is not, in the strictest sense of the word, marine. As their name suggests, most river dolphins never enter marine environments, and both the Baikal Seal and the Caspian Seal inhabit bodies of water that, despite their considerable size, are landlocked. Nonetheless, all of these animals are closely related to other species that are marine mammals in the more traditional sense.

CLASSIFICATION Biologists divide living organisms into major groups called phyla. The phylum Chordata includes the vertebrates, which encompasses mammals, fishes, birds, amphibians, and reptiles. Phyla are divided into classes, classes into orders, orders into families, families into genera (singular: genus), and genera into species. The species is the basic unit of classification and is generally what we have in mind when we talk about a "kind" of animal. Subspecies, or races, are populations of a species that differ from one another in some way or are isolated geographically and that may be developing into new species.

Marine mammals, as well as terrestrial mammals, are in the class Mammalia, which includes more than 4,000 species in 25 orders. Mammals are warm-blooded and have hair; female mammals give birth to live young and have mammary glands that secrete milk for the nourishment of offspring.

The classification of marine mammals, like that of all living things, follows the general rules of taxonomy. In this system, begun by the 18th-century Swedish naturalist Carl Linnaeus, species are grouped or separated by the degree to which they are thought to be related in terms of

The Sperm Whale is one of nature's many remarkable creations. It has managed to occupy most of the planet, occurring in the tropics as well as in cold polar waters. It dives to the ocean floor and lives in a complex, highly structured society.

evolutionary descent. For many years, the determination of such relationships was largely dependent upon similarities or differences in anatomy and morphology (form and structure). In recent years, however, DNA analysis has become increasingly important in assessing the evolutionary relatedness of species.

It is important to recognize that the current taxonomy of marine mammals is not set in stone. Central to this issue is the question of what is a "species." The traditional definition, called the "biological species concept," states that two animals are of the same species if they can mate and produce fertile offspring. More recently, many biologists have adopted the "phylogenetic species concept," according to which populations of animals that have been separated for long enough to evolve significant genetic differences are considered separate species, even though they might produce fertile offspring if they were to breed.

New genetic data have suggested taxonomic revisions for many animal groups, and marine mammals have not been exempt from such debates. For example, recent genetic analysis has indicated strong support for the separation of the right whales into three species—the North Atlantic, North Pacific, and Southern Right Whales; previously, right whales were regarded as either two species (northern and southern) or three subspecies within a single, closely related group. This new classification is a result of the phylogenetic definition of a species. Although

OPPOSITE: *The two main groups of cetaceans are the mysticetes, or baleen whales, represented here by a Humpback Whale calf (left), and the odontocetes, or toothed whales, represented by Common Bottlenose Dolphins (right).*

ABOVE: *Among the smaller whales are Cuvier's Beaked Whale (left) and the Beluga, or White Whale (right).*

North Atlantic and North Pacific Right Whales might well be able to successfully reproduce if they came into contact, it is clear from genetic analysis that they do not, and indeed that they have not mixed genetically for a very long time. The implication is that these two populations are now on separate evolutionary paths and thus should be considered separate species.

The taxonomic status of some species covered in this guide will undoubtedly change, to a greater or lesser extent, with additional genetic and other research. It is almost certain that Bryde's Whale will shortly be split into at least two species, based upon both morphological and genetic analyses. In the pinnipeds, DNA studies have recently shown remarkable similarity among the Southern Hemisphere fur seals, calling into question their division into different species. Many other examples of taxonomic uncertainty exist for marine mammals, and any field guide will inevitably become outdated quickly in this respect.

EVOLUTION All marine mammals are believed to have evolved at various times from land-dwelling ancestors. We can only speculate on the reasons behind the movement of terrestrial animals into aquatic environments and their eventual evolution into species that spend some or all of their lives in the water. Such animals were almost certainly taking advantage of new food sources and were perhaps also seeking escape from terrestrial predators. The major groups of marine mammals are united much more by their shared use of a similar environment than by a common evolutionary descent. While cetaceans, seals, and sirenians have different evolutionary origins, they have developed similar adaptations to life in the aquatic realm. Among the most important of these adaptations are the modification of limbs into flippers, the development of thick fatty insulation in the form of blubber, and a range of complex diving abilities. In addition, the cetaceans, in particular, have evolved powerful tails for propulsion.

Among marine mammals, the cetaceans are the most completely adapted to an aquatic existence. Their transformation began 55 to 60 million years ago, and the cetaceans were already well diversified by 53 million years ago. The two modern major cetacean groups (suborders) are the baleen whales (mysticetes) and the toothed whales (odontocetes). Most scientists accept that these groups share a common origin in a third, extinct suborder, the archaeocetes, which disappeared about 30 million years ago. The closest living relatives of the cetaceans are the artiodactyls (even-toed ungulates such as cows and camels), and recent genetic and paleontological studies suggest that cetaceans are more closely related to hippopotamuses than to any other terrestrial mammal. Whether cetaceans evolved from ancient artiodactyls or diverged (with the hippo) from another group of ancient animals is still the subject of much debate.

The origin of the pinnipeds is more recent than that of the cetaceans. Pinnipeds comprise three major groups of carnivores (order Carnivora): true seals (phocids), eared seals (otariids), and the Walrus (odobenids). There has long been

LEFT: *The three families of pinnipeds, or "fin-footed" carnivores: Phocidae, represented by a Weddell Seal mother and pup under a breathing hole in the Antarctic (opposite left); Otariidae, such as this young South American Fur Seal (opposite right); and Odobenidae, the Walrus (left).*

BELOW: *The three other major groups of marine mammals: order Sirenia, one member of which is the West Indian Manatee (left); carnivore family Mustelidae, represented here by the Sea Otter (middle); and carnivore family Ursidae, whose sole marine member is the Polar Bear (right).*

disagreement about whether these groups share a common evolutionary origin. One view is that all pinnipeds evolved from a common carnivorous, bear-like ancestor, from which they diverged approximately 25 million years ago. Another view maintains that the phocids and the otariids and odobenids evolved separately about 20 million years ago, the phocids from an early otter-like ancestor and the otariids and odobenids from an early bear-like form. Certainly, there are striking similarities in the morphology of otariids and bears. However, while pinnipeds and bears may represent "sister" groups within the order Carnivora, recent DNA analysis has strongly suggested that the phocids and otariids are genetically much closer to each other than to any other carnivore.

The origin of the sirenians is not entirely clear, but the earliest fossils date back approximately 50 million years and suggest that this group evolved on the shores of the prehistoric Tethys Sea between Africa and Eurasia. Both morphological and molecular studies indicate that sirenians share a common evolutionary origin with elephants, aardvarks, and hyraxes.

The otters comprise some 13 or 14 species worldwide, but only two (the Marine and Sea Otters) feed exclusively in marine waters. Otters are mustelids, members of a diverse family of small carnivores that includes, among many others, weasels, skunks, and badgers. The Sea Otter appears to have evolved from an ancestor named *Enhydritherium*, which lived (perhaps in fresh water) in Europe and North America between about 5 and 10 million years ago.

Finally, the Polar Bear appears to have a relatively recent origin, and DNA analysis has shown it to be much more closely related to the Brown Bear than to any of the other bear species in the world. Indeed, the Polar Bear and the Brown Bear are known to hybridize occasionally, a fact that further confirms their close relationship. It has been suggested that the Polar Bear evolved quite rapidly from an isolated high-latitude population of Brown Bears, probably no earlier than 250,000 years ago.

Range and Habitat

Marine mammals are remarkably successful animals that have colonized a broad range of habitats, from major rivers and coastal areas to the deep ocean. Some species, such as river dolphins and the Hawaiian Monk Seal, have a very restricted distribution. Others, like Humpback and Sperm Whales, occur worldwide. In general, the distribution of any species is determined by the resources it needs to survive, the occurrence of which may vary by season. With the exception of the Polar Bear, the Amazonian Manatee, and some baleen whales, all marine mammals need to eat for most or all of the year; thus their own distribution is closely related to that of their food. In addition, some species have specific habitat requirements for giving birth; in particular, pinnipeds and otters need to give birth on land or on ice.

Several species have a very restricted distribution. For example, as their name implies, the river dolphins occur largely or exclusively in some of the world's major rivers. Similarly, the Amazonian Manatee is confined to the freshwater Amazon River and its tributaries. The Caspian and Baikal Seals occur only in those two large inland bodies of water. Monk seals, which once were wide-ranging, today have limited ranges along the coasts of Hawaii, the Mediterranean Sea, and northwestern Africa.

Other species have a broader distribution but are confined to one hemisphere or ocean basin. The Bowhead Whale occurs only in arctic waters, and the Gray Whale and Sea Otter exist only in the North Pacific Ocean. No pinniped occurs in both the Northern and Southern Hemispheres, and many cetaceans are confined to a range either north or south of the equator, but not both.

DISTRIBUTION TYPES Several terms are used to describe different types of distribution. Animals with a circumglobal, or *cosmopolitan,* distribution are those that are found all over the world. Many of the great whales have a cosmopolitan distribution. The Humpback Whale, for example, occurs in all major oceans and at various times of the year can be found from the tropics to the poles. *Circumpolar* refers to a distribution in high latitudes around one of the poles. Marine mammals that are circumpolar, in either the Northern or the Southern Hemisphere (but not in both) include the Bowhead Whale, Narwhal, Beluga, Southern Right Whale Dolphin, Hourglass Dolphin, Arnoux's Beaked Whale, Polar Bear, and Ringed, Crabeater, Weddell, Southern Elephant, and Ross Seals. Some populations of cetaceans have a circumpolar distribution during only part of the year; these

LEFT: *Some Northern Hemisphere seals, such as these Baikal Seals in Lake Baikal, Russia, have adapted to living in landlocked bodies of water.*

RIGHT: *Amazonian Manatees are confined entirely to freshwater lakes and rivers in South America.*

The Atlantic Spotted Dolphins on the Bahama Banks (left) and Indo-Pacific Bottlenose Dolphins in coastal shallows of Shark Bay, Australia, (above) experience an underwater world that is much more varied than it appears from the surface.

include populations of the Humpback, Fin, Killer, and male Sperm Whales.

Species that occur in tropical waters on both sides of the equator are said to have a *pantropical* distribution. These include such cetaceans as the Melon-headed and Bryde's Whales and Fraser's Dolphin. In Sperm Whales, the distribution of females and juveniles is centered in tropical (and sometimes subtropical) waters, whereas mature males wander widely and penetrate even polar waters. A largely or exclusively tropical distribution that does not span the globe is characteristic of some other species, including the Hawaiian Monk Seal and all of the sirenians.

A *coastal* distribution denotes an occurrence close to coasts and often includes adjacent waters over the continental shelf. Highly coastal species may also utilize river estuaries, occasionally foraging in fresh water or around mudflats. The Sea Otter and the Marine Otter occur almost exclusively in coastal waters. Many marine mammals have a coastal distribution for part or all of their lives; these include many species of dolphins, porpoises, and pinnipeds, as well as some of the baleen whales. For other species, coastal and shelf waters may represent only a portion of their distribution, based on the seasons or other factors; for example, while many Humpback Whales spend much of the year close to coasts, they also undertake long migrations across deep ocean waters, and some feed in these more remote habitats.

Species that occur in the open sea, either year-round or for only a portion of the year, are said to be *pelagic*. The Sperm Whale and many beaked whales are truly pelagic animals, rarely coming near land except in places where the continental shelf is narrow and deep waters abut the coastline. Elephant seals and some other pinnipeds spend much of their year in the pelagic realm, as do many dolphins and baleen whales.

Any marine mammal whose distribution is partly or exclusively tied to ice is said to be *pagophilic*, or "ice-loving." Many of the pinnipeds breed and/or feed on or around ice. The Bowhead Whale, which spends much of its life in partly frozen waters, can travel considerable distances under ice, using its huge head to break through sea ice that is several feet thick. The Beluga and Narwhal also spend much time in ice (and occasionally become trapped and die there). It is common to find aggregations of several polar marine mammal species in semipermanent areas of open water known as polynyas. As noted above, the Polar Bear spends much of its life on sea ice; its remarkable insulation allows it to swim considerable distances in the frigid open water between ice floes.

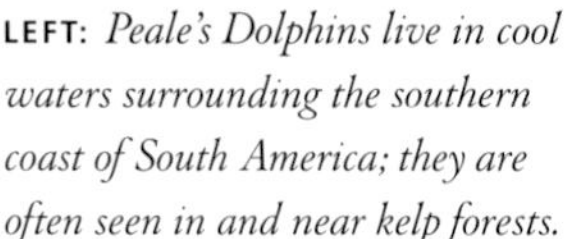

LEFT: *Peale's Dolphins live in cool waters surrounding the southern coast of South America; they are often seen in and near kelp forests.*

RIGHT: *The Polar Bear is pagophilic, or "ice-loving," with a circumpolar distribution in the Arctic.*

Within a single species of marine mammal it is not unusual to see marked variation in patterns of distribution or migration among different populations. For example, there are several types of Common Bottlenose Dolphins that inhabit very different habitats; some occur in oceanic waters, others in coastal or even estuarine areas. While as a rule Humpback Whales make extensive seasonal migrations between high and low latitudes, one population of this species appears to live year-round in the warm waters of the Arabian Sea.

EXTRALIMITAL RECORDS Confirmed observations of a species well outside its normal range are referred to as extralimital records. For example, in the spring of 2001 there was a sighting of a young Humpback Whale in the Mediterranean Sea off the Greek coast, an area where, with very few exceptions, Humpbacks are not known to occur. Recent records of Hooded Seals from the Caribbean and southern California are even more remarkable. The animals involved are sometimes referred to as vagrants or strays. In most cases, extralimital records are assumed to represent "lost" animals, although in the case of some poorly understood species they may indicate that the "normal" distribution is actually broader than we think.

MIGRATION For many marine mammals, the year is divided into distinct breeding and feeding seasons, during which the animals' distribution is quite different. In some cases, these two periods of their annual cycle are joined by a seasonal migration. Some of the baleen whales migrate huge distances from summer feeding areas in high latitudes to winter breeding and calving grounds in tropical or subtropical waters; in the case of Gray and Humpback Whales, these migrations may be 5,000 miles (8,000 km) each way. The movements of other migratory marine mammals are less dramatic and, in many cases, rather poorly understood.

The function of migration is also often unclear. Most animals migrate to take advantage of greater seasonal availability of food resources at their migratory destinations, and this is a good explanation for the extensive seasonal movements of some pinnipeds (such as elephant seals). However, food is clearly not the reason behind the extraordinary transoceanic treks of the migratory baleen whales, which fast during the winter. Their migrations may relate to the thermodynamic advantages of calving and overwintering in warm water where animals may have to devote less energy to keeping warm. However, there remains much dispute about this issue.

Behavior

The social organization of marine mammals ranges widely, from the highly complex societies found in many of the toothed whales to the largely unstructured, solitary ways of baleen whales, some pinnipeds, sirenians, the Marine Otter, and the Polar Bear. Marine mammal behavior is similarly varied.

SOCIAL ORGANIZATION Scientists use various terms to describe the different groupings of marine mammals. An aggregation may be thought of as an otherwise unassociated gathering of animals brought together around a common resource, such as food or breeding habitat. By contrast, the terms group, pod, or school usually imply some degree of association and coordination of movement among the animals. It is often difficult to assess the social structure of marine mammals. An observer on a ship may at one moment pass a school of oceanic dolphins thousands strong and nearby see a lone Blue Whale, apparently isolated from any other member of its kind. Yet appearances may be deceptive. The dolphin "school" is probably a temporary gathering that will break down into more discrete, enduring groups, while the Blue Whale may be in acoustic communication with one or more other Blue Whales many miles away.

As is the case with mating systems (see page 23), the social organization of any species has evolved to meet the specific needs of the animals concerned. In understanding animal societies, it is important to recognize two major principles. First, all social behavior has developed as a trade-off between the costs and the benefits of group living versus solitary living. Second, any social system is the sum of the behaviors of the system's individual members. This is because, as a species evolves, selection works upon individuals, not upon groups or societies. The social organization of any animal population reflects the influence of various ecological factors. In particular, group size and social behavior are strongly influenced by both pressure from predators and the distribution of resources. If animals of a particular species tend to be found in stable groups that remain together for a long time—or, conversely, if the animals are usually found alone—there is always a good reason, even though we may not know what it is.

The societies of Killer and Long-finned Pilot Whales are quite extraordinary, in that groups seem to remain together for the entire lives of the individuals in them. Evidently both of these species have developed highly efficient cooperative hunting strategies that depend upon long-term bonds among the group members. In these and perhaps some other toothed whales, most or all members of a group are closely related, and presumably breeding occurs when

Pilot whales, such as these Long-finned Pilot Whales near the Azores in the eastern North Atlantic Ocean, are among the most social cetaceans, living in stable groups that are tightly bonded. Unfortunately, hunters at the Faeroe Islands exploit this behavior by driving large numbers of pilot whales ashore.

Many pinniped species, especially fur seals and sea lions, aggregate in large numbers on limited stretches of beach during the breeding and molting seasons. These South African Fur Seals form rookeries where "harem bulls" maintain territories and females give birth and nurse their pups.

unrelated pods come together. Toothed whales in general tend to have more complex societies involving stable or semi-stable bonds, often between related individuals; this reflects not only cooperation in hunting, but also the advantages of group defense against predators such as sharks. It is not uncommon in some toothed whale societies, such as that of Sperm Whales, for animals (notably males) to leave their natal group when they reach sexual maturity.

Social bonds in other marine mammals tend to be brief. Pinnipeds, sirenians, and otters usually forage alone, although some species aggregate in large numbers for breeding or other purposes. Among baleen whales, the only longer-term relationship is that of a mother and her calf, which typically lasts six months to a year. Although large aggregations of baleen whales may be found in prime feeding areas, individual whales typically spend much of their time alone, or they associate with other animals for relatively short periods. This fluid social organization reflects the nature of a baleen whale's prey: Schools of fish or krill or patches of plankton are variable in size, and their exploitation usually does not require long-term cooperation. Furthermore, since baleen whales are not under constant threat of predation, there is little value (and probably considerable cost) to maintaining large stable groups.

COMMUNICATION Marine mammals live in a world in which it is difficult to see. Even close to the surface in the clearest tropical water visibility is restricted, and sunlight does not penetrate below depths of a few hundred feet. Consequently, visually based communication is ineffective under the water except at close range. Instead, many marine mammals communicate primarily by sound, which travels some five times faster in water than in air. When observing marine mammals at sea, it is well worth investing in an inexpensive hydrophone (underwater microphone), in order to listen in.

The sounds made by marine mammals vary considerably in type, frequency, and strength. A hydrophone dropped into the water near a school of dolphins will immediately pick up a barrage of sounds. These include long trains of rapid clicks made as the animals echolocate, as well as a variety of whistles used for communication within the group. Some dolphins have "signature" whistles that uniquely identify individuals. A similar phenomenon exists in Sperm Whales, which produce patterns of sound known as "codas" that serve as acoustic signatures for each whale in a group. Killer Whales have distinctive "dialects" that distinguish a particular family group; in well-studied populations of this species, biologists can often tell which pod is passing by without actually seeing it.

The great whales make the loudest sounds. Male Humpback Whales sing long, complex songs for hours at a time that are so loud they can sometimes be heard above the water's surface, an unusual phenomenon given that the interface between water and air generally presents an effective barrier to sound. In deep water, Humpback songs can be heard from several miles away,

LEFT: *Mother fur seals and their pups use individually distinctive calls to identify one another in the midst of what appears, from a human perspective, to be acoustic bedlam. This female Northern Fur Seal may be calling for her pup.*

BOTTOM: *Beaches inhabited by fur seals and sea lions can be noisy places. Two Guadalupe Fur Seals threaten each other (left). A Galápagos Sea Lion announces his presence (right).*

but this broadcasting pales in comparison to the vocalizations of the Blue and Fin Whales. Although both of these giants make sounds that are below the range of human hearing, with the right equipment their low-frequency booms can be detected over distances of up to 2,000 miles (3,200 km). It is not yet clear whether these extraordinary vocalizations represent long-distance communication or, as some have suggested, a sonar navigation system that allows a whale to detect underwater features like seamounts from hundreds of miles away.

Acoustic communication is also an important part of the social repertoire for marine mammals that spend time on land. For example, female pinnipeds use vocalizations to find their pups on crowded breeding beaches, and the roars, grunts, or other sounds of male pinnipeds are important parts of their threat and territoriality displays.

Communication need not be acoustic in nature to be effective. Visual displays are often used, especially by pinnipeds, to communicate aggression or defense, and many cetaceans employ postural displays to communicate various states. Male Humpback Whales frequently fight over females during the breeding season, contests that often involve threat displays in which the animal inflates its ventral pouch to make itself seem larger, or blows long streams of bubbles to warn or confuse other males. As in other animals, territoriality is found in those species for which defense of a specific resource is likely to increase reproductive success. Therefore, cetaceans are usually not strictly territorial, since their "resources" consist primarily of mobile prey that is not confined to a specific, defensible space. Territoriality is much more common in pinnipeds; females will stake out prime spots on breeding beaches, and males will aggressively defend those same areas to prevent other males from gaining access to the females within them.

SURFACING, DIVING, AND HAULING OUT Our views of most marine mammals represent no

However many times one might read of breaching or fluking on a printed page, one is never prepared for the thrill of seeing close up a Humpback Whale leap (left) or a Southern Right Whale wave its majestic tail high above the surface (right).

more than brief glimpses of the animals as they surface to breathe. Because they are mammals like we are, we sometimes think of them as surface-living creatures that dive from time to time; in fact, cetaceans and some seals are quite the opposite, passing most of their lives in an aquatic world that is largely foreign to humans, and returning to the surface just long enough to replenish their oxygen supply before descending once more into the depths. We see a blow, a rolling back, perhaps a tail raised high in the air; then the water closes once more, leaving nothing more than some swirls at the surface.

When not diving or hauled out (that is, on land), seals usually show no more than their heads above the surface. Dolphins, porpoises, and whales typically take a number of breaths at the surface, separated by shallow dives of a few seconds or longer, and then make a deep dive. Most large whales arch their backs quite prominently just before they dive. Some (like the Humpback) usually raise their flukes when diving, while others (like the Fin Whale) do not. All whales leave a "footprint" at the surface, a circular swirl made by turbulence from the movement of the flukes. It is sometimes possible to follow a whale moving invisibly just below the surface by watching for the trail of these swirls.

Most large whales and even some of the smaller ones create a visible "blow" when they breathe. The blow is not water, but rather a mixture of condensed air and atomized water droplets that varies in shape and height according to the species and the size of the individual. Right whales have a distinctly V-shaped spout, while large Blue Whales can throw up a columnar blow that may be up to 50 feet (15 m) tall and easily visible from 3 or 4 miles (5–6 km) away. However, variations in the blow among individuals of a species make it a rather unreliable means of species identification.

The speed at which marine mammals swim varies considerably, from the slow dog paddle of a Polar Bear at the surface to the feeding lunges of the Fin Whale, which can reach burst speeds exceeding 20 knots. Pinnipeds use their flippers for swimming in a variety of styles, while the cetaceans' efficient tails propel them forward with up-and-down movements rather than the side-to-side movements of fish.

Surface behavior varies widely. Many cetaceans breach (jump out of the water), a feat that in large whales can be truly spectacular. Some dolphins and whales will repeatedly slap their tails or flippers on the water surface, behaviors known as lobtailing and flipper slapping, for

reasons that are usually not clear. Many cetaceans will raise their heads above the water, presumably to look around, a behavior known as spyhopping. Some marine mammals seem to play in the water. Dolphins, pinnipeds, and even some large whales are adept at surfing in large waves, and some dolphin species are well known for their tendency to ride on the bow waves of ships.

As one would expect from animals that forage in the water, marine mammals have evolved the ability to dive with considerably more efficiency than terrestrial mammals. Of the various groups, the otters are probably the least accomplished in this regard, making short, shallow dives. Similarly, many of the smaller seals and toothed whales (such as dolphins and porpoises) typically dive for only a few minutes and to no great depth. Baleen whales, on the other hand, can remain submerged for as long 40 minutes and, in some cases, may reach depths of as much as 1,000 to 1,300 feet (300–400 m) while foraging, although they more typically exploit prey within a few hundred feet of the surface. The most remarkable examples of diving physiology are found in the Sperm Whale, the beaked whales, and some of the larger pinnipeds (such as the Weddell and elephant seals). In the case of the Sperm Whale, dives of more than two hours' duration to depths exceeding 6,500 feet (2,000 m) have been recorded.

Unlike cetaceans, most pinnipeds frequently haul out of the water onto beaches, ice, or rock ledges. Brief haulouts allow the animal to rest for short periods, while extended haulouts are required for mating, pupping, and molting activities. Pinnipeds molt at least annually to renew the skin, fur, and hair that provide them with critical waterproofing and insulation. During this time, they must spend long periods out of the water to keep warm and promote blood circulation; in some species, animals will huddle together during the molt to conserve energy.

The way in which pinnipeds move across land or ice varies by family. Phocid pinnipeds hunch their bodies or wriggle from side to side, sometime using their foreflippers. The eared seals and the Walrus can walk on their foreflippers.

ABOVE: *The whirling jumps of Spinner Dolphins, such as this individual off the Hawaiian Islands, enliven the seascape.*
RIGHT: *Gray Whales are not deep divers, even though most of their prey are captured on or near the sea bottom. They spend most of their time in relatively shallow water, whether migrating along continental coastlines or spreading out over the continental shelf to feed.*

Unlike true seals, eared seals are capable of walking, albeit somewhat awkwardly. This Juan Fernández Fur Seal hurries across a stretch of rocky ground on all four flippers.

STRANDING The appearance on shore of any living cetacean is not natural; such events are termed strandings. The stranding of live individual animals or groups is relatively common among some species of cetaceans. Without human intervention, strandings are usually fatal, since cetaceans are not adapted to life on land. Once out of the water, they overheat. Furthermore, since a cetacean's light bones cannot support the body when it is no longer in the gravity-free medium of the sea, the animal is sometimes literally crushed by its own weight.

Strandings can be divided into two types, those caused by sickness and those caused by navigational errors. Single strandings usually involve animals that are very ill and often close to death. Healthy baleen whales never strand; nearly all mysticete beachings involve whales that are on the verge of death from illness or serious injury. As a result, and also because baleen whales do not travel in stable groups, such events almost never involve more than a single whale. Similarly, terminally ill dolphins and porpoises usually strand alone, although in some cases a dependent calf will follow its sick mother up onto a beach.

Mass strandings, which usually involve three or more animals, occur more often than most people realize. The cetaceans that most commonly strand en masse are pilot whales, False Killer Whales, Sperm Whales, and some smaller odontocetes. Mass strandings probably do not result from illness, although some scientists have theorized that a sick lead animal may sometimes lead the rest of a group's members to their deaths. Most mass strandings are probably the result of several factors, including bad weather and the interaction between abnormally low tides and shallow, confusing topography. Whales foraging in deep water at high tide may suddenly find themselves in a maze of sandbars as the tide rapidly goes out. In some cases, this scenario is sufficient to cause a stranding, and if a storm is added to the mix it becomes even more difficult for the animals to reorient themselves and find their way back to the safety of deep water.

The annual molt is a critical time for pinnipeds. They haul out for long periods to conserve energy and promote rapid replacement of their pelage. Some, like this female Northern Elephant Seal on San Benito Island, Mexico, lose their hair in large clumps, called a "catastrophic" molt.

Reproduction

Marine mammals exhibit considerable variation in reproductive biology and mating strategies, which is not surprising given the diversity of their habitats. In general terms, the mating system of a species is determined by the distribution of the females, which in turn is based upon environmental constraints. For example, the limited availability of suitable pupping habitat forces pinniped females to aggregate in relatively confined areas, giving males the opportunity to monopolize groups of females. This has led to the evolution of mating systems that are heavily based upon competition among males, with the larger, stronger males tending to father a disproportionate number of offspring.

As is the case in most mammals, female marine mammals have evolved to invest more heavily in their young than the males of the species. Since male parental care is not required for the successful rearing of marine mammal offspring, males usually contribute nothing in this regard and instead attempt to maximize their mating opportunities. As a result, the most common mating system among marine mammals is polygyny, in which successful males mate with more than one female.

An extreme example of this type of mating system is found in elephant seals. A few large males may monopolize and inseminate the majority of breeding females on a pupping beach, often engaging in intense male-to-male competition and territoriality. Because the potential payoff in terms of male reproductive success is so great, large size in males has been strongly favored in the evolution of these species, leading to a huge difference in mass between male and female elephant seals.

By contrast, female baleen whales do not rely on such restricted areas for calving, nor are they tied to areas of food resources, since many baleen whales fast during the breeding season. The females tend to be widely distributed during the breeding season, and males must usually compete over single females rather than groups. Not surprisingly, therefore, far more male baleen whales contribute to the next generation of offspring than is the case in elephant seals. Furthermore, baleen whales are not territorial during the breeding season, as it is pointless to defend a territory if females are not tied to a restricted area.

In some species, males tend to compete with each other not by fighting, but rather by producing large volumes of sperm, a system called sperm

Humpback Whales migrate long distances to reach their preferred breeding grounds in shallow tropical waters. This mother, shown here with her calf near the Turks and Caicos Islands, would have spent the preceding summer on one of several feeding grounds in the northern North Atlantic Ocean.

ABOVE: *A male Spinner Dolphin swims upside-down beneath a female to mate. A lengthy research project on Spinner Dolphins in Hawaii revealed that multiple males may mate with a single female in very short sequences.* **OPPOSITE:** *This Northern Sea Lion bull appears to be in his reproductive prime, surrounded by his harem of females on a beach in Prince William Sound, Alaska.*

competition. Right whales provide the most dramatic example of this. At close to a ton, the testes of male right whales are the largest in the animal kingdom. Females of these species frequently mate serially or even simultaneously with more than one partner.

Male mating strategies in other marine mammal species include visual or acoustic displays to attract females; the best-known example of this is probably the complex song sung by male Humpback Whales. In some species (such as bottlenose dolphins), males cooperate in coalitions to compete for mates. In some marine mammals (for example, right whales), female mating strategies may include efforts to incite competition among males.

True monogamy (the repeated pairing of two animals to the exclusion of other partners) is virtually unknown in any mammal. In other animals (such as some birds), monogamy tends to occur in cases where partner bonding and parental care by both parents greatly improve the chance of offspring survival. Genetic analysis of Gray Seals has shown that some females will seek out the same mate in successive years, although the males themselves may be promiscuous. It has been suggested that in some pinniped species, partners remain together for a single season, a system known as serial monogamy.

Of all the marine mammals, pinnipeds tend to have the highest degree of synchrony in female

The transfer of energy from mother to pup must take place rapidly in the seals that breed on ice; the milk of Harp Seal mothers, like this one nursing her pup near the Magdalen Islands in Canada's Gulf of St. Lawrence, is more than 50 percent fat.

receptivity; indeed, in the Harp and Hooded Seals, all mature females in a particular breeding population are believed to become receptive within a period of only 10 to 15 days. This probably relates to the fact that the pupping habitats (on land or ice) required by female seals are spatially separated from their food sources in the water. By contrast, the peak of sexual receptivity of female sirenians is spread over about six months, reflecting the fact that they can both breed and feed in the water, and that their food source (plant matter) is available year-round. In baleen whales, breeding is also spread over a few months, although reproduction is usually strongly seasonal in winter, allowing females to time the most demanding periods of pregnancy and lactation so that they coincide with seasonal peaks in the productivity of their prey.

Like other large animals, marine mammals tend to produce single offspring (rather than multiple offspring as in smaller mammals) and to invest heavily in each. Gestation is also relatively long, varying from eight months in the case of some porpoises to as long as 16 months in the Sperm Whale. Most of the baleen whales are pregnant for close to a year, and the fetal growth rates of this group are the fastest in the animal kingdom (20 times that of primates). Pinnipeds and otters undergo an interesting phenomenon known as delayed implantation, in which the fertilized ovum does not begin development until well after mating has occurred. This probably relates to the need to synchronize mating and pupping; delayed implantation is common in species in which males and females are separated outside the breeding season, and where resource availability during the mating season is not optimal for actual development of the fetus.

The duration of lactation varies widely, from only four days in the Hooded Seal to several years in the case of some toothed whales. Prolonged lactation occurs only in species characterized by extended familial associations, such as the Sperm and Killer Whales and many dolphins, and seems to serve a social bonding function. By contrast, baleen whales, otters, and many sirenians usually wean their offspring after six to 12 months. The composition of marine mammal milk tends to be associated with the length of lactation: Pinnipeds that breed on ice, such as the Harp Seal, must raise their young rapidly in a matter of days and have milk with a fat content as high as 60 percent, while toothed whales, which have prolonged periods of lactation, have the lowest milk fat content, at between 10 and 30 percent.

Marine mammals become sexually mature (capable of reproducing) at a wide range of ages, with pinnipeds tending to mature relatively early, between about three to seven years of age. Many cetaceans mature somewhat later, although there is considerable variation in this. The Humpback Whale can mature by four years of age, while the Bowhead Whale typically does not reach maturity until 15 to 20 years old. In highly polygynous speçies, males may not become sexually mature until they are large enough to successfully compete with other males.

Food and Foraging

As a group, marine mammals have evolved to feed upon an astonishing range of prey species at virtually every level in the marine ecosystem. At one end of the scale, some of the baleen whales present one of the great paradoxes of the natural world, in which the largest animals on our planet subsist on some of the smallest. Right whales feed exclusively upon zooplanktonic organisms, which are typically the size of a grain of rice; that they can subsist on such tiny fare is testament to the extraordinary productivity of the ocean. Similarly, the Blue Whale, which is the largest animal ever to have lived on Earth, eats small, shrimp-like creatures called krill (more properly termed euphausiids). At the other end of the scale, the Killer Whale is a fearsome predator whose diet can include huge fast fish like bluefin tuna, as well as seals, dolphins, and even larger marine mammals. Indeed, Killer Whales sometimes successfully prey on the great whales, including even Blue and Sperm Whales.

It is difficult to generalize about the prey exploited by the different groups of marine mammals, but a few trends can be noted. Within groups, some species have a broad, catholic diet, while others specialize on a very limited range of prey. For example, the Humpback Whale feeds on krill and many different kinds of fish, while the Blue Whale feeds almost exclusively on krill. The term for such specialist foraging is stenophagy; thus the Blue Whale is said to be stenophagous on krill.

The baleen whales have evolved baleen, instead of teeth, to feed upon the most abundant food in the ocean, zooplankton and small schooling fish. Commonly eaten organisms include krill, copepods, herring, sand lance, capelin, menhaden, anchovies, and pelagic crabs; in addition, Gray and Bowhead Whales in particular feed upon a wide variety of invertebrates such as amphipods. The plates of the baleen form an efficient filtration system that separates prey from vast volumes of water taken into the mouth. The design of the plates and of the fringing hair on their inner surface varies from species to species according to the type of prey captured. Despite its capacious mouth, a baleen whale's throat is no more than a few inches across, a dimension that reflects the size of its prey. Baleen whales typically forage in the upper layers of the water column, preying on species that are found from the surface to a few hundred feet down.

As a rule, the odontocete diet consists of larger prey than that taken by the mysticetes. Toothed whales prey upon fish of varying sizes, including small species (such as herring and sand lance), moderate-size species (cod, salmon, and halibut), and large species (sharks and tuna). Cephalopods,

LEFT: *The Killer Whale's versatility as a predator seems to know no bounds. Here, an adult male, identifiable by its tall, spike-like dorsal fin, rushes into the surf along the Argentine coast in pursuit of a young South American Sea Lion.* **OPPOSITE:** *Unlike baleen whales, toothed cetaceans usually catch their prey one by one. Here, an Atlantic Spotted Dolphin catches a razorfish off the Bahamas.*

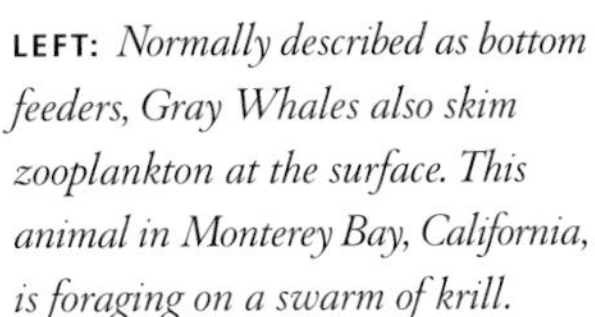

LEFT: *Normally described as bottom feeders, Gray Whales also skim zooplankton at the surface. This animal in Monterey Bay, California, is foraging on a swarm of krill.*

RIGHT: *Sea Otters sometimes use rocks to break open the shells of hard-bodied prey, such as clams. They usually float on their backs while dining.*

particularly squid, constitute the principal diet of many toothed whales, including sperm, pilot, and beaked whales. Some populations of Killer Whales feed largely upon marine mammals such as seals and cetaceans, as well as on birds such as penguins. Toothed whales feed at varying depths: Dolphins and porpoises are relatively shallow feeders, while beaked and sperm whales routinely hunt at depths of hundreds or even thousands of feet. Unlike baleen whales, which often engulf large patches of prey and ingest thousands or even millions of organisms at once, toothed whales usually feed by taking one item (such as a single fish) at a time. They often swallow prey whole, and their teeth function to grip rather than to chew. It is likely that suction plays an important role in capturing prey for many toothed whales.

Like toothed whales, pinnipeds are excellent hunters, taking a wide variety of fish (including larger species such as cod and salmon), squid, and other prey. Many seals forage close to the surface, but some of the larger phocids, notably Weddell and elephant seals, can dive to considerable depths in search of food. Perhaps the most fearsome of the pinnipeds is the Leopard Seal, which will attack almost anything available, from krill and squid to penguins and other seals. The Walrus feeds primarily upon mollusks, although it is also known to take fish, and some "rogue" Walruses kill and eat seals.

Unlike other marine mammals, the sirenians are herbivores. Manatees and the Dugong subsist on various aquatic plants, including water hyacinth and sea grasses. In some parts of the world, sirenians have even been known to invade flooded rice fields to feed. Marine and Sea Otters forage in coastal waters and are generalist feeders, although individual Sea Otters frequently seem to exploit a very limited range of prey. Their prey species, which vary by population, include mollusks, sea urchins, crustaceans, and fish. Finally, while the Polar Bear will feed on various fish species and even sometimes on Belugas or Narwhals trapped in ice, its diet is characteristically dominated by Ringed Seals.

The foraging behavior of marine mammals is as varied as their prey. Sirenians are simple grazers (and sometimes browsers) on vegetation,

The wide-open mouths of these Humpback Whales signify the late stages of a feeding run. Momentarily, their jaws will close and the huge volumes of water in their distended throats will be expelled, leaving behind fish and invertebrates trapped against the baleen fringes.

while other marine mammals employ a variety of senses and behaviors to catch their prey. Probably all of the toothed whales possess some kind of biological sonar, or echolocation, that they use routinely as their principal means of finding prey. Although most cetaceans probably have reasonably good eyesight, it is usually of limited use in the opaque environments in which many species forage. The river dolphins generally have poor vision, and instead use echolocation to forage and navigate through the turbid river waters in which they live.

There is no evidence that baleen whales possess the sort of sonar used by odontocetes, and how they find food is largely a mystery. Certainly they are adept at taking advantage of oceanographic cues, such as water temperature gradients, to locate areas of high biological productivity. They probably also use sound passively to locate (hear) prey. It is also quite likely that mysticetes have a well-developed sense of smell, which could be helpful in detecting patches of plankton or schools of fish at the surface.

Since most pinnipeds as well as Marine and Sea Otters feed in relatively shallow water, vision is probably an important sense for these groups, none of which are thought to have developed echolocation. How the deep-diving otariids (such as the Weddell Seal) locate prey is unclear, but their eyes appear designed to make maximum use of any available light. Also, some deeper-living prey of pinnipeds are likely bioluminescent.

Among the more interesting foraging techniques of marine mammals are the bubble nets or clouds that Humpback Whales blow to trap fish. The whales often work together in these efforts. In some species, cooperative foraging is well developed. Killer Whales are highly proficient group hunters, often working together to herd and capture fish or marine mammals. Baleen whales often feed alone, but in some species (such as Humpback and Fin Whales), animals work together to exploit schools of fish. By contrast, pinnipeds, otters, sirenians, and Polar Bears are usually solitary feeders. There are even a few well-documented examples of cooperative hunting by humans and cetaceans. These include a human-dolphin "cooperative" in Brazil in which bottlenose dolphins herd fish into fishing nets, as well as a long-term cooperative relationship between 19th-century whalers in New South Wales, Australia, and a local group of Killer Whales.

Status and Conservation

Few groups of animals have been as relentlessly exploited by humans as marine mammals. For millennia, humans have hunted cetaceans, pinnipeds, otters, and sirenians for their fur, meat, or oil, and a few aboriginal hunts continue today in places as far apart as the Arctic and the tropics. Historically, the exploitation of marine mammals was not greatly different in scale or in method from hunts for terrestrial mammals, although some small isolated populations of a few species had already been reduced or wiped out by the beginning of the industrial age.

Large-scale commercial hunts were in most cases initiated in the last two or three hundred years, although sustained commercial exploitation of some whales began much earlier. For example, right whales were taken by Basques in the Bay of Biscay at least as early as the 11th century. With improved navigational techniques and the development of faster, more efficient ships, sealing and whaling became profitable enterprises on a huge scale. There was a massive slaughter of fur seals in the 19th century, and as a result of the well-known excesses of the commercial whaling industry in the 20th century, more than 2 million whales were killed in the Southern Hemisphere alone. Unregulated exploitation reduced many populations of marine mammals to small fractions of their original sizes, while others were extirpated altogether. While the scale of exploitation has been greatly reduced from that of the last two centuries, hunting continues to negatively affect population sizes of many marine mammals.

Today fishing is undoubtedly the single greatest human-related cause of mortality in marine mammals. The proliferation of synthetic nets and other fishing gear has wreaked havoc on many marine mammal populations. It is likely that deaths of cetaceans, pinnipeds, sirenians, and otters from incidental entanglement collectively run into the millions worldwide each year. Fishermen in many areas exacerbate the problem by intentionally killing marine mammals, either for bait or to eliminate what they perceive as competition for fish.

In addition to hunting and fishing, marine mammals are threatened by such human-related factors as pollution, boat traffic, noise, and river damming. Among the most critically endangered species are the river dolphins, whose constrained environments make them particularly vulnerable. The fragmentation of small populations of river dolphins into even smaller isolated units further threatens this group of cetaceans. Indeed, it seems inevitable that some of the river dolphins will become extinct in the 21st century.

Habitat degradation is a particularly serious concern for marine mammals that live in coastal areas or in any environment that brings them

The Harbor Porpoise rescue program on Grand Manan Island, New Brunswick, Canada, depends on private sources of support to ensure that animals trapped in herring weirs are returned safely to the open sea. No similar program is possible for extricating cetaceans from gillnets; most entangled animals die before the nets are hauled.

into close contact with human activities. Noise from ship traffic and industrial activity may displace marine mammals from feeding or breeding areas; at the very least it can disrupt critical behaviors such as foraging or the nursing of young. Collisions between ships and marine mammals (notably cetaceans and sirenians) are common in some areas and frequently result in death or serious injury to the animal. Among North Atlantic Right Whales, ship strikes are the largest single cause of mortality and have clearly inhibited the recovery of this small remnant population.

The impact of pollution on marine mammals is difficult to assess but may be significant. Widespread industrial pollutants such as PCBs are known to have profound effects on the reproduction and health of some terrestrial species, and it appears increasingly likely that marine mammals that feed relatively high on the food chain (such as seals, toothed whales, and the Polar Bear) may be compromised by contaminants in areas of particularly heavy industrial development. For example, pollutants have been implicated in recent mass mortalities of seals and dolphins in European waters. The increasing accumulation of certain pollutants with each new generation is a further concern. The complex manner in which contaminants interact with organisms, and the many uncertainties involved, make specific cause-and-effect links very difficult to demonstrate. However, a good argument can sometimes be made for reduced reproduction or immune function as a result of contaminants.

On the broadest scale, human-induced climate change has the potential for devastating effects on some marine mammal populations. While long-term impacts of global warming are difficult to predict, one possible problem is loss of sea ice in polar regions, a phenomenon that will undoubtedly have profound consequences for the eco-

Blue Whales were prime targets of the modern whaling industry, and their numbers were rapidly depleted beginning in the late 19th century. Although Blue Whales have been protected from commercial hunting for several decades, views such as this one are still rare in all but a few parts of the species' historic range.

Many marine mammals are threatened with extinction, including almost all populations of Dugongs (opposite left) and Bowhead Whales (opposite right), as well as the Hawaiian Monk Seal (left) and the Mediterranean Monk Seal. A third species of monk seal in the Caribbean Sea and Gulf of Mexico was exterminated by humans within the last two centuries.

systems concerned. For example, the production of krill is known to be heavily dependent upon sea ice, the undersurface of which provides habitat for the algae upon which krill feed. Thus significant loss of ice cover results in greatly reduced krill abundance. Given that krill are the foundation of the ecosystem of the Antarctic, the consequences for the many species of marine mammals, birds, and fish that prey on krill, as well as for predators higher in the food chain, may well be disastrous.

A number of populations of marine mammals are critically endangered. Among the great whales, these include the North Atlantic Right Whale, the North Pacific Right Whale, Atlantic populations of the Bowhead Whale, the western North Pacific population of the Gray Whale, and most populations of the Blue Whale. As noted, the most endangered of the odontocetes are the river dolphins, together with the smallest of the cetaceans, the Vaquita. The Hawaiian and Mediterranean Monk Seals are both highly endangered, while a third species, the Caribbean Monk Seal, is almost certainly extinct. Sirenians have been espcieally vulnerable to hunting and other anthropogenic problems, and the Dugong in particular has declined precipitously through most of its range. Many Dugong populations are either extinct or critically endangered, and it seems likely that only the populations in Australia and New Caledonia remain large today.

In addition to the Caribbean Monk Seal, one or two species of marine mammals are extinct. The Atlantic Gray Whale (which may or may not have been a separate species from that found today in the North Pacific) was still extant around 1700, and its subsequent demise remains a mystery. It may have been a victim of unrecorded early hunts; alternatively, whalers may have finished off a small population that was already on its way to extinction. The remnant population of Steller's Sea Cow in the Commander Islands was extinct by 1768, the victim of fur hunters in search of seal pelts who killed these huge sirenians for food.

Many marine mammals are protected by international agreements such as the Convention on International Trade in Endangered Species (CITES) and the Convention for the Conservation of Antarctic Seals (CCAS). Bodies such as the International Whaling Commission (IWC) and various bilateral or multilateral commissions either protect marine mammals or attempt to regulate their exploitation. Many countries have national laws governing human interactions with marine mammals, although enforcement of these laws is too often inadequate or nonexistent. In the United States, all marine mammal species are protected under the Marine Mammal Protection Act, and many are also covered by the Endangered Species Act.

Watching Marine Mammals

Many species of marine mammals live in remote environments that are essentially inaccessible to a casual observer, while others are surprisingly easy to find. In many areas it is not even necessary to board a boat to watch these animals. Along the California coast, sea lions frequent harbors and haul out on rocky beaches, while Gray Whales pass within easy sight of shore on their annual migrations between Alaska and Mexico.

During spring in Cape Cod, Massachusetts, Gray and Harbor Seals use certain beaches for giving birth, while Humpback, Fin, and North Atlantic Right Whales may be seen offshore. Indeed, in some locations it is virtually impossible to miss marine mammals, even when not actively looking for them: Gaze out to sea from many places in the Hawaiian Islands in winter, and rarely more than a minute will pass before Humpback Whales appear.

IDENTIFICATION Identifying marine mammals can be simple or almost impossible, depending on the species. While it would be difficult to mistake a Polar Bear or a male Narwhal for any other animal, correctly identifying a particular seal species is often quite challenging. It becomes even more difficult to make accurate identifications of animals in the water, as seals, whales, and dolphins can move rapidly, giving only brief glimpses of their backs at the surface. Most species do, however, have unique characteristics that, either alone or in combination, help to identify them. This guide highlights key identification features for each species.

USEFUL EQUIPMENT Binoculars are probably the most useful equipment for watching marine mammals. Although it is sometimes possible to get very close to marine mammals, more often than not they will be seen from a distance of anywhere from several hundred feet to more than a mile. Binoculars (even inexpensive ones) will thus allow a much better look at the animals and greatly enhance the experience.

A camera is also standard equipment, although observers should be sure to reserve some time around marine mammals for simply watching and enjoying them without the interference of photographic equipment. A good camera with a telephoto lens (at least 200 mm) is recommended; at several hundred feet, even a huge Blue Whale will take up only a fraction of the image in a photo taken with a standard 50 mm lens. Any film is acceptable, keeping in mind that faster-speed film works better in lower light. If the camera allows it, select a fast shutter speed, since marine mammals are frequently in motion; a speed of at least 1/500

Recovery of the eastern Pacific population of Gray Whales is a conservation success story. Today, people have the privilege of observing these whales up close each winter in their breeding lagoons along Mexico's Baja California peninsula.

LEFT: *Children play on Casa Beach in La Jolla, California, within sand-tossing distance of a group of resting Harbor Seals.*
ABOVE: *A somnolent gang of California Sea Lions crowd a buoy off San Diego, California.*

of a second is recommended for sharp, focused images. Keep in mind also that many marine mammals are relatively dark-colored animals; accordingly, metering the light off the water (out of any glare) is often a good way to set the camera's exposure before photographing the animal.

ETIQUETTE Marine mammal watching should be conducted respectfully and without disturbance to the animals. In many places (such as the United States) there are specific regulations governing marine mammal watching; be aware of what these are before heading into the field. Whether on a beach or in a boat, maintain a reasonable distance, and do not interfere with the behavior of the animal. When operating a boat, do not move until it is possible to see the animal clearly; marine mammals often stay just below the surface, and many carry propeller scars on their bodies because of careless boat operation. By the same token, while waiting for an animal to reappear, keep the engine in neutral so that the propeller is disengaged. Never attempt to "cut off" a swimming marine mammal by placing the boat in its path; always stay to the side and somewhat behind the animal, and go slowly enough to easily stop or turn if something unexpected happens. Finally, Polar Bears are extremely dangerous predators and should be watched only in the company of a knowledgeable guide.

STRANDED MARINE MAMMALS Any stranded marine mammal, alive or dead, should be reported as soon as possible. In the United States and some other countries, there is a well-established network of scientific or governmental groups that are responsible for dealing with stranded marine mammals. Call either a local institution involved in this work or the local Coast Guard or police station, as in most cases these services will know whom to contact.

It is not advisable to touch a dead marine mammal, since any carcass can transmit disease; indeed, in some countries (such as the United States) it is illegal for anyone without a scientific permit to handle a marine mammal, either living or dead. Furthermore, keep well clear of any live stranded animal. The tail of a stranded whale or dolphin is a powerful weapon that can inflict serious injury, and even very sick seals or otters can give nasty bites.

In a remote location where no professional assistance is available, taking photographs of the stranded animal can be very useful. Copies of the photos, together with information on the date, location, and any other details, can then be sent to appropriate experts for species identification.

Organization of the Guide

The main part of this guide consists of the species accounts, which are descriptions of individual species of marine mammals. There are 118 species accounts divided into 10 groups, each with its own introduction.

GROUP INTRODUCTIONS The group introductions provide an overview of the marine mammals covered in the species accounts that follow and describe the families to which those species belong. Some introductions cover a single large family, while others combine two or more smaller, related families. Following the group text are illustrations of the species showing them at relative size to one another.

SPECIES ACCOUNTS The species accounts provide detailed information about individual species of marine mammals. A small number of accounts cover two or three closely related species. With a few exceptions, this guide follows the taxonomic order given in *Marine Mammals of the World: Systematics and Distribution,* by Dale W. Rice (Society for Marine Mammalogy, 1998).

Names Each species account begins with the currently accepted common name of the species. Some species have widely used alternate common names, and those are given in the introductory text of the account. Common names can vary from one part of the world to the next. Each mammal, however, has only one scientific name. The scientific name, shown below the common name, is italicized. Each scientific name consists of two parts: the genus name, which is always capitalized, and the specific name, which is always lower case. The scientific name is followed by the name of the author (or authors) who named the species and the date when it was first described in the literature. If the author and date are in parentheses, it means that the species has been transferred to a genus other than the one to which it was originally assigned.

Key Features At the top of each species account is a list of key identifying features to help the reader recognize animals in the field. The information given varies by species and may include physical characteristics, typical behavior, and range.

Illustrations and Photos Each account includes at least one portrait of the species. For species in which males and females differ markedly, separate illustrations are included. Some accounts have illustrations depicting animals at various life stages or from different populations or subspecies, or special features of the species, such as distinctive flukes. Photographs supplement the illustrations. For those species that are little known and seldom photographed, there may be no photographs in the species account.

Introductory Text A brief overview of the featured species begins each account. This section lists alternate or colloquial names, often gives the derivation of the scientific name, explains important taxonomic issues involving the species, and highlights interesting facts about the animal.

Description This section presents the animal's physical characteristics, focusing on the attributes that are most useful for identification. The description begins with the general body shape, followed by details of specific body parts, and tooth (or baleen) count, which can be used to identify stranded animals and skeletal remains. The coloration pattern is described, and significant differences between males and females or between populations are noted.

Measurements and Life Span Each species account includes a box with length and weight at birth, maximum length and weight, and life span. Obtaining measurements and age estimates of marine mammals in the wild can be very difficult and, for some species, the data is necessarily incomplete. If they differ significantly, lengths and weights are given separately for males and females. Where noted, some measurements are taken from particular geographical populations.

Pinnipeds and sirenians are measured in a straight line from the tip of the snout to the tip of the tail or flukes, respectively. In cetaceans, length is measured from the tip of the jaw to the notch between the tail flukes. Cetaceans, in particular, are rarely weighed alive, and post-mortem weighing often involves the loss of tissue, and thus under-estimation.

Life span, as used in this guide, refers to maximum longevity in the wild. Most individuals die well before reaching that age. Ages for pinnipeds, toothed cetaceans, and Dugongs are determined by sectioning the teeth to reveal light and dark bands of dentine or cementum. One light and one adjacent dark band comprise a GLG (growth layer group), and for most species one GLG is deposited per year. The ages of pinnipeds may also be determined by counting bands on the claws. Estimating age is less straightforward for the toothless baleen whales; methods include examining the alternating light (horny) and dark (waxy) laminae in the "ear plugs" or the layers in ear bones and baleen. Ages of manatees are estimated by examining the layers in ear bones.

Range and Habitat Each account includes a description of the species' geographical distribution and habitat preferences. This text provides, where appropriate, ranges of subspecies, breeding and feeding ranges, haulout locations (pinnipeds), and migration information.

A range map accompanies the text and shows where the species normally occurs. Other elements depicted include possible range, areas where there are no records of the species' occurrence but where researchers suspect they occur; extralimital records of animals outside their usual range (vagrants); and pinniped breeding sites. As the maps portray simplified sketches of distribution, it is important to read the range and habitat text in conjunction with the maps.

Similar Species This section lists other marine mammals that share all or part of the featured species' range and with which it may be confused. Similar species are not necessarily closely related to the main species. Physical characteristics that help to distinguish the similar species from the main species are provided.

Behavior This section begins by describing the social organization of the species, including group sizes and whether it is solitary or social. It covers typical behavior of the species, such as swimming and diving, molting (pinnipeds), vocalizations, and associations with other marine mammals. This section may also mention other animals that prey on the featured species.

Reproduction This section describes the species' mating system and behaviors, breeding season, length of gestation, frequency of birth, lactation and weaning, and interactions between parents and offspring. The reproductive habits of some species are little known, and in some cases no information particular to the species is available.

Food and Foraging With the exception of the herbivorous sirenians, all marine mammals prey on other animals. This section lists the main animals and plants that make up the species' diet and describes foraging behavior. Other information sometimes given in this section includes the location of the prey in the water column and how this relates to the species' diving behavior.

Status and Conservation This section explains the species' current status in the wild, giving population estimates when known. It lists historic and current threats, including deliberate persecution, such as hunting. and indirect threats, such as habitat degradation. The text also describes efforts to protect the species.

APPENDICES The Appendices, which follow the accounts, include lists of marine mammals organized by the region of the world in which they are found; an illustrated glossary of terms used in the book; diagrams showing parts of a pinniped, a baleen whale, and a toothed cetacean; photo credits; and an index.

Polar Bear and Otters

Seals, sea lions, fur seals, Walruses, Polar Bears, and Marine and Sea Otters are grouped together as marine carnivores, an informal category within the order Carnivora. Unlike most of their terrestrial kin, marine carnivores are well adapted for aquatic life. They are excellent swimmers and divers, and all but the Polar Bear forage exclusively in the water. While most terrestrial carnivores prey on other mammals, marine carnivores eat mainly fish, squid, and other aquatic invertebrates. Two key exceptions are the Polar Bear, which eats primarily seals, and the Leopard Seal, which preys on other seals, penguins, and seabirds.

The marine carnivores are often grouped into two categories, known casually as pinnipeds (see page 49) and marine fissipeds. The distinction between the two groups is based on the structure of the toes, with "fissiped" designating animals with separate toes and "pinniped" those with digits joined together either by skin webbing or cartilage. The marine fissipeds include representatives of two carnivore families, the Ursidae (bears) and the Mustelidae (weasels, otters, and relatives).

FAMILY URSIDAE The Polar Bear is the only marine species in the otherwise terrestrial bear family. All bears have a large head, small eyes, rounded ears, and a small tail. The Polar Bear's unique white pelage, waterproof fur, and excellent swimming ability are adaptations to its

Sea Otter

arctic environment. The Polar Bear has a circumpolar distribution in the Arctic, primarily near the coasts of North America and Asia, although individuals at times wander far from land on the drifting pack ice of the Arctic Ocean while searching for Ringed Seals, their primary prey.

FAMILY MUSTELIDAE Weasels, otters, and their relatives comprise the largest family in the order Carnivora. Most members of the family have long thin bodies with short legs. Adapted for pushing through dense underbrush or tunnels on land, their legs also function well for swimming. While many mustelids are at least partly aquatic, only Sea and Marine Otters dwell exclusively in marine waters, the Marine Otter along the west coast of South America, and the Sea Otter in the North Pacific Ocean. Sea Otters are unique among carnivores in that their teeth have no sharp cutting edges and even the canines are blunt and rounded. This adaptation allows them to crush their invertebrate prey, many of which have strong exoskeletons. Marine Otters have sharply pointed teeth that they use for cutting or shearing flesh.

Sea Otter *page 42* Marine Otter *page 46*

Polar Bear

Ursus maritimus
Phipps, 1774

- **THE LARGEST BEAR SPECIES**
- **PELAGE ENTIRELY WHITE, CREAM-COLORED, OR YELLOWISH WHITE**
- **MALE MUCH LARGER THAN FEMALE**
- **STRONG SWIMMER**
- **GENERALLY SOLITARY; MAY CONGREGATE IN SOME AREAS, SUCH AS REFUSE DUMPS**
- **CIRCUMPOLAR DISTRIBUTION IN ARCTIC REGIONS**

The Polar Bear is the largest land carnivore in North America and the top predator in the arctic marine ecosystem. Closely related to the Brown Bear, the Polar Bear is thought to have diverged from a Siberian population of Brown Bears relatively recently, evidently less than 1 million years ago. Matings between Brown Bears and Polar Bears in zoos have produced fertile offspring, evidence of this close evolutionary relationship. The distribution of Polar Bears largely parallels that of their primary prey, the Ringed Seal, of which they eat mainly the blubber. Polar Bears have a remarkable ability to store large amounts of fat, which enables them to fast for long periods during the winter denning season and at other times when food may be scarce. Large numbers of Polar Bears congregate in summer near Churchill, in southeastern Manitoba, Canada, where a local tour industry allows people to get fairly close to the bears. Polar Bears are also known in Norway as "ice bears."

DESCRIPTION The Polar Bear is the largest species of bear. Its ears are smaller than those of other bears, an adaptation to the cold arctic environment, and the neck is substantially longer and lacks the shoulder hump that is characteristic of other bear species. Polar Bears have partially webbed forepaws, an adaptation for swimming. The five digits on each paw are armed with strong, retractable claws. Males are about twice as big as females. Head size varies geographically, with the largest-headed bears in the Bering and Chukchi Seas and the smallest around southeastern Greenland. There are six to eight pairs of teeth in both the upper and lower jaws. The back cheek teeth are much smaller than those of other bear species.

The pelage coloration may vary from white to yellow or even light gray, while the nose, lips, eyes, and bare footpads are black.

RANGE AND HABITAT Polar Bears have a circumpolar distribution along coastal areas and islands of the Arctic, ranging north in some areas to at least 88°N and south to southern Labrador. They have a discontinuous distribution with regionally discrete populations. Denser concentrations occur near the northern coastlines of North America, Baffin Island, northern Greenland, Europe, and Russia. Scientists often recognize two subspecies, based on concentrations of denning females. *Ursus maritimus maritimus* occurs from the Canadian Northwest Territories east to western Russia near Novaya Zemlya. *U. m. marinus* ranges from the Taymyr Peninsula in Russia eastward to the Beaufort Sea and Banks Island in Canada. Vagrants occur as far south as Newfoundland, Iceland, and Norway. Occasionally Polar Bears may wander up to a few hundred miles inland into tundra, taiga, and even coniferous forest habitats.

Polar Bears occupy fast-ice and pack-ice habitats where their principal prey, Ringed Seals, occur year-round. In the Canadian Arctic, their density and reproductive success appear to correlate with densities of Ringed Seals, as well as the quality of the sea-ice habitat. Denning is most common along the coast, near areas of good spring hunting. In some areas, Polar Bears may excavate dens in multiyear pack ice, although dens are less common there, especially where the water is relatively deep and cold. One of the three largest maternity denning sites is in the Wapusk National Park in Churchill, Manitoba. Other major denning areas are in King Karl's Land in the Svalbard Islands (Norway) and at Wrangell Island (Russia).

SIMILAR SPECIES The Polar Bear is easy to distinguish from other bears by its striking, white to yellowish pelage. Also, the Polar Bear has a substantially longer neck than do Black Bears and Brown Bears, and it lacks the characteristic shoulder hump of other adult bears.

BEHAVIOR Polar Bears are generally solitary, although females may remain and travel with their cubs for two to three years. Great endurance walkers and swimmers, these bears make seasonal migrations of between 1,200 and 2,500 miles (2,000–4,000 km) across the ice and occasionally swim 60 miles (100 km) or more in open water. On the ice, Polar Bears are fast runners, reaching speeds of 25 to 30 miles per hour (40–50 km/h) during short sprints, especially when chasing Ringed Seals. Though Polar Bears are highly mobile, they are faithful to their home ranges, large areas of up to 200 square miles (500 sq km). Males often have larger home ranges than females. Females use dens throughout the winter, and both males and females may construct dens for short-term shelter at other times, especially in autumn and winter. These shelter dens allow the

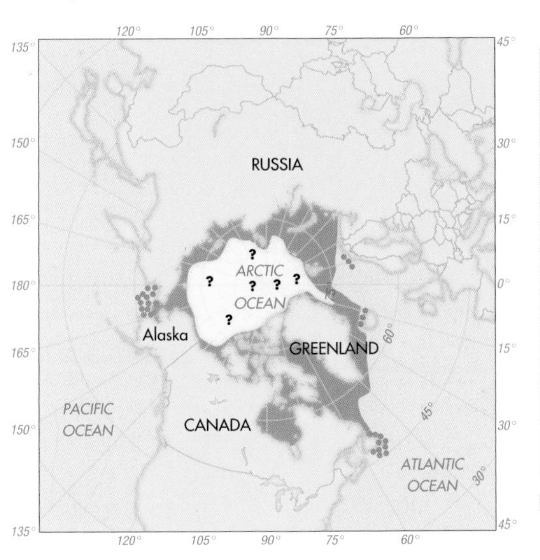

POLAR BEAR
FAMILY URSIDAE

MEASUREMENTS AT BIRTH

LENGTH	Unavailable
WEIGHT	1¼ lb (0.6 kg)

MAXIMUM MEASUREMENTS

LENGTH	**MALE**	8′6″ (2.6 m)
	FEMALE	6′11″ (2.1 m)
WEIGHT	**MALE**	1,800 lb (800 kg)
	FEMALE	660 lb (300 kg)

LIFE SPAN
MALE 29 years
FEMALE 32 years

■ RANGE ● VAGRANTS ? POSSIBLE RANGE

bears to conserve energy when food is scarce and are more often used in northern areas.

REPRODUCTION Mating occurs from late March through early May. Overall, gestation lasts about 5½ to 8 months. In November and December, pregnant females dig maternity dens in snowdrifts on fast ice, drifting pack ice, and land along the coast. In the southwestern Hudson Bay, some females excavate dens up to 60 miles (100 km) inland from the coast. Females give birth to up to three cubs in December or January. They remain in the dens with their cubs until spring, at which time the cubs are able to travel about on the ice. In most populations, cubs are weaned at about 2½ years old.

FOOD AND FORAGING The primary food of Polar Bears are Ringed Seals and, to a lesser extent, Bearded Seals, principally the blubber of both species. In spring, Polar Bears search out and excavate seal birthing lairs to prey on the young seal pups. They also stalk seals at their breathing holes or when the seals are hauled out on the ice. Polar Bears are expert hunters, capable of remaining motionless on the ice for several hours while stalking seals at their breathing holes. Polar Bears also prey on Walruses, Belugas, Narwhals, Hooded Seals, Harp Seals, some seabirds, and even reindeer in some areas, and they scavenge on whale carcasses and in refuse dumps near human settlements.

STATUS AND CONSERVATION The Polar Bear's global population has been estimated at between 25,000 and 40,000, and the denning population in southern Hudson Bay may number around 1,200. Polar Bears are hunted for food and clothing throughout the Arctic. On October 16, 2000, the United States and Russia signed a bilateral agreement to establish quotas on these subsistence harvests of Polar Bears in Alaska and northeastern Siberia. The agreement also prohibits all commercial hunting, as well as hunting in denning areas and the killing of females with cubs or bears less than one year old. Moreover, it prohibits hunting with aircraft, traps, or snares. Polar Bear reserves have been established at Polar Bear Provincial Park in Ontario, Canada, the Russian Federation's Wrangel Island Republic Reserve, several sites in the Svalbard Islands of Norway, and in the Northeast Greenland

OPPOSITE: *Juvenile Polar Bears jousting near Churchill, Manitoba.* **RIGHT:** *Female Polar Bears generally give birth to two or three cubs every three or four years. The cubs remain close to their mothers until weaned two to three years later.*

National Park, where most of the eastern Greenland population lives.

In the 1990s, Polar Bears in Hudson Bay were 10 percent thinner than they were two decades earlier and had 10 percent fewer offspring. This evidently resulted from long-term warming trends that caused the ice to melt early and substantially decreased the amount of time the bears were able to prey on Ringed Seal pups. In the late 1990s, scientists found that a number of Polar Bears near Svalbard and in the Barents Sea had both male and female reproductive organs and also had high levels of persistent organic pollutants, especially PCBs, in their blood and tissues. These pollutants, which are up to 20 times higher in that area than in other parts of the Polar Bear's range, may be affecting the species' development and immune functions.

Polar Bears travel and hunt mostly on fast ice or pack ice. Although they are adept swimmers and may cover 60 miles (100 km) or more in a day, they rarely hunt in the water.

Sea Otter

Enhydra lutris
(Linnaeus, 1758)

- LARGE BODY SIZE RELATIVE TO OTHER MUSTELIDS
- LUXURIOUS PELAGE WITH EXTREMELY DENSE FUR
- FLOATS AND SWIMS ON BACK, GROOMING OFTEN AT SEA SURFACE
- FORAGES EXCLUSIVELY IN NEARSHORE MARINE ENVIRONMENTS
- DISTRIBUTION IN COASTAL NORTH PACIFIC AND SOUTHERN BERING SEA

The Sea Otter is the most aquatic species in the order Carnivora. These otters infrequently haul out on land or ice; females even give birth and nurse their pups in coastal waters. Sea Otters spend their time either floating on their backs at the surface or diving for food at the seafloor. While at the surface, they groom almost continuously to maintain the cleanliness and insulating properties of their fur. Their luxurious pelage has the greatest density of fur of any mammal. Scientists recognize three subspecies of the Sea Otter, one in California and two in Alaska. Otters in California are called Southern or California Sea Otters, whereas those that range farther north are called Alaskan Sea Otters. The genus name is from the Greek word *enhydris,* meaning "otter"; the specific name is from the Latin word *lutra,* which also means "otter."

DESCRIPTION The Sea Otter, especially in Alaska, is large relative to most other mustelids and has a short broad head and a short blunt snout. The upper lip and cheeks are well developed and densely covered by stiff whiskers. The hindpaws are large, flipper-like, and webbed. The forepaws are rounded; Sea Otters use them like hands to manipulate food, groom, and hold various tools for breaking open shellfish. The tail is long, flattened, and oar-like. Adult males are more robust at the head and neck than females. Alaskan Sea Otters are substantially larger-bodied and more robust than California Sea Otters. There are eight pairs of teeth in both the upper and lower jaws.

The Sea Otter's body is completely covered by fur, except for the pads on the bottom of the fore- and hindpaws and the tip of the nose. Adult males and females have a dense coat of dark brown to reddish-brown underfur. The guard hairs are less dense and may be lighter brown to blond, particularly on the face and head. Newborn pups have a light, buff-colored pelage with the guard hairs becoming yellowish several weeks after birth.

RANGE AND HABITAT Sea Otters occur in near-shore waters of the North Pacific, from Japan north to the coast of the Kamchatka Peninsula, east throughout the Aleutians, and south through the Gulf of Alaska and along the Pacific coast of North America to Baja California. The California, or Southern, Sea Otter *(Enhydra lutris nereis)* ranges from northern California south to Baja California. The Alaskan Sea Otter *(E. l. lutris)* ranges from the Commander Islands in the western Bering Sea along the southeast coast of the Kamchatka Peninsula, through the Kuril Islands, and south to northern Japan. Another subspecies also called the Alaskan Sea Otter *(E. l. kenyoni)* ranges in the Aleutian and Pribilof Islands east to the Alaska Peninsula, into Prince William Sound, and along the coasts of southeastern Alaska, British Columbia, Washington, and Oregon.

Sea Otters live in a variety of coastal marine habitats, ranging from rocky sea bottoms and shorelines, where marine communities are most diverse, to areas where mud or sand is the primary sea-bottom substrate. They mostly occur in areas that are shallower than 130 feet (40 m), though they may travel through much deeper waters when moving between foraging areas or during seasonal dispersal.

SIMILAR SPECIES River Otters may occasionally occur in marine habitats. Sea Otters differ from River Otters in their large, flipper-like hind limbs, dorso-ventrally flattened tail, and flattened molars. River Otters generally swim or float on their bellies when in the water, whereas Sea Otters almost always float and mostly swim at the surface on their backs.

BEHAVIOR Sea Otters may occur alone or in small groups. Sometimes a dozen or more may aggregate into floating rafts in near-shore waters or in kelp beds, where they may wrap themselves or even tie themselves up in the kelp. Although they may occur in groups of several to dozens or more in areas where food is abundant and kelp beds are thick, Sea Otters are not very social. Adult males often segregate from the rest of the population during most of the year. Most Sea Otters spend virtually all of their lives in the water, although some individuals occasionally haul out on rocky coastlines or sandy, cobble, or snow-covered beaches in winter. Pups are quite vocal, and their high-pitched, piercing calls to their mothers are loud and often audible for a hundred yards or more, even in areas of heavy, noisy surf. Sea Otters are easily distinguished at the water's surface by their habit of floating belly-up when sleeping or while rubbing and grooming their fur. Grooming is an important activity, as it maintains general cleanliness and ensures the waterproof, insulating quality of the underfur, a key property for thermoregulation and heat conservation in the cold waters of the North Pacific and Bering Sea.

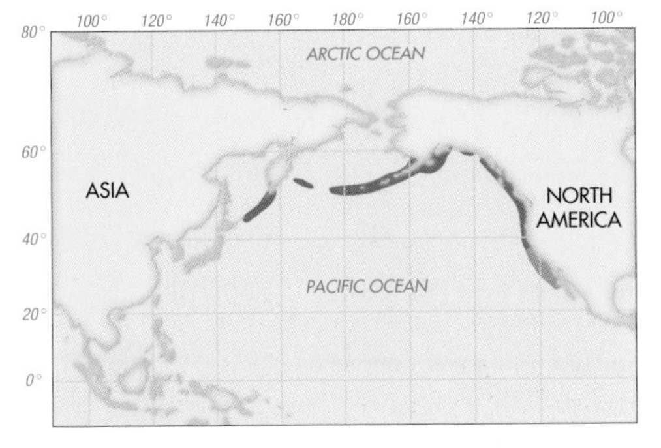

■ RANGE OF ALASKAN SUBSPECIES
■ RANGE OF CALIFORNIA SUBSPECIES

SEA OTTER
FAMILY MUSTELIDAE

MEASUREMENTS AT BIRTH
LENGTH 22–24″ (55–60 cm)
WEIGHT 4–5 lb (2–2.25 kg)

MAXIMUM MEASUREMENTS
LENGTH **MALE** 4′10″ (1.48 m)
FEMALE 4′7″ (1.4 m)
WEIGHT **MALE** 99 lb (45 kg)
FEMALE 73 lb (33 kg)

LIFE SPAN
MALE 15 years
FEMALE 20 years

REPRODUCTION Sea Otters have a polygynous mating system. Males establish aquatic territories in areas near females and their pups and may mate with several or more females during a breeding season. During mating attempts, males often bite females on their noses; thus sexually mature females may have bloody noses during the breeding season, and older females may have distinguishing scars. Females give birth throughout the year. In California, most pups are born between December and February, while in Alaska most are born from May to June. Gestation lasts about 9 to 10 months. Females nurse their pups for about six months, sometimes as long as a year, before they abruptly wean and abandon them. Females continue to forage while nursing their pups, and pups may begin to forage in shallow habitats at about six weeks old.

FOOD AND FORAGING The diet of Sea Otters varies with the physical and biological characteristics of the habitats in which they live. In rocky-bottom habitats, Sea Otters generally search out and eat large-bodied prey (including lobsters, sea urchins, and abalone) that offer the greatest caloric reward. In soft-bottom habitats, where prey are often smaller and more difficult to find, Sea Otters also eat a variety of burrowing invertebrates, such as pismo clams. Throughout their range, Sea Otters mostly forage in habitats where water depth is less than 130 feet (40 m), though

ABOVE: *Sea Otters often anchor themselves in kelp while resting between foraging trips. Mothers leave their pups in secure kelp bed areas while foraging.* **BELOW:** *Sea Otters use their chests and bellies as tables on which to rest the crabs, lobsters, abalone, and sea urchins they catch at the sea bottom.*

A Sea Otter pup rests on its mother's belly. While at the surface, Sea Otters spend most of their time floating on their backs.

juvenile males may forage in somewhat deeper habitats. Sea Otters are known to use tools while feeding. They often carry rocks to the surface to use as hammers for crushing the shells of sea urchins and other shellfish. Balancing the prey on their chest and belly, they use their forepaws to hold and manipulate the rock.

STATUS AND CONSERVATION Before commercial hunting began in 1741, the range-wide population of Sea Otters may have numbered between 150,000 and 300,000. By 1911, when the United States, Japan, Russia, and Great Britain negotiated an international treaty to prohibit further commercial hunting, only several thousand Sea Otters were thought to remain at scattered locations throughout their vast historic range. The species recovered substantially in most areas, and scientists reintroduced them into others. However, there have been recent unexplained declines in several populations. The Alaskan populations, estimated to number about 100,000 in the early 1900s, declined rapidly and substantially throughout the Aleutian Islands during the mid- and late 1990s. While reasons for the decline are not certain, some scientists have suggested that predation by Killer Whales was the primary, if not the sole, cause. The California population has also declined, from an estimated 2,377 in 1995 to about 1,700 in 2000. The California Sea Otter is listed as "threatened" under the U.S. Endangered Species Act and "depleted" under the Marine Mammal Protection Act of 1977; its status is due to both its small population and potential threats, including entanglement in fishing nets and oil spills by tankers along the heavily traveled central California coast. Efforts in the late 1980s to reestablish populations of Sea Otters in southern California have essentially failed, with the relocated otters either returning to their original homes in central California, dying from human-related causes, or disappearing. In the meantime, a small colony established itself naturally at San Miguel Island in the early 1990s, and others moved south along the California coastline south of Point Conception in the late 1990s.

Marine Otter

Lutra felina
(Molina, 1782)

- **SMALL BODY WITH WELL-WEBBED FEET, STRONG CLAWS**
- **LONG POINTED TAIL, RELATIVELY SHORTER THAN IN FRESHWATER OTTERS**
- **ACTIVE MOSTLY DURING DAY, ESPECIALLY NEAR DAWN AND DUSK**
- **RANGE RESTRICTED PRINCIPALLY TO PACIFIC COAST OF SOUTH AMERICA**

The Marine Otter is the smallest member of the genus *Lutra* and is the only otter of its genus that lives almost exclusively in marine habitats. Scientists know very little about the behavior of this secretive creature, which occurs in small, geographically fragmented populations. Marine Otters range along the coasts of central Peru south to Cape Horn and along the Atlantic coast of Tierra del Fuego. They occasionally enter estuaries and, rarely, the lower reaches of freshwater streams or rivers. The genus name is from the Latin for "otter," and the specific name is from the Latin for "cat." The Marine Otter is also called the "sea cat," or *gato marino*, and is known as *chungungo* and *chinchmen* in parts of Chile and Peru.

DESCRIPTION The Marine Otter is relatively small, rarely exceeding 3 feet (1 m) in total length. The head is relatively broad and flat, with a slight forehead and a blunt nose. The moderate-size fore- and hindpaws are webbed and have robust claws. The tail is pointed and relatively short for a lutrid otter. There are eight to nine pairs of teeth in the upper jaw and nine pairs in the lower jaw.

The short, brownish-gray underfur is covered by longer, dark brown guard hairs, and the pelage appears uniformly dark brown. The tips of the guard hairs may, however, fade to lighter brown several months after molting occurs. The ventral pelage is lighter, with the cheeks, chin, and throat often lightest of all.

RANGE AND HABITAT Marine Otters range along the west coast of South America, from central Peru, near Chimbote, south to Cape Horn and Le Maire Strait in southern Argentina. They occasionally enter estuaries and rarely enter the lower reaches of freshwater streams and rivers. Historically, Marine Otters ranged up the east coast of South America into southern Argentina, and small numbers may still live near Staten Island off Argentina's southern tip. Their marine habitat is characterized by rocky shorelines where strong winds and heavy seas prevail and where there are thick growths of seaweed and other algae.

SIMILAR SPECIES No other otters occur in marine habitats within the range of the Marine Otter. Marine Otters do, however, occasionally enter freshwater rivers, where they may be confused with the freshwater River Otters, which range from central Chile to Cape Horn and north into Argentina and may sometimes occur along the ocean coast. River Otters are larger, with a proportionally longer tail, and prefer more protected habitats compared to the more open and wave-exposed coastal areas inhabited by the Marine Otter.

BEHAVIOR Marine Otters have a secretive nature, and little is known about their social behavior. They appear to be most active in the water during the day. When swimming, Marine Otters submerge all but the head and a small area of the back, although adult females sometimes swim on their back, carrying their pups on their belly in a fashion similar to Sea Otters. Observations of fights between males suggest that they may be territorial. Aggressive interactions between individuals may also occur over competition for food and include active, high-pitched vocalizing and physical combat.

REPRODUCTION Little is known about reproduction in Marine Otters. Breeding evidently occurs in summer (December and January), and gestation may last from two to four months. Pups are born in autumn and winter and remain with their parents for up to 10 months. Females have litters of two pups on average, but occasionally give birth to three or four pups.

FOOD AND FORAGING Marine Otters are thought to prey mainly on shellfish, nearshore marine fish, and freshwater prawns.

STATUS AND CONSERVATION The present abundance of this species is unknown. It is protected by various measures throughout its range. A number of Marine Otters are killed each year by poachers hunting for their pelts; this poaching has caused reduced abundance in many areas and eradication from some. In addition, Marine Otters

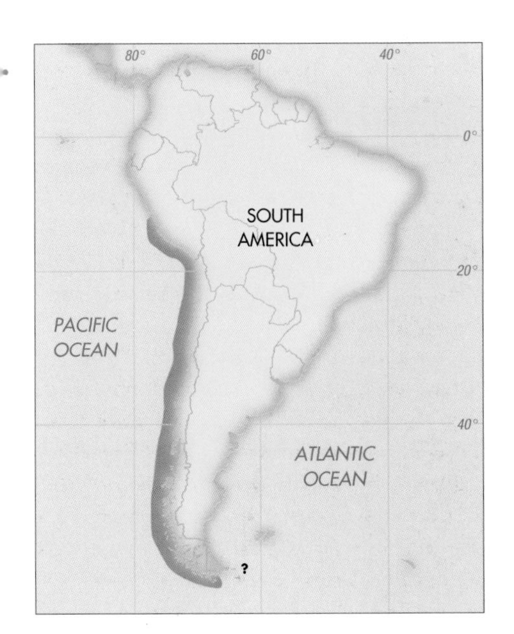

MARINE OTTER
FAMILY MUSTELIDAE

MEASUREMENTS AT BIRTH
LENGTH Unavailable
WEIGHT Unavailable

MAXIMUM MEASUREMENTS
LENGTH 3′9″ (1.15 m)
WEIGHT 10 lb (4.5 kg)

LIFE SPAN Unavailable

■ RANGE
? POSSIBLE RANGE

are killed by incidental entanglement and drowning in fishing nets and are deliberately killed by fishermen, who perceive them as competitors for fish. The reduction of shellfish populations caused by kelp harvesting, as well as direct harvests of fish and shellfish, also adversely affect this species. As a result of reduced abundance, Marine Otter populations have become fragmented into widely separated small populations, further threatening this species' survival.

ABOVE: *Little is known about the behavior of the rather secretive Marine Otter. These otters are active mostly during the day and forage almost exclusively in the marine environment.*

LEFT: *Marine Otters are dark dorsally and lighter ventrally. Their webbed feet are well adapted for swimming, and the strong claws are ideal for catching fish and shellfish.*

Pinnipeds

Although formerly classified in their own separate order, the Pinnipedia, the consensus now is to consider pinnipeds as belonging to three closely related families within the order Carnivora: Otariidae (the "eared" seals), Odobenidae (the Walrus), and Phocidae (the "true" seals), with a total of more than 30 species. The term "pinniped" is derived from the Latin *pinnipes* for "wing- or fin-footed." The name refers to the family's modified appendages, which have a fin-like appearance.

All pinnipeds have relatively large bodies that are well adapted to their aquatic lifestyle. The forelimbs and hindlimbs have been modified into paddle- or fin-like structures, the major limb bones are all enclosed in the body, and the pelvis, shoulders, and spine are adapted for efficient swimming while still allowing effective locomotion on land. All pinnipeds have a tiny tail, and some have small external ear flaps. Most species also have a short snout and relatively large eye sockets and eyes; they depend on good underwater vision. Their bodies are insulated by a layer of fat, also called blubber, situated immediately beneath the skin. The blubber helps the animal conserve heat. Members of the family Phocidae have a particularly thick layer of blubber that also serves as an energy reserve during long periods of fasting (up to a month in some species).

In general, marine carnivores have fewer teeth than terrestrial carnivores. Pinnipeds have comparatively fewer incisors, and most have large,

Harbor Seal, San Diego, California

conical canines that are often larger in males than in females. The cheek teeth are similar in shape and modified for a soft flesh diet, with little or no crushing surface.

All pinnipeds shed their hair, or molt, once a year, usually during a brief period in summer or autumn, though the timing varies among species and populations. Hairs are shed singly except in a handful of species that shed in large sheets with the upper layer of epidermis still attached to the hair. This type of molt is called catastrophic.

The breeding social structure varies among species from extreme polygyny in the elephant seals and some otariids to loose polygyny in Weddell Seals and possibly other ice-breeding phocids to perhaps monogamy or serial monogamy in Harbor Seals and monk seals. Those species that are most polygynous are also the most sexually dimorphic in size, with adult males reaching several times the mass of adult females. Otariids and some phocids form breeding rookeries where large numbers of individuals congregate each year on a predictable schedule; the ice-breeding phocids tend to be solitary or form much more dispersed breeding aggregations. Pinnipeds give birth once a year to a single pup. Twins (confirmed recently in the Weddell Seal using molecular genetic analyses) are rare, likely owing to the large size of the newborn—often about one-quarter to one-third the size of the mother in members of the family Phocidae—and to the large investment of energy required to produce fat-rich milk during the short lactation period. A pinniped female becomes receptive and may mate again within one to several weeks after giving birth. In all pinnipeds studied in detail to date, the embryo develops for only a short period after the egg is fertilized and then remains quiescent until it attaches to the uterine wall about two to three months later, after which growth resumes. This phenomenon is called embryonic diapause or delayed implantation, and allows the mother to accomplish parturition, early lactation, and mating within a single season, while still getting a period of respite from the energy demands of pregnancy.

Pinnipeds have been hunted by indigenous peoples throughout the world for several thousand years or more. The meat of these animals was used for food, their fat for fuel, their furs and skins for clothing and shelter, and their teeth and bones for tools and weapons. In some areas, local populations may have been hunted to extinction. Modern commercial sealing, which began in earnest in the late 1700s and continued into the 19th and early 20th centuries, decimated many species of pinnipeds, particularly virtually all fur seals. Since the early or mid-20th century, when some degree of regulation finally began, most pinniped populations worldwide have been recovering, and many are now robust. Today, five principal international treaties and a number of bilateral and multilateral international agreements provide for the protection and conservation of pinnipeds. Threats to the vitality and persistence of pinnipeds include pollution, overharvesting of fish, squid, and shellfish stocks, and outbreaks of disease.

FAMILY OTARIIDAE The family Otariidae comprises the sea lions and fur seals, also known as eared seals because they have external ear flaps, or pinnae. There are at least 15 species in seven genera. One of the two fur seal genera has a single species, the Northern Fur Seal, which occurs only in the North Pacific and the Bering Sea. The other has eight species, one in the Northern Hemisphere along the coast of Baja California and southern California (the Guadalupe Fur Seal) and seven in the Southern Hemisphere. The

Phocid pups, like this young Gray Seal, are born on land or ice, and they grow rapidly. In most species, pups are weaned within days or weeks after birth.

sea lions are divided into five genera, each having a single species, except *Zalophus,* which some researchers consider as having two species. Sea lions occur only in the Pacific and western South Atlantic Oceans and along portions of the Indian Ocean coast of Australia.

In all otariid species the male is significantly larger than the female. Eared seals can walk on land on both their fore- and hindflippers. The foreflippers are long with the digits entirely enclosed by skin and cartilage, forming a fin-like appendage that provides their primary means of propulsion when swimming. Fur seals have a short dense layer of hair close to the skin, which traps air and provides waterproof insulation. They also have an outer layer of longer, less dense hair. The closely related sea lions lack the fur seals' dense, waterproof underfur.

FAMILY ODOBENIDAE The Walrus is the only living representative of the family Odobenidae. Its huge body, sparse hair, heavy and coarse mustache, and robust canine teeth modified into long tusks make it unmistakable. This species lacks external ear flaps. The foreflippers are like those of the fur seals and sea lions, while the hindflippers are closer to those of phocids. Unlike the phocids, however, the Walrus can rotate its hindflippers underneath its body and walk on them when on ice or land. In water it uses either the fore- or hindflippers for propulsion.

The Walrus lives throughout most of the Arctic and Subarctic, principally on or near the pack ice, though large numbers, especially males,

Caribbean Monk Seal

Monachus tropicalis
(Gray, 1850)

The Caribbean Monk Seal once lived throughout much of the Caribbean Sea and Gulf of Mexico. It was an easy target of hunters, who killed it mainly for oil. The last confirmed sighting of Caribbean Monk Seals was made in the 1950s, and the species is almost certainly extinct. Although many species of pinniped were hunted to presumed or near extinction in the 18th and 19th centuries, the Caribbean Monk Seal is the only one to have been exterminated in recent times.

This Antarctic Fur Seal (left) is dwarfed by a Southern Elephant Seal (right), not only the largest species in the family Phocidae, but also the largest living pinniped.

aggregate on beaches in some areas in summer. The ancestors of the Walrus were evidently abundant in the Pacific Ocean, where a diverse group of forms lived along the North American coast as far south as southern California. They then colonized the Atlantic, apparently became extinct in the Pacific, and then recolonized the Bering Sea in more recent times.

FAMILY PHOCIDAE Members of the family Phocidae, called true or earless seals because they lack external ear flaps, include 19 species (one of them extinct) in 13 genera. Five species live exclusively in the Southern Hemisphere and have a circumpolar distribution around the Antarctic Continent or a little farther north in the Southern Ocean and on subantarctic islands. Four species live exclusively in the North Pacific or the Bering Sea; three live exclusively in the North Atlantic; and two are found mostly in the Arctic. Three others have highly localized populations.

Unlike eared seals and Walruses, phocids cannot rotate their hindflippers under their bodies to walk. On land they use their foreflippers to pull themselves along, while their hindflippers trail passively behind. On ice they can move very quickly and glide with sinusoidal movements across the surface. In water they use their hindflippers for propulsion, moving them from side to side while spreading the digits widely. The foreflippers are used mostly for steering and to help the seal maintain its position in the water column. The tips of the digits of phocid foreflippers are not completely enclosed in skin, but there is some short webbing between them and they have well-developed claws. Phocids have short hair, and some also have a very sparse second layer of hair closer to the skin.

Eared Seals and Walrus

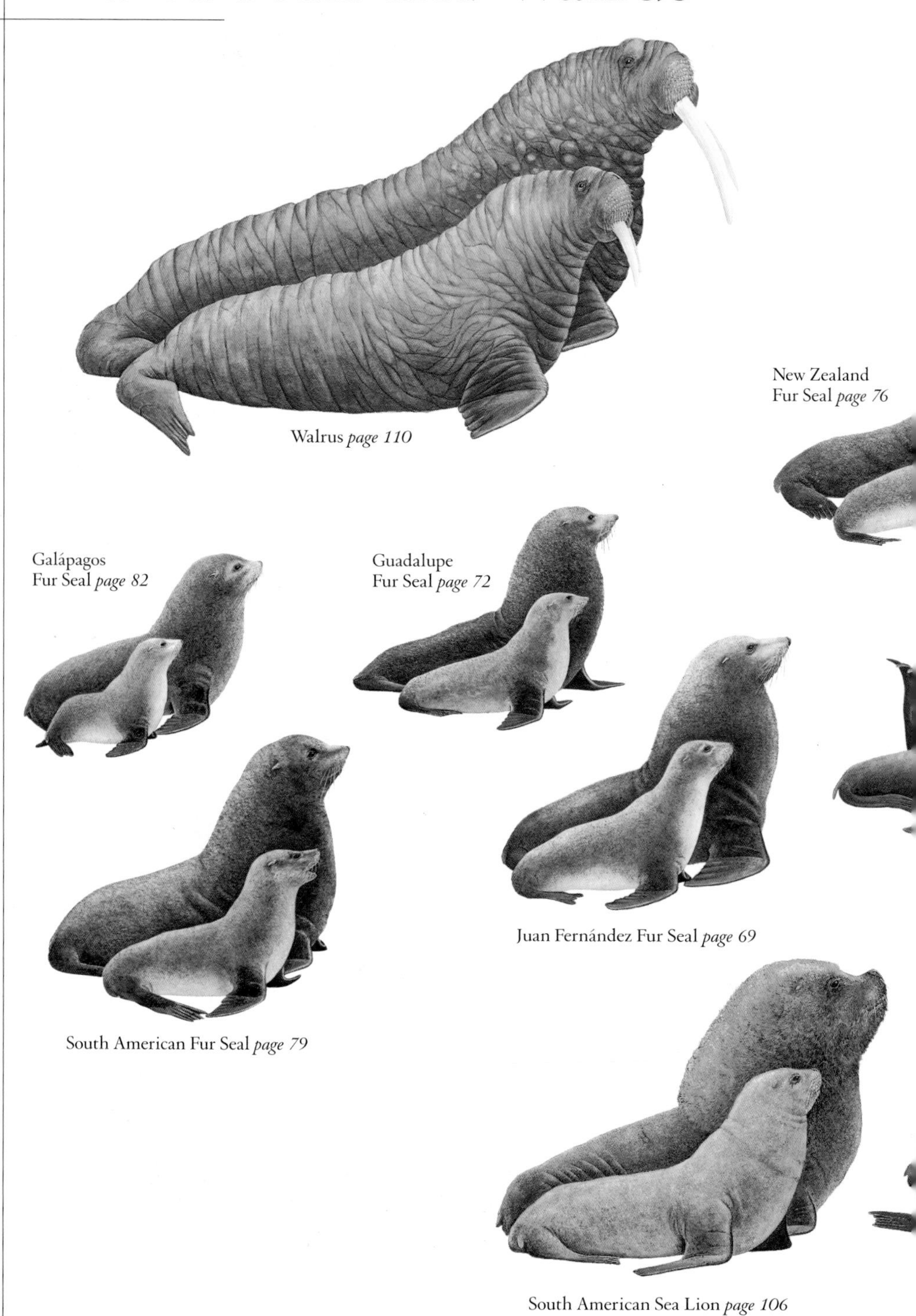

Antarctic Fur Seal
page 62
Subantarctic Fur Seal
page 66
South African and
Australian Fur Seals page 58
New Zealand
Sea Lion page 102
Australian
Sea Lion page 98
Northern Fur Seal page 86
California and Galápagos
Sea Lions page 90
Northern Sea Lion page 94

True Seals

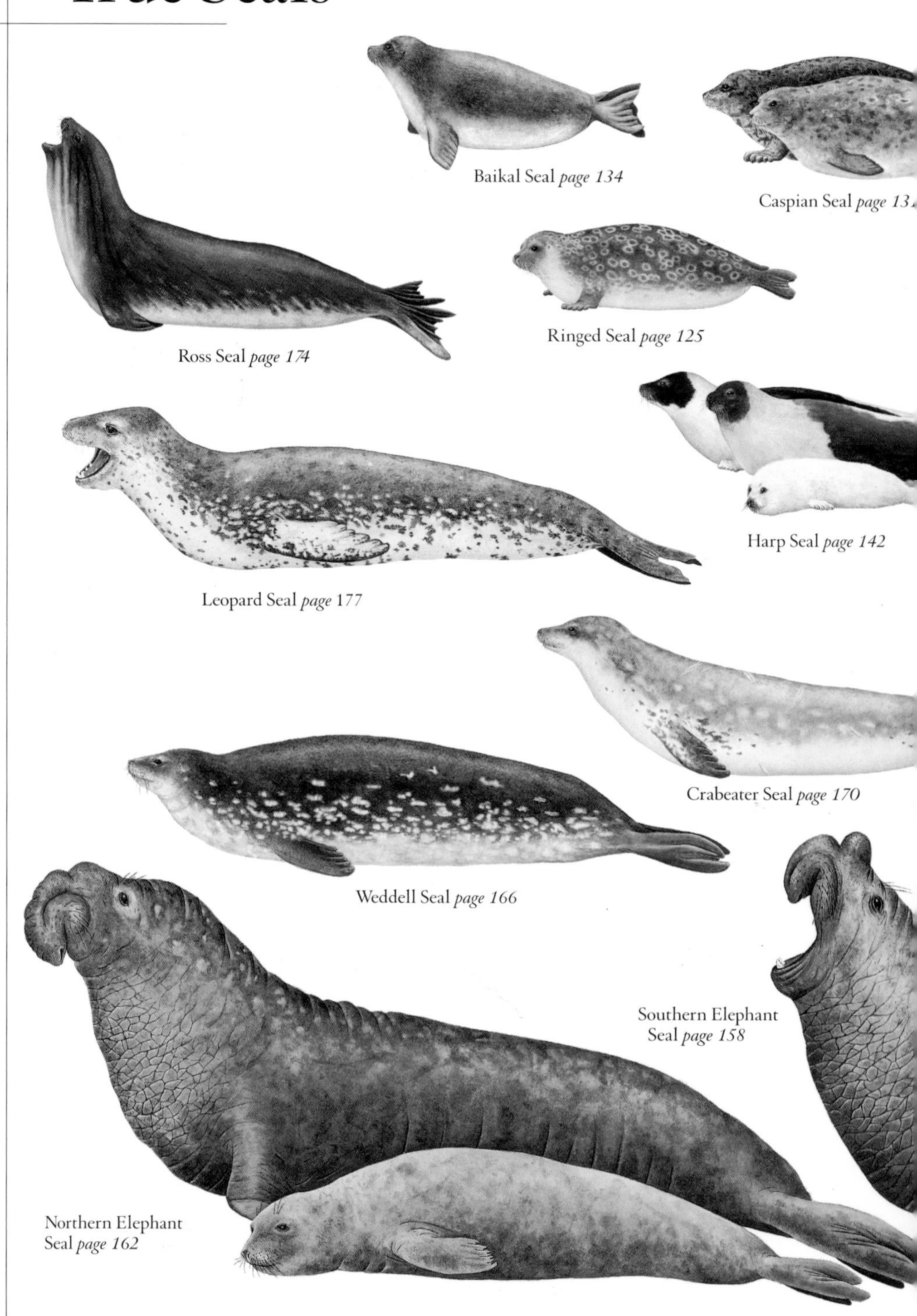
Baikal Seal *page 134*
Caspian Seal *page 13*
Ross Seal *page 174*
Ringed Seal *page 125*
Harp Seal *page 142*
Leopard Seal *page 177*
Crabeater Seal *page 170*
Weddell Seal *page 166*
Southern Elephant Seal *page 158*
Northern Elephant Seal *page 162*

Harbor Seal page 118
Gray Seal page 138
Largha Seal page 122
Bearded Seal page 114
Ribbon Seal page 128
Hooded Seal page 146
Mediterranean Monk Seal
page 150
Hawaiian Monk Seal
page 154

South African and Australian Fur Seals

Arctocephalus pusillus
(Schreber, 1775)

- LARGEST OF ALL FUR SEALS
- SNOUT POINTED, ROBUST, EITHER FLAT OR SLIGHTLY CURVED UPWARD
- MODERATELY LONG FACIAL WHISKERS THAT EXTEND PAST LONG PINNAE
- BREED IN SOUTHERN AUSTRALIA AND SOUTHERN AFRICA

Since 1971, scientists have recognized two subspecies of *Arctocephalus pusillus:* the South African Fur Seal, also called the Cape Fur Seal, and the Australian Fur Seal, also commonly called the Tasmanian or Tasman Fur Seal. Virtually identical in appearance, they are distinguished principally by their widely disjunct distributions. The species was originally described from an illustration of a pup, thus the specific name *pusillus,* Latin for "little." In fact, these are the largest of the fur seals. The colonies along the coast of Namibia, in southwestern Africa, are especially large, due to a highly productive nearshore marine habitat, as well as limited disturbance of the coastline, which is near inland diamond mines and has strictly controlled access.

DESCRIPTION South African and Australian Fur Seals are the largest and most robust of all the fur seals. The head is large and broad, with a long pointed snout that may appear flat or upturned slightly at the tip. The pinnae and vibrissae are long, and the vibrissae may extend backward past the pinnae, especially in adult males. The foreflippers are covered by sparse hair over about three-quarters of their length. The hindflippers are short relative to the large body, with short fleshy tips on the digits. Males are substantially larger and more robust than females. There are no obvious differences in body shape between the subspecies, although scientists have reported slight differences in skull morphology. There are 10 pairs of teeth in the upper jaw and eight pairs in the lower jaw.

Adult males are dark gray to brown with a darker mane of short coarse hairs and a light belly, while adult females are light brown to gray with a light throat and darker back and belly. The foreflippers are dark brown to black. Pups are born black and molt to a gray pelage with a pale throat within three to five months.

RANGE AND HABITAT Australian Fur Seals *(Arctocephalus pusillus doriferus)* breed in Bass Strait, at four islands off Victoria in southeastern Australia and five islands off Tasmania. The largest colonies are off Victoria at Lady Julia Percy Island and Seal Rocks, and off Tasmania at Reid and Judgement Rocks. Vagrants have been reported as far north as Port Stephens in New South Wales. Historically, Australian Fur Seals ranged from South Australia to southern Tasmania and to Jervis Bay, New South Wales. Breeding colonies of the South African Fur Seal *(A. p. pusillus)* extend along the coast of southern Africa, from Cape Cross (Namibia) and around the Cape of Good Hope to Black Rocks, Cape Province. Vagrants have been reported north along Africa's western coast to Angola and south to Marion Island.

Both Australian and South African Fur Seals prefer to haul out and breed on rocky islands, rock ledges and reefs, and pebble and boulder beaches. However, some large breeding colonies of South African Fur Seals occur on sandy beaches along the southern coast of South Africa, and a large aggregation of nonbreeding fur seals regularly hauls out on a sandy beach at Cape Fria in northern Namibia.

SIMILAR SPECIES New Zealand Fur Seals and Australian Sea Lions may overlap with Australian Fur Seals in Bass Strait. New Zealand Fur Seals are smaller, with a darker brown pelage, and mostly occur on sandy beaches. Australian Sea Lions are lighter brown (especially compared with female Australian Fur Seals) and have a more robust body and a wider, less pointed snout; they commonly occur on sandy beaches. Subantarctic Fur Seals may rarely overlap with South African Fur Seals in southern South Africa or at Marion Island. Subantarctic Fur Seals are smaller and have a wider, less pointed snout and occur farther south, mostly at and near subantarctic islands.

BEHAVIOR South African and Australian Fur Seals may travel alone at sea. Females nursing pups go to and from offshore foraging areas in groups of a dozen or more. When resting in nearshore waters, large groups sometimes raft in kelp beds, and sometimes anchor themselves to kelp fronds at the surface. Around Australia in Bass Strait, most dives of lactating females were reportedly to depths of 210 to 280 feet (65–85 m), which in that area is close to the ocean depth.

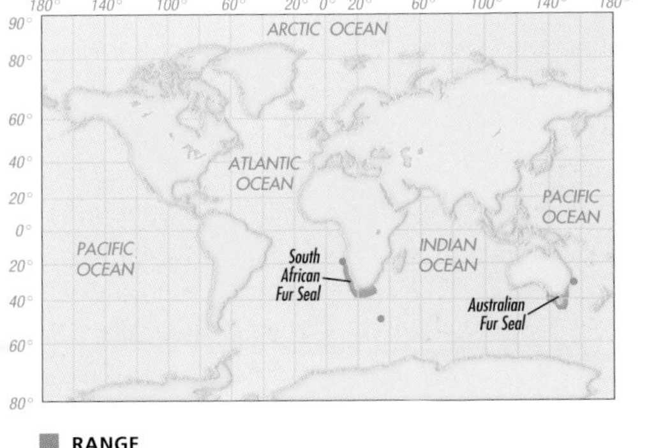

SOUTH AFRICAN AND AUSTRALIAN FUR SEALS
FAMILY OTARIIDAE

MEASUREMENTS AT BIRTH
LENGTH 24–31″ (60–80 cm)
WEIGHT 11–26 lb (5–12 kg)

MAXIMUM MEASUREMENTS
LENGTH **MALE** 7′5″ (2.27 m)
FEMALE 5′7″ (1.71 m)
WEIGHT **MALE** 790 lb (360 kg)
FEMALE 260 lb (120 kg)

LIFE SPAN
MALE 18 years
FEMALE 21 years

This suggests that those fur seals forage mostly for prey that live at the sea bottom or in bottom sediments. Dives of one male studied were to around 45 feet (14 m) and lasted about 2½ minutes, with the deepest dive to 335 feet (102 m) and lasting almost seven minutes. Lactating females depart the colony to forage around sunrise and arrive back at the rookery between dusk and early morning several days later. Most pregnant females spend a substantial period of time (about 7 weeks) away from land just before the beginning of the breeding season. South African and Australian Fur Seals molt in late summer to early autumn, and there may be differences in the timing of the molt according to age and sex. In some areas, Killer Whales and white sharks occasionally eat these fur seals.

REPRODUCTION South African and Australian Fur Seals have a polygynous breeding system. Adult males establish breeding territories beginning in late October. While males are sexually mature at six to eight years old, most are not physically able to establish or maintain breeding territories until around 10 to 12 years old. They defend access to females against other males with ritualized threats and sometimes with direct aggression until females come into estrus. Females give birth to a single pup between late October and late December, with peak pupping in late November and early December. They are ready to mate about 6 to 10 days later. After mating, females begin alternating brief periods foraging at sea with several days ashore nursing their pups. Foraging trips last about seven days in winter and about four days in summer and autumn. Pups are usually weaned at four to six months old, although some may suckle for a year or longer. Pup mortality during the first two months is about 15 to 20 percent.

FOOD AND FORAGING These fur seals eat squid, octopus, and a variety of bony fish. In Australia, they often forage over the continental shelf at the sea bottom. Off South Africa, they may feed just off the continental shelf on schooling fish and squid that live in the cool, productive upwelling waters of the Benguela Current.

STATUS AND CONSERVATION The Australian Fur Seal is still recovering in both abundance and distribution from the commercial sealing for pelts conducted principally from 1798 through 1825 (some hunting continued at a few sites through 1899). Historic populations may have been as high as 200,000, but in 1991 abundance figures were estimated at about 47,000 to 60,000 off southeastern Australia and 15,000 to 20,000 off Tasmania. Most breeding and haulout sites are now protected from disturbance by Australian federal, state, and territorial law. Australian Fur Seals may become entangled in fishing nets and drown, and fishermen may also illegally shoot fur seals that damage their nets. In October 2000, the Tasmanian government authorized the killing of "nuisance" fur seals around fisheries after failed efforts to move them away from areas of fishing activity. The South African Fur Seal population is robust, numbering several hundred thousand or more. Commerical harvesting and scientific collecting of the species continued through the 20th century and into at least 2001.

OPPOSITE: *Australian Fur Seals prefer to haul out on rocky beaches and reefs along the mainland coast of southeastern Australia or nearby islands.*

ABOVE: *South African Fur Seals, also known as Cape Fur Seals, along the Skeleton Coast of Namibia. These fur seals often enter the surf in late morning and may spend most of the day resting there to escape the intense midday heat.*

RIGHT: *Adult male Australian Fur Seals are substantially larger than females and have more robust chests and necks.*

Antarctic Fur Seal

Arctocephalus gazella
(Peters, 1875)

- GRAY TO BROWN DORSALLY, ADULT FEMALE PALER VENTRALLY
- BROAD SHORT SNOUT; MALE HAS DISTINCTLY CONVEX FOREHEAD
- MATURE MALE HAS LONG DARK MANE, THICK NECK AND CHEST
- LARGEST BREEDING COLONIES JUST SOUTH OF ANTARCTIC CONVERGENCE
- SOME SEALS HAUL OUT ON PACK ICE IN WEDDELL SEA AND OFF ANTARCTIC PENINSULA IN SUMMER

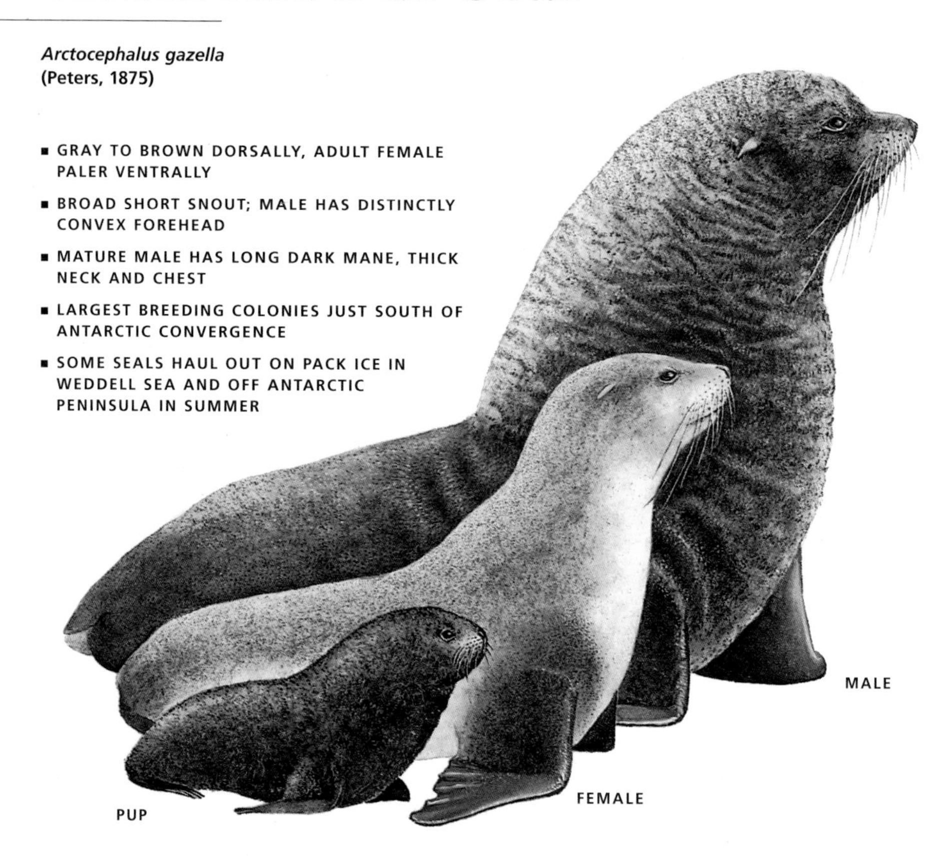

Antarctic Fur Seals were the target of massive commercial seal harvests beginning in the 1790s, when U.S. sealers hunted them extensively at South Georgia Island. They were also hunted at the South Shetland Islands when those colonies were discovered in 1819, and at the South Orkney and South Sandwich Islands. By 1820, records indicated that more than 1 million Antarctic Fur Seals had been killed for their pelts. From the few hundred seals that survived, the species has recovered to its current population of about 2 to 3 million and today may be found at many islands around the Southern Ocean. The specific name is attributed to the *SMS Gazelle*, a German vessel that collected the type specimen from Kerguelen Island. The Antarctic Fur Seal is sometimes also called the Kerguelen Fur Seal.

DESCRIPTION Relative to other fur seals, the Antarctic Fur Seal has a short, broad, and blunt snout, and adult males have a distinctively convex forehead. The snout appears flat from the forehead to the nose, especially in adult males. Adults have quite long facial vibrissae that extend beyond the pinnae. There are nine pairs of teeth in the upper jaw and eight pairs in the lower jaw.

Adult males are uniformly dark brown to charcoal, with some lighter bleaching at the ends of the long guard hairs on the mane. Females and juveniles are gray dorsally and lighter ventrally, particularly on the chest and neck, with a light blaze on the flanks (absent in adult males) that extends toward the tail. Pups are usually black at birth. A small percentage may be cream-colored or whitish.

RANGE AND HABITAT Breeding colonies are scattered around Antarctica, from about 70°W near the Antarctic Peninsula to about 80°E in East Antarctica. About 95 percent of the population breeds at South Georgia, principally off the northern coast of Bird Island, and farther south at the South Orkney and South Shetland Islands. Other key breeding colonies are mostly south or slightly north of the Antarctic Convergence, at Macquarie, McDonald, Heard, Bouvet, Marion, Kerguelen, Prince Edward, and Crozet Islands.

On some islands, such as Macquarie Island, breeding seals prefer cobble and rocky beaches, whereas nonbreeding seals often wander far inland onto grassy meadows or tussock grass. At other sites, such as Heard Island, Antarctic Fur Seals haul out on broad meadows within about 200 feet (60 m) of the beach. In late summer, some adult and subadult males haul out on ice floes, particularly in the western Weddell Sea, and in autumn they haul out in relatively large numbers to molt at island sites along the Antarctic Peninsula, at Heard Island, and on the continent at Mawson and Davis Research Stations. The whereabouts of most seals during winter are poorly known; they are at sea continuously and virtually none are seen ashore.

SIMILAR SPECIES Antarctic Fur Seals are similar in appearance to Subantarctic Fur Seals, and the two species occur together at several sites, including Marion and Macquarie Islands, although they generally segregate into different habitats or areas. Subantarctic Fur Seals have a distinctive orange-yellow face and chest and shorter flippers. Adult males have a tuft of hair on the top of the head, and adult females and juveniles are darker. New Zealand Fur Seals, which overlap on Macquarie Island, are lighter, with a larger snout and shorter foreflippers. Some Antarctic Fur Seals have been observed hauled out in recent years at the Juan Fernández Islands among Juan Fernández Fur Seals; Antarctic Fur Seals may be distinguished by their darker pelage and shorter snout.

BEHAVIOR Adult and subadult males may form groups while molting along the Antarctic Peninsula in late summer and early autumn. Adult females are gregarious but relatively asocial other than the strong bond they establish with their pups, although there are occasional aggressive encounters with nearby females or other pups and brief interactions with adult males to mate. These seals appear to be solitary when foraging and migrating. Females evidently remain at sea continually between breeding seasons, and juveniles may spend several years at sea before returning to natal sites to mate for the first time. The deepest recorded dive for the species is to

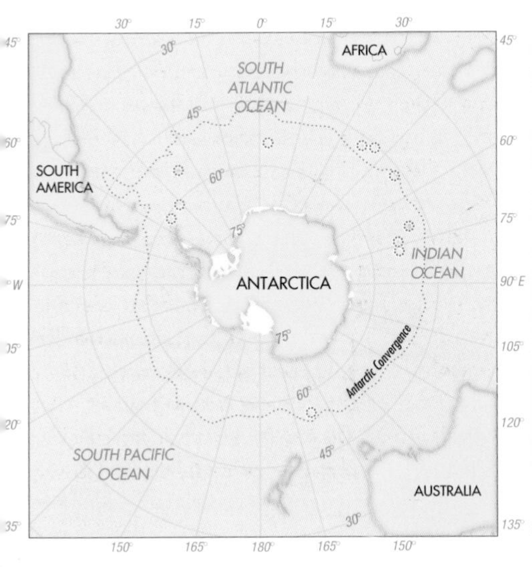

ANTARCTIC FUR SEAL
FAMILY OTARIIDAE

MEASUREMENTS AT BIRTH
LENGTH 24–28″ (60–70 cm)
WEIGHT 9–13 lb (4–6 kg)

MAXIMUM MEASUREMENTS
LENGTH **MALE** 6′7″ (2 m)
FEMALE 4′5″ (1.35 m)
WEIGHT **MALE** 440 lb (200 kg)
FEMALE 88 lb (40 kg)

LIFE SPAN
MALE 14 years
FEMALE 23 years

about 590 feet (180 m); the longest dive lasted 10 minutes. The diving ability of pups substantially improves during the first few months of life, and by about four months old their diving patterns are similar to those of adult females. Leopard Seals eat Antarctic Fur Seal pups, especially at Heard Island, Seal Island near the Antarctic Peninsula, and South Georgia. Survival of suckling pups may be particularly low in years when krill abundance near a colony is insufficient to allow lactating females to forage effectively. Pup mortality also increases in highly congested breeding colonies, owing to the difficulties of pups reuniting with their mothers, harassment by subadult males, and disease. At some sites where similar fur seal species occur together, interbreeding has become more common in recent years as populations have increased. This has been particularly noted at Macquarie Island, where Antarctic, Subantarctic, and New Zealand Fur Seals may all hybridize.

REPRODUCTION The breeding system of the Antarctic Fur Seal is polygynous, and dominant breeding males mate with as many as 20 females during a successful season. Adult males establish breeding territories on beaches in late October to mid-November, preferably just along the shoreline. They are fiercely territorial during the breeding season and aggressively defend access to estrous females from other males, mostly with stereotyped physical displays, lunges, and vocalizations. Occasionally they resort to physical combat, biting at an opponent's chest, neck, and foreflippers. Males may fast during the breeding season for six to eight weeks, losing up to 3½ pounds (1.5 kg) a day. The gestation period lasts about a year. Females give birth to a single pup between mid-November and late December. They mate about 7 to 10 days later and then begin a series of foraging trips at sea that last several days each. In between, they are ashore for one to several days to nurse their pups. Pups are weaned at about four months old.

FOOD AND FORAGING Antarctic Fur Seals feeding near South Georgia and the Antarctic Peninsula appear to forage exclusively on krill, at least in summer. At Heard and Macquarie Islands they eat pelagic deepwater lanternfish and squid, the latter more often in autumn and winter. Males also kill and eat gentoo, king, and macaroni penguins at Bird and Marion Islands. Lactating females forage mostly at night. Nighttime dives of lactating females are to depths of about 100 feet (30 m) or less, while daytime dives are to 130

LEFT: *Very little is yet known about the foraging behavior of young Antarctic Fur Seals. Pups begin entering the water when three to four months old and start to develop their swimming skills in shallow nearshore waters.* **OPPOSITE TOP:** *At South Georgia, Antarctic Fur Seals often haul out on tussock grass, also a key nesting habitat of the endemic water pipit. The increasing population of fur seals on some beaches has had detrimental effects of that rare bird's population.* **OPPOSITE BOTTOM:** *A small percentage of Antarctic Fur Seal pups have cream or yellowish-white colored fur.*

to 245 feet (40–75 m). These depths appear to correspond with the daily vertical distribution of the krill. Nothing is known of the foraging behavior of juveniles or adult males.

STATUS AND CONSERVATION During the 18th and 19th centuries, Antarctic Fur Seals were virtually exterminated by commercial sealers hunting for their fine pelts. However, a little more than a century after these large-scale harvests ended, a small remnant colony at Bird Island off South Georgia had increased to nearly 2 million seals. This colony may have been the source of recruits for most other extant colonies. In the late 1980s, the population at Heard Island was estimated at 15,000, and that at Macquarie Island at about 400. At Marion Island the total population numbered around 1,200 in 1995. The Antarctic Fur Seal may be threatened by the developing commercial trawl fisheries near many of its breeding colonies, including those at Macquarie and South Georgia Islands, and around the Antarctic Peninsula, particularly during summer when lactating females are dependent on abundant nearby prey resources.

The species is protected on Macquarie Island by the Tasmanian Parks and Wildlife Service and on Heard Island by the Australian Antarctic Division. On Prince Edward and Marion Islands, the South African Department of Environment and Tourism prohibits visitors, except for researchers and weather station personnel on Marion Island. At sea, these seals are protected from harvesting in the Antarctic Treaty area south of 60°S by the Convention for the Conservation of Antarctic Seals. The Falkland Islands Dependencies Conservation Ordinance protects them at South Georgia and the South Sandwich Islands.

Subantarctic Fur Seal

Arctocephalus tropicalis
(Gray, 1872)

- DISTINCTIVE, LIGHT ORANGE OR CREAM-COLORED FACIAL MASK
- SHORT, POINTED, FLATTENED SNOUT
- LONG FACIAL VIBRISSAE
- SHORT BROAD FOREFLIPPERS
- ADULT MALE HAS PROMINENT TUFT OF HAIR ON HEAD
- BREEDS MAINLY ON ISLANDS IN SOUTHERN INDIAN AND ATLANTIC OCEANS

The Subantarctic Fur Seal was first described from a specimen mistakenly thought to have come from the northern coast of Australia, thus the specific name, from the Greek word *tropikos,* meaning "tropical." Most colonies of these fur seals, however, occur on subantarctic islands, just north of the Antarctic Convergence. In the early 1800s, commercial sealers supplying fur to European, Asian, and North American markets significantly reduced the number of Subantarctic Fur Seals. Only a few thousand of these survived in remnant populations at Gough, Amsterdam, and Marion Islands. The species recovered substantially during the past century. Researchers first believed the Subantarctic Fur Seal to be the same species as the Antarctic Fur Seal but later designated the two as distinct species, and recent genetic data have confirmed the distinction. Subantarctic Fur Seals are also known as Amsterdam Fur Seals or Amsterdam Island Fur Seals.

DESCRIPTION The Subantarctic Fur Seal has a medium-size but robust body. The head is short with a short broad snout and long facial vibrissae. The foreflippers are broad and relatively short. Mature males are substantially larger than females, with a thick broad chest and a tuft of hair on top of the head. There are 10 pairs of teeth in the upper jaw and eight pairs in the lower jaw.

On both males and females, the chest, snout, and face are cream-colored or rusty orange, and the belly is tan to light brown, sometimes with a dark band between the foreflippers. The area around the insertion of the foreflippers is dark brown, in contrast to the somewhat lighter dorsal pelage. On males the back is gray or chocolate brown and darkens with age, while on females the back is a lighter gray. Pups are black at birth. They molt at about three months old to a pelage that is dark brown dorsally and light gray to cream-colored on the chest and belly.

RANGE AND HABITAT These are among the most widely distributed of the fur seals, breeding at sites in the South Atlantic, Indian, and Pacific Oceans. Most seals breed north of the Subtropical Front on the temperate islands of Gough in the South Atlantic and Amsterdam in the Indian Ocean. Smaller colonies occur farther south near the Antarctic Convergence, at Marion and Prince Edward Islands in the Prince Edward Island group, and at Crozet and Macquarie Islands. Small numbers haul out seasonally at Heard Island, and one pup was born there in 1987. Breeding seals prefer rocky coastal habitats, while nonbreeders often haul out on tussock slopes above these beaches.

SIMILAR SPECIES Antarctic Fur Seals occur with Subantarctic Fur Seals at some islands, such as Marion Island, although they mostly occur south of the Antarctic Convergence, whereas Subantarctics mostly occur north of it. Antarctic Fur Seals generally breed in more open vegetated areas. They lack the orange-yellow face and chest of Subantarctic adults, and males lack the tuft of hair on top of the head; females and juveniles are lighter in color. South African Fur Seals, which may overlap with Subantarctic Fur Seals at sea and occasionally at subantarctic islands, are substantially larger and have longer, pointed snouts. Males have a dark face and chest, and females have a gray rather than cream or orange throat and chest. New Zealand Fur Seals, which overlap at Macquarie Island, lack the Subantarctic's orange facial mask.

BEHAVIOR Subantarctic Fur Seals occur in small groups along rocky coastlines. Lactating females are relatively asocial, keeping away from other seals while ashore with their pups. Males may associate in small groups after the breeding season; however, their social encounters consist primarily of threat postures, vocalizations, and spars. Adult males make three main vocalizations: barks to advertise territorial status, guttural growls or puffs to affirm territorial boundaries, and high-intensity calls to warn or challenge other males. The primary call of females is a loud, tonal honk to attract pups upon their return from foraging trips. Pups respond to these calls with a bleat. Seals begin molting in March and April, and there may be some differences in molt pattern based on age, sex, and reproductive status. Subantarctic Fur Seals evidently hybridize with

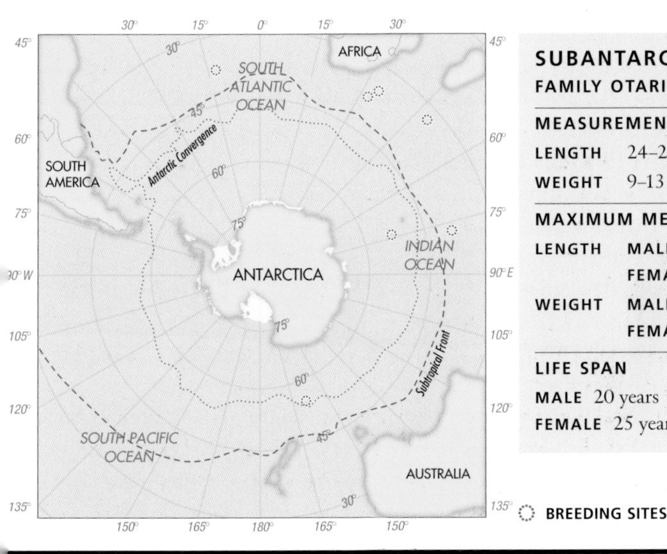

SUBANTARCTIC FUR SEAL
FAMILY OTARIIDAE

MEASUREMENTS AT BIRTH
LENGTH 24–28″ (60–70 cm)
WEIGHT 9–13 lb (4–6 kg)

MAXIMUM MEASUREMENTS
LENGTH **MALE** 6′7″ (2 m)
FEMALE 4′7″ (1.4 m)
WEIGHT **MALE** 350 lb (160 kg)
FEMALE 110 lb (50 kg)

LIFE SPAN
MALE 20 years
FEMALE 25 years

LEFT: *Subantarctic Fur Seals have a distinct orange face mask and chest, long vibrissae, and relatively short narrow foreflippers.*
BELOW: *Adult Subantarctic Fur Seals have a tuft of hair on the top of the head. Open-mouth vocalizations are threats used to exclude other breeding males from access to receptive females.*

Antarctic Fur Seals at the Prince Edward Islands and Crozet and Macquarie Islands, and perhaps also with South African Fur Seals at Marion Island. A small colony that occurs at Macquarie Island also hybridizes with New Zealand Fur Seals.

REPRODUCTION The breeding system is polygynous. Adult males begin establishing breeding territories in late October and defend them with physical posturing, vocal threats, and occasional physical battles. Males may fast for up to two months while defending their territories. Females give birth from early November through December, with most pups born in early December. Females mate about six to seven days later, and then begin a series of foraging trips to sea lasting 5 to 12 days each, alternating with one to several days ashore nursing pups. Foraging trips in winter are longer and may last around 18 to 28 days. Pups are weaned at about 10 months old. Gestation lasts about 11 months.

FOOD AND FORAGING Near Macquarie, Marion, and Amsterdam Islands, Subantarctic Fur Seals eat mostly mesopelagic lanternfish, components of the deep scattering layer that migrate toward the surface at night. Lactating females at Amsterdam Island forage primarily at night. They dive continually to depths of around 50 to 65 feet (15–20 m) in summer and around 100 feet (30 m) in winter, each dive lasting about 1 to 1½ minutes. The deepest recorded dive for the species was to 682 feet (208 m), and the longest recorded dive lasted about 6½ minutes. At other sites, nocturnal diving is also the rule.

STATUS AND CONSERVATION Fur seals were discovered at Macquarie Island in 1810, and by 1820 about 200,000 had been harvested for their pelts and the population exterminated. It is not certain how many of these were Subantarctic Fur Seals. The world population of Subantarctic Fur Seals numbers around 280,000 to 350,000, and several colonies are increasing. The population at Macquarie Island was around 110 in 1995. There were about 50,000 Subantarctic Fur Seals at Amsterdam Island in the early 1990s, and at Marion Island the population numbered about 49,000 and annual births had increased about 2 percent since 1989. The population at Gough Island numbered around 200,000 in 1978 and was increasing. Subantarctic Fur Seals are protected from harvesting by the Convention for the Conservation of Antarctic Seals. The population at Macquarie and Heard Islands is considered to be endangered. Despite great reductions in population during the 1800s, including extinction at several sites, the genetic variability of the Subantarctic Fur Seal is remarkably high.

Juan Fernández Fur Seal

Arctocephalus philippii
(Peters, 1866)

- LONG POINTED SNOUT
- MALE'S THICK YELLOWISH MANE CONTRASTS WITH DARK PELAGE
- HAULS OUT PRIMARILY ON ROCKY COASTS
- DISTRIBUTION RESTRICTED TO JUAN FERNÁNDEZ AND SAN FÉLIX ISLANDS

These fur seals are named for the Spanish navigator Juan Fernández, who discovered them in the mid-1500s. William Dampier, a pirate and explorer, visited the Juan Fernández Islands in the late 1600s and wrote about large numbers of fur seals there, attracting sealers who quickly reduced the population. In fact, the species was thought extinct until a Chilean scientist discovered about 200 fur seals at the islands in 1965. The population has gradually increased and was estimated to number about 12,000 in the late 1990s. Because of their isolation along the rocky coasts of the Juan Fernández Islands, little is known about these animals.

DESCRIPTION The Juan Fernández Fur Seal is the second-smallest fur seal (after the Galápagos Fur Seal). The body is relatively robust. The snout is long, slender, and pointed; adult males have a more bulbous nose than females and juveniles. The foreflippers and hindflippers are stubby, and the hindflippers have fleshy tips on the digits. Adult males are longer than adult females and substantially more robust at the chest, neck, and shoulders. They have a mane of long coarse guard hairs that extends from the top of the head to the shoulders. There are 10 pairs of teeth in the upper jaw and eight pairs in the lower.

Adult males are dark brown to black, with the guard hairs on the mane tipped golden yellow to tan. Adult females are chocolate brown overall, though the tips of the guard hairs may fade to yellow or tan. Pups are black at birth, and the pelage becomes lighter during the first few years.

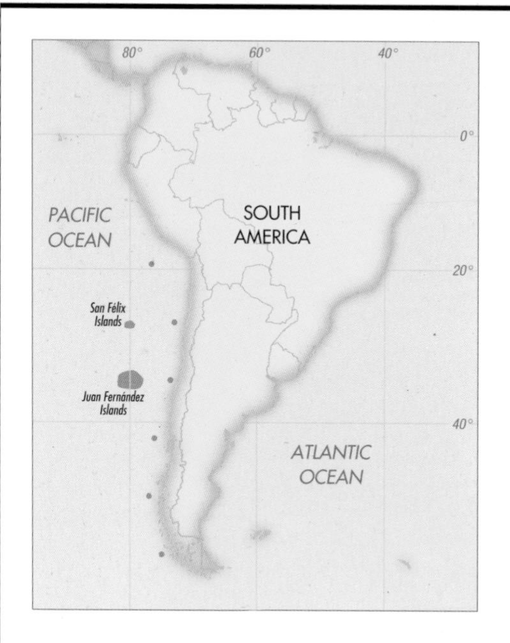

JUAN FERNÁNDEZ FUR SEAL

FAMILY OTARIIDAE

MEASUREMENTS AT BIRTH

LENGTH	26–28″ (65–70 cm)
WEIGHT	13–15 lb (6–7 kg)

MAXIMUM MEASUREMENTS

LENGTH	**MALE** 6′7″ (2 m)
	FEMALE 4′7″ (1.4 m)
WEIGHT	**MALE** 310 lb (140 kg)
	FEMALE 110 lb (50 kg)

LIFE SPAN Unavailable

RANGE
VAGRANTS

RANGE AND HABITAT Juan Fernández Fur Seals breed only at the Juan Fernández and San Félix Islands off the coast of Chile. Their distribution at sea is poorly known. In autumn and winter, vagrant seals have been recorded as far south as Punta San Juan, Chile, and as far north as Peru. When hauled out ashore, Juan Fernández Fur Seals prefer caves and beaches with rocky, volcanic substrates, especially at the base of bluffs. At Más a Tierra Island, they occur on the barren, dry south shore and have so far avoided the island's wetter, heavily vegetated north coast.

BELOW: *Juan Fernández Fur Seals haul out and breed along beaches of volcanic boulders at the Juan Fernández and San Felix Island groups off central Chile.*

OPPOSITE: *Male Juan Fernández Fur Seals have more bulbous noses than females. This difference begins appearing when seals are two to four years old. The foreflippers of these fur seals are rather stubby and partially covered by hair on the dorsal surface and have barely noticeable nails.*

SIMILAR SPECIES Antarctic Fur Seals may haul out among Juan Fernández Fur Seals at the Juan Fernández Islands, and South American Sea Lions might occur with transient Juan Fernández Fur Seals along the Chilean coast. Both species have a shorter, less pointed snout than the Juan Fernández Fur Seal. South American Sea Lions have a shorter pelage that is lighter brown to tan.

BEHAVIOR When foraging, Juan Fernández Fur Seals appear to be solitary, though very little is known about their habits at sea. On land, females seem to be very picky about where they haul out, mostly choosing rugged, rocky coastlines and caves. Though gregarious, they are not particularly social and tend to keep at least a few feet from each other when ashore. These preferences determine where adult males are attracted and where the most intense interactions occur during the breeding season between competing territorial males. Juan Fernández Fur Seals tend to limit their onshore activities to the morning, later afternoon, and evening because of hot midday temperatures. They may enter the water often to cool off during the day as temperatures rise, particularly when winds are light. While in the water they often rest with the head pointed down and the hindflippers waving in the air.

REPRODUCTION Juan Fernández Fur Seals are polygynous breeders, and males compete aggressively to establish and maintain territories along areas of the coastline where females prefer to haul out. Their territories may extend into tide pools or areas just offshore. Pups are born from mid-November through December. Females remain with their pups for seven to nine days before mating, keeping near the water to stay cool during the hot summer days. After mating, females begin a series of foraging trips at sea, coming ashore to nurse their pups in between trips. Female Juan Fernández Fur Seals engage in the longest foraging trips, as well as periods spent ashore, of any otariid species, remaining at sea an average of about 12 days, sometimes up to 25 days, and spending about five days ashore with their pups. Pups are weaned at about 10 months old.

FOOD AND FORAGING Lactating females eat mostly lanternfish and squid, diving to depths of 35 to 300 feet (10–90 m) and ranging as far as 300 miles (500 km) offshore. They mostly dive and feed at night, when the prey migrate upward in the water column as part of the deep scattering layer. Nothing is known about the diet or foraging behavior of males, nonlactating females, or juveniles.

STATUS AND CONSERVATION Beginning in the late 1600s, these seals were hunted intensively by commercial sealers for their pelts, with up to perhaps several million killed through the 1700s and early 1800s. The species was presumed extinct from 1824 until 1965, when a small colony of about 200 seals was discovered in the Juan Fernández Islands. The Chilean government has protected Juan Fernández Fur Seals from hunting or disturbance since 1978, although fishermen on Más Afuera Island evidently poach them for bait, food, or barter. The population was estimated at 12,000 in the early 1990s.

Guadalupe Fur Seal

Arctocephalus townsendi
Merriam, 1897

- THICK, UNIFORMLY DARK PELAGE
- NARROW POINTED SNOUT, ESPECIALLY IN MALE
- RELATIVELY SHORT HINDFLIPPERS AND FLIPPER DIGITS
- ON FOREFLIPPERS, HAIR ON DORSAL SURFACE EXTENDS BELOW WRIST
- BREEDS ALMOST EXCLUSIVELY ON GUADALUPE ISLAND

The Guadalupe Fur Seal is the only member of the genus *Arctocephalus* that occurs in the Northern Hemisphere. The specific name refers to C. H. Townsend, an American scientist who collected several fur seal skulls from Guadalupe Island in the late 1800s. Commercial sealing had reduced the population to only a few dozen animals by then. Surprisingly, molecular evidence so far suggests that the gross reduction in population size may not have substantially affected the genetic variability of the species. A small colony was discovered at San Benito Island (Baja California) in 1997, and in the same year a pup was born at the Channel Islands off southern California. In addition, several subadult males have been seen in the Gulf of California in recent years.

DESCRIPTION The Guadalupe Fur Seal has a slender pelvis and hind end relative to the rest of the body. The head is long and narrow and fairly flat, especially in females. The snout is long, narrow, and pointed. The hindflippers are short, relative to those of most other fur seals, and the foreflippers are broad, with hair extending below the wrist on the dorsal surface. Adult males are considerably longer and larger-bodied than adult females, and they have a thicker chest and neck, and a thick, uniform mane that extends from the forehead to the shoulders. There are 10 pairs of teeth in the upper jaw and 8 pairs in the lower jaw.

The pelage of both sexes ranges from uniform dark brown to black. On males, the guard hairs on the mane are tipped with light tan or yellowish.

Pups are black at birth, lightening to dark brown after several years. Pups may molt during the first few weeks and again when several months old.

RANGE AND HABITAT The Guadalupe Fur Seal's current breeding range is limited almost exclusively to Guadalupe Island off the Pacific coast of Baja California. Researchers discovered another small colony at San Benito Island off Baja California in the 1997, and one pup was born at San Miguel Island, in the Channel Islands off southern California, the same year. In late summer, after the breeding season, some seals range north to the Channel Islands and occasionally as far as central California. The southernmost sightings of this species have been at Los Islotes in the southern Gulf of California. The whereabouts of Guadalupe Fur Seals during the nonbreeding season, from autumn through spring, are generally not known. Historically, Guadalupe Fur Seals bred from the Channel Islands south at least to San Benito and Cedros Islands, and they may also have ranged seasonally as far north as the Farallon Islands (although the fur seals that bred there were evidently Northern Fur Seals). The validity of historic reports of the species as far south as Socorro and Clipperton Islands, off Mexico, has been questioned.

When ashore during the breeding season, Guadalupe Fur Seals favor rocky habitats near the water's edge and caves at windier sections of coastlines, as these areas offer the most relief from the summer heat. The territories of some breeding males may be centered around tide pools, which attract adult females.

SIMILAR SPECIES Northern Fur Seals occasionally occur with Guadalupe Fur Seals on land at San Miguel Island in the Channel Islands, and their ranges at sea may overlap from southern to central California. Northern Fur Seals have a shorter, less pointed snout, longer digits on the hindflippers, and hair on the dorsal surface of the foreflippers that ends abruptly at the wrist. Female Northern Fur Seals can also be distinguished by their more contrasting, dark-and-silver color pattern. California Sea Lions may occur with Guadalupe Fur Seals at Guadalupe Island and in waters off Baja California and southern California to central California. Juveniles and adult female sea lions can be distinguished by their light tan to yellow pelage and shorter snout; similarly, adult males are larger and chocolate brown to black, but they have a shorter snout, a more robust hind end, and a light tuft of hair on the top of the head.

BEHAVIOR Researchers know little about the whereabouts of Guadalupe Fur Seals during the nonbreeding season from September through May, but they are presumably solitary when at sea. On land they are mostly asocial and space out to avoid body contact and interactions with other seals. On exceptionally hot and calm days,

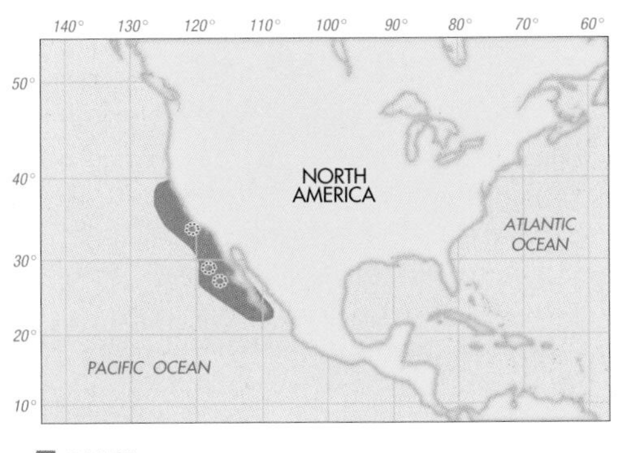

GUADALUPE FUR SEAL
FAMILY OTARIIDAE

MEASUREMENTS AT BIRTH
LENGTH 24″ (60 cm)
WEIGHT 4–9 lb (2–4 kg)

MAXIMUM MEASUREMENTS
LENGTH **MALE** 7′3″ (2.2 m)
FEMALE 6′3″ (1.9 m)
WEIGHT **MALE** 490 lb (220 kg)
FEMALE 121 lb (55 kg)

LIFE SPAN
MALE 18 years
FEMALE 23 years

they may regularly enter the surf or tide pools to cool down. In nearshore waters, they appear to spend most of their time grooming at the water's surface. At other times, they rest with the head pointed downward and the hindflippers protruding above the water, or float on their sides with the tips of one foreflipper and one hindflipper touching above the water. At San Nicolas Island in the Channel Islands off southern California, male Guadalupe Fur Seals have occasionally established territories among breeding California Sea Lions. While they have been observed attempting to mate with the sea lion females, no obvious hybrids have yet been encountered.

REPRODUCTION Guadalupe Fur Seals are polygynous breeders; males may mate with up to a dozen females during a single breeding season. Males establish small territories along rocky coastlines or in caves. They may bark quietly and continually while patrolling their territories. When confronting other males, they roar or loudly

ABOVE: *Guadalupe Fur Seals cool off during a hot day by hanging almost motionless in the water with their heads pointed down.*
RIGHT: *Adult male Guadalupe Fur Seal with females. Males are substantially larger and have more robust chests and necks than females. A male Guadalupe Fur Seal may mate with up to a dozen females in a season.*

cough to assert dominance. Females give birth from early June through July, with a peak in late June. They mate about a week after giving birth, and then begin a series of foraging trips lasting two to six days, sometimes up to 13 days when warm-water conditions prevail and prey is less abundant and more dispersed. They come ashore for four to six days between foraging trips to nurse their pups. Pups are weaned at about nine months old.

FOOD AND FORAGING The Guadalupe Fur Seal's diet is poorly known but appears to consist of pelagic squid, lanternfish, and mackerel. Lactating females may travel a thousand miles or more from the breeding colony to forage. They appear to feed mostly at night, at depths of about 65 feet (20 m), with dives lasting about 2½ minutes.

STATUS AND CONSERVATION Commercial hunting by sealers, whalers, and Sea Otter hunters during the 1800s as well as subsistence hunting by aboriginals in the Channel Islands nearly exterminated Guadalupe Fur Seals; in fact, the species was thought to be extinct by the early 1900s. In 1949, a single male was observed at San Nicolas Island, and soon afterward a small colony of several dozen seals was discovered at Guadalupe Island. In 1975, the Mexican government designated Guadalupe Island a pinniped sanctuary (although some poaching and disturbance to the breeding sites continue there). North of Mexico, the Guadalupe Fur Seal is listed as a threatened species under the U.S. Endangered Species Act. The population was estimated at about 6,000 in 1987 and 10,000 in the late 1990s. The severe reduction of the Guadalupe Fur Seal has evidently had a less substantial effect on its gene pool compared to other similarly depleted pinniped species, as relatively high levels of genetic variability have been reported.

Guadalupe Fur Seals occasionally haul out at the Channel Islands off California in summer. The male fur seals attempt to herd and breed with California Sea Lion females and may also displace resident California Sea Lion males from their breeding areas.

New Zealand Fur Seal

Arctocephalus forsteri
(Lesson, 1828)

- LONG, STRAIGHT, POINTED SNOUT
- LARGE FLESHY NOSE
- LONG PINNAE
- PREFERS ROCKY AND COBBLE BEACHES FOR HAULING OUT AND BREEDING
- BREEDS MAINLY AT SOUTH ISLAND, NEW ZEALAND, AND SOUTHEASTERN AUSTRALIA

There are two geographically separated breeding populations of New Zealand Fur Seals: one around South Island, New Zealand, and the other along the coast of southeastern Australia. While some genetic differences exist between the two populations, there are no obvious morphological distinctions. The specific name refers to the naturalist J. G. A. Forster, who described these seals during his voyage around the world with Captain James Cook in the late 1700s. The species was first included in the genus *Otaria,* but recent classifications include it in *Arctocephalus.*

DESCRIPTION The New Zealand Fur Seal is thick-bodied with a long, flat head. The snout is long and pointed, and the nose large and fleshy. The pinnae and facial vibrissae are relatively long. Both the fore- and hindflippers are rather short and narrow, with barely noticeable claws. Adult males develop a thick chest and neck, and have a mane of long guard hairs on the back of the neck. There are 10 pairs of teeth in the upper jaw and eight pairs in the lower jaw.

Adult males are generally dark grayish brown, sometimes appearing slightly paler ventrally. Females and juveniles are grayish brown, with a lighter belly and throat. Pups are black at birth, molting at about two to three months old to a gray pelage that is lighter below.

RANGE AND HABITAT New Zealand Fur Seals prefer jumbled rocky and cobble beaches for hauling out and breeding. Principal breeding colonies occur in New Zealand at South Island and Stewart Island, at scattered locations on the coast of Western and South Australia, and off Tasmania at Maatsuyker Island. There are also breeding colonies at subantarctic Chatham, Campbell, Antipodes, Bounty, Auckland, and Macquarie Islands and at Kangaroo Island off South Australia. Nonbreeding seals occasionally range as far west as Perth on the west coast of Western Australia and northeast to Queensland (Australia) and New Caledonia. Prior to hunting by commercial sealers in the early 1800s, populations were evidently abundant in eastern Bass Strait at the Furneaux Island Group.

SIMILAR SPECIES Australian Fur Seals and Australian Sea Lions overlap with New Zealand Fur Seals along the coast of southern Australia. The sea lions are lighter, with a light gray to tan pelage, and have a blunter snout, a broader head, and a more robust hind end. Australian Fur Seals have a darker brown pelage. Subantarctic and Antarctic Fur Seals occur with New Zealand Fur Seals at Macquarie Island. Subantarctics are easily distinguished by their orange facial mask and pale chest. Antarctics are darker, with longer vibrissae, a shorter snout, and longer flippers.

BEHAVIOR The social structure of New Zealand Fur Seals is most ordered during the summer breeding season. Competing males establish and defend territories along rocky coastlines and in or near coastal caves, where females congregate to give birth and nurse their pups. Adult males fast for up to two months during this time and remain mostly inactive, except for occasional threatening and fighting or when they are mating. Females are mostly asocial and avoid close contact with other females. Females and their pups and some juveniles may be ashore throughout the year, though their numbers on shore are lowest in winter. Adult males are mostly absent from the breeding sites from March through September, but they may show up at more southern islands, such as Macquarie, Campbell, and the Antipodes Islands in late autumn when they are molting.

REPRODUCTION New Zealand Fur Seals have a polygynous breeding system. Adult males establish breeding territories in mid- to late November, and for the next one to two months they defend groups of pre-estrous females against other males. Females give birth to a single pup anywhere from late November through mid-January, with most births occurring from early December to early January. Females come into estrus about eight days after giving birth. They stay with their pups for about 10 days, and then depart on the first of repeated foraging trips at sea. These trips last several days and alternate with one to three days spent ashore nursing. About 1 percent of pups die during the first six weeks of life, and about another 8 percent die during the next 10 weeks.

FOOD AND FORAGING New Zealand Fur Seals breeding at Kangaroo Island eat mostly cephalopods in summer and bony fish in winter,

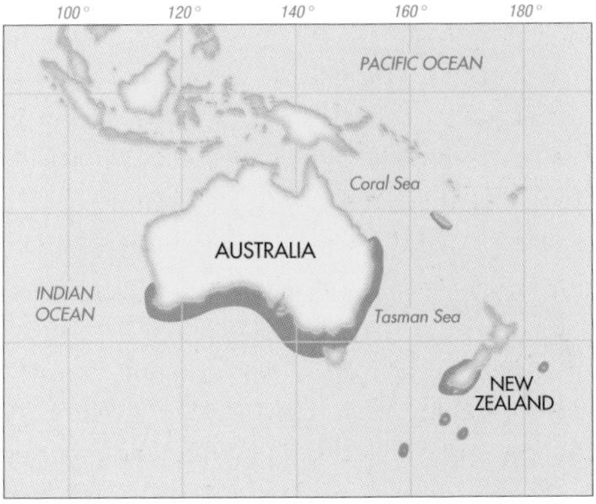

100° 120° 140° 160° 180°
PACIFIC OCEAN
Coral Sea
AUSTRALIA
INDIAN OCEAN
Tasman Sea
NEW ZEALAND

■ RANGE

NEW ZEALAND FUR SEAL
FAMILY OTARIIDAE

MEASUREMENTS AT BIRTH

LENGTH	24–28″ (60–70 cm)	
WEIGHT	9–13 lb (4–6 kg)	

MAXIMUM MEASUREMENTS

LENGTH	**MALE**	8′2″ (2.5 m)
	FEMALE	4′11″ (1.5 m)
WEIGHT	**MALE**	400 lb (180 kg)
	FEMALE	110 lb (50 kg)

LIFE SPAN Unavailable

and males occasionally eat seabirds, including little and rockhopper penguins. They forage at night, diving to about 35 to 50 feet (10–15 m), with deeper dives near dawn and dusk. Dives are shallowest and shortest during summer, becoming deeper and longer through autumn and winter. Lactating females dive up to 900 feet (275 m) for as long as 11 minutes.

STATUS AND CONSERVATION The population of New Zealand Fur Seals in New Zealand totaled about 50,000 in the early 1980s and appears to be increasing. Estimates of the species' abundance in Australia in 1991 totaled about 35,000. The population at Macquarie Island, principally nonbreeding seals hauled out in April and May, was estimated at 2,000 in the late 1980s. Approximately 10,000 New Zealand Fur Seals have been incidentally killed in nets of trawl fisheries around South Island between 1989 and 1998. Recently, aquaculture facilities have developed along the coast at South Island, attracting the fur seals to the net pens and generating covntroversy over potential damage and loss of fish. A deep-sea fishery for Patagonian toothfish occurs within this species' range near Macquarie Island, though the rate of entanglement of seals in those trawl nets appears to be much lower.

ABOVE LEFT: *New Zealand Fur Seals forage mostly at night, though they may spend part of the day in the water to cool down.*
ABOVE RIGHT: *When away from shore on foraging trips, New Zealand Fur Seals rest at the surface during the day and forage mostly at night. The "jug handle" pose of the hindflippers arching to anchor against a foreflipper is characteristic of many fur seal species.*
BELOW: *During the breeding season, male New Zealand Fur Seals use vocal and visual threats to maintain territories. Occasionally these threats escalate into physical combat, with males pushing against each other's chests and slashing with their canine teeth.*

South American Fur Seal

Arctocephalus australis
(Zimmermann, 1783)

- MODERATELY LONG POINTED SNOUT
- LONG PINNAE
- STOCKY BODY
- OCCURS ALONG ROCKY COASTS OF SOUTH AMERICA AND FALKLAND ISLANDS

Perhaps the largest continuous commercial harvest of any pinniped has been of South American Fur Seals in Uruguay. Hunting began there in 1515 and continued virtually without interruption through the 1980s, when it finally began to decrease as the demand for furs declined. South American Fur Seals occur mainly along the coasts of South America, with a smaller population at the Falkland Islands, which is reported to be larger-bodied and which some researchers consider to be a subspecies (*Arctocephalus australis australis*). These seals have also been called Southern Fur Seals and Falkland Fur Seals.

DESCRIPTION The South American Fur Seal is relatively stocky with a long pointed snout, a prominent forehead, and rather long pinnae. The flippers are long and slender. Adult males are substantially more robust than females at the chest, head, and neck. During puberty, males develop a mane of long guard hairs that extends from the top of the head to the shoulders. There are 10 pairs of teeth in the upper jaw and eight pairs in the lower jaw.

Adult males are dark brown to black, with the guard hairs of the mane tipped light gray to yellow. Females and juveniles are dark brown to dark gray dorsally and paler brown or gray on the snout, neck, chest, and belly. Pups are black at birth and gradually lighten as they molt during the first month of life and perhaps again several months later.

RANGE AND HABITAT These seals occur along the coasts of South America, north to central Peru in the Pacific Ocean, and north along the

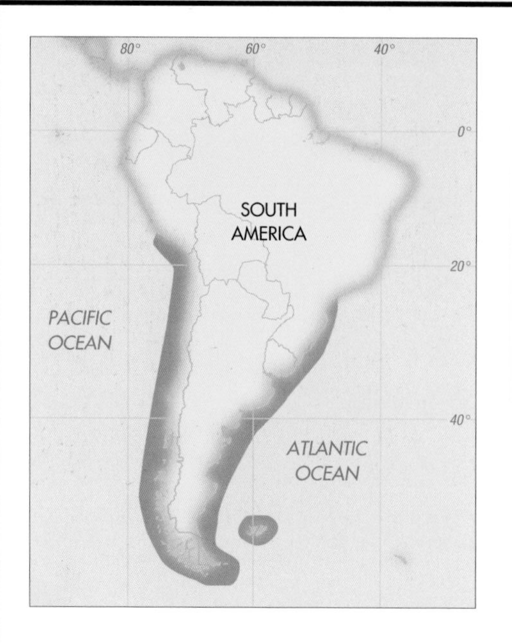

SOUTH AMERICAN FUR SEAL
FAMILY OTARIIDAE

MEASUREMENTS AT BIRTH
LENGTH 24–26″ (60–65 cm)
WEIGHT 8–12 lb (3.5–5.5 kg)

MAXIMUM MEASUREMENTS
LENGTH **MALE** 6′3″ (1.9 m)
FEMALE 4′7″ (1.4 m)
WEIGHT **MALE** 440 lb (200 kg)
FEMALE 110 lb (50 kg)

LIFE SPAN Unavailable

■ RANGE

coast of Argentina to Uruguay in the Atlantic. Small colonies also occur throughout the Falkland Islands (at Volunteer Rocks, Elephant Jason Island, and New Island). South American Fur Seals prefer rocky beaches for hauling out and breeding. In autumn and winter, after the breeding season, some seals range northward along the Atlantic coast to southern Brazil.

SIMILAR SPECIES South American Sea Lions occur throughout the range of South American Fur Seals. They are much larger, have a lighter tan pelage and a shorter, broader snout, and lack the dense underfur of fur seals.

BEHAVIOR Researchers know little about the whereabouts and behavior of South American Fur Seals during the nonbreeding season, when they spend more time at sea. Groups of up to 20 individuals have occasionally been seen traveling together. Near shore, they may raft together in small groups with their hindflippers waving above the sea's surface and their heads pointing downward below the water. At colonies in Peru and Uruguay, South American Sea Lion males abduct, kill, and eat fur seal pups. In some years, this mortality may account for up to 13 percent of all pups born. Along the coast of Peru, vampire bats attack these fur seals, especially the pups.

REPRODUCTION South American Fur Seals have a polygynous breeding system. Males are territorial and may engage in intense fights when

Female South American Fur Seals give birth to a single pup from late October through mid-December and then nurse them for 6 to 12 months before abruptly weaning them. When females return to colonies after feeding at sea for several days, they quickly relocate their pups—first by their vocalizations and then by scent to confirm.

RIGHT: *Adult male South American Fur Seals are stocky and robust-bodied with a long pointed snout, which may appear to be slightly upturned, and a prominent forehead. They do not have a tuft of hair on the top of the head as do males of many other species of eared seal.*

BELOW: *South American Fur Seals prefer to haul out on rocky beaches and ledges like this one at New Island in the Falkland Islands.*

defending access to estrous females against challenging males. Pups are born from late October through mid-December. Females mate about 7 to 10 days after giving birth and then begin foraging, spending three to six days at sea and returning to land for one or two days between trips to nurse their pups. Pups are weaned at about six months old at breeding colonies in Peru; in Uruguay, they continue nursing for up to 12 months.

FOOD AND FORAGING South American Fur Seals eat a variety of fish, cephalopods, and invertebrates. Their key prey species are anchovy in the Pacific Ocean and lobster and krill in southern Chile. When foraging, they typically dive to about 130 feet (40 m), although dives of up to 560 feet (170 m) lasting for seven minutes have been recorded.

STATUS AND CONSERVATION Beginning at least 6,000 years ago, aboriginals hunted South American Fur Seals for their meat and fur. Europeans began harvesting these seals soon after first encountering them in the early 1700s. Harvesting intensified in the late 1700s and continued at substantial levels into the 1980s, when the demand for furs began to decline. South American Fur Seals continue to be hunted in Uruguay; about 12,000 males were harvested annually between 1987 and 1991, and about 5,000 annually since then. South American Fur Seals are also killed incidentally in fishing nets and traps, and are illegally hunted for their meat, for use as bait, and to reduce competition for fisheries, especially in Peru. In early 1997, several thousand pups may have perished in an oil spill off the coast of Uruguay. The largest population of South American Fur Seals occurs in Uruguay and may number about 250,000. In the early 1990s, the population in Peru numbered at least 20,000 but declined to about 6,000 by 1998. The decline was evidently caused by increased mortality or relocation of fur seals due to warming of coastal waters during the 1997–1998 El Niño.

Galápagos Fur Seal

Arctocephalus galapagoensis
Heller, 1904

- SMALLEST OF THE EARED SEALS
- SHORT POINTED SNOUT
- MALE LACKS SAGITTAL CREST
- RESTRICTED TO THE GALÁPAGOS ISLANDS

The Galápagos Fur Seal, endemic to the Galápagos Islands, is the smallest member of the otariid family and the only fur seal that breeds in tropical latitudes. It is closely associated with the cool upwelling waters that circulate around the islands and create conditions for greater productivity of prey populations than generally occurs in tropical waters. Galápagos Fur Seals have the lowest reproductive rate of any pinniped, presumably because productivity of prey communities in these tropical waters is more variable compared to that in temperate and polar environments. During El Niño years, the upwelling systems fail and the near-surface waters warm substantially, disrupting and impoverishing the regional food chain. As a consequence, pup survival is extremely poor in some years.

DESCRIPTION The Galápagos Fur Seal has a small body and a short pointed snout. The foreflippers are long and broad, and the hindflippers short and slender. Adult males are two to three times larger than females, with a thicker neck and broader chest and shoulders. Males lack the sagittal crest on the cranium that is typical of other male otariids and have no distinct mane. There are 10 pairs of teeth in the upper jaw and eight pairs in the lower jaw.

Adult males are dark brown on the back and sides, with pale tan heads and bellies, and dark brown to black flippers. Adult females are similarly patterned, but lighter, especially on the ventral surfaces. Pups are black at birth and become lighter over the next several months. They may molt during the first few weeks and again at several months old.

RANGE AND HABITAT Galápagos Fur Seals breed only in the Galápagos Islands. They have been reported breeding on at least 15 of the islands in the archipelago. Most breeding sites are on west-facing beaches, which are closer to the productive upwelling and cool offshore waters. The largest colonies are at Isabela Island, which accounts for about one-third of the population, and at Fernandina Island. The Galápagos Fur Seal is thought to be nonmigratory, remaining in the Galápagos as a year-round local resident. Nothing is known of their whereabouts when they are foraging at sea.

SIMILAR SPECIES Galápagos Sea Lions occupy the same range as Galápagos Furs Seals and can be distinguished by their larger size and much lighter pelage, and by their vocalizations, which are generally louder. Adult male sea lions bark, compared with the coughs and growls of the fur seal males. The pup-attraction call of female sea lions is siren-like compared with the sheep-like bawl of fur seal females. In addition, Galápagos Sea Lions prefer to haul out on open beaches, whereas the fur seals favor rocky substrates.

BEHAVIOR The tropical heat has a great influence on the social organization and behavior of Galápagos Fur Seals. When ashore, they often find shelter from the sun behind large boulders or in caves, spacing out from one another to avoid body contact. They cool off by lying close to the water's edge and frequently entering the surf or tide pools. The territories of adult males are constrained to smaller areas along the splash zone, where females aggregate for thermoregulatory purposes. There is a strong bond between mothers and pups, maintained by joint recognition of vocalizations and the mother's recognition of her pup's scent.

REPRODUCTION Galápagos Fur Seals have a polygynous breeding system. Males establish territories on beaches and remain there for up to 27 days at a time. Because of the warm climate in the Galápagos, females can give birth over a period of several months; thus males may have several periods of successful breeding during a season. After a brief period feeding at sea to recover weight lost while defending their territory, they may return to establish a new territory. Females give birth in September and October, when temperatures are coolest and nearshore abundance of prey is greatest. Females come ashore about two or three days before giving birth. They mate about eight days later, and then begin an alternating cycle of one to three days foraging at sea and about one day ashore nursing their pups. Pups are usually not weaned until they are two or three years old. These mother-pup bonds are the longest of any otariid pinniped.

FOOD AND FORAGING Galápagos Fur Seals eat lanternfish and squid, components of the deep scattering layer that migrate toward the ocean surface at night. They appear to spend more time foraging during new moons, perhaps because

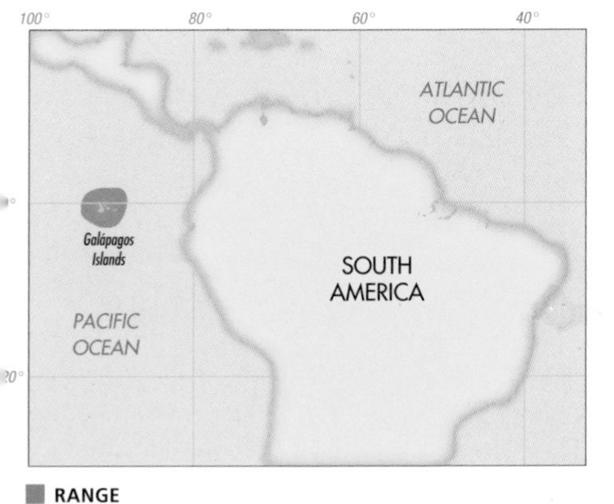

GALÁPAGOS FUR SEAL
FAMILY OTARIIDAE

MEASUREMENTS AT BIRTH
LENGTH Unavailable
WEIGHT 7–9 lb (3–4 kg)

MAXIMUM MEASUREMENTS
LENGTH **MALE** 5′3″ (1.6 m)
FEMALE 4′3″ (1.3 m)
WEIGHT **MALE** 154 lb (70 kg)
FEMALE 88 lb (40 kg)

LIFE SPAN Unavailable

Galápagos Fur Seal females nurse their pups for about one day at a time, alternating with one to three days at sea foraging. Most pups are nursed for at least two years but occasionally three, the longest period of maternal care for any pinniped except the Walrus.

their vertically migrating prey comes closer to the surface then. Foraging depths increase as the seals age. Six-month-old seals dive to about 20 feet (6 m) for 50 seconds; one-year-olds dive up to 150 feet (47 m) for 2½ minutes; and at 18 months old seals dive to 200 feet (61 m) for 3 minutes. The longest and deepest recorded dive for this species was a female's dive to 555 feet (169 m) for 6½ minutes.

STATUS AND CONSERVATION During the early 1800s, Galápagos Fur Seals were exploited by commercial whalers who occasionally landed on the islands, and later they were hunted by commercial sealers. At least 22,000 seals were killed during this period, and by the early 20th century the species was almost extinct. Nonetheless, scientific collectors continued to kill Galápagos Fur Seals until 1934, when the Ecuadoran government prohibited hunting of the species. The prohibition has been well enforced since 1959, when most of the Galápagos archipelago was established as a national park. Feral dogs were eliminated from the coastlines of some islands, and in 1998 the government established a no-fishing zone around the islands ranging from 15 to 40 miles (25–65 km), although this has

Galápagos Fur Seal in typical "jughandle" pose, adopted when resting at the sea surface.

LEFT: *Galápagos Fur Seals spend time in nearshore waters to get relief from the intense daytime tropical heat.* **BELOW:** *Adult male Galápagos Fur Seals establish breeding territories on rocky beaches near the water's edge. They maintain these territories with visual and vocal threats toward other males and occasionally brief physical combat.*

been substantially violated by local fishermen. The population grew from an estimated 27,000 seals in 1978 to about 40,000 in 1988. Extremely warm-water El Niño conditions in 1982 and 1983 and again in 1997 and 1998 greatly reduced prey abundance and substantially affected pup and perhaps adult survival. In 1982, 33 percent of pups died within a month of birth, and none survived more than five months. Once waters cooled again in 1983, however, about 90 percent of newborn pups survived their first month. Of the pups born during the 1997–1998 El Niño event, virtually none survived their first year. In January 2001, some fur seals may have become oiled and died from a spill of about 240,000 gallons of oil that occurred when a boat ran aground at San Cristóbal Island, but the impact of this spill on the fur seal population is not clear.

Northern Fur Seal

Callorhinus ursinus
(Linnaeus, 1758)

- DARK BROWN OR BLACK PELAGE

- SMALL HEAD, SHORT SNOUT, BLUNT NOSE

- HAIR ON DORSAL SURFACE OF FOREFLIPPERS STOPS ABRUPTLY AT WRIST

- RELATIVELY LONG HINDFLIPPERS, ESPECIALLY LONG TIPS OF DIGITS
- DISTRIBUTION LIMITED TO NORTH PACIFIC, BERING SEA, SEA OF OKHOTSK

The only representative of the genus *Callorhinus,* the Northern Fur Seal differs substantially in its morphology from other fur seals and sea lions. The most distinctive differences include a small head, short snout, long hindflippers, and an abrupt termination of hair on the foreflippers at the wrist. The Northern Fur Seal is one of only two fur seals that live in the Northern Hemisphere, the other being the Guadalupe Fur Seal. The species has been extensively hunted for its fine pelt. Revenues from harvests of these seals on the Pribilof Islands for just a few years in the 1860s matched the price paid to Russia by the United States for the purchase of Alaska. The genus name is from the Greek words *kallos,* for "beautiful object," and *rhinos,* for "skin" or "hide," and refers to the fine quality of the Northern Fur Seal's pelage. The specific name means "bear-like" in Latin.

DESCRIPTION The Northern Fur Seal has a stocky body and a small head with a very short snout. The hindflippers are the largest of any otariid pinniped, with the tips of the digits particularly long and extending well beyond the small claws. The foreflippers are relatively broad, and the hair on the dorsal surface ends abruptly in a clean line across the wrist, a characteristic that distinguishes this species from Guadalupe Fur Seals. Adult males are substantially longer and heavier than females. Males develop massive chests, shoulders, and necks, and a short stiff mane that extends from the lower neck to the shoulders; some adult males have an obvious short tuft of lighter guard hairs on top of the head. There are 10 pairs of teeth in the upper jaw and eight pairs in the lower jaw.

Adult males are uniformly dark brown to black. Adult females are dark brown or gray dorsally and light gray, silver, or cream on the throat and lower chest and in a narrow stripe along the belly. Pups are born with a uniform black pelage that they gradually molt during the first two to four months of life. Their dorsal surface lightens to lighter black or dark brown, and the ventral pelage lightens to silver, gray, or cream on the lower chest and belly.

RANGE AND HABITAT The distribution of the Northern Fur Seal is limited to the North Pacific Ocean, the Bering Sea, and the Sea of Okhotsk. Primary breeding colonies are in the Bering Sea at the Pribilof Islands and the Commander Islands. Smaller colonies occur at Robben Island in the Sea of Okhotsk, the Kuril Islands off northern Japan, Bogoslof Island near the Aleutian Islands, and San Miguel Island off southern California. Adult males begin leaving colonies in September, and most are thought to remain in the Bering Sea and North Pacific along the Aleutian Islands in winter and early spring. Females have returned to sea by late October, migrating into the central North Pacific and south along the California coast to feed until spring. Some seals may occur at San Miguel Island year-round.

Breeding seals mostly occupy rocky beaches, though young males may move several hundreds of feet inland to rest and spar. At San Miguel Island, one colony occurs on a broad sandy beach, whereas another occupies a small rocky islet off the island's northwestern shore.

SIMILAR SPECIES Northern Sea Lions occur in the northern part of the Northern Fur Seal's range, and California Sea Lions and Guadalupe Fur Seals occur in the southern part, off the coast of North America. The sea lions are readily distinguished by their tan to yellowish pelage and longer, dog-like snouts. Northern Sea Lions are lighter in color and much larger and more robust, and the adults are especially massive. California Sea Lions have much shorter fleshy tips on the digits of their hindflippers and have hair on the foreflippers extending beyond the wrist. Guadalupe Fur Seals have larger foreflippers, with hair extending past the wrist on the dorsal surface, smaller hindflippers with short terminal flaps, and a much longer, pointed snout.

BEHAVIOR Northern Fur Seals form relatively dense aggregations during the breeding season but appear to be solitary when foraging at sea. After pups are weaned in October, they remain ashore for several weeks, spending brief periods of time in the water developing swimming skills before migrating south by late November. Northern Fur Seals may forage at sea continuously for several months or more from autumn through spring. Some may remain at sea for several years

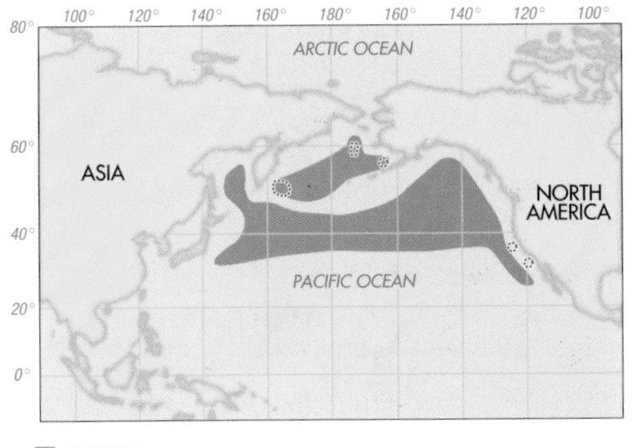

NORTHERN FUR SEAL
FAMILY OTARIIDAE

MEASUREMENTS AT BIRTH
LENGTH 24–26″ (60–65 cm)
WEIGHT 9–13 lb (4–6 kg)

MAXIMUM MEASUREMENTS
LENGTH **MALE** 6′11″ (2.1 m)
FEMALE 4′11″ (1.5 m)
WEIGHT **MALE** 600 lb (270 kg)
FEMALE 132 lb (60 kg)

LIFE SPAN
MALE 18–20 years
FEMALE 20–25 years

before returning to the colonies. Vocalizations are an important component of the mother-pup bond during the three to four months of nursing. Females regularly leave their pups in large aggregations for several days at a time to forage at sea; upon their return, females quickly recognize their pups by their highly individualized vocalizations. Recent findings indicate that Northern Fur Seals may even recognize one another by vocalizations after a separation of several months to years.

REPRODUCTION Northern Fur Seals are polygynous breeders. Adult males establish territories in early and mid-May. They aggressively herd females within their territories until mating occurs. Males vigorously defend the females in their territories against other males with ritualized postures, vocalizations, and occasional combat. Females give birth from late May through late June. They stay with their pups for about eight days, at which time they are ready to mate. After mating, females begin a series of foraging trips lasting two to 10 days each; they spend about one or two days ashore between trips nursing their pups. While foraging, females leave their pups in large aggregations. Upon their return, mothers recognize their pups primarily by vocalization and, when closer, by smell. After three to four months, females abruptly wean their pups, return to sea, and migrate south until the next breeding season.

FOOD AND FORAGING Northern Fur Seals eat a variety of nearshore and pelagic squid and fish. Most data on their foraging habits come from lactating females foraging near rookeries in the Bering Sea in summer. They make mostly shallow dives, to depths of 50 to 165 feet (15–50 m), but some dives may reach 820 feet (250 m) and last about 1½ to 3 minutes. Seals foraging in pelagic habitats tend to dive at night, while those over the continental shelf dive mostly during the day.

ABOVE: *Adult male Northern Fur Seals may be several times larger than females. During the breeding season, the males actively herd females and prevent them from leaving their territories until they are receptive for mating about eight to 10 days after giving birth.*
LEFT: *Virtually all Northern Fur Seals depart island colonies by late October and spend the next seven to eight months at sea foraging far from land.*

Young Northern Fur Seal pups remain ashore for one to several days while their mothers are at sea feeding. Their mothers will return to nurse them for about a day and then go to sea to forage again. This cycle repeats until the pups are around four months old, when the mothers abruptly wean and abandon them before migrating south for the winter.

STATUS AND CONSERVATION Native Russians and Americans hunted Northern Fur Seals for several thousand years, and commercial harvests beginning in the 1700s greatly reduced the species' abundance. In the early 1900s, Russia, Japan, Canada, Great Britain, and the United States formed an international treaty to prohibit the harvests of Northern Fur Seals at sea (which were mostly females) in exchange for managed harvests of subadult males on the Pribilof Islands. Even this managed harvest was ended in 1984. Some subsistence harvests continue on the Pribilof Islands, which are governed by regulations published under authority of the U.S. Fur Seal Act and the U.S. Marine Mammal Protection Act. About 1,400 seals were harvested there in 1997 and about 1,550 in 1998, about 75 percent of which were on St. Paul Island. There are about 1.2 million Northern Fur Seals across the species' entire range. Substantial numbers have become entangled and died in fragments of derelict fishing net in the Bering Sea and North Pacific Ocean.

On some beaches at St. Paul Island in the Bering Sea, bachelor males take a chance at intercepting and mating with females as they depart on foraging trips. The females, who have likely already mated with a more dominant male, must run this gauntlet several or more times during the summer breeding season.

California and Galápagos Sea Lions

Zalophus californianus
(Lesson, 1828)

- ADULT MALE DARK BROWN TO BLACK, FEMALE AND JUVENILE TAN TO YELLOWISH
- ADULT MALE HAS WELL-DEVELOPED SAGITTAL CREST, TUFT OF HAIR ON HEAD
- COMMONLY VISIT BOATING MARINAS
- RESPOND WELL TO TRAINING EFFORTS
- RESTRICTED TO COASTAL WATERS IN EASTERN NORTH PACIFIC AND GALÁPAGOS ISLANDS

California and Galápagos Sea Lions are perhaps the most familiar pinnipeds in the world. Highly gregarious and easily trained, these are the sea lions commonly seen in zoos, circuses, and oceanariums. They are quite active, both behaviorally and vocally, and even large adult males are easy to tame and train. In the wild, they have a similarly gregarious reputation. Along North America's west coast, for example, they are frequent visitors to boating marinas, sometimes hauling out by the dozens on yachts, bait barges, and buoys, and occasionally damaging or sinking them. Moreover, some California Sea Lions have made a habit of lounging around small dams and locks, particularly in downtown Seattle, Washington, where they feast on migrating salmon returning to their spawning grounds upriver. Scientists have generally recognized three subspecies of *Zalophus californianus,* based predominantly on their geographic isolation from each other: the California, Galápagos, and Japanese Sea Lions (the latter is now presumed extinct). One taxonomist has recently suggested that these populations should be considered distinct species. Further comparative research will undoubtedly help to resolve these taxonomic questions. The genus name, derived from the Greek prefix *za* and *lophos* for "crest," refers to the large sagittal crest on the skull of adult males.

DESCRIPTION Adult females and juveniles are slender-bodied; adult males are robust at the shoulders, chest, and neck, and are rather slender at the hind end. The snout is long, straight, and narrow. The foreflippers are broad, with hair on the upper surface extending past the wrist and tiny claws. The short hindflippers have short fleshy tips at the end of the digits and short narrow claws. Adult males have a pronounced forehead, exaggerated by the sagittal crest, that is crowned by a light brown to blond tuft of hair. Males do not have the well-developed mane typical of other sea lion species. There are nine pairs of teeth in the upper jaw and eight pairs in the lower jaw.

Adult males are mostly dark brown to black, with areas of light tan on the face. Adult females and juveniles are light gray to silver just after the annual molt, but their pelage quickly fades to blond or tan. Pups are born dark brown to black and fade to light brown within a few weeks. Pups molt at about four or five months old to a light gray pelage that darkens to chocolate brown, then fades to light tan or yellowish over several months. Beginning at about puberty, males generally do not fade to lighter tan after molting. A small number of males, however, remain light yellow to blond throughout their lives.

RANGE AND HABITAT California Sea Lions *(Zalophus californianus californianus)* breed at the Channel Islands off southern California; at islands along the northern Pacific coast of Baja California, including the offshore Guadalupe Island; and on islands along the east coast of Baja California in the middle and southern Gulf of California. In most areas, they prefer to haul out and breed on sandy beaches. Although they may move far inland or up coastal slopes at night or during cool weather, they stay closer to the water's edge during warm days. After the breeding season, large numbers, particularly males, migrate north in the Gulf of California and along the Pacific coast of North America as far as British Columbia. Galápagos Sea Lions *(Z. c. wollebaeki)* are generally restricted to the Galápagos Islands and occur on virtually all of the islands in the archipelago. A small colony became established in 1986 near the coast of Ecuador at La Plata Island. Vagrants have been reported along the coasts of Ecuador and Colombia. The Japanese Sea Lion *(Z. c. japonicus)* formerly ranged from along the southern coast of Kamchatka, into the southern Sea of Japan, and south along the west coasts of southern Korea and Japan, with vagrants ranging to the east coast of Korea. Scientists believe that the Japanese Sea Lion was exterminated throughout its range by the early 1950s.

SIMILAR SPECIES Northern Sea Lions and Northern Fur Seals overlap with California Sea Lions from southern British Columbia south through central California, including breeding colonies at Año Nuevo and the Farallon Islands. Northern Sea Lions are substantially larger, with a more robust head and a broader snout, and males are tan to blond with a conspicuous mane on the neck,

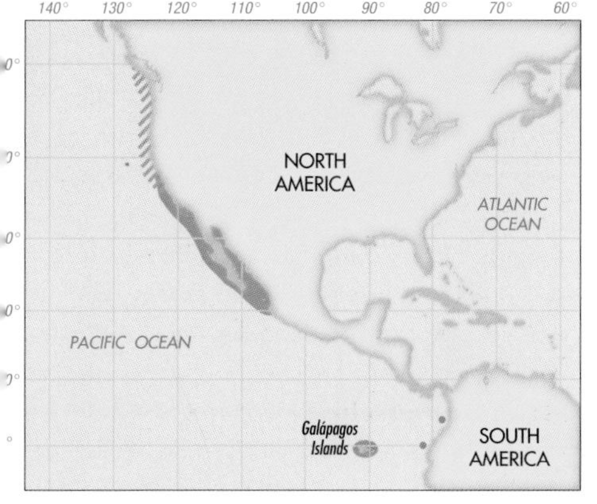

RANGE OF CALIFORNIA SEA LION (BREEDING)
RANGE OF CALIFORNIA SEA LION (NONBREEDING)
RANGE OF GALÁPAGOS SEA LION
VAGRANTS

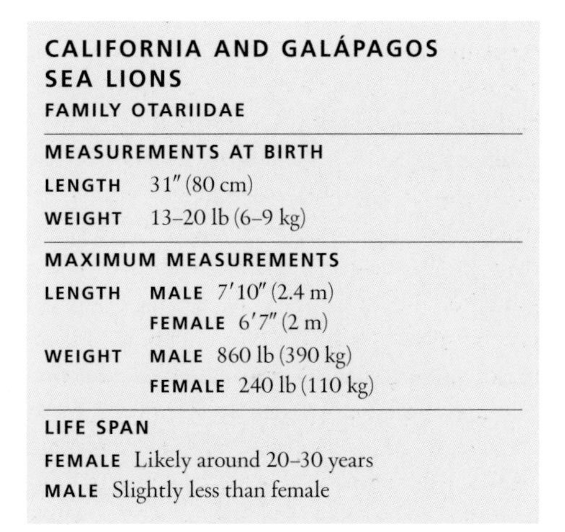

CALIFORNIA AND GALÁPAGOS SEA LIONS
FAMILY OTARIIDAE

MEASUREMENTS AT BIRTH
LENGTH 31″ (80 cm)
WEIGHT 13–20 lb (6–9 kg)

MAXIMUM MEASUREMENTS
LENGTH **MALE** 7′10″ (2.4 m)
FEMALE 6′7″ (2 m)
WEIGHT **MALE** 860 lb (390 kg)
FEMALE 240 lb (110 kg)

LIFE SPAN
FEMALE Likely around 20–30 years
MALE Slightly less than female

throat, and chest. The fur seals are darker, with larger flippers and a shorter snout. Guadalupe Fur Seals may overlap from central California to Baja California. They have a longer snout, and adult males have a less pronounced forehead with no tuft of hair; females are lighter. The Galápagos Fur Seal, whose range overlaps with that of the Galápagos Sea Lion, is smaller and darker.

BEHAVIOR California and Galápagos Sea Lions form large aggregations when ashore, particularly on the Channel Islands, where thousands may haul out in dense groups. They may also travel and cavort in groups of a dozen or more when at sea near haulout and breeding beaches, and perhaps when en route to offshore foraging areas. In summer, females dive mostly to depths of about 245 feet (75 m) for about four minutes. Their dives become deeper and longer the rest of the year. The maximum recorded diving depth for the species is 1,760 feet (536 m), and the maximum duration is 12 minutes. California Sea Lions may spend several days at a time at sea, and as much as one to two weeks in some seasons. While at sea, they dive almost continuously, resting only briefly at the surface. Adult females and juveniles molt from early autumn through winter, and adult males appear to molt in January and February, after they have migrated north. There are an increased number of subadult and adult males ashore at the Channel Islands in winter, likely hauled out to molt before returning to breeding colonies in Mexico.

REPRODUCTION California and Galápagos Sea Lions are polygynous breeders. Adult males establish breeding territories along the water's edge or near tide pools, nearshore ledges, or reefs, where females may congregate. Males compete with ritualized physical displays, including aggressive posturing, threat vocalizations, and occasional physical combat. Females give birth from late May through July; most births occur in late June. They remain ashore with their pups for about seven to 10 days, and then begin foraging at sea for one to three days at a time, spending about a day ashore nursing their pups. Pups may accompany their mothers on some foraging trips. Those that remain behind congregate in large groups to socialize and play; their mothers quickly find and reunite with them, recognizing them by vocalizations and smell. Pups are abruptly weaned and abandoned at about six months old, although some may continue to nurse for up to a year, and rarely longer. Females become estrous and are ready to mate about three weeks after giving

ABOVE: *California and Galápagos Sea Lions are gregarious and may form dense groups of several hundred or more when ashore.*
LEFT: *A Galápagos Sea Lion female and her pup seek shade from the intense midday tropical sun.*

California Sea Lions "rafting" near the breakwater at Monterey, California. These rafts loosely form in midmorning as seals return from feeding trips and then break up in midafternoon as the sea lions either haul out or move offshore to forage again.

birth, by which time they are already foraging at sea. Consequently, females exercise some choice over which and how many males they mate with.

FOOD AND FORAGING California and Galápagos Sea Lions have a diverse diet, with northern anchovy, market squid, sardines, Pacific and jack mackerel, and rockfish among their favored prey. They feed mostly in cool, upwelling waters near the mainland coast, along the continental shelf edge, and around seamounts, and may also sometimes forage on the sea bottom. California Sea Lions are notorious for approaching commercial and sport-fishing boats and seizing fish from lines and nets. In northern California, Oregon, and Washington, males position themselves at the mouths of streams and rivers to intercept lampreys, salmon, and eels that become densely concentrated there during annual migrations.

STATUS AND CONSERVATION California Sea Lions were hunted along the west coast of North America and at the Channel Islands for at least several thousand years before the arrival of Europeans in the 16th century, and Japanese Sea Lions were probably also hunted for several millennia. Early whalers, sealers, and Sea Otter hunters likely killed them for food during the 1800s and 1900s. There was small-scale commercial harvesting of California Sea Lions in Baja California and southern California from the 1920s through the early 1940s. In the early 1900s, bounties were paid out for them in Washington, Oregon, and California because of their perceived competition with commercial fisheries. California Sea Lion populations have increased at least fourfold since the 1970s, when the United States and Mexico limited killing and harassment of the species. The population numbered around 175,000 in 2001, with perhaps two-thirds of the sea lions living in U.S. waters. Galápagos Sea Lions may number around 30,000, but there have been no thorough population surveys for several decades. Japanese Sea Lions have been presumed extinct since the early 1950s, as there have been no reliable sightings since then.

California Sea Lions are agile swimmers. They are also curious and often approach divers and snorkelers, sometimes nipping at their snorkels and fins.

Northern Sea Lion

Eumetopias jubatus
(Schreber, 1776)

- **LARGEST EARED SEAL**
- **LIGHT BROWN TO BLOND PELAGE WITH DARK BROWN FLIPPERS**
- **ADULT MALE HAS WELL-DEVELOPED MANE EXTENDING TO SHOULDERS**
- **DISTRIBUTION RESTRICTED TO NORTH PACIFIC OCEAN AND SOUTHERN BERING SEA**

The Northern Sea Lion is the largest member of the eared seals. The only sea lions that occur in the Bering Sea and the Gulf of Alaska, they can be easily distinguished from Northern Fur Seals, with which they occur, by their more robust body and lighter pelage. The genus name, from the Greek words *eu* ("typical" or "well") and *metopion* ("broad forehead"), refers to this species' well-developed forehead. The specific name, from the Latin word *jubatus* ("having a mane"), refers to the extensive mane of adult males. The Northern Sea Lion is also known as the Steller Sea Lion after Wilhelm Steller, the German surgeon and naturalist who described sea lions he encountered while shipwrecked aboard the Russian ship *Vitus Bering* at Bering Island in 1742.

DESCRIPTION The Northern Sea Lion is large, with a robust body and head. The snout is short, blunt, and broad. The foreflippers are broad, with hair sparsely covering about three-quarters of the dorsal surface. The hindflippers are short and slim, with the first and fifth digits longer than the middle digits. Adult males are substantially larger than females, especially at the head, neck, and chest. They have a mane of long hairs that extends from the back of the head down the neck to the shoulders and, to a lesser extent, onto the throat and chest. There are nine pairs of teeth in the upper jaw and eight pairs in the lower jaw.

Adults are light brown to blond and are generally darker ventrally than dorsally. The foreflippers and hindflippers are dark brown to black, contrasting with the lighter pelage color.

Females are often lighter than males. Pups are born with a dark brown to black pelage. They molt at about four months old to a lighter brown coloration and become lighter with each successive annual molt. Juveniles attain the adult coloration when about three years old.

RANGE AND HABITAT Northern Sea Lions are restricted to the North Pacific Ocean and southern Bering Sea. They range from near Hokkaido, Japan, north to the Kuril and Commander Islands and along the Kamchatka Peninsula; east through the Aleutian and Pribilof Islands into the Gulf of Alaska; and south along the Pacific coast of North America to central California. They are most abundant now in the Gulf of Alaska, southeastern Alaska, and British Columbia. Historically, Northern Sea Lions bred at the Channel Islands off southern California; however, they have not been seen there since the early 1980s. Northern Sea Lions occur in coastal waters when feeding and migrating. They haul out on rocky reefs, ledges, and beaches, and in some areas on sandy beaches.

SIMILAR SPECIES Northern Fur Seals overlap with Northern Sea Lions throughout most of their range. They are substantially smaller, with a proportionally smaller, less robust head and long hindflippers with long fleshy tips at the ends of the digits, and they have a dark brown to black pelage with dense underfur and longer guard hairs. California Sea Lions, principally adult and subadult males, overlap with Northern Sea Lions from southern British Columbia south to central California. They are substantially smaller, with a dark brown to black pelage and a relatively longer, narrower snout. Adult males have a well-developed sagittal crest, with a tuft of light hair on top of the head, and lack a well-developed mane.

BEHAVIOR Northern Sea Lions aggregate in modest numbers on rocky and gravel beaches and rocky reefs throughout the year. They space out from each other during the breeding season, although pups may gather in close groups to socialize and play while their mothers forage at sea. While lactating females are generally intolerant of physical contact during the breeding season, they may have brief social interactions with other females. Females bond strongly with their newborn pups and later recognize them by vocalizations and smell. After the breeding season, Northern Sea Lions may be more gregarious and tolerant of close contact. Molting occurs from late summer through early winter, depending on age and sex. Northern Sea Lions often haul out on sea ice in the Bering Sea and the Sea of Okhotsk, an unusual behavior for otariid pinnipeds, which generally avoid ice. When foraging, juveniles dive to average depths of 70 feet (21 m), with a maximum depth of 655 feet (200 m). Most of their dives last less than one minute. Adult females dive deeper than

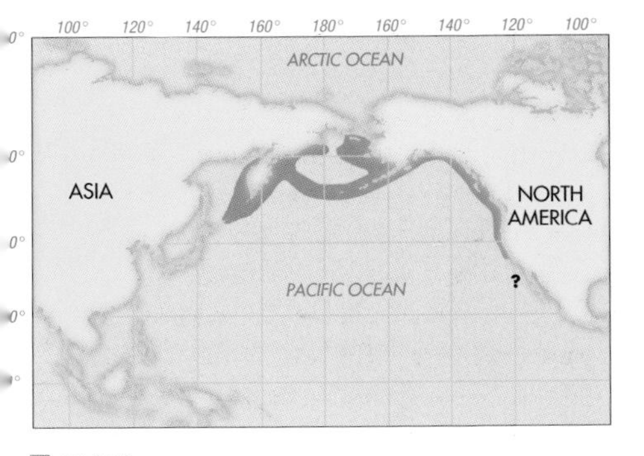

RANGE
? POSSIBLE RANGE

NORTHERN SEA LION
FAMILY OTARIIDAE

MEASUREMENTS AT BIRTH
LENGTH 3′4″ (1 m)
WEIGHT 35–51 lb (16–23 kg)

MAXIMUM MEASUREMENTS
LENGTH **MALE** 11′ (3.3 m)
FEMALE 9′6″ (2.9 m)
WEIGHT **MALE** 2,400 lb (1,100 kg)
FEMALE 770 lb (350 kg)

LIFE SPAN
FEMALE About 18–25 years
MALE Perhaps slightly less than female

Northern Sea Lions may form dense aggregations during the summer breeding season and again in autumn and winter when they molt. The abundance of this species declined precipitously from the late 1970s through 2001, especially in the western Aleutian Islands and in southern California waters, from which they had disappeared by the early 1980s.

juveniles. Dives are generally shallower at night than during the day and deeper in spring and summer than in winter.

REPRODUCTION The mating system of the Northern Sea Lion is polygynous. Adult males arrive at colonies in early to mid-May and establish territories with stereotyped visual and vocal threats. Occasionally these threat displays escalate to physical combat, though such fights are usually brief. Females give birth from late May through early July, with most pups born in late June. Mothers remain with their newborn pups for about nine days and then alternate about one to three days at sea foraging with several hours to two days ashore nursing their pups. Females are ready to mate about 11 to 14 days after giving birth. Gestation lasts about a year. Most pups are weaned by the time they are one year old.

FOOD AND FORAGING Scientists in Alaska and at the U.S. National Marine Fisheries Service are conducting studies to better identify the foraging habits and habitats of Northern Sea Lions in order to improve conservation and management policies for the species and for commercial fisheries. Northern Sea Lions are presumed to forage mostly close to continental and island coastlines. Past data have shown seasonal changes in diet that evidently reflect changes in local abundance and distribution of prey. Walleye pollock is an important fish prey for Northern Sea Lions in the Gulf of Alaska and Bering Sea. Herring, rockfish, cod, squid, and octopus are key prey in coastal waters of British Columbia. Northern Sea Lions may range into some freshwater rivers in Washington, Oregon, and California to catch lampreys and adult salmon returning from the sea to spawn or young salmon migrating out to sea. Adult and

OPPOSITE: *Northern Sea Lions occasionally raft together for several hours during the day. Recent studies have also shown that small numbers of these sea lions may gather together while hunting herring and other schooling fish in the northern Gulf of Alaska.* **LEFT:** *Adult male Northern Sea Lion (left) and California Sea Lion (right) showing the differences in size, color, and body shape.*

subadult males have been reported killing and eating Northern Fur Seal pups at the Pribilof Islands and small Harbor Seals, Ringed Seals, and Sea Otters in other areas.

STATUS AND CONSERVATION For several thousand years, aboriginals in the North Pacific and Bering Sea hunted Northern Sea Lions for food and clothing. In the early 1900s, as a result of fishermen's complaints that the sea lions were affecting their catches, substantial numbers were killed for bounty from British Columbia south to California. In addition, large numbers of pups were commercially harvested in Alaska from 1959 through 1972, and Native Americans in the Aleutian and Pribilof Islands harvested hundreds annually in the early 1990s.

The abundance of Northern Sea Lions has greatly declined across the species' range, from several hundred thousand in the 1970s to about 60,000 to 70,000 by the late 1990s. The decline may be due to substantial increases in commercial fishing activities in the Bering Sea and Gulf of Alaska, or to long-term natural environmental changes in marine communities in the North Pacific, or it may be due to a combination of these factors. Elevated levels of copper, mercury, and selenium were detected in Northern Sea Lions that foraged along the coast of central California. Much scientific research is needed to sort out the causes of the population decline and to determine the best policies to conserve Northern Sea Lions and other components of the marine ecosystem in the North Pacific, Gulf of Alaska, and Bering Sea. Two populations are recognized in the northeastern Pacific Ocean. The U.S. Endangered Species Act lists the eastern population, east of Cape Suckling (144°W), as threatened and the western population, west of Cape Suckling, as endangered. Northern Sea Lions are also protected by the Canadian Fisheries Act.

Australian Sea Lion

Neophoca cinerea
(Péron, 1816)

- RELATIVELY LARGE HEAD WITH LONG NARROW SNOUT
- ADULT MALE HAS ROBUST CHEST, SHOULDERS, HEAD, AND NECK
- LIGHT RING AROUND EYE IN JUVENILE MALES
- RELATIVELY NEARSHORE DISTRIBUTION OVER CONTINENTAL SHELF WHEN FORAGING
- ENDEMIC TO WESTERN AND SOUTHERN AUSTRALIA

The Australian Sea Lion is endemic to Australia, with most of its largest breeding colonies located at offshore islands in eastern South Australia. These sea lions were an important source of food for aboriginal Australians, early European explorers, and shipwrecked sailors. They were hunted in historic times for their oil and in prehistoric times for their meat, and perhaps also their skins. Though the breeding colonies have been protected from human disturbance and harvest for some time, the species' range-wide abundance still remains fairly low, at about 10,000 seals. While Australian Sea Lions appear to prefer haulouts on sandy beaches or smooth rocky reefs, they regularly wander inland, sometimes up to several miles. The genus name, which means "new seal," was first applied in 1866. The specific name is derived from the Latin *cinereus,* for "ash-colored," in reference to the pelage of juveniles and adult females.

DESCRIPTION The Australian Sea Lion has a robust stocky body (particularly adult males), a relatively large head, and a long narrow snout. The flippers are short and narrow, with small claws on the hindflippers and barely visible claws on the foreflippers. Adult males are up to twice the length and three times the mass of females. They have a robust chest, shoulders, head, and neck, and long rough hair on the neck that often looks like a mane. There are nine pairs of teeth in the upper jaw and eight pairs in the lower jaw.

Adult males are dark brown with yellow to white areas on the nape of the neck and the top of the head, and sometimes with spotting on the chest. Adult females and juveniles are silver to gray dorsally and cream-colored ventrally. Pups are born with a pelage of soft, dark brown hair, which fades until it is shed after about two months. Juvenile males often have a light-colored ring around the eyes.

RANGE AND HABITAT The breeding range of the Australian Sea Lion is limited to southern coastal areas of Australia, from Houtman Abrolhos (28°S, 112°E) in Western Australia east to near Kangaroo Island (34°S, 138°E) in South Australia. While most of the largest colonies occur at offshore islands in eastern South Australia, at least 28 have been identified in Western Australia and 38 in South Australia, most of them relatively small and located on sheltered beaches at isolated islands. The best-known mainland breeding colonies are at Point Labatt in South Australia and near Twilight Cove (Thundulda) in Western Australia. During the nonbreeding season, adult and subadult males range more widely along Western Australia's coast. Vagrants may appear during the nonbreeding season as far north as Shark Bay and as far east as Portland, Victoria.

Australian Sea Lions appear to prefer to haul out and breed on sandy beaches or smooth rocky reefs. However, they regularly wander inland, sometimes up to several miles.

SIMILAR SPECIES Australian Sea Lions overlap with Australian Fur Seals in south-southeastern Australia and with New Zealand Fur Seals in southwestern Australia. The fur seals have a thicker, shorter pelage that is a darker gray to black, a relatively smaller head, and a more pointed snout. Occasionally Australian Sea Lions may be seen along the southern coast of South Island, New Zealand, with New Zealand Sea Lions. Male New Zealand Sea Lions lack the pale areas on the head and neck, and have a longer snout; females and juveniles are difficult to distinguish.

BEHAVIOR These sea lions form small breeding colonies of several hundred or fewer. They are mostly asocial, except during the breeding season, when social interactions are generally limited to mothers and their pups and brief encounters between mature males and females to mate. Australian Sea Lions may spend more time ashore when molting, which occurs from August through September at the Dangerous Reef colony, but perhaps in other months at other colonies. Australian Sea Lions evidently do not migrate and may forage locally for most of their lives. When returning from foraging trips, they are often observed bodysurfing onto the beach. Because the sea lions at the colony at Seal Bay on Kangaroo Island are relatively undisturbed by close human approach, tourists have been allowed to move about the beaches where the sea lions haul out after the breeding season. However, sea lions at other colonies appear to be less tolerant of human presence.

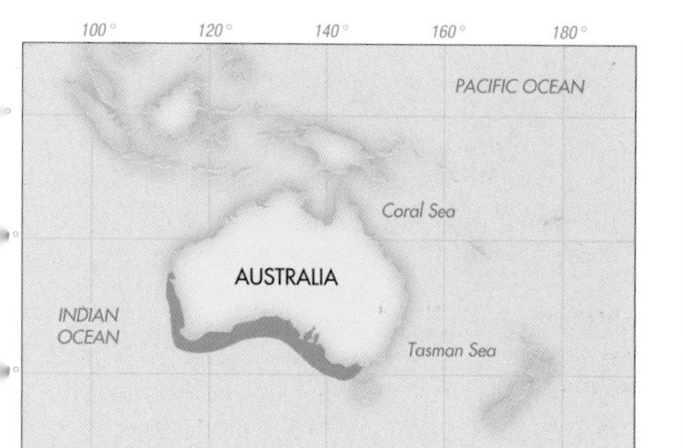

RANGE

AUSTRALIAN SEA LION
FAMILY OTARIIDAE

MEASUREMENTS AT BIRTH

LENGTH	24–27" (62–68 cm)
WEIGHT	15–18 lb (6.5–8 kg)

MAXIMUM MEASUREMENTS

LENGTH	MALE	8′2″ (2.5 m)
	FEMALE	5′11″ (1.81 m)
WEIGHT	MALE	660 lb (300 kg)
	FEMALE	220 lb (100 kg)

LIFE SPAN Unavailable

REPRODUCTION Although Australian Sea Lions are substantially sexually dimorphic in size, they do not appear to be polygynous breeders. Rather, they seem to engage in what has been called serial monogamy, in which a male maintains a short-term territory (for two to four weeks) near a solitary female until she comes into estrus, mates with her, then evidently returns to sea to feed before locating another female to guard and mate. Adult males aggressively herd females during the peak of the breeding season, which may vary by several months across the range. Breeding does not take place at the same time among the various rookeries, even when they are adjacent to each other. Pups are born from January to October, with most births occurring in June. Females give birth about one to two days after hauling out; they are ready to mate about 10 days later. Gestation lasts about 12 months. A female remains with her pup for about 10 to 14 days, then begins a series of short feeding trips at sea that last about two days each, coming ashore in between for about 1½ days to nurse. Females may nurse pups for up to two or three years if they do not give birth in consecutive seasons. Pup mortality during the first six months is about 23 percent for pups born in winter and about 7 percent for pups born in summer; it may be as great as 56 percent at some sites, owing to aggression of adult and subadult males and predation by sharks.

FOOD AND FORAGING The diet of Australian Sea Lions is poorly known. What has been so far described suggests that it may include a variety of fish, small sharks, octopus, squid, and occasionally penguins. Some lactating females were recorded foraging on the seabed of the continental shelf within 20 miles (30 km) of shore, at depths of 490 feet (150 m).

STATUS AND CONSERVATION Aborigines and early European visitors hunted Australian Sea Lions for food, oil, and, to some extent, clothing. Commercial sealing during the first quarter of the 19th century exterminated colonies at most accessible sites. While there are not adequate data to determine trends in abundance, populations are thought to be far less numerous than they were prior to this commercial harvesting.

ABOVE: *Female Australian Sea Lion nursing pup while seeking shade from the intense midday summer heat.*
LEFT: *An Australian Sea Lion "porpoising" as it travels quickly from a terrestrial haulout to an offshore foraging site.*

In 1992, the species' population was estimated to be about 11,200, with about 2,600 pups born each year. Three colonies in central South Australia—Dangerous Reef, Seal Bay on Kangaroo Island, and The Pages Islands—account for about 42 percent of the total population, and three colonies in Western Australia account for about 9 percent. In South Australia, all of the species' breeding sites occur within conservation parks managed by the Department of Environment, Heritage and Aboriginal Affairs. To promote recovery of the Australian Sea Lion population in South Australia, the National Parks and Wildlife Act, passed in 1972, restricts access to some colonies during the breeding season in order to limit disturbance by humans. Australian government agencies are studying the impact of tourists on the colony at the popular Seal Bay site. There has also been some concern that tour operators promoting viewing and interacting with white sharks may inadvertently be drawing the sharks to some breeding colonies, thereby increasing pup mortality. Several protected areas have been established near colonies in South Australia, and two marine parks were established within the species' range in Western Australia in the late 1990s.

A group of Australian Sea Lions foraging. These sea lions feed mostly on the seafloor within 20 miles (30 km) of shore.

New Zealand Sea Lion

Phocarctos hookeri
(Gray, 1844)

- MALE HAS MASSIVE NECK, CHEST, AND HEAD
- PREFERS SANDY BEACHES
- BREEDS MAINLY AT AUCKLAND ISLANDS, NEW ZEALAND

New Zealand Sea Lions have a fairly restricted range around the Auckland Islands, where they crowd together on sandy beaches during the breeding season. They are also known as Hooker's Sea Lions, after Sir Joseph D. Hooker, the British botanist who collected the first specimens of the species in the early 1800s. The genus name is derived from the Greek word *arktos,* meaning "bear," and refers to this sea lion's bear-like skull, especially in males. Some adult males engage in the unusual behavior of killing, and sometimes even eating, pups of their own species. They may also kill and eat the pups of other seal species. In 1997, one adult male killed about 43 percent of the Antarctic and Subantarctic Fur Seal pups born at Macquarie Island that year.

DESCRIPTION The New Zealand Sea Lion has a robust head (massive in adult males). The head is flat and the snout blunt and short. The flippers are rather short and narrow with tiny claws. Adult males are substantially larger-bodied and heavier than females. As males age, they develop a robust neck and chest, and a mane of coarse long hair that extends from the back of the head and down the neck to the shoulders. There are 10 pairs of teeth in the upper jaw and eight pairs in the lower jaw.

Adult males are uniformly dark brown. Adult females are mostly yellowish to creamy white, with darker patches on the snout and upper foreflippers. Pups are dark brown at birth, but within a month male pups have a light area on top of the head and neck, while female pups have a mostly light brown to tan pelage.

RANGE AND HABITAT New Zealand Sea Lions are generally restricted to the Auckland Island group, with about 95 percent of all pups born at Enderby, Dundas, and the Figure of Eight Islands. Other small colonies exist at Campbell Island to the south and the Snares Islands to the north. Some sea lions, mostly adult males, haul out in autumn and winter during the nonbreeding season at Port Pegasus on Stewart Island. Males of various ages may occur year-round on the Otago Peninsula at Papanui Beach, and sometimes elsewhere on the mainland of South Island, New Zealand. Males also regularly appear at Macquarie Island, some 375 miles (600 km) south of the Aucklands, presumably migrating there to forage during the nonbreeding season. Remains of New Zealand Sea Lions have been found in Maori middens on New Zealand's North Island, at Cape Kidnappers and the Coromandel Peninsula, suggesting that this species may once have occurred there.

New Zealand Sea Lions seem to prefer to haul out on soft sandy beaches with easy access to the water, although many wander inland to rest on grass-covered cliffs or even deep in forests.

SIMILAR SPECIES Australian Sea Lions occasionally occur with New Zealand Sea Lions on the southern coastline of South Island, New Zealand. While females and juveniles are hard to tell apart, Australian Sea Lion adult males have more rounded, blunter faces with shorter snouts, as well as a characteristic pale area on top of the head and neck, compared with the uniformly dark pelage of adult male New Zealand Sea Lions.

BEHAVIOR Adult females are gregarious and often crowd together on beaches during the breeding season. Once they give birth, however, they are aggressive toward other females and pups that approach them too closely. Several adult males have been observed abducting, killing, and occasionally eating pups at Dundas Island. In 1998, two males killed and ate at least 38 pups, about 2 percent of the total born that season. New Zealand Sea Lions can be habitual wanderers and may sometimes be found several miles inland from the coast. Pups at Enderby Island, where rabbits are particularly abundant, may starve to death after they wander into rabbit burrows and become trapped. Though these sea lions may spend more time at sea in autumn and winter between breeding seasons, there is as yet no evidence of migrations, and they may forage fairly near breeding sites during this time. Adults molt in March and April; juveniles may molt slightly earlier.

REPRODUCTION The mating system of New Zealand Sea Lions is polygynous, with dominant males maintaining harems of about 12,

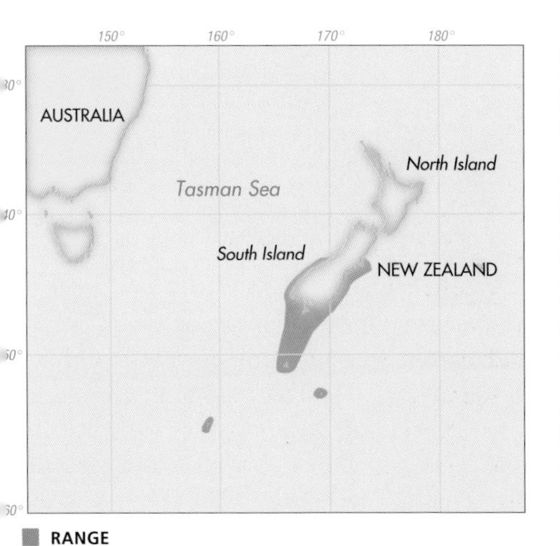

NEW ZEALAND SEA LION
FAMILY OTARIIDAE

MEASUREMENTS AT BIRTH

LENGTH	30–31″ (75–80 cm)	
WEIGHT	**MALE**	18 lb (8 kg)
	FEMALE	15 lb (7 kg)

MAXIMUM MEASUREMENTS

LENGTH	**MALE**	11′ (3.3 m)
	FEMALE	6′7″ (2 m)
WEIGHT	**MALE**	990 lb (450 kg)
	FEMALE	350 lb (160 kg)

LIFE SPAN
MALE 23 years
FEMALE 18 years

sometimes as many as 25, females. Males establish small territories on sandy beaches in late October and early November and defend them against nearby males with ritualized threat postures and vocalizations. Occasionally serious fights occur, which the larger male usually wins. Males may maintain tenure on their territories for several months, fasting throughout the period. Pregnant females arrive at the breeding sites in late November and early December and give birth about 10 days later. Most pups are born in mid-December. Females are ready to mate about seven to 10 days after giving birth. They then begin a series of foraging trips to sea that last one to three days each, coming ashore for about one or two days between trips to nurse their pups. Most pups are weaned at about eight months old, although some may suckle for a year or longer.

ABOVE: *A breeding group of New Zealand Sea Lion adult males, females, and pups. Males establish and defend small breeding territories on sandy island beaches and compete with each other for opportunities to mate.* **RIGHT:** *A New Zealand Sea Lion clears the surface of the water by using its foreflippers to generate great propulsive force. This behavior is known as porpoising.*

Female New Zealand Sea Lion. Females and juveniles have a tan to yellow pelage and a flat head, with a robust, rounded snout. Adult males are larger and darker and have thicker chests and necks than females.

FOOD AND FORAGING These sea lions eat mostly flounder, octopus, small squid, and crustaceans. Adult males also eat seabirds and penguins, including gentoo penguins at Macquarie Island. They occasionally kill and eat fur seal pups. Lactating females forage at depths of about 400 feet (120 m). They dive continuously, apparently to the seafloor, for about four minutes at a time. The deepest recorded dive for the species is 1,555 feet (474 m).

STATUS AND CONSERVATION Maori natives hunted New Zealand Sea Lions during prehistoric times, and later shipwrecked sailors and early settlers to New Zealand hunted and killed them. After the discovery of the Auckland Islands in 1806, commercial sealers rapidly reduced the species' population, selling the pelts in Asian markets and rendering oil from the blubber for use in lighting and as a lubricant. In 1894, a government decree provided some protection for the species, and in 1946 commercial sealing was prohibited. New Zealand Sea Lions numbered between 13,000 and 17,000 in the mid-1990s; however, a viral or bacterial disease that broke out in 1998 resulted in the deaths of about 60 percent of pups born that year, as well as about 20 percent of adult females. New Zealand Sea Lions also suffer mortalities from incidental entanglement in trawl nets set for squid, predation by sharks and Killer Whales, and the killing and eating of pups by adult males.

South American Sea Lion

Otaria flavescens
(Shaw, 1800)

- **ADULT MALE'S HEAD MASSIVE, WITH SHORT, BLUNT, UPTURNED SNOUT**
- **ADULT MALE HAS LONG MANE EXTENDING TO SHOULDERS**
- **LIGHT YELLOW TO ORANGISH-BROWN PELAGE**
- **NEARSHORE DISTRIBUTION MOSTLY ALONG SOUTHERN SOUTH AMERICA**

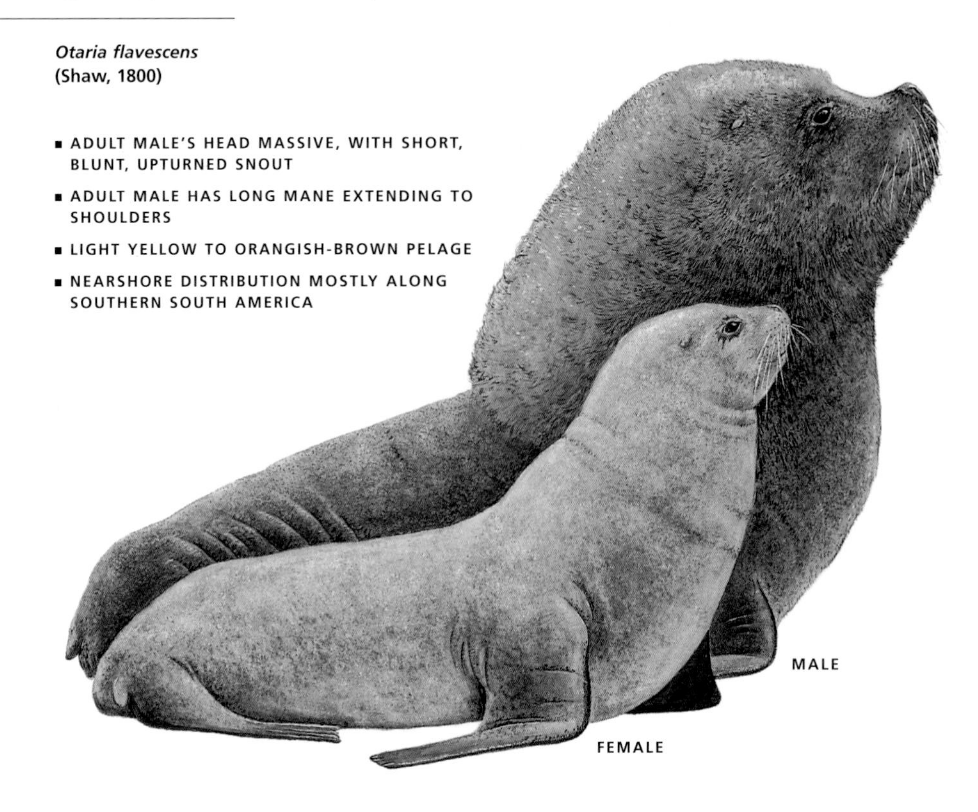

The South American Sea Lion is the most abundant and commonly seen marine mammal along the coast of Argentina. The species also occurs elsewhere along the South American coast, and a small colony resides at the Falkland Islands. South American Sea Lions are large-bodied animals, and males are particularly striking, with a massive head and neck and a robust mane that distinguishes them from all other Southern Hemisphere fur seals and sea lions. The genus name means "little ear" in Greek *(otarion)* and refers to the small pinnae. The specific names applied in the past have been either *byronia,* referring to Commodore John Byron, who presumably collected the type skull specimen, or *flavescens,* from the Latin *flavus,* for "yellowish." Taxonomists have recently argued most convincingly for the use of *flavescens.* The South American Sea Lion is also commonly called the Southern Sea Lion and, in southern parts of its range, the Patagonian Sea Lion.

DESCRIPTION South American Sea Lions are large-bodied, with large robust heads. The snout is short, broad, blunt, and upturned. The pinnae are small relative to the size of the head. The foreflippers are long and broad, and the hindflippers appear short against the bulky torso. Adult males are substantially larger than adult females, with a massive head that seems to dwarf the rest of the body, and a more robust throat and chest. They have a well-developed mane of guard hairs that extends from the forehead and chin over the neck, throat, and chest to the shoulders. There are 10 pairs of teeth in the upper jaw and eight pairs in the lower jaw.

Adults have a light yellow to orangish-brown pelage. Males are generally darker than females and juveniles, although some males may be pale gold. The male's mane may be slightly lighter than the rest of the body, especially on the top of the head. Pups are born with a curly natal pelage that is dark brown or black dorsally and dark gray to orange ventrally. They molt at about one month old to a shorter, chocolate brown coat. This coat fades during the first year, so that juveniles vary in color from dark brown or reddish brown to orange.

RANGE AND HABITAT South American Sea Lions occur in coastal waters of South America from Peru south to southern Chile in the Pacific and then north to southern Brazil in the Atlantic. They are among the most abundant and commonly seen marine mammals in the southwestern Atlantic along the coast of Argentina. A small colony also occurs at the Falkland Islands. Generally, they breed in the more southern parts of their range and then travel northward in winter and spring, evidently following the cold Falkland Current in the Atlantic.

SIMILAR SPECIES South American Fur Seals occur throughout the range of the South American Sea Lion. They are smaller-bodied, with a much smaller head, a narrow, pointed snout, and a dark gray to black pelage with a distinct layer of fur beneath the longer guard hairs. South American Fur Seals stay mainly at sea during the nonbreeding season from late summer through early spring, coming ashore primarily during the summer breeding season, compared with the more sedentary South American Sea Lions, which often haul out year-round. Transient Juan Fernández Fur Seals along the Chilean coast have a longer, more pointed snout and a darker pelage.

BEHAVIOR These sea lions aggregate in small scattered colonies, the density of which varies with topography and temperature. They are more dispersed along rocky coastlines than on sandy or gravel beaches, and will also space out more when it is warm and sunny, often just at the water's edge to cool down. Vocalizations are key elements of territorial establishment and maintenance. Males bark often when establishing and maintaining territories and herding females, and they make high-pitched, directional calls during initial encounters with other males. The vocalizations of South American Sea Lions vary enough that individuals may recognize each other for at least short periods, particularly

80° 60° 40° 0° 20° 40°
SOUTH AMERICA
PACIFIC OCEAN
ATLANTIC OCEAN

SOUTH AMERICAN SEA LION
FAMILY OTARIIDAE

MEASUREMENTS AT BIRTH
LENGTH 33″ (85 cm)
WEIGHT 24–33 lb (11–15 kg)

MAXIMUM MEASUREMENTS
LENGTH **MALE** 9′2″ (2.8 m)
FEMALE 7′3″ (2.2 m)
WEIGHT **MALE** 770 lb (350 kg)
FEMALE 310 lb (140 kg)

LIFE SPAN
About 20 years

RANGE

A group of South American Sea Lions hauled out on the coast of Patagonia, Argentina.

during the breeding season. Females and juveniles molt in late summer and autumn; adult males molt slightly later. Pups are vulnerable to predation by Killer Whales, which have been known to temporarily beach themselves in order to catch pups lounging at the water's edge. Vampire bats attack some sea lions along the coast of Chile and Peru.

REPRODUCTION South American Sea Lions are polygynous breeders. The breeding strategies of males are influenced by the physical characteristics of the habitat. On rocky beaches, males establish and defend small territories where females aggregate to cool off, sometimes in the shade of coastal bluffs or boulders or near tide pools. They herd and contain groups of females within these territories until the females are ready to mate. On longer and broader cobble or sandy beaches, males stay along the surf line where it is cooler and attempt to herd and mate with females going to sea to cool down or to forage. Males defend their territories or access corridors to the sea with vigorous posturing, vocal threatening, and occasional brief physical fights that usually last less than 15 seconds. Pups are born from September through March, depending on location, and females are ready to mate soon after giving birth. Females remain with their pups for about a week and then begin a series of foraging trips to sea that last about three days each, spending two days ashore nursing their pups between trips. Lactating females evidently do not range far from the colonies to forage. Pups remain ashore for the first month and then gradually spend more time in the nearshore surf developing swimming skills. Pups may be nursed for a year or longer.

OPPOSITE: *A juvenile South American Sea Lion returning from a nighttime foraging trip in nearshore waters.*

LEFT: *Male South American Sea Lion with females, Ushuaia Bay, Argentina. Adult males have massive heads and well-developed manes.*

FOOD AND FORAGING Along the southern coast of Chile, South American Sea Lions tend to forage for slow-swimming fish that live on the seafloor, rather than for the more abundant but faster-swimming, schooling pelagic species. Throughout their range South American Sea Lions eat a diversity of fish and cephalopods, particularly small squid that are also harvested by commercial fisheries. Their diet overlaps substantially with the diets of resident gentoo and Magellanic penguins. Off southern Argentina, they eat a variety of invertebrates and fish, principally Argentine hake and anchovy, as well as cephalopods such as red octopus, Argentine shortfin squid, and Patagonian squid. South American Sea Lions off Patagonia mostly forage at the seafloor in nearshore habitats. Lactating females at the Falkland Islands forage mostly at night at depths to 820 feet (250 m). They usually stay within 30 miles (45 km) of their breeding beaches, though some females range up to 95 miles (150 km) away.

STATUS AND CONSERVATION South American Sea Lions had been hunted for several thousand years for subsistence before Europeans began commercial harvests in the early 1500s, and the commercial sealing industry in Uruguay continued until the late 1970s, longer than anywhere else in the world. At Mar del Plata in southern Argentina, South American Sea Lions occurred in large numbers until coastal development and direct hunting had exterminated them by the late 1800s. Small numbers of nonbreeding sea lions began to reappear in the area in the 1960s. South American Sea Lions were also harvested commercially in Chile until the early 1900s, and in 1976 government-sanctioned harvests resumed, with annual harvests of several thousand pups as well as some adults and juveniles. The population in the Falkland Islands has declined substantially, from about 350,000 in 1938 to about 30,000 in 1965, and it continues to dwindle despite protection, possibly as a result of increased commercial fisheries for squid and fish in the area. Population estimates in the 1970s and early 1980s were about 100,000 along the coast of Chile, about 20,000 in Peru, about 50,000 in Argentina, and about 30,000 in Uruguay. South American Sea Lions are currently protected from harvests in Argentina and the Falkland Islands.

Walrus

Odobenus rosmarus
(Linnaeus, 1758)

- LARGE, ROBUST, AND SPARSELY HAIRED
- DENSE, SHORT, STIFF WHISKERS ON UPPER LIP
- LONG EXTERNAL TUSKS
- MALE HAS MASSIVE CHEST, WART-LIKE NODULES ON NECK AND CHEST
- DISTRIBUTION CIRCUMPOLAR, PRINCIPALLY LIMITED TO SHALLOW ARCTIC REGIONS

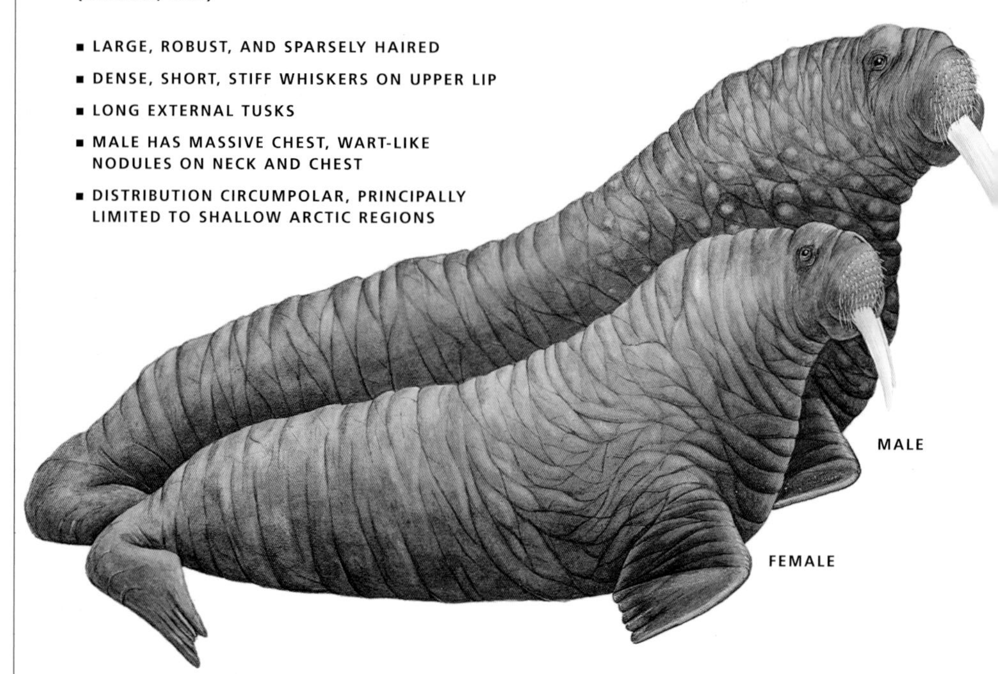

Perhaps the most distinguishing feature of the Walrus are its long tusks, which are enlarged upper canine teeth that grow continually throughout the animal's life. Walruses use their tusks mainly in social interactions, such as when males compete for females during the breeding season, but also as an aid in hauling out and moving on ice floes. This is reflected in the genus name, derived from the Greek words for "tooth" and "I walk." The specific name comes from *rossmaal* or *rossmaar,* the Scandinavian name for the Walrus. Walruses have a dense bristly assemblage of vibrissae, or whiskers, growing from thick mustacial pads on their upper lip and cheeks. The highly sensitive whiskers and pads are key adaptations for finding, identifying, excavating, and consuming prey from the seafloor sediment. Walruses can be exceptionally dangerous and have on occasion attacked and punctured inflatable boats with their tusks. Taxonomists recognize three subspecies of Walrus.

DESCRIPTION The Walrus has a large, robust torso, massive in adult males, that dwarfs its relatively small head. The head is flat and broad, and the snout short and blunt. The thick broad cheeks and upper lip are densely covered by short, robust vibrissae. While the Walrus lacks external pinnae, its flippers are similar in general shape to those of the otariids. The foreflippers are short and broad with very small claws. The hindflippers are short, with longer inner and outer digits and small claws. The Walrus is unique among extant pinnipeds in having external tusks. These upper canine teeth grow gradually as the Walrus ages, sometimes reaching 3 feet (1 m). Adult males are substantially larger than females, mostly in the neck, chest, and shoulders. Males develop wart-like nodules on the neck and chest when mature. There are five pairs of teeth in the upper jaw and four pairs in the lower jaw.

Adults and juveniles are sparsely covered with short, light brown to blond hair. In males, the hair becomes sparser with age, and from a distance males may appear to be bare-skinned. Hair density appears to remain constant in females throughout their lives, although it is still sparse. The skin, up to 2½ inches (6 cm) thick, is visible through the sparse pelage. Skin color varies, appearing pale in the water or when hauled out in cold temperatures, and sometimes bright pink when hauled out in warmer temperatures. The color differences are caused by the amount of blood being circulated to the skin—less when Walruses are in cold temperatures, in order to conserve heat, and more when they are hauled out in warm conditions and are thermoregulating to dump excessive internal body heat. Newborns (usually called calves) have short, gray to grayish-brown hair.

RANGE AND HABITAT Walruses live in relatively shallow water and coastal habitats of the North Atlantic, Bering Sea, and Arctic Ocean near North America and Eurasia. They mostly occur in pack ice during the spring breeding season. In summer and early autumn, they may haul out in large dense groups on sandy and pebble beaches to molt and rest.

The Atlantic Walrus *(Odobenus rosmarus rosmarus)* ranges from the eastern Canadian Arctic east to the Kara Sea and occurs in four relatively geographically isolated populations: one from Lancaster and Jones Sounds south to Hudson and Ungava Bays and east to the west coast of Greenland; another along the east coast of Greenland; a third at Svalbard, Norway; and a fourth from the eastern Barents Sea east to the Kara Sea near Novaya Zemlya. Vagrants have been reported along the European coast south to the Bay of Biscay and the British Isles. The Pacific Walrus *(O. r. divirgens)* ranges in the Bering Sea from the Kamchatka Peninsula east to Bristol Bay, Alaska, and in the western Arctic from the Chukchi Sea and along the coast of northeastern Siberia east to Point Barrow, Alaska. Vagrants have been reported south to the Alaska Peninsula and into the North Pacific near Kodiak Island, Cook Inlet, and Yakutat Bay. The Laptev Sea Walrus *(O. r. laptevi)* ranges from the eastern Kara Sea east to the western East Siberian Sea.

SIMILAR SPECIES The Walrus's tusks, distinctive body shape, and thick mustacial pads covered with short stiff vibrissae set it apart from all other pinnipeds that may overlap with it, including Northern Sea Lions, and Largha, Ringed, Hooded, Harp, Gray, and Ribbon Seals.

BEHAVIOR Walruses occur in small loose groups on the sea ice during the breeding season and occasionally form larger groups at the pack-ice edge when they haul out to molt. At other times, large

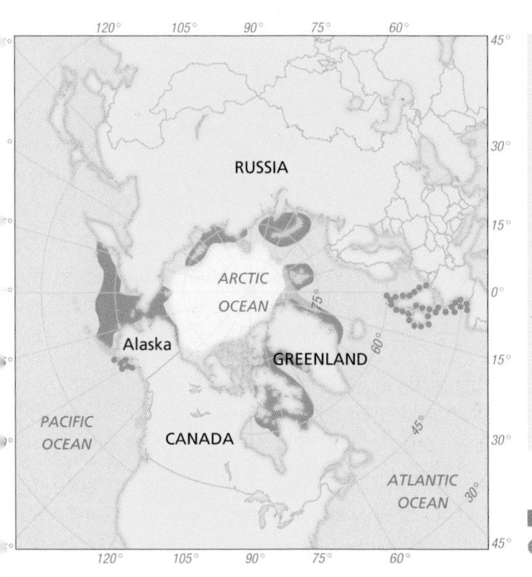

WALRUS
FAMILY ODOBENIDAE

MEASUREMENTS AT BIRTH		
LENGTH	3′4″–3′11″ (1–1.2 m)	
WEIGHT	99–165 lb (45–75 kg)	
MAXIMUM MEASUREMENTS		
LENGTH	**MALE**	12′ (3.6 m)
	FEMALE	10′ (3 m)
WEIGHT	**MALE**	4,200 lb (1,900 kg)
	FEMALE	2,600 lb (1,200 kg)
LIFE SPAN		
40 years		

■ RANGE
● VAGRANTS

dense groups of Walruses hauling out to rest or to molt may form on island beaches. From late spring through autumn, adult males form groups at several coastal sites at Bristol Bay in Alaska and the western Bering Sea in Russia. These groups are highly congested, with Walruses often lying on top of each other. At that time females, pups, and juveniles occur mostly in the pack ice. Females develop strong bonds with their calves and vigorously defend them, often using their tusks against other Walruses that come too close. Walruses may use their tusks for several reasons. Both sexes use them as an aid in hauling out on the ice, and males may use them in territorial battles or to attract females. Walruses apparently do not use their tusks for foraging on the sea bottom, as scientists previously believed. While at sea, Walruses may be submerged about 80 percent of the time, presumably diving and foraging. The diving patterns of Walruses are not well studied. Most dives presumably last less than 10 minutes; the longest recorded dive lasted about 25 minutes. Most dives are likely shallower than about 260 feet (80 m), as the benthic mollusks Walruses prey upon generally occur in shallow coastal or continental shelf habitats. The deepest recorded dive for the species was to 436 feet (133 m).

REPRODUCTION Because of the ice-loving, or pagophilic, habits of the Walrus, scientists know relatively little about its breeding behavior. Some believe the Walrus is a polygynous breeder and that males establish small aquatic territories, like leks, into which they attempt to attract receptive females. This system differs from the territorial and group breeding systems of other pinnipeds, in which males either actively herd small groups of females or simply position themselves near females and aggressively exclude other males. Male Walruses may physically and vocally threaten each other and may occasionally engage in intense combat. Mating evidently occurs in the water. Gestation lasts about 15 months. Females give birth on pack ice from mid-April through early June and nurse their pups for 2 to 2½ years. Because of the long nursing period, females give birth only every two to three years at most.

FOOD AND FORAGING Walruses prey principally on mollusks, other invertebrates, and small

OPPOSITE: *A female Walrus and her young pup in the Bering Sea pack ice. Walrus mothers are intensely protective of their pups and may charge other animals or boats that come too close.*

RIGHT: *A group of molting male Walruses in summer. On warm days, these aggregations can be seen from several miles away because the animals' skin becomes bright pink as blood is shunted to the skin surface to dump excess heat.*

fish that live at the seafloor and in bottom sediments. They can create powerful suction power with their well-developed cheeks and are able to suck soft-bodied clams from their shells even when the clams are burrowed deep in bottom sediments. Some Walruses have been observed killing and eating seabirds, seals, and Northern Sea Lions.

STATUS AND CONSERVATION For thousands of years or more, native arctic peoples have hunted Walruses for their meat, blubber, skin, bones, and tusks. Walruses have also been hunted commercially for at least 1,000 years. Commercial hunting was most extensive from the 1600s through the 1800s in the northeastern Atlantic, where many local populations were exterminated. Russian commercial hunting of Pacific Walruses accounted for the killing of up to 8,000 animals annually from 1931 through 1960. Russia continues to hunt Walruses commercially in the Bering and Chukchi Seas. Commercial hunting of Walruses was banned in the United States in the 1940s and in Canada in 1931. Populations of the Pacific Walruses recovered in most areas, possibly doubling between 1960 and 1980 to around 200,000 but appeared to start declining in the 1990s, owing to reduced birth rates and increased juvenile mortality from as yet unidentified causes. Populations of the Atlantic Walrus have remained relatively small, estimated at around 2,000 in the North Atlantic in the early 1990s, 600 near Svalbard, and 5,000 in the Laptev Sea. Subsistence harvests continue in Alaska, eastern Canada, Greenland, and Russia.

Sparring male Walruses at an island haulout site in summer. Most encounters at this time are generally brief and harmless as the males just jostle for some resting space. During the breeding season in late winter and spring, the duels may be more intense, and males may become seriously wounded as they compete with each other for a female's attention.

Bearded Seal

Erignathus barbatus
(Erxleben, 1777)

The Bearded Seal, one of the largest pinnipeds in the Arctic, is named for the dense "beard" of whiskers on its upper lip and cheek, which is so abundant that it appears to obscure the mouth. Its beard, long large body, and disproportionately small head readily distinguish it from other true seals. The Bearded Seal has a circumpolar distribution in the Northern Hemisphere in association with the edge of the pack ice. Taxonomists recognize two subspecies of Bearded Seal: one in the Atlantic sector of the Arctic and one in the Pacific sector. The distinction between the two subspecies is based on apparent geographic separation; however, the two subspecies mingle in the Canadian and Russian Arctic, and morphological differences between them are slight. Bearded Seals are noticeably vocal underwater. Regional vocal dialects have been reported, suggesting that there may be some isolation among local populations. The genus name is from the Greek word *gnathòs*, for "jaw," and refers to this seal's large cheek cavity and robust lower jaw. The specific name is derived from the Latin word *barba*, for "beard." Norwegian sealers called Bearded Seals "square flippers," for the squared-off shape of the foreflippers.

DESCRIPTION The Bearded Seal has a large body, and its relatively small head and short fore-flippers give it an even more long-bodied and robust appearance. The head is broad and the snout short and blunt. Compared to other true seals, the cheek cavity is quite large, and the lower jaw is deep and robust. There is a dense array of thick vibrissae on the upper lip, short in the middle and quite long on the sides of the mouth and cheeks. The longer whiskers are straight when wet but curl up noticeably when dry. The vibrissae are smooth, like those of monk seals, rather than beaded as in other true seals. Monk seals and Bearded Seals have four abdominal teats, unlike other phocid pinnipeds, which have two. The foreflippers have a squared-off appearance, with the five digits all about equal length (this feature is quite different from other phocid seals, which have a substantially longer first digit). The hind-flippers have a typical phocid shape but may appear small due to the adult's long thick body. The claws on the foreflippers are short and robust, while those on the hindflippers are small and pointed. Adult females may be slightly larger than adult males. There are nine pairs of teeth in the upper jaw and eight pairs in the lower jaw.

- SQUARED-OFF, SHORT FOREFLIPPERS
- LONG AND LARGE-BODIED WITH SMALL BROAD HEAD
- DENSE ARRAY OF THICK WHISKERS ON UPPER LIP AND CHEEKS
- TWO PAIRS OF RETRACTABLE ABDOMINAL TEATS
- CLOSELY ASSOCIATED WITH DRIFTING SEA ICE IN THE ARCTIC

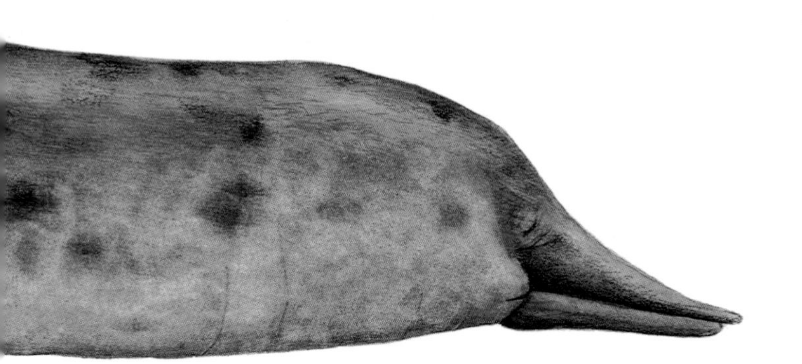

Adults have a gray or dark brown pelage, often slightly darker dorsally, with scattered dark spots and light rings. Pups shed their white lanugo coat in utero and are born with a pelage of long, wavy, bluish to brown hair. Bearded Seals that occur near Svalbard, Norway, often have rust-colored faces and, occasionally, foreflippers, evidently a discoloration caused by iron-oxide deposits. Some researchers think these seals become discolored from foraging in soft-bottom sediments that are rich in iron monosulfides.

RANGE AND HABITAT Bearded Seals have a circumpolar distribution in the Northern Hemisphere south of 80°N, in the Arctic and rarely into subarctic regions. They are generally associated with drifting sea ice in shallow-water areas. Their whereabouts in winter are poorly known;

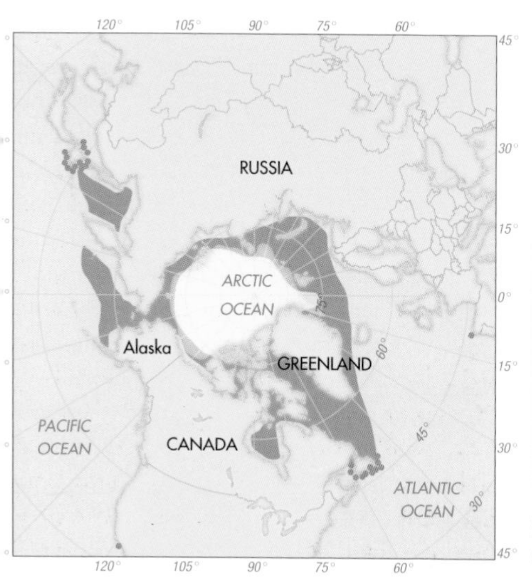

BEARDED SEAL
FAMILY PHOCIDAE

MEASUREMENTS AT BIRTH
LENGTH 4′3″ (1.3 m)
WEIGHT 75 lb (34 kg)

MAXIMUM MEASUREMENTS
LENGTH **MALE** 6′11″–7′10″ (2.1–2.4 m)
FEMALE May be slightly longer than male
WEIGHT **MALE** 570 lb (260 kg)
FEMALE 600–790 lb (270–360 kg)

LIFE SPAN
About 31 years

RANGE
VAGRANTS

many are thought to occur in open-water polynyas, particularly in the North Water Polynya in Baffin Bay. There are two subspecies of Bearded Seal. *Erignathus barbatus barbatus* occurs from the eastern Canadian Arctic and the Gulf of St. Lawrence east through the North Atlantic Ocean and into the Barents and Laptev Seas. *E. b. nauticus* ranges from the Russian Arctic, the Sea of Okhotsk, and the Bering Sea east into the western Canadian Arctic. The highest densities of Bearded Seals have been observed in the northern Bering Sea near St. Lawrence and St. Matthew Islands, the eastern Beaufort Sea near Banks Island, the eastern Canadian Arctic in the western Hudson Bay and Ungava Bay, and around Svalbard, Norway. Vagrants have been recorded as far south as northeastern Newfoundland in the western Atlantic, Portugal in the eastern Atlantic, and Hokkaido, Japan, in the western Pacific.

SIMILAR SPECIES The Bearded Seal may overlap in the southern portions of its range with Harbor Seals, throughout the Arctic with Ringed Seals, in the Bering Sea with Largha and Ribbon Seals, and in the Atlantic sector of the Arctic with Harp, Gray, and Hooded Seals. The Bearded Seal is readily distinguished by its larger size, longer body, dense beard, and squared-off foreflippers. Harbor Seals are distributed farther south and rarely occur in pack-ice habitats. Ringed Seals have a dark pelage with light-colored rings. Largha Seals have a light pelage with dark spots. Ribbon Seals have a distinct white banding pattern. Harp Seals have a light pelage with a distinctive dark saddle pattern. Gray Seals are more heavily spotted. Hooded Seals have a darker pelage with white patches.

BEHAVIOR Bearded Seals are not generally gregarious, though they may occur locally in high densities where drifting sea ice becomes concentrated. In spring, they may also concentrate in nearshore pack-ice habitats, where females give birth on the most stable areas of ice. Rarely, during times when the sea ice melts completely, they may haul out on land in the Okhotsk, White, and Laptev Seas. While molting between April and August, Bearded Seals spend substantially more time hauled out than at other times of the year. In some areas of heavy ice, Bearded Seals maintain breathing holes with the robust claws of their foreflippers. Bearded Seals are especially vocal underwater, particularly males during the breeding season. Their songs can be heard for dozens of miles or more, and native hunters often locate them by listening for these sounds. Adult

ABOVE: *A Bearded Seal breathes at the surface after foraging at the seafloor for several minutes.*
LEFT: *From a distance, Bearded Seals can be distinguished from other Northern Hemisphere seals by the small head in relation to the long robust body.*

The dense long facial whiskers and uniformly colored pelage help to distinguish Bearded Seals from other pinnipeds in the Arctic.

females near the coast of Spitsbergen, Norway, dive mostly at night, to depths of around 65 feet (20 m), with each dive lasting two to four minutes. The deepest recorded dive was to 945 feet (288 m), and the longest recorded dive lasted 19 minutes. Pups typically dive to about 30 feet (10 m) for about one minute; the maximum recorded dive was to 275 feet (84 m) and lasted 5½ minutes. Bearded Seals are important prey of Polar Bears in some areas and may also be eaten by Killer Whales. Pups are occasionally killed and eaten by Walruses.

REPRODUCTION Scientists know little about the breeding system of Bearded Seals. They may be monogamous, or adult males may mate with a few females sequentially. In spring, pregnant females seek out the most stable pack-ice areas, usually near coasts. Unlike Ringed Seals, which excavate lairs, female Bearded Seals give birth on the open surface of the sea ice. Most pups are born in early to mid-April in the southern Bering Sea and Sea of Okhotsk, late April in the Bering Strait area, and early May near Svalbard, Norway. The nursing period is rather short, and pups are usually weaned by the time they are about 15 days old. Lactating females spend about 8 percent of their time hauled out on the ice nursing their pups and the rest of the time in the water, evidently foraging. Adult females and their pups usually haul out on small ice floes away from other seals, where adult males may visit them for short periods to check on the receptiveness of females for mating. Mating is thought to occur in the water.

FOOD AND FORAGING Bearded Seals mostly hunt on the seafloor in shallow continental shelf areas of the Arctic. Their diet mainly consists of crabs, shrimp, mollusks, arctic and saffron cod, flatfish, sculpins, and octopus that live in or just above the seafloor sediment. They may also eat marine algae in some regions. Locally, this algae, found in the intestines of the killed seals, is considered a rare vegetarian delicacy.

STATUS AND CONSERVATION The total population of Bearded Seals has been estimated at more than 500,000, with perhaps 300,000 living in the Bering Sea. Several years of exceptionally heavy ice coverage in the eastern Beaufort Sea in the early 1970s may have led to a substantial decline of Bearded Seals there. Some scientists think that commercial sealers hunting mostly for Harp and Hooded Seals exterminated the local population of Bearded Seals around Labrador and Newfoundland. Bearded Seals have been hunted for subsistence throughout the Arctic for several centuries or more, harvested for meat and clothing. Russian sealers harvested thousands of Bearded Seals annually in the Sea of Okhotsk during the 1960s and in the Bering and Chukchi Seas from 1976 through 1985. Some harvests continued in the Bering Sea through at least the late 1980s.

Harbor Seal

Phoca vitulina
Linnaeus, 1758

- VARIABLY TAN TO SILVER WITH SCATTERED DARK SPOTS, OR DARK WITH LIGHT RINGS
- FREQUENTS COASTAL LAGOONS AND BAYS
- HAULS OUT ON COBBLE AND SANDY BEACHES, TIDAL FLATS, SANDBARS, ROCKY REEFS
- WIDESPREAD ALONG COASTS AND ISLANDS OF NORTH ATLANTIC AND NORTH PACIFIC

Widely distributed in coastal habitats of both the North Pacific and North Atlantic Oceans, Harbor Seals are often seen swimming in the surf off sandy beaches of the northern United States. There are five recognized subspecies, distinguished principally by their geographical distribution, and there does not appear to be much long-distance movement among regional populations. Populations vary in their coloration, becoming darker with lighter spots in more southerly latitudes of their range in the Pacific. The Harbor Seal is known by at least several dozen names through various parts of its range, including the Common Seal in the North Atlantic and the Kuril Seal in the western North Pacific. The specific name is derived from the Latin words *vitula,* meaning "calf," and *innus,* meaning "-like."

DESCRIPTION The Harbor Seal is a medium-size phocid pinniped with a short, spindle-shaped body. The head is robust and the snout rather broad and long. The flippers are relatively short, with sturdy claws on the foreflippers. Adult males may be slightly larger than females. There are nine pairs of teeth in the upper jaw and eight pairs in the lower.

The pelage pattern varies substantially with latitude. Some seals are light tan or silver with scattered dark spots; others are black with scattered, light, incomplete rings. Lighter morphs are often less spotted ventrally. In the Pacific, the dark morph may be more common in southern areas, whereas light and intermediate morphs may predominate in northern areas. Seals in the North Atlantic are mostly dark with light rings. In some regions, Harbor Seals may have red or rust-colored areas, evidently owing to deposition of iron oxide on the hair shafts. This has been particularly noted in San Francisco Bay, where strong seasonal winds cause sediments to become suspended in the water in shallow areas where the seals forage. Pups generally shed their long, woolly, grayish lanugo pelage in utero and resemble their parents at birth; pups born early in the season, however, may retain their lanugo coat for several days.

RANGE AND HABITAT Harbor Seals range widely in coastal areas of the North Pacific and North Atlantic. Five subspecies are recognized, based principally on geographic distribution. *Phoca vitulina stejnegeri* ranges throughout the western Aleutian Islands to the Commander Islands, and along coasts in the western North Pacific from the southeastern area of the Kamchatka Peninsula southward through the Kuril Islands to the east coast of Hokkaido, Japan. *P. v. richardii* ranges throughout the eastern Aleutians and along the southern coast of the Alaska Peninsula to Prince William Sound, and southward along the coast to Baja California. In the Bering Sea they occur in southern Bristol Bay, and small numbers occur at the Pribilof Islands. Several thousand seals live at the Channel Islands off southern California. Vagrants have occurred in the Gulf of California and at Guadalupe Island, several hundred miles off the Pacific coast of Baja California. Three subspecies are found in the North Atlantic. *P. v. vitulina* occurs in the Barents Sea, the southern Baltic Sea, and in the eastern North Atlantic from the British Isles south to Portugal. A small population of this subspecies is isolated in the Arctic Ocean at Svalbard, Norway. *P. v. concolor* occurs in the western North Atlantic along the coast of northeastern Canada east to northwestern and northeastern Greenland and Iceland and south to New Jersey. Vagrants have been recorded as far south as Florida, as well as far up the St. Lawrence River to Lake Champlain. *P. v. mellonae* occurs in several freshwater lakes and rivers in northeastern Canada on the Ungava Peninsula.

Harbor Seals forage in a variety of marine habitats, including deep fjords, coastal lagoons and estuaries, and high-energy, rocky coastal areas. They may also forage at the mouths of freshwater rivers and streams, occasionally traveling several hundred miles upstream. They haul out on sandy and pebble beaches, intertidal rocks and ledges, and sandbars, and occasionally on ice floes in bays near calving glaciers.

SIMILAR SPECIES Harbor Seals overlap with Ribbon, Largha, and Ringed Seals in the North Pacific and with Harp, Hooded, Gray, Ringed, and Bearded Seals in the North Atlantic. Harbor Seals generally avoid areas of pack ice where pagophilic Hooded, Harp, and Ringed Seals occur. Largha Seals are similar in color to the light-morph Harbor Seals but rarely occur in the ice-free habitats preferred by Harbor Seals. Gray Seals are larger, with longer, less-tapered snouts, and a pelage that is white with mostly black blotches. Ribbon and Bearded Seals are distinctive and easy to distinguish from Harbor Seals.

BEHAVIOR Except for the strong bond between mothers and pups, Harbor Seals are generally intolerant of close contact with other seals. They are nonetheless gregarious, especially during the molting season, which occurs between spring and autumn, depending on geographic location. While molting, groups of several hundred seals may haul out at the same tide bar, sandy or cobble beach, or exposed intertidal reef. When hauled out they spend most of their time sleeping,

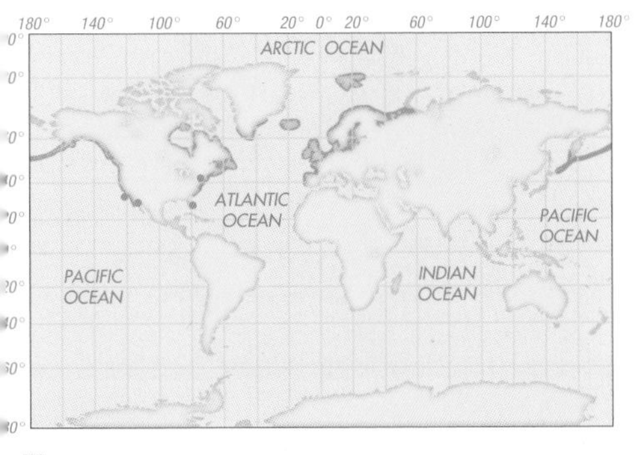

HARBOR SEAL
FAMILY PHOCIDAE

MEASUREMENTS AT BIRTH
LENGTH 28–40″ (70–100 cm)
WEIGHT 18–26 lb (8–12 kg)

MAXIMUM MEASUREMENTS
LENGTH **MALE** 6′3″ (1.9 m)
FEMALE 5′7″ (1.7 m)
WEIGHT **MALE** 370 lb (170 kg)
FEMALE 290 lb (130 kg)

LIFE SPAN
MALE 25 years
FEMALE 35 years

OPPOSITE: *Harbor Seals may spend some of their time at sea sleeping on the seafloor and drifting with currents and tidal surge.* **LEFT:** *Harbor Seals haul out on a variety of habitats throughout their range in the North Pacific and North Atlantic Oceans, including sandy and sometimes vegetated or rocky beaches.* **BELOW:** *A Harbor Seal forages in one of its varied nearshore marine habitats.*

scratching, yawning, and scanning for potential predators, including humans, foxes, coyotes, bears, and raptors. In late autumn and winter, Harbor Seals may be at sea continuously for several weeks or more, presumably feeding to recover body mass lost during the reproductive and molting seasons and to fatten up for the next breeding season.

Harbor Seals are solitary when at sea. They are rather curious around human divers and may nip at the fins of scuba divers and snorkelers. These are the least vocal of all pinnipeds. Pups have a sheep-like call to which their mothers respond with a call or by navigating toward them. Other vocalizations on land are brief grunts and growls, used as warnings to neighbors to keep at least a flipper's length distance away. Some low-frequency, pulsed sounds that have been recorded underwater are thought to be threat displays of patrolling males. Harbor Seals are preyed on by large sharks in some areas, particularly near Sable Island, Nova Scotia, and perhaps in some areas along the California coast. They are also hunted by transient Killer Whales in British Columbia and the Gulf of Alaska. Brown bears and coyotes have been reported killing and eating Harbor Seals in some areas of the Pacific Northwest and Gulf of Alaska.

REPRODUCTION Harbor Seals appear to be serially monogamous; males generally mate with one or a few females during a season. Males may display and vocalize, both to attract females and threaten challenging males. A primary display during the breeding season is to slap the surface of the water with the fore- or hindflippers. On land, males display with aggressive vocalizations, mostly growls or coughs, accompanied by thrusts of the head. They also vigorously wave their foreflippers at nearby seals and may scratch other seals with the foreflipper claws or make open-mouth threats. Females give birth in spring and summer, generally earlier the farther south they live. As yet, twins have not been observed in this species. Mothers nurse their pups for three to four weeks, then abruptly wean and abandon them. Mating takes place in the water soon after pups are weaned.

Unusually precocial for pinnipeds, Harbor Seal pups are able to swim and dive within minutes of birth. In the water, they often ride on their mothers' backs by holding on with their fore-

flippers. In some areas, such as Sable Island, Nova Scotia, lactating females may begin to forage when their pups are only a day or two old. Pups may remain on the beach during these times, but they more often accompany their mothers to sea, and both mother and pup spend substantial amounts of time diving.

FOOD AND FORAGING Harbor Seals eat a highly varied diet, feeding on demersal fish, pelagic schooling fish, octopus, and squid, depending on prey availability. In the western Gulf of Alaska, newly weaned Harbor Seal pups forage for slow-moving shrimp, switching to faster-moving capelin as their swimming and foraging skills improve. Harbor Seals spend about 85 percent of the day diving, and much of their diving is presumed to be active foraging in the water column or on the seabed. They dive to depths of about 30 to 500 feet (10–150 m), depending on location; seals in southern California sometimes dive to about 1,500 feet (450 m). Dives generally last a few minutes, although the longest recorded dive (in the eastern North Atlantic) lasted 31 minutes.

STATUS AND CONSERVATION Harbor Seals may number 500,000 or more throughout their range. Historically, native inhabitants have hunted these seals to various extents for several hundred to several thousand years. In the early 1900s, bounties were paid out in some areas for killing Harbor Seals because of their supposed competition with commercial fishermen, particularly in Alaska, British Columbia, eastern Canada, Massachusetts, and England. Harbor Seals are still killed legally in Canada, Norway, and the United Kingdom to protect fish farms or local fisheries. In the western Gulf of Alaska, Harbor Seal populations declined substantially from the mid-1970s through the mid-1980s. This decline correlated with an increase in commercial trawl fisheries, particularly for shrimp. Long-term ecosystem changes owing to natural changes in the environment may also partially explain the decline. Some seals died as a result of the *Exxon Valdez* oil spill in Prince William Sound in 1989, but the consequences to the population have been debated. Substantial numbers of Harbor Seals, perhaps more than 20,000, died from a distemper virus in the eastern North Atlantic in the late 1980s, and an influenza virus also killed several hundred seals in New England in 1979 and 1980. The population in the Baltic Sea has remained at a low of about 600 seals since around 1994 owing to hunting by humans and perhaps reproductive and immunological suppression from chronic exposure to chemical pollution in the Baltic Sea.

Largha Seal

Phoca largha
Pallas, 1811

Scientists know little about the Largha Seal, which spends its time either in the open ocean or in pack-ice habitats that are difficult to access. During the breeding season, these seals are known to form small family groups consisting of an adult female and her pup and an adult male waiting to mate with the female. The Largha Seal was named by the coastal Tungas people, who lived along the western Sea of Okhotsk. It is also known as the Spotted Seal.

DESCRIPTION The Largha Seal is a relatively small-bodied phocid pinniped, with a round head and a narrow, moderately long snout. The nostrils point more forward than upward. The flippers are short and narrow, with short sturdy claws on the foreflippers and small narrow claws on the hindflippers. The first and fifth digits on the hindflippers are slightly longer than the inner digits, and the first digit on the foreflipper is slightly longer than all the others. There are no obvious differences in size, shape, or coloration between males and females, though males may be slightly longer and heavier. There are nine pairs of teeth in the upper jaw and eight pairs in the lower jaw.

Adults are generally light gray to silver dorsally and ventrally, with dark spots scattered over the entire pelage. The spotting is perhaps somewhat denser dorsally, with small rings occasionally appearing in areas of concentrated pigment. The dorsal pelage is often slightly darker and may sometimes appear dull brownish several months after the molt. Pups are born with a long, woolly, white lanugo coat, which is shed around two to four weeks after birth and replaced by an adult-like, patterned pelage.

RANGE AND HABITAT Largha Seals range from the Seas of Okhotsk and Japan, east throughout the Bering Sea and along the coast of Alaska to Bristol Bay, and have occasionally been seen in the western Beaufort Sea. They haul out and breed on sea ice in all areas, and more rarely on sandbars and gravel beaches. Some seals, mostly weaned pups and juveniles, appear around the coast of northern Hokkaido, Japan, in spring and occasionally range to the southern end of Hokkaido. After the ice melts in spring and summer, Largha Seals evidently forage in the open ocean.

- RELATIVELY SMALL-BODIED
- PELAGE MOSTLY LIGHT GRAY WITH SCATTERED SPOTS
- NEWBORN HAS LONG, WOOLLY, WHITE COAT
- PREFERS ICE HABITATS, RARELY OCCURS ON BEACHES
- DISTRIBUTION RESTRICTED TO THE BERING, OKHOTSK, AND JAPAN SEAS

SIMILAR SPECIES Largha Seals may share habitats in the Bering, Okhotsk, and Japan Seas with Ribbon, Ringed, and Harbor Seals, and in the western Beaufort Sea with Ringed and Bearded Seals. Harbor Seals, most similar in appearance, are larger and generally use terrestrial rather than ice habitats. Ribbon and Ringed Seals are readily distinguished by their pelage patterns (distinctive bands on the Ribbon Seal and rings and denser spotting on the Ringed Seal), and Bearded Seals have a much longer and larger body and dense long vibrissae sprouting from the plump upper lip and cheeks.

BEHAVIOR Because Largha Seals live mostly in pack-ice habitats that are difficult to access, or

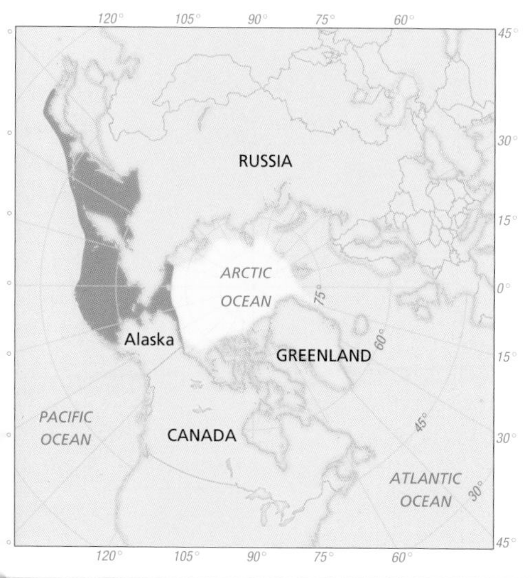

RANGE

LARGHA SEAL
FAMILY PHOCIDAE

MEASUREMENTS AT BIRTH

LENGTH 30–36″ (77–92 cm)

WEIGHT 15–26 lb (7–12 kg)

MAXIMUM MEASUREMENTS

LENGTH **MALE** 5′7″ (1.7 m)
FEMALE 5′3″ (1.6 m)

WEIGHT 290 lb (130 kg)

LIFE SPAN

30–35 years

LEFT: *Largha Seals, also called Spotted Seals, are similar in body shape, size, and coloration to Harbor Seals, but Largha Seals haul out and breed principally on pack ice, compared to the terrestrial habitats preferred by Harbor Seals.* **RIGHT:** *Newborn Largha Seals also have a thick white lanugo pelage, which is shed when they are about three to four months old, revealing the adult spotted color pattern.*

in the open ocean, their behavior is virtually unstudied. They may occur in well-spaced family groups on the sea ice during the breeding season in spring.

REPRODUCTION Largha Seals are thought to be seasonally monogamous. During the breeding season, they are most often seen well spaced out on the ice in triads consisting of an adult female, her pup, and an adult male. Females give birth on the surface of ice floes from January through mid-April, with a peak in mid- to late March. Males are thought to join a female and her pup about a week after pupping, and the group remains together until the pup is weaned at three to four weeks old, at which time mating occurs and the male leaves the group. This system limits the mating opportunities of males during a breeding season; however, males that mate early in the season may later find an unattended female-pup pair or may displace another male from a triad. Mating evidently takes place in the water.

FOOD AND FORAGING Adults and juveniles eat a variety of schooling fish (pollock, capelin, arctic cod, and herring), epibenthic fish (especially flounder, halibut, and sculpin), and crabs and octopus at depths of up to 1,000 feet (300 m). Weaned pups apparently mostly eat amphipods, krill, and other small crustaceans.

STATUS AND CONSERVATION Native peoples along the eastern Russian coast and in Alaska have traditionally killed small numbers of Largha Seals for subsistence. The Soviet Union made some commercial harvests from the 1930s through the 1980s in the Sea of Okhotsk and the western Bering Sea, and Japan also commercially hunted these seals in the Sea of Okhotsk at times. Largha Seals occasionally drown in fishing nets set in coastal waters of northern Hokkaido, Japan. Population abundance is poorly known but has been estimated at around 350,000 to 400,000, with about half of the seals living in the Bering and Chukchi Seas.

Ringed Seal

Pusa hispida
(Schreber, 1775)

- DARK DORSAL PELAGE WITH SCATTERED, IRREGULAR LIGHT RINGS AND DARK BACKGROUND
- SMALLEST TRUE SEAL NEXT TO BAIKAL SEAL
- EXCAVATES BIRTH LAIRS BENEATH ICE SURFACE
- DISTRIBUTION CLOSELY ASSOCIATED WITH LANDFAST AND PACK ICE
- WIDELY DISPERSED IN ARCTIC BASIN AND BERING, OKHOTSK, JAPAN, AND BALTIC SEAS

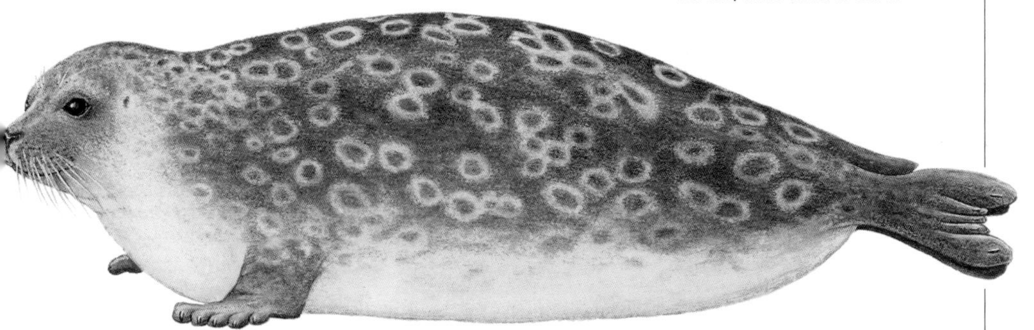

The smallest and most common seals in the Arctic Ocean and the Bering and Baltic Seas, Ringed Seals have long been important prey for native inhabitants of the Arctic. They are also the top prey of Polar Bears, and during the breeding season, Ringed Seals excavate birth lairs in ice and snow to protect themselves against this predation. The Baikal Seal is the only other pinniped known to use such structures for giving birth and raising pups. Scientists recognize five subspecies of the Ringed Seal, including two freshwater populations. The Ringed Seal was formerly included in the genus *Phoca* along with the Baikal and Caspian Seals. However, taxonomists have recently reinstated these three species to the genus *Pusa*, which is derived from the common name for the Ringed Seal used by the Inuit of Greenland and various eastern North Atlantic cultures. The specific name is derived from the Latin word *hispidus*, meaning "hairy" or "bristly," and refers to the adult pelage, which is often stiffer than that of other phocid pinnipeds. The common name refers to the scattered irregular rings on the pelage.

DESCRIPTION The Ringed Seal has a small plump body and a small head. The snout is narrow, short, and cat-like. The flippers are small, with short slender claws on the hindflippers and robust claws on the foreflippers that may be more than an inch long. There are nine pairs of teeth in the upper jaw and eight pairs in the lower jaw.

The pelage of adults is dark dorsally with scattered irregular rings, and lighter and less ringed ventrally. Newborn pups have a woolly, white lanugo coat that they shed at about six to eight weeks old to reveal an unspotted pelage that is uniformly dark silver or gray dorsally and light silver ventrally. The ringed pattern develops at the first annual molt when seals are a little more than one year old.

RANGE AND HABITAT Ringed Seals have a circumpolar distribution throughout the Arctic Ocean, Hudson Bay, and Baltic and Bering Seas. They are closely associated with sea ice. In summer they often occur along the receding ice edge and farther north in denser pack ice. Five subspecies are recognized. The most widely dispersed form, *Pusa hispida hispida*, occurs in the Arctic Basin. *P. h. ochotensis* occurs in the Sea of Okhotsk and the Sea of Japan, and *P. h. botnica* occurs in the Baltic Sea. Freshwater populations

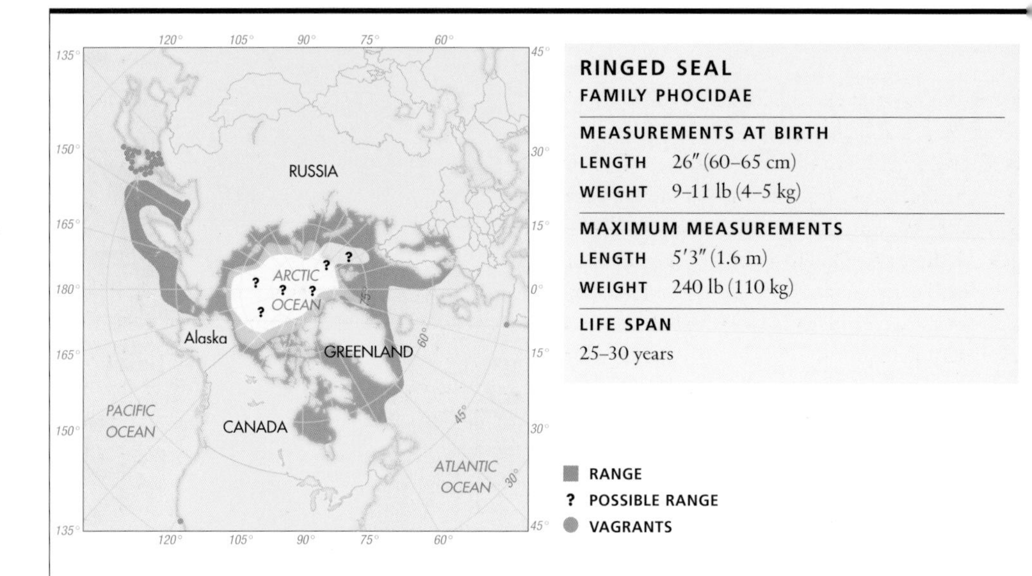

RINGED SEAL

FAMILY PHOCIDAE

MEASUREMENTS AT BIRTH

LENGTH 26″ (60–65 cm)

WEIGHT 9–11 lb (4–5 kg)

MAXIMUM MEASUREMENTS

LENGTH 5′3″ (1.6 m)

WEIGHT 240 lb (110 kg)

LIFE SPAN

25–30 years

RANGE
? POSSIBLE RANGE
VAGRANTS

include *P. h. saimensis* in Lake Saimaa in eastern Finland and *P. h. lagodensis* in Lake Ladoga, Russia. Vagrants from the marine populations have ranged as far south as Portugal in the Atlantic Ocean and California in the Pacific.

SIMILAR SPECIES Harbor, Harp, Hooded, Gray, Bearded, Ribbon, and Largha Seals may occupy similar habitats in various parts of the Ringed Seal's range. All but Harbor and Largha Seals can be readily distinguished by their body and head morphology and pelage patterns. Largha Seals, which may overlap in the Bering, Okhotsk, and Japan Seas, have a spotted rather than a ringed pelage pattern and are larger but more slender than Ringed Seals, with relatively longer, wider snouts. Harbor Seals prefer ice-free habitats and are rarely seen in ice.

BEHAVIOR Though there are areas of high density of Ringed Seals through the Arctic, these seals do not aggregate in large groups. Rather, they are largely solitary and space out from one another by hundreds of yards or more. During the breeding season, triads of an adult female, her pup, and an adult male form short-term family groups. These groups are not easily observed, however, as the seals remain in lairs in the ice and snow excavated by the females for pupping and nursing. The excavation of lairs in and under sea and lake ice is unique to Ringed Seals and is evidently an adaptation for escaping predation by Polar Bears. Some lairs are quite complex, with several chambers. Females evidently leave pups in the lairs for short periods while they forage nearby. Throughout winter, Ringed Seals maintain breathing holes by chewing away newly formed ice. Individuals may favor particular breathing holes, perhaps excluding other seals from loosely associated underwater territories. Ringed Seals molt in June and July; while molting, they spend more time basking on the surface of the ice than in other seasons. Ringed Seals are the primary prey of Polar Bears, and are also occasionally eaten by Walruses and Killer Whales.

REPRODUCTION The breeding system of the Ringed Seal is thought to be either mildly polygynous or serially monogamous, but is not well

known because of the difficulty in finding and observing seals during the breeding season. Females excavate lairs in the pressure ridges or accumulated snow on sea or lake ice, and in Lake Saimaa in snowdrifts along the shoreline. They give birth in March and April in most areas, a little earlier in the Baltic Sea. Pups are weaned and mating occurs between April and May. Males evidently patrol under the ice searching for receptive females. They may stay with a female for several days until they mate, and then return to the water to patrol for another potential mate.

FOOD AND FORAGING When feeding along the sea-ice edge in summer, Ringed Seals eat mostly polar cod, even though the potential prey biomass there is dominated by pelagic crustaceans. The seals evidently selectively choose these prey, which represent about only 1 percent of the fish and crustacean biomass. In these areas, Ringed Seals eat smaller cod, evidently at shallower depths than the sympatric Harp Seals. Most dive depths for *P. h. hispida* are 35 to 150 feet (10–45 m) for sexually mature males, and 330 to 475 feet (100–145 m) for subadult males and postpartum females. Most dives last about 4 minutes for adult males and 7½ minutes for adult females. The longest dive recorded is about 23 minutes, although the seal may actually have been resting on the sea bottom rather than feeding.

STATUS AND CONSERVATION Ringed Seals have been key subsistence prey for native arctic peoples, who hunt them for food for humans and dogs as well as for skins to make clothing. Levels of PCBs are higher in seals taken by subsistence hunters in the European and Russian Arctic than in other arctic regions. These higher levels are thought to be due to continued use of PCBs in Russian electrical equipment. Though never completely surveyed, the species may number as many as 4 million. Ringed Seals in the Baltic Sea are considered to be at risk because of heavy pollution, which affects the seals' immune systems and reproductive success. Although about half of the Ringed Seals in Lake Saimaa breed in coastal areas located within national parks, poaching and threats associated with fisheries in other parts of the lake seriously threaten this small population.

OPPOSITE: *Ringed Seals have a robust body and small head and foreflippers. The dark pelage background with scattered light rings is characteristic of the species.* **RIGHT:** *Ringed Seals are the primary prey of Polar Bears and so are extremely wary when surfacing in their breathing holes, which may be staked out by patient, hungry bears. Ringed Seals maintain these breathing holes by abrading the ice with their canine teeth.*

Ribbon Seal

Histriophoca fasciata
(Zimmermann, 1783)

The Ribbon Seal is named for its distinctive pattern of ribbon-like bands that circle the head and neck, the foreflippers, and the hips. The specific name is derived from the Latin word *fascia,* meaning "band." Scientists have recorded Ribbon Seals making several types of underwater sounds, presumably for social interaction during the breeding season. A unique air sac that branches off the trachea in adults may help produce these sounds. The air sac is well developed in males and poorly developed or absent in females, suggesting that males may use vocalizations to compete with other males or to atrract females. Ribbon Seals live in the pack ice of the Bering and Okhotsk Seas, and perhaps the Chukchi Sea, during most of the year, but some scientists think they are pelagic in summer, perhaps ranging south to the coast of Japan and into the central North Pacific Ocean.

DESCRIPTION The Ribbon Seal is more slender and streamlined than other phocid pinnipeds in the North Pacific. The head is short and broad, and the snout short and narrow at the tip. The flippers are short and narrow; the foreflippers have a longer outer digit and robust claws, while the hindflippers have longer first and fifth digits and small, narrow claws. There are nine pairs of teeth in the upper jaw and eight pairs in the lower.

The adult pelage is distinct and unmistakable. The body is dark brown in females to near black in males, interrupted by distinct light stripes that circle the neck, each foreflipper, and the hips. These stripes are often less distinct in adult females. Pups are born with long white hair, which they shed shortly after weaning to reveal a pelage that is grayish dorsally and silver-gray ventrally. Juveniles are gray to light brown; faint striping starts to develop when the seals are one or two years old.

RANGE AND HABITAT Ribbon Seals live principally throughout the offshore pack ice of the Bering Sea, the southern Chukchi Sea, and into the Sea of Okhotsk along the east coast of Russia. There appear to be two principal breeding grounds, one in the Sea of Okhotsk and one in the Bering Sea. The whereabouts of Ribbon Seals during the nonbreeding season are poorly

- STRIKINGLY PATTERNED (PARTICULARLY MALE) WITH LIGHT-COLORED BANDS ON A DARK BODY
- OCCURS IN OKHOTSK, BERING, AND CHUKCHI SEAS, AND NORTHERN PART OF NORTH PACIFIC OCEAN

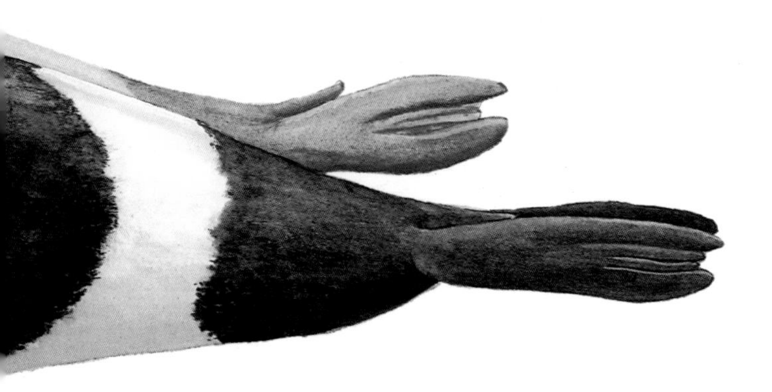

known. Adults generally occur in pack-ice habitats from January to May, and they may be pelagic during the rest of the year. At least some move north with the receding ice edge in summer to the Chukchi and Beaufort Seas. Others may range south to the Aleutian Islands, northern Hokkaido, and the central North Pacific Ocean. In 1962, an adult male hauled out and was captured near Morro Bay, California.

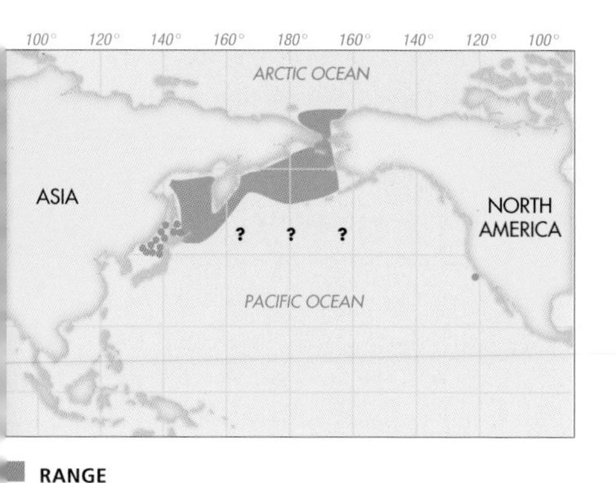

RANGE
? POSSIBLE RANGE
VAGRANTS

SIMILAR SPECIES The unique pelage pattern of adult Ribbon Seals makes them unmistakable throughout their range, although juveniles may be difficult to distinguish from Largha Seals and Harbor Seals from a distance. The two species may overlap in the Bering, Okhotsk, and Japan Seas. Harbor Seals avoid pack-ice habitats, preferring instead to haul out on coastal beaches, tidal flats, and sandbars.

RIBBON SEAL
FAMILY PHOCIDAE

MEASUREMENTS AT BIRTH
LENGTH 31–35″ (80–90 cm)
WEIGHT 20–22 lb (9–10 kg)

MAXIMUM MEASUREMENTS
LENGTH 6′3″ (1.9 m)
WEIGHT 220 lb (100 kg)

LIFE SPAN
26 years

TOP LEFT: *A Ribbon Seal pup at St. Lawrence Island in the northern Bering Sea.*
BOTTOM LEFT: *Adult male Ribbon Seal. Pups and juveniles lack the distinct banding pattern of adults.*
BELOW: *The species' unique pattern is more distinct in the adult male than in the female, shown here.*

BEHAVIOR Scientists know very little about the social behavior of Ribbon Seals, due to the inaccessibility of their offshore pack-ice breeding habitats in winter and early spring and their unknown whereabouts during summer and autumn. They appear to occur in scattered small groups or pairs. Juveniles evidently molt in April and May, and adults from May through July. Ribbon Seals produce at least two underwater sounds, one a long whistle that sweeps down from a high frequency to a low one, the other a short puffing noise that covers a wide range of frequencies.

REPRODUCTION Ribbon Seals appear to be monogamous, and mating evidently takes place in the water during May; however, these behaviors have never been observed. Females give birth in April and May, primarily on sea ice in the Bering Sea and the Sea of Okhotsk. They nurse their pups for about three to four weeks, then abruptly wean and abandon them.

FOOD AND FORAGING The diet of Ribbon Seals is poorly known but appears to consist principally of pelagic and demersal fish, especially pollock and arctic cod, and squid. Young seals may feed mainly on crustaceans.

STATUS AND CONSERVATION The abundance of Ribbon Seals is poorly known, but scientists estimate the species to number around 250,000. Between 1961 and 1967, Russians harvested around 13,000 Ribbon Seals each year in the Bering Sea, possibly overexploiting some local populations. Annual quotas of around 3,000 were afterward imposed; however, since the breakup of the Soviet Union in the early 1990s kills have been less well regulated and documented. While commercial hunting evidently no longer occurs in the Bering Sea, there is poaching in some areas. Other threats to the species include drowning from incidental entanglement in commercial fishing nets in the North Pacific Ocean and Bering Sea.

Caspian Seal

Pusa caspica
(Gmelin, 1788)

- RELATIVELY LIGHT PELAGE WITH DARK SPOTS, ESPECIALLY ON DORSAL SURFACE
- PUP BORN WITH WHITE WOOLLY COAT
- DISTRIBUTION RESTRICTED TO SALINE, LANDLOCKED CASPIAN SEA

This seal is endemic to the Caspian Sea, the world's largest inland body of salt water. It is closely related to the Ringed Seal and the two species are similar morphologically, except that the Caspian Seal has a generally light pelage with scattered spots, in contrast to the Ringed Seal's light rings scattered over a dark background. The history of the relationship between the two species is not yet settled. Some scientists believe the Caspian Seal is a direct descendant of Ringed Seals that migrated southward during an early glaciation event and then were isolated when the ice sheets retreated; others think the ancestral stock for Caspian Seals may have originated instead in the large, brackish, inland Paratethys Sea during the Miocene and Pliocene epochs (roughly 2 to 24 million years ago) and that the Ringed Seal evolved from seals that dispersed north to the Arctic and were later isolated. There is concern over the long-term vitality and persistence of the Caspian Seal because of its confined isolation in an increasingly polluted environment.

DESCRIPTION The Caspian Seal is relatively small-bodied, similar in size to the Ringed Seal, which is thought to be its closest relative. Its head is delicate with a relatively long narrow snout, like that of Ringed and Baikal Seals. The flippers are short, with moderate claws on the foreflippers and short, narrow claws on the hindflippers. There are 10 pairs of teeth in the upper jaw and eight pairs in the lower jaw.

Adults have a grayish-yellow to dark gray pelage and are darker dorsally than ventrally. There are numerous, mostly dark spots scattered over the pelage, especially on the dorsal surface. Pups are born with a long white lanugo coat that is shed and replaced by a short darker pelage about three to four weeks after birth.

RANGE AND HABITAT This seal lives only in the Caspian Sea, a large inland sea about 100 feet (30 m) below sea level, located between Europe and Asia. Most breeding occurs in the northern part of the sea, where ice forms in winter and early spring and provides substrate for females to haul out, give birth, and nurse their pups. A small breeding colony also occurs farther south on islands along Turkmenistan, especially Ogurchinsky Island. Later in the year, seals haul out on a number of islands in the northern Caspian Sea, and some portion of the population may

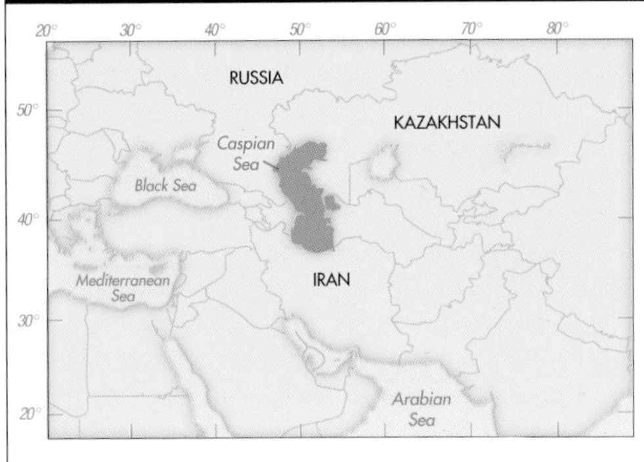

CASPIAN SEAL

FAMILY PHOCIDAE

MEASUREMENTS AT BIRTH

LENGTH 25–31″ (64–79 cm)

WEIGHT 11 lb (5 kg)

MAXIMUM MEASUREMENTS

LENGTH **MALE** 4′11″ (1.5 m)
FEMALE 4′7″ (1.4 m)

WEIGHT Unavailable

LIFE SPAN

MALE 26 years
FEMALE 29 years

migrate south to feed. However, the whereabouts of most seals during the ice-free period from late spring through autumn is relatively unknown.

SIMILAR SPECIES There are no other pinniped species that occur within the range of the Caspian Seal.

BEHAVIOR Scientists know little about the social, reproductive, or foraging behaviors of the Caspian Seal. Though Caspian Seals may congregate on the sea ice in winter, they are relatively asocial. Larger groups may form on islands in the northern end of the sea in summer when the seals are molting; however, social interactions during these times largely consist of aggressive snorts or flipper waving intended to maintain distance between individuals. Eagles prey on young Caspian Seal pups, and wolves may kill seals hauled out on islands in the northern part of the sea.

REPRODUCTION The breeding system is evidently one of serial monogamy. Females give birth in January and early February, mostly on the sea ice in northern areas of the Caspian Sea. They nurse their pups for four to five weeks. Mating occurs while or just after pups are weaned in late February and early March.

FOOD AND FORAGING Caspian Seals have a relatively broad diet that includes a variety of fish and crustaceans. In autumn and winter, their primary prey are evidently sculpins, gobies, and crustaceans. In spring, they eat mostly herring, roach, carp, sprat, and smelt.

OPPOSITE: *Caspian Seals may crowd together on small sandy islands in the northern Caspian Sea when molting.*

RIGHT: *A Caspian Seal rests at the surface after a foraging dive. The recent invasion in the Caspian Sea by a comb jelly, an exotic species that devastated the marine food web of the Black Sea, has raised some concerns about the conservation of this isolated seal.*

STATUS AND CONSERVATION Caspian Seals have been hunted for at least several centuries. From the mid-1800s through the early to mid-1900s, annual harvests typically numbered 115,000 seals, although 227,000 were evidently killed in 1935. Since the late 1960s, annual catches have been regulated at 65,000 or fewer. The population was estimated at about 500,000 in the early 1990s and was thought to be stable. However, increasing pollution and declining fish populations caused by overfishing have raised concerns about long-term population trends for this species. Mass mortalities of several hundred seals were reported in the northern part of the Caspian Sea in the spring of 1997, the autumn of 1999, and throughout the spring of 2000. The most recent mass mortality was reportedly caused by canine distemper, when a lack of ice in winter evidently forced the seals to crowd together on island beaches, increasing social contact and facilitating the spread of the virus. Another mortality event appeared to be developing in autumn 2000. In addition, Caspian Seals are threatened by climatic warming, which has resulted in poor ice formation (a key requisite for pup survival) and early breakup of ice in some years. The recent invasion from the Black Sea of a comb jelly, a voracious predator on fish larvae, is another concern.

Baikal Seal

Pusa sibirica
(Gmelin, 1788)

- SMALLEST TRUE SEAL, THOUGH QUITE ROTUND
- LARGE, FORWARD-FACING EYES, SET CLOSE TOGETHER
- LONG VIBRISSAE
- UNIFORMLY COLORED PELAGE, DARKER DORSALLY
- ISOLATED IN LAKE BAIKAL IN SOUTHEASTERN SIBERIA

The Baikal Seal is the smallest of the true seals and the only exclusively freshwater pinniped species. Far removed from its nearest relatives, the Ringed and Caspian Seals, the Baikal Seal is endemic to Lake Baikal in southeastern Siberia, the world's oldest and deepest freshwater lake. The species is thought to be derived from Ringed Seals that were isolated in Lake Baikal about a half million years ago, after the expansion and retreat of the Eurasian ice sheets. It was formerly included with the Ringed and Caspian Seals in the genus *Phoca;* however, taxonomists have recently reinstated these three species to the genus *Pusa*, which is the common name given to the Ringed Seal by the Inuit of Greenland and various eastern North Atlantic cultures. The Baikal Seal is known locally as *nerpa*, the Russian word for "seal."

DESCRIPTION The Baikal Seal is small and quite rotund, especially during the nonbreeding season, when large amounts of fat are stored in subcutaneous areas of the midsection. Its rotund appearance is exaggerated by its short body length relative to other pinnipeds. The head is broad, especially around the large eyes, and the snout is rather long and narrow. The large eyes are set closer together than in most other true seals and are forward facing. The flippers are short but broad. The foreflippers have robust claws, used to excavate birth lairs and to keep breathing holes open in winter. There are nine pairs of teeth in the upper jaw and eight pairs in the lower jaw.

Adults and juveniles have a pelage of rather long soft fur that is silver-gray to dark gray dorsally and light gray to yellowish white ventrally. They are uniformly colored, lacking the spots and rings of their closest relatives, the Caspian and Ringed Seals. Newborn pups have a long white lanugo coat that they shed at about four weeks old to reveal a silver-gray pelage.

RANGE AND HABITAT Baikal Seals occur only in Lake Baikal in southeastern Siberia, a large deep freshwater lake 1,500 feet (450 m) above sea level, about 350 miles (550 km) long, and up to 5,000 feet (1,500 m) deep. Most seals live year-round in the northern half of the lake, and virtually all breeding takes place there. A portion of the population migrates south in autumn and winter as the ice forms and spreads southward, and then north in spring and summer as the ice melts and the ice edge recedes. Seals have been reported to occasionally travel short distances up tributary streams and rivers along the lake's northern and western shores, and also into the lower Angara River, which is the only outlet river at the southwestern end of the lake.

SIMILAR SPECIES Baikal Seals are the only pinnipeds that live in Lake Baikal and so are unlikely to be confused with any other animals.

BEHAVIOR Baikal Seals are relatively asocial and generally solitary. Small groups may form to molt on island and some lakeshore beaches in spring and summer if the lake ice has melted. Baikal Seals may dive continuously from September through May, when the lake is frozen over, hauling out only infrequently. In the middle of the lake, where the water depth is about 3,300 to 5,250 feet (1,000–1,600 m), they dive mostly to depths of 35 to 160 feet (10–50 m). Occasionally they descend to more than 1,000 feet (300 m). Dives usually last two to six minutes, although some recorded dives have lasted more than 40 minutes.

REPRODUCTION Pregnant females excavate subsurface lairs in the ice or in pressure ridges, where they give birth and nurse their pups. Pups are born from mid-February through March, with a peak in mid-March. Pups emerge from their lairs in April and may lie basking on the surface on mild days. Mothers nurse their pups for about two months and then abruptly wean and abandon them at about the time they are ready to mate again. The nursing period may be shorter in the more southern areas of the lake, owing to early breakup of the ice in some years. Mating occurs in the water but has never been witnessed. Scientists have estimated that around 80 to 90 percent of sexually mature females give birth each year. The percentage is somewhat less in years of unfavorable, reduced ice conditions.

FOOD AND FORAGING Baikal Seals eat a large variety of fish. Their most important year-round prey are sculpins and a fish known commonly in Russia as *golomyanka.* They also prey upon omuls, fish that are heavily exploited by commercial fisheries, although omuls are evidently not a key component of their diet.

STATUS AND CONSERVATION Baikal Seals were hunted for thousands of years by native peoples who lived around Lake Baikal. Commercial harvests by the Russians began in the early 1900s, with up to 10,000 seals killed annually. The actual harvests since the dissolution of the Soviet Union in 1991 are less well known, as unregulated poaching may be substantial; local villagers have likely hunted them for subsistence food. There are three primary hunting seasons: April, when pups are harvested on the shore-fast ice;

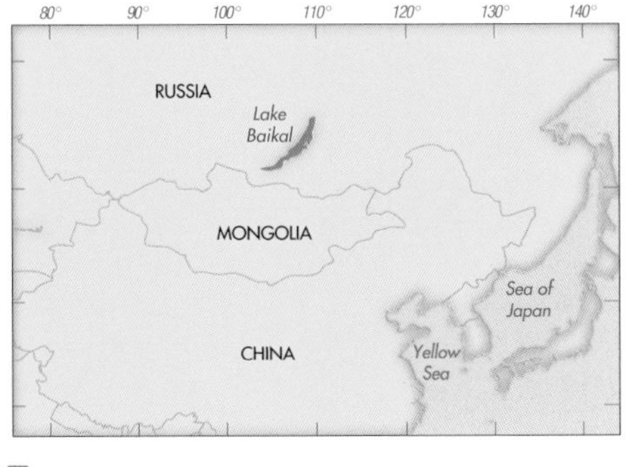

RANGE

BAIKAL SEAL
FAMILY PHOCIDAE

MEASUREMENTS AT BIRTH
LENGTH 25–26″ (64–66 cm)
WEIGHT 9 lb (4 kg)

MAXIMUM MEASUREMENTS
LENGTH 4′7″ (1.4 m)
WEIGHT 176–198 lb (80–90 kg)

LIFE SPAN
FEMALE 43 years
MALE Slightly less than female

May to June, when molting seals are harvested in northern areas before the ice breaks up and melts; and October to November, when nets are set along the lake's eastern shores as new ice begins to form in sheltered coves and bays.

There were an estimated 58,000 Baikal Seals in 1986 and perhaps as many as 100,000 in 1994. There is concern, however, that the population is in decline, as the number of pre-reproductive seals was remarkably small in the late 1980s, resulting in few seals being recruited into the breeding population, while older breeding seals continued to die. Moreover, a distemper virus killed at least several thousand seals in 1987 and

ABOVE LEFT: *Baikal Seals are the smallest of the true seals and are confined to the freshwater Lake Baikal in southern Siberia. They are short but rotund, which makes the head look particularly small.*

ABOVE RIGHT: *Baikal Seals generally haul out on seasonal ice that covers most of Lake Baikal's surface in winter and spring, but in summer they may have to crowd onto small rocky islets to molt.*

BELOW: *Baikal Seals feed intensively in summer and autumn in order to build up a thick layer of fat in preparation for the winter and the next breeding season.*

ABOVE: *The Baikal Seal's eyes are large and well adapted for foraging and navigating in poorly lit conditions under the ice that covers the lake in winter.*

1988, and there was an unusual mortality of several dozen seals at the southern end of the lake in 1997. Several hundred or more seals may also die each year from entanglement in fishing nets. A survey in 2000 reported an estimate of about 85,000 seals in the lake. Scientists have recently reported high levels of organochlorine compounds in the tissues of Baikal Seals, evidently a result of runoff into the lake from agricultural activities and industries. Levels are higher in males because females transfer some of their burden to pups before birth and when nursing, while males tend to accumulate the compounds in their tissues throughout life.

Gray Seal

***Halichoerus grypus* (Fabricius, 1791)**

During the breeding season, these gregarious seals spend several weeks ashore or on nearshore fast ice, making them accessible for observation and handling. The species has been well studied around the United Kingdom and Newfoundland. The spotting pattern of Gray Seals appears to be individually unique and has been used in population studies to identify individuals. There are three distinct populations, all located in the North Atlantic Ocean. The scientific name translates to "hook-nosed sea pig" and is derived from the Greek words *halios* ("of the sea"), *khoiros* ("pig"), and *grupos* ("hook-nosed"). In eastern Canada the Gray Seal is also known as "horsehead," in reference to the distinctive shape of the male's head.

DESCRIPTION The Gray Seal is large-bodied and robust, rotund at the torso and slender toward the hind end. The head is conspicuously long, broad, and flat with no obvious forehead. The flippers are short and rather thick. The foreflippers are blunt at the end, with digits all roughly the same length and with long slender claws. Adult males are up to three times larger than adult females, with a proportionally larger head and a longer, fleshier snout. Mature males develop a robust neck and chest with prominent folds or wrinkles. The chest may become heavily scarred from fighting with other males. There are nine to ten pairs of teeth in the upper jaw and eight pairs in the lower jaw.

Adult males are generally a uniform dark gray, brown, or black with scattered light spots and blotches over most of the body. Adult females and juveniles are mostly light silver or gray with dark brown, olive, or black blotches. The ventral coloration, especially of females and juveniles, may be lighter. Pups are born with a long thick lanugo coat that they shed at about two to four weeks to attain a muted adult pelage.

RANGE AND HABITAT Scientists recognize three primary populations of this species, all in the northern North Atlantic Ocean. One population occurs in the Baltic Sea and is now concentrated mostly in the northeastern Gulf of Finland, having been substantially reduced in size during the past several centuries. Another broadly distributed population ranges around Iceland, Norway, and the British Isles, with principal colonies at the Faroe Islands, the Hebrides, North Rona Island, and the Orkney, Shetland, and Farne

- SUBSTANTIAL DIFFERENCES BETWEEN MALE AND FEMALE IN COLOR, SIZE, AND HEAD AND NECK SHAPE
- MALE DARK WITH LIGHT SPLOTCHES, FEMALE LIGHT WITH DARK SPLOTCHES
- MALE'S SNOUT LONG AND BROAD
- MALE'S NECK AND CHEST ROBUST WITH FOLDS AND WRINKLES, HEAVILY SCARRED IN OLDER MALES
- OCCURS IN NORTHERN NORTH ATLANTIC OCEAN

Islands. Vagrants occasionally stray south to Germany, France, and Portugal. A third population is centered in the Gulf of St. Lawrence and along the Atlantic coasts of Labrador, Newfoundland, and Nova Scotia. A key colony near Nova Scotia is at Sable Island. Others are at various islands in the Gulf of St. Lawrence and Northumberland Strait. Gray Seals form colonies on rocky island or mainland beaches, though some seals give birth in sea caves and on sea ice, especially in the Baltic Sea.

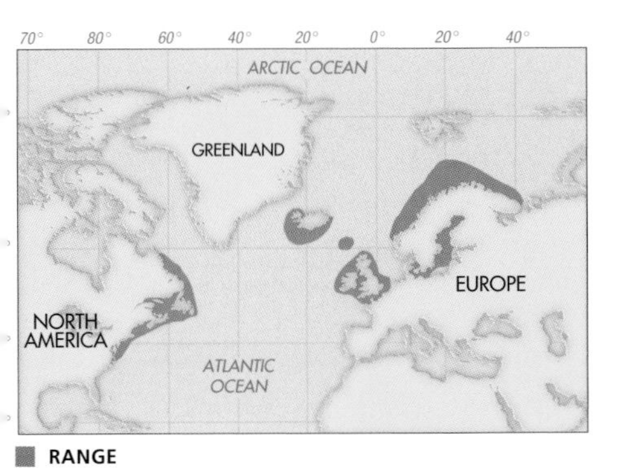

RANGE

SIMILAR SPECIES Harbor, Harp, Hooded, and Bearded Seals occur in many areas of the Gray Seal's range in the North Atlantic. Harbor Seals, which may be confused with young Gray Seals, have nostrils set farther apart and eyes located more forward on the face, closer to the nostrils. Harp Seals are substantially smaller and have a distinct dark saddle pattern and a black face. Hooded and Bearded Seals have a pelage similar to that of female Gray Seals. Juvenile and female Hooded Seals have a short blunt snout with

GRAY SEAL
FAMILY PHOCIDAE

MEASUREMENTS AT BIRTH

LENGTH		35–43″ (90–110 cm)
WEIGHT	MALE	35–40 lb (16–18 kg)
	FEMALE	29–35 lb (13–16 kg)

MAXIMUM MEASUREMENTS

LENGTH	MALE	8′6″ (2.6 m)
	FEMALE	6′7″ (2 m)
WEIGHT	MALE	770 lb (350 kg)
	FEMALE	440 lb (200 kg)

LIFE SPAN

35–40 years

LEFT: *Gray Seals forage in nearshore waters during the summer breeding season but may migrate to offshore foraging areas distant from breeding sites in autumn and winter.*

OPPOSITE: *Adult females are fiercely protective of their pups and will threaten and chase away any other seals, including male suitors, until the pups are weaned.*

larger, more forward-pointing nostrils, while adult males are easy to distinguish by their large, flaccid, inflatable nose. Bearded Seals have a short broad snout and a "beard" of densely packed vibrissae on the upper lip and cheeks.

BEHAVIOR Gray Seals are gregarious, gathering to breed, molt, and rest in groups of several hundred or more at island coasts and beaches or on land-fast ice and pack-ice floes. They space out at these sites and generally avoid social contact, except when females are nursing young or for mating. Little is known of their social behavior or whereabouts at sea, but they are thought to be solitary when feeding. Telemetry data indicate that at least some seals may forage seasonally in waters close to colonies, while others may migrate long distances from their breeding areas to feed in pelagic waters between the breeding and molting seasons. Gray Seals molt in late spring or early summer and may spend several weeks ashore during this time. Individual seals in the Baltic Sea appear to use a single primary haulout site throughout the year. When feeding, most seals remain within 45 miles (75 km) of their haulout sites; overall home ranges vary from 420 to 2,470 square miles (1,090–6,400 sq. km). Between July and January, seals at some sites haul out for about five hours at a time and forage at sea for about 35 hours. Some seals may remain at sea continually for up to 20 days.

Pups are born with a thick, wavy lanugo pelage, which is shed about two to four weeks after birth to reveal a faint adult pelage pattern.

REPRODUCTION Gray Seals are generally polygynous breeders. Adult males compete for rights and opportunities to mate with females, mostly with vocal and physical displays, though conflicts occasionally escalate to physical combat. Females give birth from September through November around the British Isles and Iceland, in November and December along Finland, and from late December through early February in the western Atlantic. In the Gulf of St. Lawrence, where seals breed on ice, pups suckle for about 15 days. Females lose about 13 pounds (6 kg) a day while fasting and nursing their pups. Some females that give birth on fast ice or pack ice may occasionally enter the water to feed while lactating, leaving their pups on the ice for short periods. Male pups are somewhat heavier at birth, and they grow faster and weigh more at weaning age than female pups. Pups are abruptly weaned at three to four weeks old, when females mate. Mating may take place on land, on ice, or in the water.

FOOD AND FORAGING In the Baltic Sea, Gray Seals prey mainly on herring, while along the British coast they forage mostly on the seafloor for demersal and benthic fish and cephalopods. Most dives of lactating females are to the seafloor and last about 1½ to 3 minutes (maximum 9 minutes). Juveniles in the Baltic Sea dive mostly to depths of 65 to 130 feet (20–40 m).

STATUS AND CONSERVATION Gray Seals were harvested for subsistence for several centuries or more by coastal people in the northeastern and northwestern Atlantic and around the United Kingdom. They were hunted extensively during the 1900s in Canada, the northeastern United States, and the Baltic Sea for bounties paid as a means to reduce competition with fisheries for common prey. Government-sponsored harvests in Great Britain occurred from 1962 through 1979. The population in the Baltic Sea numbered around 100,000 in the early 1900s but now numbers only around 2,000 to 3,000, owing to mortalities from heavy hunting and poaching, as well as from drowning in fishing nets and perhaps disease. Seals may be more susceptible to various diseases because of the extreme pollution of the Baltic Sea. The population in the western North Atlantic was estimated at around 110,000 and increasing in the late 1980s, and that in the British Isles at about 85,000 and increasing in the early 1990s. About 20,000 Gray Seals may live in the eastern North Atlantic.

Harp Seal

Pagophilus groenlandicus
(Erxleben, 1777)

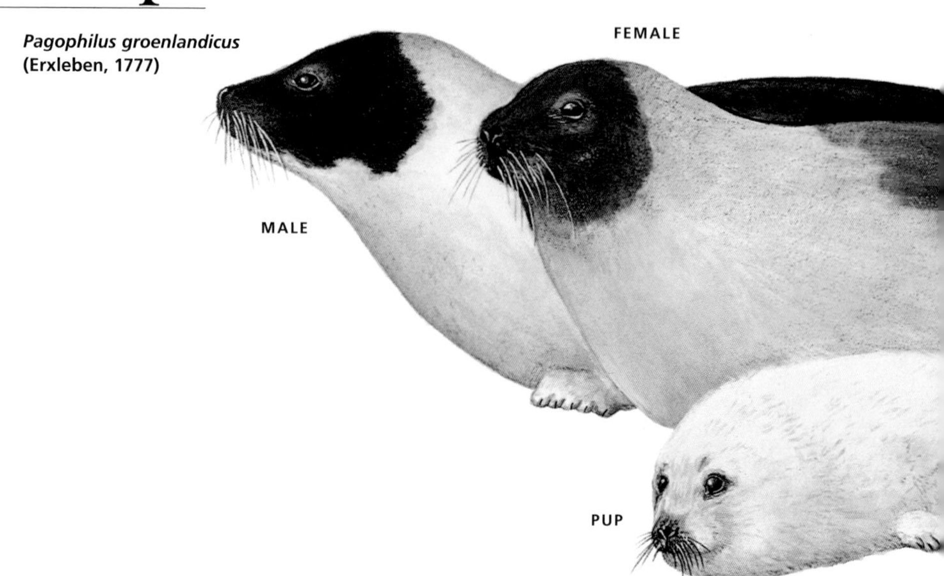

The Harp Seal is the most abundant pinniped in the Northern Hemisphere. Its common name refers to the distinctive, horseshoe-shaped saddle on the backs of adults, which resembles a lyre or a harp. The genus name means "lover of ice," and indeed these seals gather in large numbers in the drifting pack ice of the North Atlantic. Sealers call newborn Harp Seal pups "whitecoats," in reference to their long, woolly, white lanugo coat. Recently weaned pups are called "beaters," evidently referring to their awkward swimming behavior, while one-year-olds are "bedlamers," perhaps derived from a French term meaning "beast of the sea." The long-running and substantial harvests of these seals by Canadian sealers, particularly the widely publicized clubbing of young whitecoat pups for their fur, made this species well known. Prior taxonomies have included the Harp Seal in the genus *Phoca*. However, a recent comprehensive revision reinstated the species to the genus *Pagophilus*. Harp Seals are also known as Greenland Seals and sometimes as Saddleback Seals.

DESCRIPTION The Harp Seal has a rather robust body and a relatively small head, which is broad and flat with a narrow snout. The flippers are short and narrow; the foreflippers have thick strong claws, and the hindflippers have longer inner and outer digits and small narrow claws. There are eight pairs of teeth in both the upper and lower jaws.

Adults typically have a light gray pelage, a black face, and a black saddle behind the shoulders. The black saddle extends in a lateral band on both sides toward the pelvis, forming a pattern that resembles a harp. Some adults are sparsely spotted, with the harp pattern not completely developed. Development of the adult pelage correlates with sexual maturation. Newborn pups, called "whitecoats," have a long, white, woolly lanugo coat; this is replaced soon after weaning, at about three to four weeks old, by a short, silver pelage with scattered small dark spots.

RANGE AND HABITAT These seals are distributed in the pack ice of the North Atlantic Ocean, from Newfoundland to northern Russia. There are three recognized breeding populations: one in the White Sea, another near Jan Mayen Island (called the "West Ice"), and a third in the western

- BLACK FACE AND DISTINCTIVE BLACK SADDLE ON BACK
- NEWBORN PUP HAS A LONG, WHITE, WOOLLY PELAGE
- OCCURS IN LARGE GROUPS IN PACK ICE DURING BREEDING SEASON
- DISTRIBUTION LIMITED TO NORTH ATLANTIC AND WHITE SEA

North Atlantic, which is split into the "Gulf" subpopulation off northeastern Newfoundland and in the Gulf of St. Lawrence and the "Front" subpopulation off southern Labrador. The White Sea population is considered to be a separate subspecies *(Pagophilus groenlandicus oceanicus);* the "West Ice" and western North Atlantic populations are similar enough morphologically to be considered one subspecies *(P. g. groenlandicus).*

Harp Seals may travel up to 1,600 miles (2,500 km) during their migrations. In late spring after the breeding season, Harp Seals on the sea ice of the southern Barents and White Seas migrate north to summer feeding grounds, following the receding ice edge. In some years, however, large numbers have appeared near Finnmark, Norway, in winter and spring. Harp Seals in the western North Atlantic also move north to feed in late

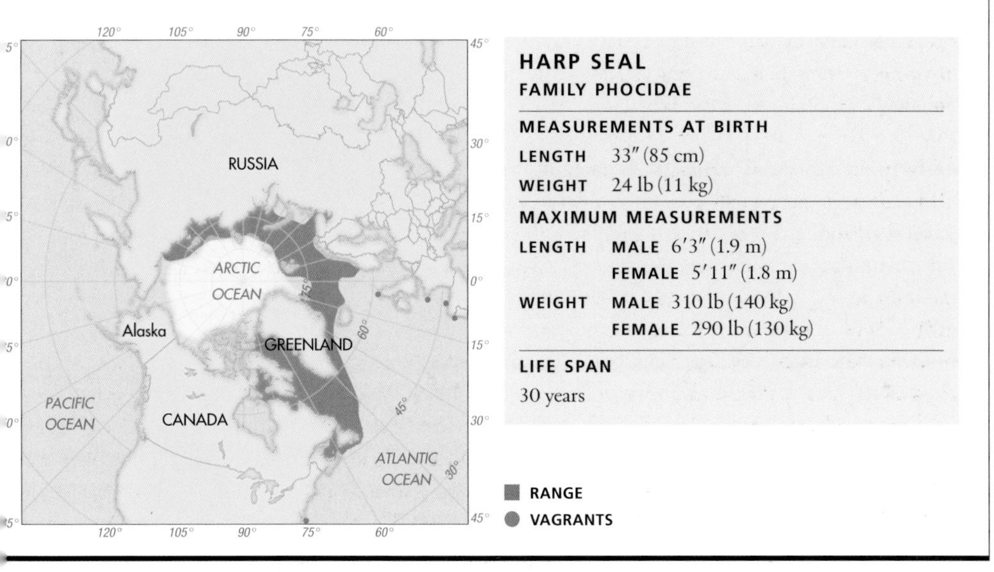

HARP SEAL
FAMILY PHOCIDAE

MEASUREMENTS AT BIRTH

LENGTH	33″ (85 cm)
WEIGHT	24 lb (11 kg)

MAXIMUM MEASUREMENTS

LENGTH	MALE	6′3″ (1.9 m)
	FEMALE	5′11″ (1.8 m)
WEIGHT	MALE	310 lb (140 kg)
	FEMALE	290 lb (130 kg)

LIFE SPAN

30 years

Harp Seals associate with pack ice in the Antarctic throughout the year, foraging under the ice in winter and spring and along the pack's receding edge in summer and autumn.

spring and summer as far as Thule, Greenland, and Jones Sound between Devon and Ellesmere Islands. Seals in the eastern North Atlantic move as far north as Svalbard. Vagrants have been reported as far south as Scotland, Germany, and France in the east and Virginia in the west.

SIMILAR SPECIES Harp Seals may overlap in distribution with Harbor, Ringed, Gray, Bearded, and Hooded Seals in the North Atlantic and Bearded and Ringed Seals in the Russian Arctic. However, adult Harp Seals are usually easy to identify by their unique, harp-shaped saddle marking.

BEHAVIOR Harp Seals aggregate in large numbers in the pack ice when breeding in February and March and later when molting in late spring. These aggregations, particularly of breeding populations, may number up to several thousand seals. Large groups may also feed and travel together during their extensive seasonal migrations. Harp Seals are very vocal, engaging in a variety of underwater calls in addition to those they give on the ice during the breeding season. In the Gulf of St. Lawrence, scientists have recorded 19 different call types during the breeding season. A "double-grunt" call is the most common. Underwater calls vary among populations, particularly between the "Gulf" subpopulation off Newfoundland and the "West Ice" population near Jan Mayen Island. After the breeding season, Harp Seals follow the ice edge north as it recedes in spring and summer. Seals may congregate in groups along the ice edge when they molt in late April and May. Adult males appear to molt first, followed by juveniles and nonbreeding seals, and finally reproductive females.

REPRODUCTION Harp Seals are thought to be serially monogamous. Breeding males search out receptive females and defend access to them against other males until mating. Though gregarious when breeding, individual females and their pups avoid close contact with other seals. Individual males investigate the female-pup pairs and may remain with one pair until the female is ready to mate. Females give birth on the pack ice from mid-February through early March and nurse their pups for about 12 days. Mating occurs in the water in late March and early April, just after weaning. The gestation period lasts about 11 months.

FOOD AND FORAGING In summer, Harp Seals feed along the ice edge as it recedes north into arctic regions, foraging mostly for polar and arctic cod. These fish represent only about 1 percent of the area's fish and crustacean biomass, which is dominated by pelagic crustaceans. Harp Seals eat larger cod and evidently dive to greater depths than Ringed Seals, with which they occur. While migrating in spring and fall, Harp Seals feed on capelin, herring, and krill. Most foraging occurs at depths of less than 300 feet (90 m), although dives may sometimes exceed 820 feet (250 m).

STATUS AND CONSERVATION The total population of Harp Seals may be 6 million, with the largest population in the western North Atlantic estimated at 4½ million seals in 1994. The "West Ice" population numbered almost 300,000 in 1994, including approximately 60,000 pups born that year. The White Sea population numbered 1½ to 2 million seals in 1998, based on an estimated 300,000 to 400,000 pups born.

Since the late 1700s, Harp Seals have been hunted in the western North Atlantic for their oil, fur, and skin, and in Newfoundland also for their meat. By 1832, over 700,000 Harp Seals were being killed annually. Kills declined to about 500,000 a year through the 1860s and to about 160,000 a year by the early 1900s. Today, commercial hunting continues throughout the species' range. The hunt along the coast of Canada was particularly controversial, with sealers clubbing to death large numbers of whitecoat pups less than three weeks old. In 1987 the commercial pup hunt was banned, with only subsistence hunting allowed to continue. From 1997 through 1999, the total mortality of Harp Seals in the western North Atlantic—including the Canadian harvest, the unregulated harvest off Greenland, and incidental deaths in fisheries—was estimated at 465,000 a year. About 245,000 seals were killed in Canada in 1999, and about 91,000 were killed in 2000, the decline in numbers possibly due to both socioeconomic factors and the early melting and breaking up of the pack ice, which dispersed the seals. Russia and Norway jointly manage harvests of weaned pups and adults near Jan Mayen Island and in the Barents Sea. Norwegian sealers killed about 15,000 seals annually between 1991 and 1996 and about 7,000 in 1999. Russian sealers killed about 35,000 whitecoat pups in 1999. Small numbers of Harp Seals are killed for subsistence in Canada and northwestern Greenland and by local hunters in Iceland, totaling about 60,000 annually.

ABOVE: *Harp Seal pups are covered with a long, thick, white lanugo pelage at birth, which is shed to reveal a light pelage with scattered black spots just after weaning.* **RIGHT:** *The Harp Seal's pelage coloration changes during each successive annual molt until the characteristic saddle, or "harp," pattern develops when seals become sexually mature.*

Hooded Seal

Cystophora cristata
(Erxleben, 1777)

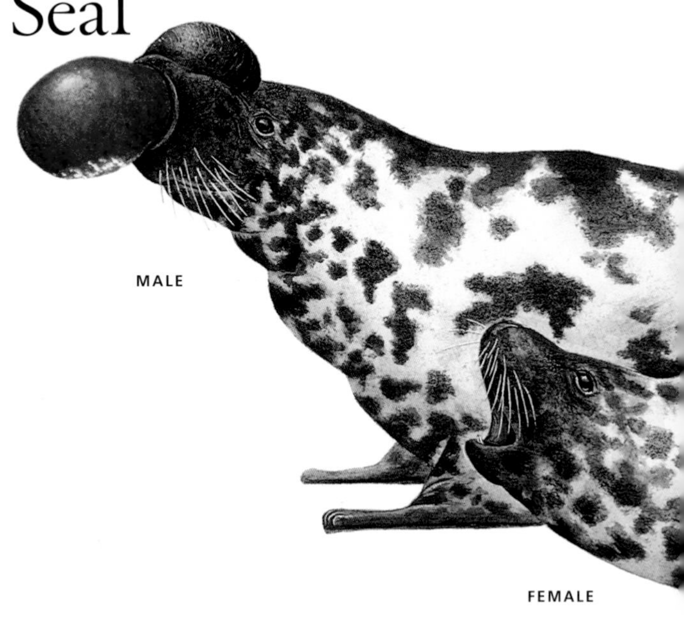

The Hooded Seal is named for the inflatable nose ("hood") and the extrudable nasal sac ("bladder") of adult males, which they use to attract females and threaten other males during the breeding season. The genus name comes from the Greek *kustis,* meaning "bladder," and *phoros,* meaning "carrying," while the specific name is from the Latin *crista,* for "crest." Another common name for the species is Bladder-nosed Seal. Young Hooded Seals are known to be great wanderers, and there have been reports of their occurrence as far away from their primary homes in the North Atlantic as the Beaufort and Bering Seas, southern California, the Caribbean, and southwestern Spain. Hopeful reports of the Caribbean Monk Seal, thought to be extinct, have recently been discounted by some researchers as actually being vagrant juvenile Hooded Seals.

DESCRIPTION The Hooded Seal is large relative to other seals, and males are larger and heavier than females. The head is broad and robust with a short snout. Flippers are short and narrow. Adult males have a flaccid, enlarged nose, the "hood," that can be inflated to make the head and face appear larger. Among other phocid pinnipeds, only the enlarged proboscis of adult male Northern and Southern Elephant Seals is similar to the "hood" structure of male Hooded Seals. Another unique secondary sexual characteristic of males is their nasal septum, which they inflate and extrude through one nostril to form a large, red, balloon-like structure, or "bladder." During the breeding season, males use both the hood and the bladder to attract females and during dominance displays with competing males. Hooded Seals have eight pairs of teeth in the upper jaw and seven pairs in the lower jaw.

Adult males are silver-gray with irregular black blotches over most of the body, and are darker on the face and dorsal surface of the foreflippers. Females are lighter and less mottled than adult males. Pups shed their lanugo coat in utero and are born with a thick pelage that is bluish gray dorsally, yellowish or creamy white ventrally, and darker on the face and foreflippers. The silvery, mottled adult pelage begins to develop in juveniles after the first or second annual molt.

RANGE AND HABITAT The breeding range of Hooded Seals is limited to the central and west-

- EXTRUDABLE MEMBRANE ("BLADDER") IN NOSTRIL OF ADULT MALE
- INFLATABLE NOSE ("HOOD") OF ADULT MALE
- MOSTLY LIGHT PELAGE, WITH SCATTERED LARGE DARK BLOTCHES
- STRIKING BLUISH DORSAL PELAGE OF YOUNG SEALS
- ADULTS GENERALLY SOLITARY WHEN NOT MATING OR MOLTING
- BREEDS IN NORTH ATLANTIC OCEAN

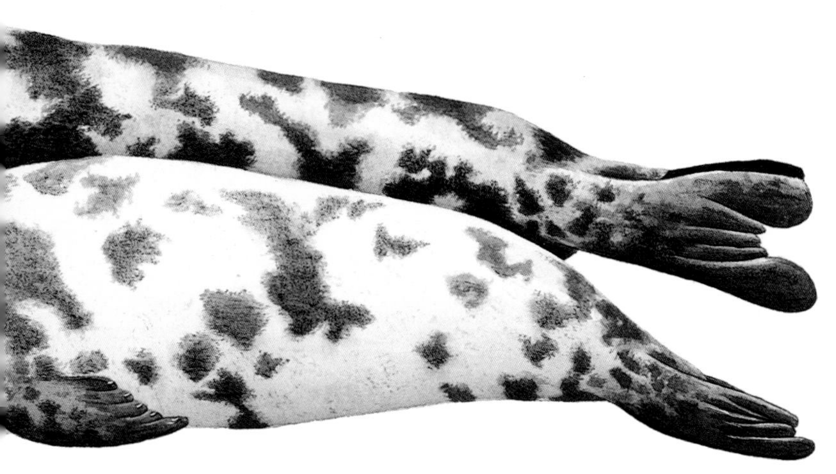

ern North Atlantic Ocean. Major breeding sites occur in the Gulf of St. Lawrence, east of Newfoundland in an area called the "Front," in the Davis Strait between northern Canada and Greenland, and near Jan Mayen Island. During the nonbreeding season, Hooded Seals appear to stay mostly along the edge of the pack ice as it recedes northward. Molting seals from all breeding populations are thought to aggregate near eastern Greenland in the pack ice in late spring and summer. Their whereabouts in autumn and winter are poorly known; however, they may be pelagic during this time, ranging away from the continental shelves into deep oceanic waters. Immature Hooded Seals may keep to these deeper waters for several years. Juveniles have sometimes wandered surprisingly far during the nonbreeding season. In the early 1990s, a young

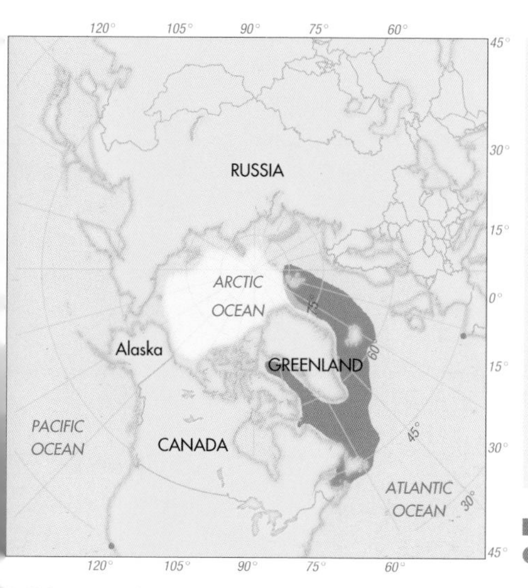

HOODED SEAL
FAMILY PHOCIDAE

MEASUREMENTS AT BIRTH
LENGTH 35–40″ (90–100 cm)
WEIGHT 25 lb (11 kg)

MAXIMUM MEASUREMENTS
LENGTH **MALE** 10′ (3 m)
FEMALE 7′10″ (2.4 m)
WEIGHT **MALE** 880 lb (400 kg)
FEMALE 660 lb (300 kg)

LIFE SPAN
FEMALE 35 years
MALE Perhaps slightly less than female

■ RANGE
● VAGRANTS

adult female in good condition came ashore near San Diego, California. She had evidently been at sea for several years, wandering away from the western North Atlantic population as a juvenile, entering the Beaufort Sea, and then traveling down through the Bering Sea and into the eastern North Pacific Ocean. Other Hooded Seals have appeared as far south as the coast of Portugal in the eastern North Atlantic and the Caribbean in the western North Atlantic.

SIMILAR SPECIES The striking color pattern of adult Hooded Seals—a light background with dark splotches scattered over the body and a dark face—is unique; no other phocid pinniped has a similar pattern. In addition, the adult male's enlarged, inflatable proboscis is a unique feature. Female Gray Seals have a similar pelage to that of Hooded Seals, although their face is lighter and their snout longer.

BEHAVIOR Hooded Seals appear to be solitary, except when breeding and molting. Both males and females vocalize on the ice during the breeding season, males to establish social dominance, and mothers and pups for recognition. Females are exceptionally aggressive when defending their pups. The mother-pup bond lasts only four days, during which pups suckle nearly continually and gain weight rapidly on the exceptionally fat-rich mother's milk.

REPRODUCTION The mating system of Hooded Seals is described as serial monogamy, in which an adult male appears to remain with a female for a short period, mate with her, and then search the area for other estrous females. Males aggressively defend females from other males with vocalizations and visual displays, including the inflation of their hood and bladder. Breeding Hooded Seals aggregate in the pack ice from mid-March through early April. Females give birth to a single pup on the middle of the ice floes throughout that period, with a peak in late March. They nurse their pups for a remarkably short four days, the shortest nursing period known for any mammal, then abruptly wean and abandon them. While nursing, pups double their birth weight, gaining about 45 to 65 pounds (20–30 kg). Females are ready to mate about five to seven days after giving birth.

An adult male Hooded Seal seeks out a female-pup pair and remains with them for several days until the female is receptive for mating. Hooded Seal females nurse their pups for a remarkable four days, the shortest lactation period of any mammal, during which the pup doubles its birth weight from around 25 pounds (11kg) to 50 pounds (22 kg).

FOOD AND FORAGING Hooded Seals appear to feed mostly on squid and fish, especially Greenland halibut, redfish, and cod. Foraging dives typically last 15 minutes or longer and reach depths of 300 to 2,000 feet (100–600 m). Some dives exceed 3,280 feet (1,000 m) and last more than 52 minutes.

STATUS AND CONSERVATION Breeding stocks of Hooded Seals in the early 1990s were estimated at 250,000 near Jan Mayen Island and about 300,000 off Newfoundland. The species is hunted in various parts of its range. In Canada, about 26,000 seals (about three times the established quota) were killed in 1996. Pups have been the primary target because of their fine pelage, more highly valued by the fur industry than that of any other phocid pinniped. Hooded Seals are also hunted in Greenland for subsistence and as an ingredient in dog food. Substantial numbers of these seals have died incidentally in commercial fishing nets, ropes, and traps along the coast of Norway. Polar Bears may prey heavily on Hooded Seals during the breeding season, and Killer Whales occasionally hunt them at sea.

OPPOSITE TOP: *Hooded Seals are named after the adult male's characteristic nose, or "hood."*
OPPOSITE BOTTOM: *The hood becomes inflated during sexual display.*
LEFT: *An inflatable red nasal membrane can also be blown out of one nostril, and is used to threaten other males and perhaps to attract breeding females.*

Mediterranean Monk Seal

Monachus monachus
(Hermann, 1779)

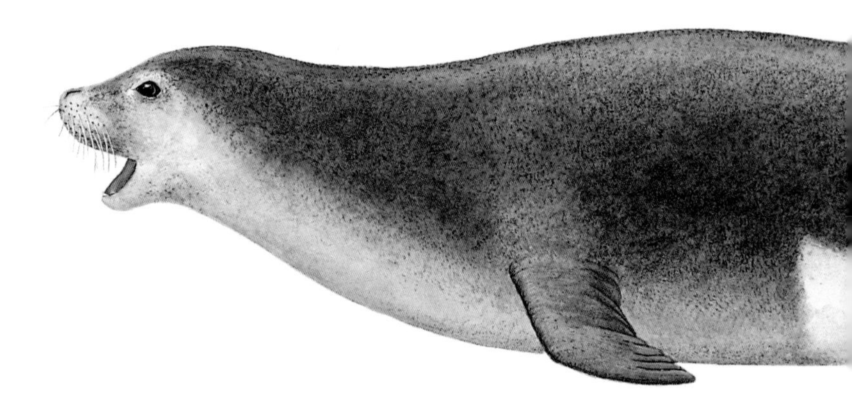

The Mediterranean Monk Seal is perhaps the most endangered pinniped in the world, owing to its small population and limited range. This seal has long been known in the Mediterranean Sea, with references to its presence dating back to the classical period. Nonetheless, scientists know very little about its biology or its historical distribution and abundance. There are differing opinions about the derivation of the scientific name. One theory is that *monachus*, from the Greek word for "monk-like, solitary," refers to the seal's smooth head and short flippers, which resemble a hooded human head and elbows protruding from a monk's frock. At one time, Mediterranean Monk Seals ranged throughout the Mediterranean Sea, into the Black Sea, and south along the Atlantic coast and near-shore islands of northwestern Africa. Substantial reductions in range have accompanied large reductions in population size in all areas.

DESCRIPTION The Mediterranean Monk Seal has a relatively long, slender body and a small, flat, broad head. The snout is short, broad, and flat, with nostrils that face upward, as in Hawaiian Monk Seals, rather than more forward, as in other Northern Hemisphere phocid pinnipeds. The vibrissae are smooth rather than beaded. The foreflippers and hindflippers are relatively short, with small slender claws. Unlike other phocid pinnipeds, monk seals and the Bearded Seal have two pairs of retractable abdominal teats, rather than one pair. There are eight pairs of teeth in both the upper and lower jaws.

Mediterranean Monk Seals are reported to have the shortest hair of any pinniped. Adult males are dark brown to black over most of the body, with a white abdominal patch around the umbilicus that extends about halfway up the sides. Adult females are countershaded, uniformly dark brown dorsally and lighter on the belly, chest, throat, and sides of the face. Some individuals may be lighter gray or darker brown to black. Pups are born with a yellowish lanugo coat that they gradually shed; by about seven weeks old they have a whitish-gray pelage.

RANGE AND HABITAT Mediterranean Monk Seals currently range from Yugoslavia south and west along the eastern and southern coasts of the Mediterranean to the Atlantic coast of northwestern Africa. The largest colony, at the Cabo Blanco

- GENERALLY A UNIFORM DARK BROWN DORSALLY, LIGHTER VENTRALLY
- LONG SLENDER BODY WITH SHORT FOREFLIPPERS, SMALL FLAT FOREHEAD, BROAD SNOUT
- TWO PAIRS OF RETRACTABLE ABDOMINAL TEATS
- HAULS OUT AND BREEDS IN CAVES
- DISTRIBUTION RESTRICTED TO MEDITERRANEAN SEA AND NORTHWESTERN AFRICA

Peninsula on the coast of Western Sahara, Africa, numbered about 100 seals in 1997. The primary haulout and breeding habitats at Cabo Blanco are sandy beaches in two coastal caves. These seals also use sea caves along rocky coastlines in other parts of their range. Some occasionally haul out on open beaches along the coast of northwestern Africa. The species was declared extinct in Italian waters in the mid-1980s; however, vagrant nonbreeding seals have occasionally been observed off northwestern Corsica, in the northeastern and southwestern islands of Sardinia, the southwestern coast and islands off Sicily, and the southeastern coasts of Puglia. Vagrants have also been observed north of Yugoslavia in Croatia, along the Atlantic coast of Portugal and France, and south along the coasts of Senegal and Gambia. Historically, Mediterranean Monk Seals ranged into the Black Sea, but none has been seen there since the early 1990s.

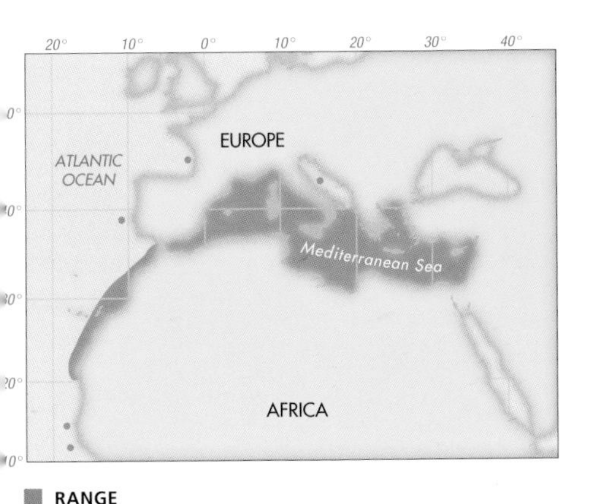

MEDITERRANEAN MONK SEAL

FAMILY PHOCIDAE

MEASUREMENTS AT BIRTH

LENGTH 33–43″ (85–110 cm)

WEIGHT 35–40 lb (16–18 kg)

MAXIMUM MEASUREMENTS

LENGTH 7′10″ (2.4 m)

WEIGHT **MALE** 710 lb (320 kg)
FEMALE 660 lb (300 kg)

LIFE SPAN Unavailable

SIMILAR SPECIES No other pinnipeds live within the range of the Mediterranean Monk Seal. Vagrant Harbor, Gray, and Hooded Seals that may occasionally wander from their North Atlantic habitats can be easily distinguished by their darker, spotted coloration.

BEHAVIOR These seals are relatively asocial. Interactions ashore are mostly limited to females and their pups during the several weeks of nursing. Because many of the breeding sites are in secluded coastal sea caves, observations of Mediterranean Monk Seals have been limited in the past. The installation of video cameras in caves in recent years has allowed some observation of the comings and goings of these seals and of the behaviors of mothers and pups.

REPRODUCTION Little is known about the reproductive biology of these seals, though some researchers have suggested that they are mildly polygynous, with males defending aquatic territories where they mate with females. Females are thought to attain sexual maturity at five or six years old. (Remarkably, one three-year-old female gave birth at Cabo Blanco in 1997, indicating that she was sexually mature and mated at about two years old.) Pups are born throughout the year, with a peak in October and November. Most pups are born in isolated caves. Mortality of young pups is generally high, particularly in October and November when frequent storms and high surf may wash the caves. At the Cabo Blanco colony, gestation may last slightly more than one year, so that females that reproduce in consecutive years give birth later each year. However, if their pup dies during the suckling period, females molt early and then give birth earlier the next year.

FOOD AND FORAGING Little is known about the diet of Mediterranean Monk Seals. They presumably eat a diversity of fish, octopus, and squid, and are thought to forage mostly at depths of 165 to 230 feet (50–70 m) or shallower.

STATUS AND CONSERVATION The Mediterranean Monk Seal is perhaps the most endangered pinniped, owing to its small, fragmented, and isolated populations that range through a number of different political jurisdictions. Beginning at least as early as the classical Greek period, these seals were hunted for their oil and hides. More recently, fishermen have deliberately killed

ABOVE LEFT: *Mediterranean Monk Seals are the rarest of all pinnipeds, numbering only around 400 to 500 throughout their range in the Mediterranean Sea and the*

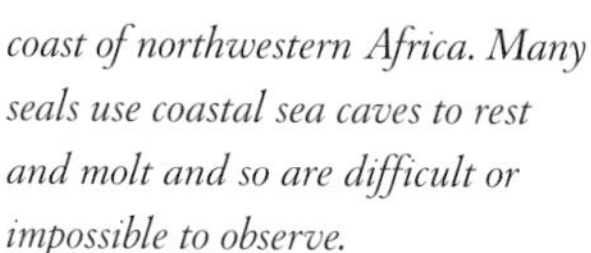

coast of northwestern Africa. Many seals use coastal sea caves to rest and molt and so are difficult or impossible to observe.

ABOVE RIGHT: *The orange markings on the head of this Mediterranean Monk Seal were applied during a study of this endangered species.*

Mediterranean Monk Seals may get entangled in fishing nets and drown, and are sometimes shot to prevent them from taking or damaging netted fish.

them, perceiving them to be competitors for fish and squid. Indeed, deliberate killing is the biggest source of mortality around Greece and is viewed as the most serious threat to the species' survival there. Seals are also occasionally entangled and drowned in fishing gear. In addition, coastal development, regional military aggression, and direct disturbance have evidently displaced colonies from many areas. Range-wide abundance for the species was estimated at around 400 seals in the mid-1990s. A large number of seals perished along the coast of northwestern Africa from an infectious viral disease. The largest colony, at the Cabo Blanco Peninsula, was reduced from about 300 seals in 1996 to about 100 by mid-1997, owing to a viral epidemic and mass mortality in the spring of 1997.

Hawaiian Monk Seal

Monachus schauinslandi
Matschie, 1905

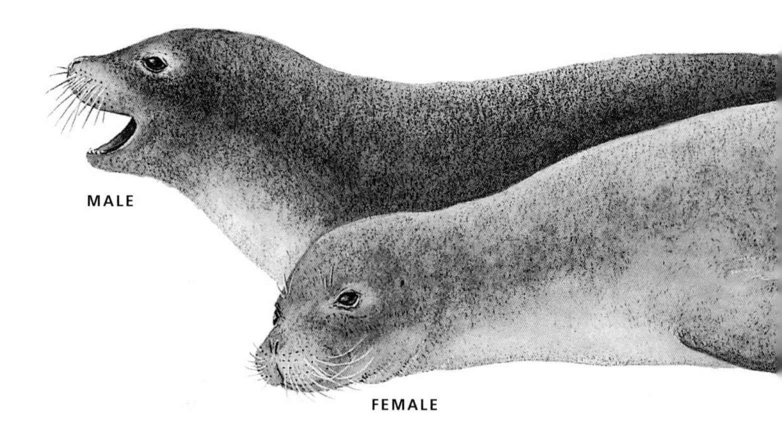

The Hawaiian Monk Seal is one of the most endangered marine mammals in the world, second only among pinnipeds to its close relative the Mediterranean Monk Seal in its risk of extinction. Scientists consider the species to be a living fossil, as some of its anatomical features are only slightly modified from those of the earliest fossil monk seals that lived along the east coast of the United States 14 to 16 million years ago. Hawaiian Monk Seals may have colonized the Hawaiian Islands as early as 10 million years ago. Today, they occur exclusively in U.S. waters, principally in the remote northwestern Hawaiian Islands. Government scientists are closely monitoring all Hawaiian Monk Seal colonies. Strategies to promote population recovery have included harboring recently weaned pups in enclosures away from potential predation by Galápagos and tiger sharks. Another strategy is to translocate seals from larger colonies, where food may be inadequate, to smaller colonies such as those at the Midway Islands and Kure Atoll.

DESCRIPTION The Hawaiian Monk Seal is a medium-size phocid seal with a slender body and short flippers. The head is relatively small, broad, and flat, and the snout is short, with the nostrils situated on top rather than pointed forward as in other North Pacific phocids. Unlike other phocid pinnipeds, monk seals and the Bearded Seal have two pairs of retractable abdominal teats, rather than one pair. Males may be slightly smaller than females. There are eight pairs of teeth in both the upper and lower jaws.

Adults have a relatively short pelage that is slightly darker dorsally than ventrally. They are gray or brown just after molting, and their pelage fades over a few months to yellowish tan. In some older seals, the pelage may darken to a deep brown after each annual molt. Pups are uniformly black at birth and throughout the first four to six weeks of life but molt soon after weaning to a silvery-gray pelage.

RANGE AND HABITAT These seals are principally restricted to the northwestern Hawaiian Islands. The main colonies are at French Frigate Shoals, Pearl and Hermes Reef, Lisianski and Laysan Islands, Kure Atoll, and the Midway Islands, with smaller numbers living at Nihoa and Necker Islands. A few seals have also been reported at Johnston Atoll over the years; two adult

- UNIFORMLY GRAYISH WHEN NEWLY MOLTED, THEN TAN OR YELLOWISH
- TWO PAIRS OF RETRACTABLE ABDOMINAL TEATS
- DISTRIBUTION RESTRICTED MAINLY TO NORTHWESTERN HAWAIIAN ISLANDS

PUP

males were translocated there from French Frigate Shoals in 1997. Increased numbers have been appearing at main Hawaiian Islands in recent years, with several pups born at Maui, Kauai, Oahu, and Molokai since the late 1980s. The fathers of at least some of those pups are likely males that were translocated to the main Hawaiian Islands in the 1990s from French Frigate Shoals, where they had been mobbing and injuring females and pups. Most seals appear to remain at their natal colonies throughout their lives, although a few move among some of the colonies, especially between Kure Atoll, the Midway Islands, and Pearl and Hermes Reef and between Laysan and Lisianski Islands.

SIMILAR SPECIES The Northern Elephant Seal is the only phocid pinniped in the North Pacific that might be confused with the Hawaiian Monk Seal. Though Northern Elephant Seals generally stay far to the west and north of the Hawaiian Monk Seal's range, two tagged juveniles were

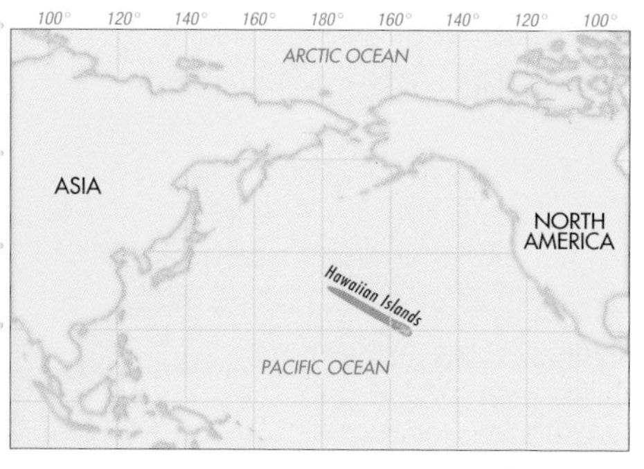

RANGE

HAWAIIAN MONK SEAL
FAMILY PHOCIDAE

MEASUREMENTS AT BIRTH
LENGTH 3′4″ (1 m)
WEIGHT 35–40 lb (16–18 kg)

MAXIMUM MEASUREMENTS
LENGTH **MALE** 6′11″ (2.1 m)
FEMALE 7′10″ (2.4 m)
WEIGHT **MALE** 510 lb (230 kg)
FEMALE 600 lb (270 kg)

LIFE SPAN
30 years

reported at the Midway Islands in the 1970s. Juvenile Northern Elephant Seals have a similar pelage color, but their heads are larger and their nostrils more forward-pointing.

BEHAVIOR Hawaiian Monk Seals are rather solitary when ashore and maintain fairly large distances from each other. Close encounters, both in and out of the water, are generally aggressive, with vocal threats common and occasional brief physical interactions. The only substantial social bonds among these seals appear to be between lactating females and their pups. Hawaiian Monk Seals also appear to be intolerant of human presence, which is likely one reason for their disappearance from Sand Island at the Midway Islands over the several decades that the U.S. military was in residence there. With the departure of the military in the early 1990s, together with strictly enforced approach limits, the monk seals have returned to Sand Island, and their numbers have increased substantially. A similar pattern of recovery occurred at nearby Kure Atoll since the recent closing of the U.S. Coast Guard Loran station that operated there from the 1960s until 1992. Molting seals shed the upper layer of skin together with the old hair in large patches. This catastrophic molt is similar to that of elephant seals but unlike all other pinnipeds, which shed single hairs. Tiger and Galápagos sharks attack, injure, and consume Hawaiian Monk Seals, especially weaned pups and juveniles, at several colonies.

REPRODUCTION Hawaiian Monk Seals evidently have a serially monogamous breeding system. From March to August, depending on the location of the colony, males patrol beaches where females are nursing, searching for estrous females to mate with. In the water, they threaten and occasionally fight with other males over receptive females. Mating evidently always takes place in the water, and the bonds between mating seals are brief. Most pups are born from March through June, with peak pupping in late March and early April. Females remain ashore nursing their pups for about four to six weeks. Lactating females are aggressive toward other seals; however, they appear to nurse pups other than their own on crowded beaches when confused about their own pup's identity, and females whose pups have died or become lost often adopt other pups. At some colonies, particularly at Laysan and Lisianski Islands, males have outnumbered females. These population skews have resulted in groups of males mobbing and grossly injuring individual females while attempting to mate with them. Resulting female mortalities have caused a further skew in the sex ratio and heightened the problem. A number of the more aggressive males have been captured and moved to the main Hawaiian Islands and a few to Johnston Atoll.

FOOD AND FORAGING Hawaiian Monk Seals eat primarily reef-dwelling fish, cephalopods, and invertebrates, especially lobsters. They appear to forage mostly on the seafloor in coral reef

OPPOSITE: *A juvenile Hawaiian Monk Seal searching for food.* **RIGHT:** *Hawaiian Monk Seals forage mostly within the coral atolls of the northwestern Hawaiian Islands, where they pursue secretive prey, such as octopus and lobsters, that hide in small caves in the coral reefs. The seals sometimes sleep in these caves, perhaps using them as temporary refuge from predatory tiger and Galápagos sharks.*

habitats within the atolls at depths of 200 feet (60 m) or less. Some seals also forage on the outer slopes of atolls or on distant banks to depths of more than 820 feet (250 m). Dives are generally short, lasting four minutes or less. Some dives last longer than 30 minutes, although it is unclear whether these are foraging dives or whether the seals are simply resting on the bottom or in coral crevices and caves. At Pearl and Hermes Reef, recent studies have shown that with minor exceptions seals remain and forage within the small atoll year-round, with adult males, females, and juveniles segregating to some extent. Seals at French Frigate Shoals, however, have ranged up to a few hundred miles to forage.

STATUS AND CONSERVATION Hawaiian Monk Seal populations have been reduced by unregulated and largely unreported commercial harvests in the early 1800s, and by subsistence hunting several hundreds of years earlier and during the late 1800s and early 1900s. By the early 1900s, sightings of these seals were rare. Military activities at the Midway Islands, Pearl and Hermes Reef, Kure Atoll, and French Frigate Shoals further impacted those colonies at various times through the end of the 20th century. The Hawaiian Monk Seal population was estimated at about 1,400 seals in 2000.

While most colonies are slightly increasing or stable, the largest colony, located at French Frigate Shoals, declined substantially during the 1980s and early 1990s. The decline was largely due to poor survival of juveniles. Adult females at French Frigate Shoals weigh less and are shorter and less rotund than females at the increasing colony at Pearl and Hermes Reef, presumably because of differences in food availability. In addition, differences in hematology and blood chemistries between the two colonies suggest that disease may be affecting seals at French Frigate Shoals. Young seals become entangled and many drown in plastic debris and in derelict fishing gear hung up on coral reefs. A U.S. multi-agency effort to remove these items recovered more than 40 tons of net debris between 1997 and 2001.

Hawaiian Monk Seals breed, molt, and rest on small coral beach islands, but they are mostly asocial and intolerant of close association or contact with each other.

Southern Elephant Seal

Mirounga leonina
(Linnaeus, 1758)

The Southern Elephant Seal is the largest seal in the world. These seals spend most of their lives at sea, staying mostly at depths of 600 to 2,000 feet (180–600 m). Recent scientific research has revealed that, like their relatives the Northern Elephant Seals, Southern Elephant Seals are remarkable divers, occasionally diving as deep as 5,000 feet (1,500 m) for as long as two hours. The species was first described from a specimen collected in 1759 from the Juan Fernández Islands off the coast of Chile. The genus name is derived from *miouroung*, the name that Australian aboriginals gave to these seals. The specific name means "lion" in Latin and may refer to the roar-like threat vocalization of adult males. Southern Elephant Seals are also commonly known as "sea elephants."

DESCRIPTION The Southern Elephant Seal, like its northern counterpart, is a bulky seal with a very large head. The eyes are exceptionally large and well adapted for detecting bioluminescent prey in the poorly lit ocean depths where these seals forage. The flippers are short relative to body size. The claws in the foreflippers are stout and broad, and may grow to 2 inches (5 cm) long, whereas those in the hindflippers are very short and slender. Adult males are massive, more than five times larger than adult females. Around puberty, males develop an elongated snout with an inflatable nose, or proboscis, that they may inflate while making vocal threats during aggressive encounters with other males. In older males, the neck becomes creased and calloused. There are eight pairs of teeth in the upper jaw and seven pairs in the lower jaw. The upper and lower canine teeth are conical and well rooted in the jaws, and can grow to more than 6 inches (15 cm) long. All other teeth are small and peg-like.

Just after their annual molt, adults have a gray to silver pelage, but the color quickly fades to tan, yellowish brown, or dark brown. Adult females and juveniles are generally lighter ventrally. Pups are born with a wavy, black pelage that they molt soon after weaning at about three weeks old to a silver or bright gray coat. This pelage darkens over the next several weeks to chocolate brown dorsally and light brown to yellowish tan ventrally.

RANGE AND HABITAT Southern Elephant Seals breed and haul out principally on oceanic islands throughout subantarctic regions of the Southern

- LARGEST SEAL IN THE WORLD
- MALE'S SNOUT MUCH LARGER THAN FEMALE'S, SOMEWHAT INFLATABLE
- MAKES DEEP LONG DIVES

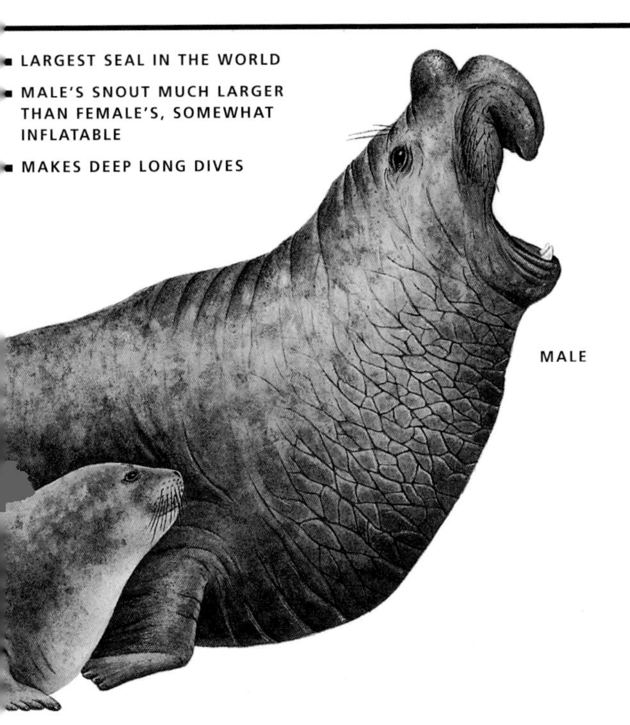

- SHEDS LARGE PATCHES OF SKIN WITH HAIR WHEN MOLTING
- UNDERGOES EXTENSIVE MIGRATIONS THROUGHOUT SOUTHERN OCEAN
- ON LAND RESTRICTED MOSTLY TO OCEANIC SUBANTARCTIC ISLANDS

Hemisphere. Large colonies have more recently developed along the coast of southern Argentina, especially at Peninsula Valdés. The principal island colonies are at the Falkland, South Orkney, South Georgia, South Sandwich, Gough, Marion, Crozet, Kerguelen, Heard, and Macquarie Islands. Small colonies have recently appeared along the Antarctic Peninsula and on the South Shetland Islands. During the nonbreeding season, Southern Elephant Seals may migrate over great areas in the Southern Ocean. Some seals travel south to forage near the Antarctic continent and may haul out and molt at Vestfold Hills and along the Antarctic Peninsula, among other places, rather than return the great distance to their subantarctic breeding sites. Vagrants have been reported in Angola, South Africa, the southern Australian coast, and New Zealand, and one individual was

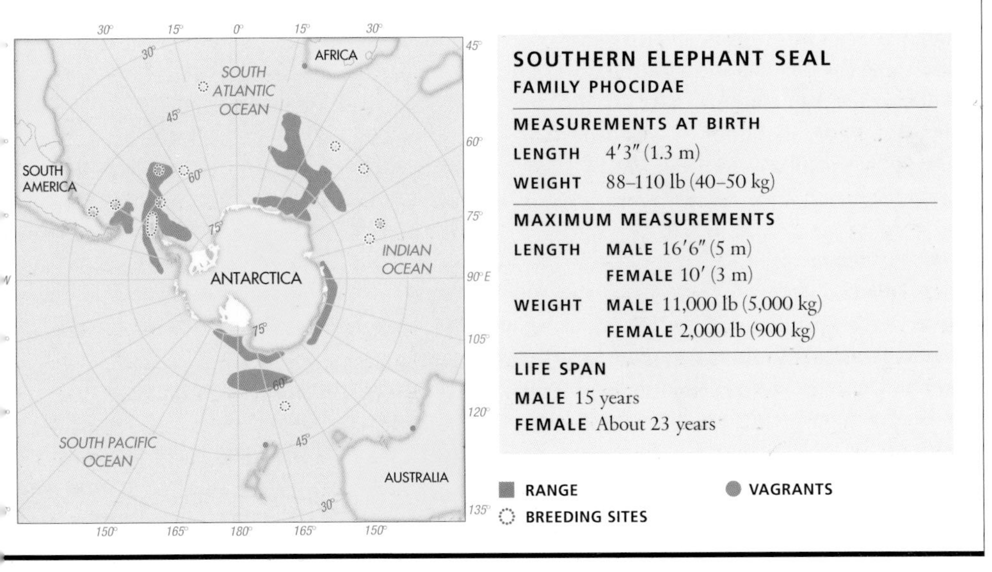

SOUTHERN ELEPHANT SEAL

FAMILY PHOCIDAE

MEASUREMENTS AT BIRTH

LENGTH 4′3″ (1.3 m)

WEIGHT 88–110 lb (40–50 kg)

MAXIMUM MEASUREMENTS

LENGTH **MALE** 16′6″ (5 m)
FEMALE 10′ (3 m)

WEIGHT **MALE** 11,000 lb (5,000 kg)
FEMALE 2,000 lb (900 kg)

LIFE SPAN

MALE 15 years

FEMALE About 23 years

Southern Elephant Seals shed their hair once each year in summer or early autumn. The upper layer of skin is also shed in large patches, leaving bare skin for a short while as the new hair emerges. Among the pinnipeds, only monk seals, and perhaps Ross Seals, have a similar kind of "catastrophic" molt.

reported from the Northern Hemisphere in 1989, on the coast of Oman. When ashore Southern Elephant Seals prefer sandy, gravel, and sometimes cobble beaches.

SIMILAR SPECIES Adult Southern Elephant Seals are easily distinguished from all other pinnipeds in the Southern Hemisphere by their large size and uniformly brown to tan pelage. At sea, their pelagic habits make it unlikely for them to be seen with any other species.

BEHAVIOR Southern Elephant Seals aggregate in large numbers during the breeding season; adult males may remain ashore continually for up to two months, while pregnant females remain ashore for about four weeks. After the breeding season, seals return to sea to forage for about two months before hauling out again to molt. Seals are most gregarious during the molting season; they may crowd together in very tight, sometimes large groups and spend much of their time sleeping. While molting, they remain ashore, fasting, for about four weeks. After molting, they return to sea to forage until the next breeding season. During their two foraging periods at sea each year, Southern Elephant Seals migrate over large areas of the Southern Ocean, covering several thousand miles or more during each migration. All evidence so far indicates that they are solitary while at sea. They dive virtually continually, spending only about three to four minutes breathing at the surface between dives. Adults forage at depths of up to 5,000 feet (1,500 m) for as long as two hours, though most dives are to around 2,000 feet (600 m) and last 20 to 30 minutes.

REPRODUCTION Southern Elephant Seals have a well-developed polygynous breeding system. Adult males arrive at breeding beaches in August, and pregnant females arrive later, in September and early October. Pups are born about a week after the females haul out. Breeding males threaten each other with stereotyped visual and vocal displays and occasionally charge and fight with each other. While most fights are brief, serious injuries to the chest, neck, and foreflippers may result from males slashing and ripping at opponents with their large canine teeth. Dominance hierarchies develop, based on size and age, and vocal signaling is the most important behavior for establishing and maintaining these hierarchies. Males make low-pitched sounds composed of a series of pulses that have little variation in frequency. They may recognize each other from these calls, at least for several weeks during the breeding season. Only the most dominant males gain mating access to females. Gestation lasts about 10 months. Pups are weaned and quickly abandoned at about three weeks old, when their mothers mate and return to sea to feed and recover mass lost while nursing.

FOOD AND FORAGING Southern Elephant Seals eat a variety of deepwater, pelagic fish and squid that live in the water column at depths of 600 to 2,000 feet (180–600 m). In some areas they forage on the bottom for species that live at the seafloor or in seafloor sediments.

STATUS AND CONSERVATION Southern Elephant Seals were hunted for hundreds to perhaps thousands of years for food and clothing by aboriginals in Australia, particularly in Bass Strait at King and Hunter Islands, and perhaps also along the coasts of southern Argentina and Chile. Commercial harvests in the 1800s and 1900s were extensive on most islands in the Southern Hemisphere, particularly in the early 1900s when the blubber, which could be rendered to oil, was in great demand for fueling lamps and sometimes also for food. The species was reduced to low numbers at many sites and exterminated at some. Commercial sealing continued at South Georgia until 1964, when shore-based whaling closed down, forcing the sealing industry also to close because of its inability to sustain itself economically. Abundance at most sites increased substantially in the early and mid-1940s. In the 1960s and 1970s, many populations began to decline from as yet unexplained causes. Colonies at South Georgia and the Kerguelen Islands, however, have remained relatively stable, and the colony along the southern coast of Argentina has steadily increased. The species may now number around 750,000, with the highest number of seals centered at South Georgia.

Young male Southern Elephant Seals sparring in practice for more serious battles when they become mature and compete for mating opportunities.

Northern Elephant Seal

Mirounga angustirostris
(Gill, 1866)

The Northern Elephant Seal is the largest phocid pinniped in the Northern Hemisphere. Commercial sealers hunted these seals so intensively during the early 1800s that the species was thought extinct by the late 1800s. Perhaps a few dozen at most survived into the 20th century, but the population has since increased steadily in number and range. Today the largest colonies are found off southern California at the Channel Islands, with smaller but increasing colonies along the coast of central California. The specific name is from the Latin words *angustus,* meaning "narrow," and *rostrum,* meaning "snout," in reference to the more slender snout of the Northern Elephant Seal compared with its closest relative, the Southern Elephant Seal. The Northern Elephant Seals are also called "sea elephants" in older literature.

DESCRIPTION The Northern Elephant Seal has a robust torso that tapers to narrow hips. The foreflippers are short, with slightly longer outer digits and long broad claws. The hindflippers are broad when fanned out, making them good for propulsion in swimming. Males begin to develop an elongated fleshy nose, or proboscis, at about puberty, and by the time they are seven or eight years old, the nose overhangs the lower lip by 6 to 10 inches (15–25 cm). Males may inflate their enlarged nose during the winter breeding season to resonate sound when they are vocally threatening each other. Adult males may be three to four times the mass of adult females, with a broad thick chest, a larger head, and much longer hindflippers. There are eight pairs of teeth in both the upper and lower jaws. The robust conical canine teeth may be up to 8 inches (20 cm) long in adult males, with about two-thirds of the tooth anchored in the jaw. Virtually all the other teeth are small and peg-like.

Just after molting adults and juveniles are bare-skinned and black. The new coat is silver to dark gray, but the pelage fades during the next several months to brown, tan, or yellow. Most of the year, adult males are uniformly dark brown, except for the chest and neck, which are mostly hairless, heavily calloused, and scarred, and appear speckled with pink, white, and light brown. Adult females and juveniles are mostly light brown to chocolate brown dorsally and lighter brown to tan ventrally; a few individuals may be uniformly blond. Pups are black at birth, molting

- ADULT MALE HAS LARGE INFLATABLE NOSE
- SHEDS LARGE PATCHES OF SKIN WITH HAIR WHEN MOLTING

- BREEDS IN EASTERN NORTH PACIFIC FROM CALIFORNIA TO BAJA CALIFORNIA
- NONBREEDING DISTRIBUTION FAR-RANGING, TO CENTRAL NORTH PACIFIC AND ALASKA

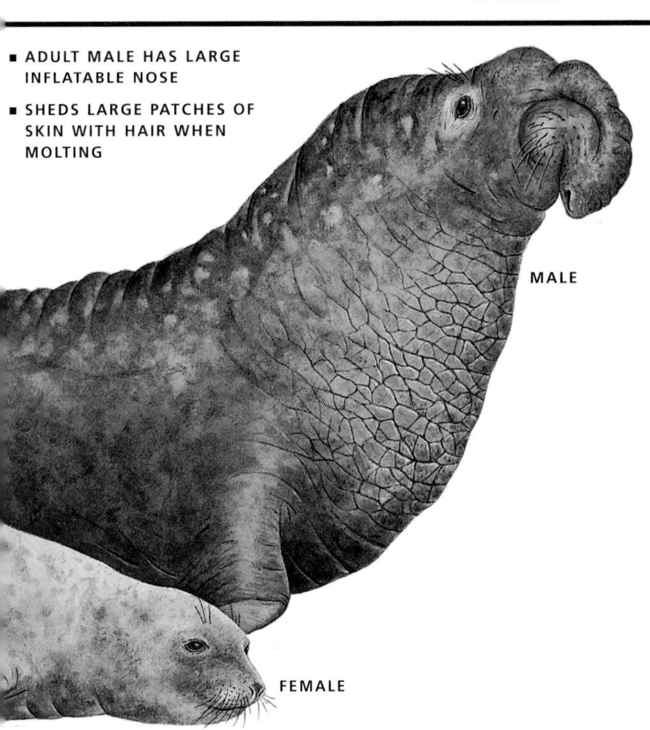

to a silver-gray pelage just after weaning at three to five weeks old. Their pelage darkens to chocolate brown dorsally, fades to light tan to yellowish ventrally by two to three months old, and then fades to lighter brown or yellow.

RANGE AND HABITAT The largest breeding colonies occur at the Channel Islands in southern California, at San Miguel and San Nicolas Islands. Others occur at Farallon and Año Nuevo Islands off central California; Santa Barbara, Santa Rosa, and San Clemente Islands in the Channel Island group; Coronado, Guadalupe, San Benito, Cedros, and Natividad Islands along the Pacific coast of Baja California; and on the central California mainland at Punta Gorda, Point Reyes, Point Año Nuevo, Cape San Martin, and Point Piedras Blancas. In spring and summer, seals also haul out at these sites for molting, as well as at some sites in Washington and British Columbia. Outside of the breeding and molting seasons, for roughly 8 to 10 months of the year,

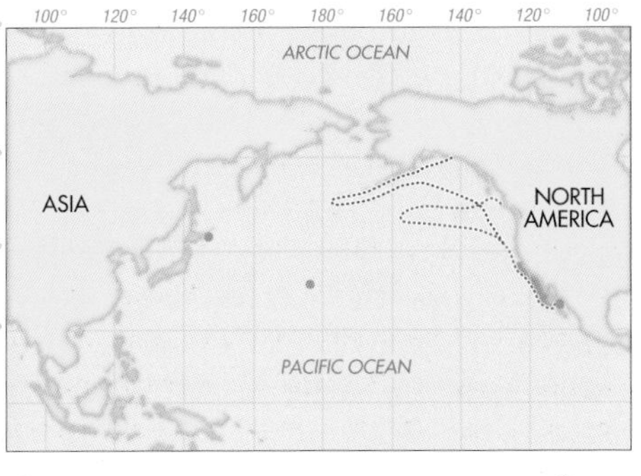

VAGRANTS
BREEDING AND MOLTING RANGE
FORAGING RANGE ♂
FORAGING RANGE ♀

NORTHERN ELEPHANT SEAL
FAMILY PHOCIDAE

MEASUREMENTS AT BIRTH
LENGTH 4′1″ (1.25 m)
WEIGHT 77 lb (35 kg)

MAXIMUM MEASUREMENTS
LENGTH **MALE** 13′6″ (4.1 m)
FEMALE 10′ (3 m)
WEIGHT **MALE** 4,400 lb (2,000 kg)
FEMALE 1,300 lb (600 kg)

LIFE SPAN
MALE 12–14 years
FEMALE 18–20 years

LEFT: *Northern Elephant Seal pup developing swimming skills before heading to sea for the first time to forage. Northern Elephant Seals spend most of their lives at sea, coming ashore for brief periods of three to four weeks when breeding in winter or molting in spring and summer.* **OPPOSITE:** *Once weaned, Northern Elephant Seal pups form dense concentrations and mostly sleep while fasting and molting for one to two months before going to sea for the first time in early spring.*

Northern Elephant Seals range widely to forage into the eastern and central North Pacific, and as far north as the Gulf of Alaska and the Aleutian Islands. Vagrant juveniles have been reported from Japan, the Midway Islands in the northwestern Hawaiian Islands, and the southern Gulf of California.

Northern Elephant Seals prefer to haul out and breed on sandy beaches, but they may use cobble and pebble beaches when their preferred habitats are not available. They are rarely seen at sea because they mostly forage away from the coast and spend about 80 to 90 percent of their time submerged, usually at great depths.

SIMILAR SPECIES Harbor Seals may occur with Northern Elephant Seals at sea and on land throughout coastal areas of the North Pacific; however, elephant seals are easy to distinguish by their large size and by the large, pendulous, inflatable nose of adult males.

BEHAVIOR During the breeding and molt seasons, Northern Elephant Seals may aggregate in groups of a few dozen to several hundred, depending on beach topography. Numbers of seals ashore are greatest in late January at the peak of the breeding season, and again in late April and early May when juveniles and adult females are molting. Males molt from May to August. When ashore to molt, these seals may crowd together in dense groups and spend most of their time sleeping. The rest of the year, Northern Elephant Seals are traveling or foraging at sea and are presumably solitary. Adults make two migrations a year, the first after the breeding season and the second after the molt season. Northern Elephant Seals dive continually when at sea, spending only brief periods of one to three minutes at the surface between dives. Most dives of adults are to depths of 1,000 to 2,500 feet (330–800 m) and last an average of 20 to 30 minutes. The deepest recorded dive, by an adult male, was to 5,141 feet (1,567 m), and the longest recorded dive, by an adult female, lasted two hours.

REPRODUCTION Northern Elephant Seals are polygynous breeders and aggregate in groups of a few dozen to several hundred during the breeding season from December through February. Adult males establish dominance hierarchies around groups of females and defend their rights of access to females until mating occurs. Dominance hierarchies are structured by the size and age of the males and are maintained by stereotyped physical and vocal threats, and occasionally by physical combat. Most contests do not escalate beyond an exchange of vocal threats. These loud, cracking vocalizations, called "clap-threats," can sometimes be heard for several miles. While vocalizing, the male rears up with its mouth wide open. Physical battles, though

rare, can result in serious injury, as males vigorously bite and rip at opponents with their large canine teeth. Females give birth about six to eight days after hauling out in December or January. They remain ashore fasting and nursing their pups for about three weeks. During this time females maintain constant contact with their pups and aggressively fend off other females and pups with threat vocalizations and bites. Females mate again about three weeks after giving birth. After mating, they abruptly wean and abandon their pups ashore and return to sea to feed. Gestation lasts about 11 months.

FOOD AND FORAGING Northern Elephant Seals mostly eat mesopelagic fish and squid, though some seals may forage on the sea bottom at the continental shelf for slow-moving skates, rays, sharks, and rockfish. Many of their prey have bioluminescent organs that Northern Elephant Seals may rely on when hunting in the deep dark oceanic regions where they usually forage. They remain at sea for 8 to 10 months each year, diving continually. They may forage during most of this time to recover the substantial body mass lost from fasting during the winter breeding and spring and summer molt seasons.

STATUS AND CONSERVATION Northern Elephant Seals were thought to be extinct by the late 1800s, the result of a long period of subsistence harvests by aboriginals over several thousand years, followed by intense commercial sealing in the early and mid-1800s. When a few seals appeared at Guadalupe Island in the 1880s, the Smithsonian Institution sent a scientific expedition to collect them for museum specimens, as the species had not yet been well described. They shot seven of the eight seals they found and again thought the species effectively extinct. Perhaps a few dozen seals survived, likely foraging far away at sea, and the population began a steady increase in the early 1900s. Northern Elephant Seals progressively colonized islands and some mainland sites in southern and then central California through the 1980s as the population increased. In the early 1990s, scientists estimated that the population numbered more than 100,000, with about two-thirds of the breeding population occurring at the Channel Islands. By 2000, the population may have numbered more than 150,000.

Most interactions between breeding Northern Elephant Seal males are settled by vocal or visual threats. Occasionally males engage in physical combat and may battle off into the surf until one or both become exhausted or one flees. The calloused, mostly hairless chest and neck develop with age and provide some protection from the vicious bites of other males.

Weddell Seal

Leptonychotes weddellii
(Lesson, 1826)

The Weddell Seal is named after Captain James Weddell, who described and illustrated the species in his narrative of a sealing voyage to Antarctica in the 1820s. Weddell Seals breed in nearshore fast-ice habitats of the Antarctic to 78°S. No other mammal, other than humans, occurs this far south. These seals are deep divers, often reaching depths of 1,800 feet (600 m) or more, and sometimes remaining submerged for more than an hour.

DESCRIPTION The Weddell Seal has a large, robust body that appears to dwarf the relatively small head, which is flat and broad. The snout is short, and the eyes large and well adapted for foraging in the deep, dimly lit waters beneath the Antarctic ice. The flippers are short and narrow, with sturdy claws on the foreflippers and short, narrow claws on the hindflippers. Females may be slightly larger than males. The robust upper canine and second incisor teeth are pointed forward slightly. There are eight pairs of teeth in both the upper and lower jaws.

Recently molted adults and juveniles are bluish black dorsally, with irregular white blotches scattered over the chest and, to a lesser extent, the belly. The hair gradually fades to brown through winter and into early summer. Pups are born with a long, woolly, light brown to gray coat, which they shed within four weeks.

RANGE AND HABITAT These seals have a circumpolar distribution around Antarctica. They typically breed close to shore along cracks in the fast ice, and have been thought to associate with land-fast ice close to the continent year-round. However, recent tracking studies indicate that some seals may regularly migrate to offshore areas after the breeding season, and juveniles may spend several years there before returning to breeding colonies in the fast ice. Other breeding colonies occur at Larsen Harbor at South Georgia Island and at Signy Island in the South Orkney Islands. Vagrants have been observed in winter and autumn in South Australia and at Heard and Macquarie Islands.

SIMILAR SPECIES Crabeater, Leopard, and Ross Seals overlap with the Weddell Seal in Antarctica. Crabeater and Leopard Seals are more

- LARGE AND ROBUST, WITH RELATIVELY SMALL HEAD
- RELATIVELY UNIFORM DARK COLOR ON BACK, WITH STRIKING, LIGHT AND DARK BLOTCHES ON CHEST
- STURDY, FORWARD-POINTING UPPER CANINE AND SECOND INCISOR TEETH
- HIGHLY VOCAL, ESPECIALLY UNDERWATER
- OCCURS IN FAST-ICE AND SOME PACK-ICE HABITATS OF ANTARCTICA

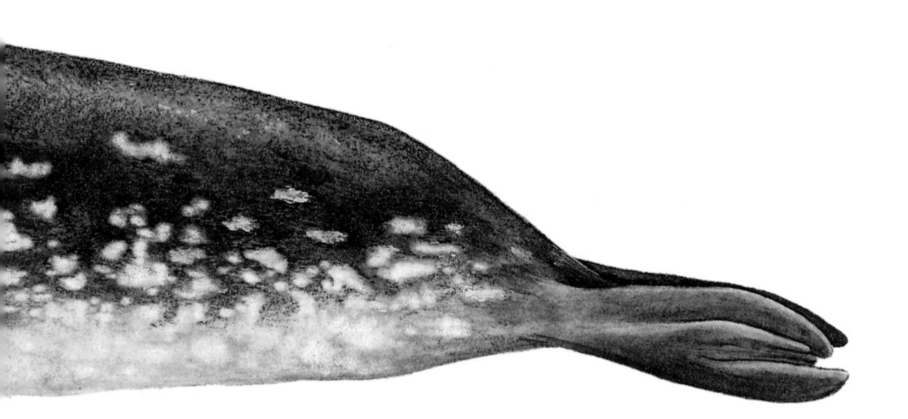

slender. The Leopard Seal's head is much larger and its snout longer, and it has small dark spots scattered along its ventral surface. The Crabeater Seal also has a few ventral dark spots, mainly on the belly and flanks. Ross Seals are uniformly dark brown dorsally and light tan ventrally before they molt, and have a blue cast to the pelage after molting, with striking dark streaks along the throat to the upper chest.

BEHAVIOR Although Weddell Seals gather along tidal ice cracks and leads during the breeding and molting season, they generally space out from one another by several feet or more. The typical social group is composed of a mother and pup. Adult males may spend most of the breeding season in the water, but they occasionally haul out toward the end of the breeding season to rest. The rest of the year, Weddell Seals are solitary,

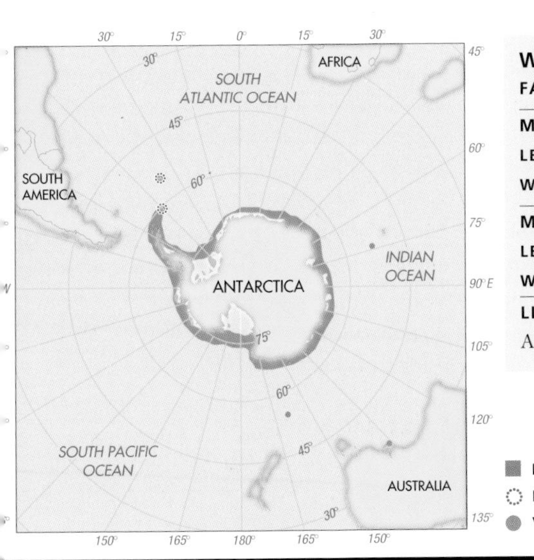

WEDDELL SEAL
FAMILY PHOCIDAE

MEASUREMENTS AT BIRTH
LENGTH 3′11″ (1.2 m)
WEIGHT 49–55 lb (22–25 kg)

MAXIMUM MEASUREMENTS
LENGTH 11′ (3.3 m)
WEIGHT 1,200 lb (550 kg)

LIFE SPAN
About 20 years

RANGE (INCLUDING BREEDING)
BREEDING SITES
VAGRANTS

although yearlings may associate with each other through the first winter. The forward-protruding canine and incisor teeth of the Weddell Seal are used to abrade the sea ice and keep breathing holes open throughout the year. The heavy wear this causes to the teeth can cause serious infection and occasionally lead to death. Weddell Seals are exceptionally vocal, particularly when they are underwater. They produce a variety of sounds, from low-frequency buzzes to higher-frequency whistles and chirps. Some pulsed sounds, which the seals often make during the breeding season, are so intense that they can be felt through the thick fast ice. Male Weddell Seals use calls to defend their aquatic territories and perhaps also to attract females. At the end of the breeding season, Weddell Seals begin to feed intensively in preparation for their annual molt, which occurs in mid-January for juveniles and in February for adults. Leopard Seals and Killer Whales prey on Weddell Seals in some areas where leads in the fast ice allow them to penetrate the nearshore zone in summer.

REPRODUCTION The unique fast-ice breeding habitat limits the availability of suitable pupping sites, resulting in heightened competition among males and a high degree of polygyny, compared to other pinnipeds that mate in the water. Recent studies have found that males may not actually father pups until they are 10 to 16 years old. Adult males appear to defend aquatic territories under the fast ice from early October through early December. They rarely haul out, and virtually all mating occurs under the ice. Females give birth from late September or October through mid-November, with peak pupping in late October. Females are relatively solitary after giving birth. They may enter the water as the pups age to make brief feeding dives. Pups may enter the water within a few weeks of birth, but their dives are fairly shallow and they may not accompany their mothers to depth on many dives. Females remain with their pups for about seven to eight weeks, then abruptly wean and abandon them. They are ready to mate again just before or after they wean their pups.

ABOVE: *Most Weddell Seals give birth and mate in nearshore fast-ice habitats around Antarctica. Recently researchers have found that offshore pack-ice and polynyas are also important foraging areas for young and, perhaps, mature but nonreproductive seals.*

Weddell Seals are deep divers. Females continue to forage to some extent during the summer breeding season when they often leave their pups on the ice surface for brief periods. Pups may also begin accompanying their mothers on short dives when around two weeks old and may be taught some important lessons in finding and capturing prey.

FOOD AND FORAGING In some areas, Weddell Seals eat mostly nototheniid fish, antarctic cod, cephalopods, and crustaceans. In other areas they may eat mostly krill and fish. They forage in a variety of habitats, from the sea bottom in shallow regions to midlevel waters in deeper and offshore zones. There have been a few observations of these seals killing, skinning, and eating chinstrap penguins. In spring and summer, Weddell Seals presumably hunt in bouts of 40 to 50 consecutive dives over several hours, followed by several hours spent hauled out on the ice resting. Adult females regularly dive to depths of 525 feet (160 m) or more from October to December and 160 to 650 feet (50–200 m) in January. Young seals make shallower dives, to 400 feet (120 m). The deepest recorded dive for the species is 2,460 feet (750 m), and the longest dive lasted 73 minutes.

STATUS AND CONSERVATION The overall abundance of the Weddell Seal is poorly known, although the population is thought to number at least 1 million. Long-term monitoring at some sites has produced estimates of local abundance; the international Antarctic Pack Ice Seals Program (APIS), completed in early 2000, may produce abundance estimates for most of the Antarctic. From the 1950s through the early 1980s, Weddell Seals were killed near a number of research stations to feed sled dogs, and the population near U.S. and New Zealand bases at Ross Island was reduced to small numbers; however, they have since recovered to their prior abundance. The Convention for the Conservation of Antarctic Seals permits limited harvesting of Weddell Seals, and Russia conducted a small-scale experimental harvest in the 1980s.

Female Weddell Seals give birth to a single pup, rarely to twins, in October and nurse it for about six weeks before weaning and abandoning it. Recent tracking of the movements of pups by satellite have shown that many pups travel rather far from the breeding sites during the first year and may remain in distant foraging areas for several years before returning to breeding colonies.

Crabeater Seal

Lobodon carcinophaga
(Hombron and Jacquinot, 1842)

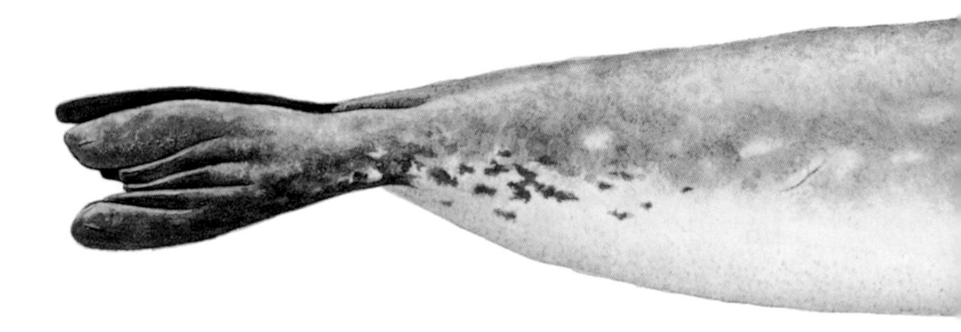

Contrary to its common name, the Crabeater Seal feeds primarily on krill. This species has highly modified lobed, interlocking cheek teeth that function to sieve invertebrate prey from gulped seawater. Its genus name is derived from the Greek words *lobos*, meaning "lobed," and *odon*, meaning "tooth," in reference to the cheek teeth. The specific name means "crabeater." Crabeater Seals are widely distributed around Antarctica. Recent studies have indicated that at least some seals may migrate over several thousand miles in autumn and winter. Groups of up to several hundred Crabeater Seals have been seen swimming and possibly foraging together in some areas in late summer.

DESCRIPTION The Crabeater Seal has a relatively slender body and so appears long for a phocid seal. The head is small relative to the body. The snout is long and narrow and appears slightly upturned at the end because of the dorsally positioned nostrils. The foreflippers are relatively long. There is no noticeable difference in body size between males and females, although females are sometimes slightly larger, perhaps because they store more fat for lactation or because they live slightly longer than males. The Crabeater Seal's teeth, the most modified of any mammal, are highly adapted to a diet of tiny krill. The postcanine teeth and molars have lobe-shaped cusps that interlock to sieve the krill from gulps of seawater. There are eight pairs of teeth in both the upper and lower jaws.

Adults and juveniles are mostly blond to gray, with some dark spotting or mottling, particularly on the sides near the hind- and foreflippers. They often have conspicuous, longitudinal paired scars on the body caused by Leopard Seal attacks. Pups have a woolly, gray pelage at birth, which they shed at or just after weaning at about four weeks old.

RANGE AND HABITAT Crabeater Seals occur principally in the pack ice around Antarctica, their year-round distribution varying seasonally as the pack ice recedes and advances. During the breeding season, they generally stay over the continental shelf. During summer and autumn, they become more widespread in the outer pack ice. Scientists do not know if these seals migrate regularly; however, satellite-linked radios have tracked some movements of several thousand miles. Crabeater Seals occasionally range north

- LONG SLENDER BODY
- SMALL HEAD WITH LONG POINTED SNOUT
- BLOND TO GRAY PELAGE WITH SCATTERED DARK SPOTS AND MOTTLING
- BODY COMMONLY SCARRED FROM LEOPARD SEAL ATTACKS
- CHEEK TEETH MODIFIED TO FORM SIEVE THAT FILTERS OUT INVERTEBRATE PREY
- CIRCUMPOLAR AROUND ANTARCTICA

to Macquarie Island, and vagrants sometimes appear on the Australian mainland in autumn and winter. They may also wander onto the Antarctic continent, far up glaciers and into dry inland valleys. One mummified carcass was found at an elevation of 3,600 feet (1,100 m) on the Ferrar Glacier.

SIMILAR SPECIES These seals share ranges and habitats with Leopard, Weddell, and Ross Seals in most areas near Antarctica. Leopard Seals are more spotted and have a massive head and extremely long foreflippers. Weddell Seals are more heavy-bodied and have a dark pelage and large, forward-directed eyes. They favor fast-ice habitats, compared with the Crabeater Seal's preference for inner and outer pack ice. Ross Seals have a thick neck, a blunt snout, and countershaded coloration, with dark streaks from chin to chest.

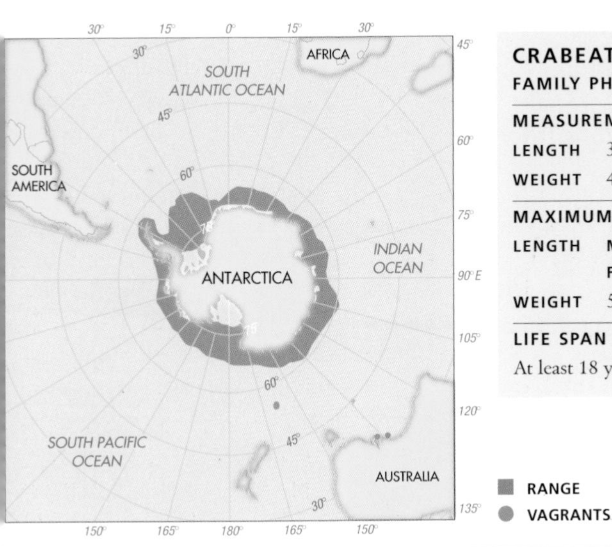

RANGE
VAGRANTS

CRABEATER SEAL
FAMILY PHOCIDAE

MEASUREMENTS AT BIRTH

LENGTH 3′11″ (1.2 m)
WEIGHT 44–66 lb (20–30 kg)

MAXIMUM MEASUREMENTS

LENGTH **MALE** 8′6″ (2.6 m)
FEMALE 8′8″ (2.65 m)
WEIGHT 510 lb (230 kg)

LIFE SPAN

At least 18 years; possibly up to 25–30 years

They favor interior pack-ice habitats. The characteristic sinusoidal, fast, and graceful movements of Crabeater Seals on the ice also distinguish them from the slower lurching movements of Leopard, Ross, and Weddell Seals.

BEHAVIOR These seals are often solitary or in small groups. However, in some areas of Antarctica, particularly along the Antarctic Peninsula, they may occur in aggregations of several hundred, and large groups of a few dozen or more are occasionally observed traveling and perhaps feeding together. On the ice, Crabeater Seals are extremely fast, often gliding quickly and gracefully, with little apparent effort. They usually propel themselves with side-to-side movements of the sides and hindflippers, in a fashion similar to swimming; sometimes they also use their foreflippers. Crabeater Seals molt in January and February, during which time they may keep to the outer reaches of the shrinking pack ice. While molting, they feed mostly at night and spend the days hauled out on the ice. The longitudinal paired scars inflicted by Leopard Seals on juvenile Crabeaters suggest that Leopard Seals may kill and eat large number of young Crabeaters.

REPRODUCTION Crabeater Seals appear to have a serially monogamous breeding system. Females give birth to pups in October and early November. The mother-pup pairs are widely dispersed over the pack ice during this time, and a male remains with a pair, defending them against competing males, until the female comes into estrus. Females nurse their pups for about three weeks, during which time the pups gain about 130 to 175 pounds (60–80 kg). Females are ready to mate within one or two weeks after weaning and abandoning their pups. Mating has never been observed but presumably takes place in the water in pack-ice habitats. After mating, the male may search for other lactating or receptive females in nearby areas, while the female begins feeding to recover lost weight before molting occurs in midsummer.

OPPOSITE: *The Crabeater Seal's teeth are the most modified of mammalian teeth. Crabeater Seals eat mostly krill, and their cheek teeth act as sieves with which to filter the small zooplankton out of gulps of seawater.*

ABOVE: *Crabeater Seals gather on pack-ice floes in large groups of up to several hundred individuals in some seasons and locations, but they are most often encountered in loose groups of ten or fewer.*

FOOD AND FORAGING Crabeater Seals eat mostly krill, but where krill is seasonally less abundant they also eat fish and squid. In summer, they feed mainly at night, diving to depths of 65 to 100 feet (20–30 m). Near twilight they make deeper dives, occasionally reaching 1,400 feet (430 m).

STATUS AND CONSERVATION Crabeater Seals have been estimated to number anywhere between 10 and 15 million. If this population estimate is correct, Crabeater Seals greatly outnumber all other pinniped species worldwide. A multinational program, completed in 2000, that surveyed seals on the pack ice around Antarctica may provide firmer estimates of the species' abundance and geographical population density. Crabeater Seals have not been commercially hunted regularly, except for small harvests of a few thousand by Norway in September and October of 1964 and by Russia in November and December of 1986. In 1955, a group of several hundred of these seals were found dead, from unknown causes, along the Antarctic Peninsula. Because krill is the Crabeater Seal's primary prey, there is some concern over the potential impact on the species of commercial harvests of krill, particularly along the Antarctic Peninsula. Crabeater Seals are protected from harvest under the Convention for the Conservation of Antarctic Seals.

Ross Seal

Ommatophoca rossii
Gray, 1844

Captain James Clark Ross of the HMS *Erebus* described the Ross Seal during a British expedition to Antarctica in the early 1840s, from two seals collected in the pack ice of the Ross Sea. This seal is also called the "singing seal," evidently owing either to its open-mouth, head-up posture when approached on the ice or its siren-like underwater vocalizations. The Ross Seal has remarkably large eyes relative to its small head, which is generally inferred to suggest that the species is a deep diver. Indeed, the genus name is derived from the Greek word *omma,* meaning "eye." The Ross Seal's small teeth and large eyes suggest that it feeds mostly on squid at substantial depths. The longitudinal streaks that run from the mouth down the chest are striking identifying features for this rarely encountered seal.

DESCRIPTION The Ross Seal is relatively small, with a thick chest and a short neck that is barely apparent. The head is small but broad, with a blunt snout and large, forward-pointing eyes. The foreflippers and hindflippers are short and have small claws. The postcanine teeth are small and peg-like, while the small canine teeth are conical and very sharp. There are eight pairs of teeth in both the upper and lower jaws.

Adults are uniformly dark brown or chestnut brown to black dorsally and light brown to tan ventrally, with dark spots or streaks on the sides. Several brown to reddish stripes or streaks run from the mouth down the throat and chest. There are no reported differences in appearance between males and females, although the species has been little studied. Pups are rarely seen.

RANGE AND HABITAT These seals have a circumpolar distribution around Antarctica. In summer they principally occur in regions of heavy, consolidated pack ice, with areas of greater densities appearing to be in the central and western Ross Sea and the King Haakon VII Sea. Recent studies have revealed that at other times some seals may range far north of the pack ice, where they forage at relatively great depths. Rare sightings have been reported from Heard Island and southern Australia.

SIMILAR SPECIES Crabeater, Weddell, and Leopard Seals share habitats with Ross Seals in

- RELATIVELY SMALL, WITH BROAD HEAD, THICK CHEST, SHORT NECK
- DARK PELAGE WITH SCATTERED SPOTS, DARK STREAKS FROM CHIN TO THROAT

- LARGE EYES, SHORT BLUNT SNOUT
- SMALL SHARP CANINE TEETH, SMALL PEG-LIKE POSTCANINE TEETH
- CHARACTERISTIC HEAD-UP, OPEN-MOUTH POSE WHEN APPROACHED
- CIRCUMPOLAR AROUND ANTARCTICA

Antarctica. Weddell Seals are most easily confused with Ross Seals. They are distinguishable at close range by their larger size, their proportionally smaller head relative to their body size, and their uniform dark pelage with light blotches. When approached, Weddell Seals usually turn on their side or back, in contrast to the typical head-up posture displayed by Ross Seals. Leopard Seals are larger, with a massive head, and have a spotted, gray dorsal coloration. Crabeater Seals are longer-bodied and more slender, with a lighter pelage.

BEHAVIOR Ross Seals are solitary and widely dispersed whenever encountered. They are known to be quite vocal when in the water, with a typically siren-like call that can be heard over several miles or more. On the ice, their threat

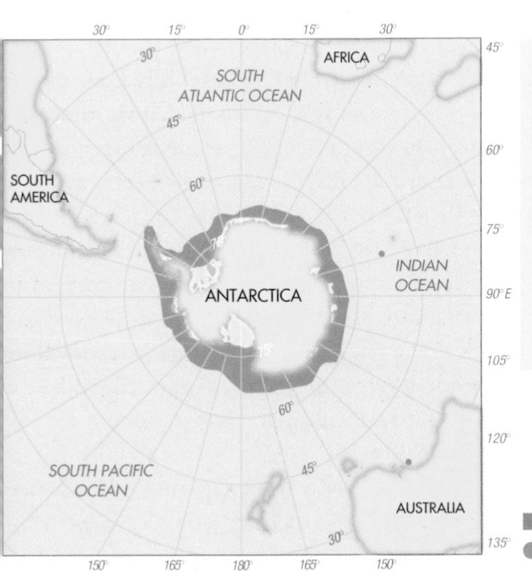

ROSS SEAL
FAMILY PHOCIDAE

MEASUREMENTS AT BIRTH
LENGTH 3′4″–3′11″ (1–1.2 m)
WEIGHT 35–60 lb (16–27 kg)

MAXIMUM MEASUREMENTS
LENGTH 7′10″ (2.4 m)
WEIGHT 440 lb (200 kg)

LIFE SPAN Unavailable

RANGE
VAGRANTS

ABOVE LEFT: *When approached closely, Ross Seals take a head-up, open-mouth pose and may make growling or gurgling threat sounds. Other trilling and chugging sounds are made with the mouth closed when the unique laryngeal air sacs are inflated, causing the chest and throat to bulge out substantially.*

ABOVE RIGHT: *The canine teeth of Ross Seals are small, conical, and sharp and are well adapted for grabbing and holding squid, which may be their main prey.*

vocalizations consist of trilling, chugging, or popping sounds that they make with the mouth fully or mostly closed. The throat and chest are greatly expanded during these vocalizations, evidently owing to the inflation of laryngeal air sacs. When humans approach, Ross Seals typically arch their necks back with their mouths wide open, and sometimes vocalize. This posture may be submissive rather than threatening. Ross Seals molt in January and February, with the timing varying according to sex and age.

REPRODUCTION Because breeding areas of Ross Seals are generally inaccessible during early summer, there is little substantive information about the species' breeding behavior and mating system. The fact that they are generally solitary when encountered suggests a mating system of serial monogamy rather than polygyny. Females evidently give birth in heavy pack-ice habitats in November and December. They may nurse their pups for three to four weeks before abruptly weaning them and mating again. Scars on the chest, neck, and foreflippers of males suggest that males may fight each other for access to estrous females.

FOOD AND FORAGING The scant data available indicates that these seals prey on deepwater squid and fish that they capture mostly at night at depths of 330 to 990 feet (100–300 m). Dives last about 6 minutes on average. When some seals are pelagic in autumn and winter, dives are longer (up to 30 minutes) and deeper (up to 2,500 feet/750 m).

STATUS AND CONSERVATION Small numbers of Ross Seals have been collected for scientific studies, and about 30 were killed by a Russian expedition in 1986 and 1987 under special permit. Killing or harassing Ross Seals is regulated by the Convention for the Conservation of Antarctic Seals.

Leopard Seal

Hydrurga leptonyx
(Blainville, 1820)

- PELAGE GENERALLY SILVER TO DARK GRAY, DARKER DORSALLY, VARIOUSLY SPOTTED
- BODY RELATIVELY SLENDER BEHIND TORSO
- HUGE HEAD AND JAW, LONG FOREFLIPPERS
- NOSTRILS POSITIONED DORSALLY AT END OF LONG, NARROW NOSE
- DISTRIBUTION MOSTLY ALONG EDGE OF FAST ICE AND PACK ICE NEAR ANTARCTICA

With its long sinuous body, heavy flat head, and huge gape, the Leopard Seal is often described as reptilian. Its sinister appearance when stalking prey—and sometimes humans walking on ice floes or near the edge of fast ice—further enhances this image. Leopard Seals are top predators in the Antarctic, preying on virtually everything in the food chain of the Southern Ocean, including krill, squid, fish, penguins, and other seals. Their large canine and postcanine teeth reflect their largely carnivorous diet. The Leopard Seal was described from a seal killed near the Falkland Islands in 1820. The genus name is from the Greek word for "water" and the Greek or Latin words for "I work" or "I drive," and refers to this seal's mostly aquatic life. The specific name means "small claw," or "slender claw."

DESCRIPTION The Leopard Seal has a slender, sinuous body behind a bulky torso and a massive head and jaw with a huge gape. The head and snout are long, narrow, and flat dorsally, and the nostrils point upward rather than forward as in other antarctic phocids. The facial vibrissae are short, clear, and barely noticeable. The foreflippers are very long and capable of powerful swimming, as well as graceful steering. There is no remarkable sexual dimorphism in body size, shape, or coloration, although mature females may be slightly larger than similarly aged males. The canine teeth are large and the postcanine teeth robust and complex, adapted for eating a variety of prey, including penguins and other seals, as well as filtering krill out of the water column. The postcanine teeth are multilobed like those of Crabeater Seals but have only three cusps instead of four or five. There are eight pairs of teeth in both the upper and lower jaws.

Adults have a gray pelage that is dark dorsally and light ventrally, with dark blotches scattered over the chest, belly, and flanks. They are silvery just after molting in January and February but fade to gray over the next several months. The pelage of newborns is similar to that of adults, although the hair is longer and softer.

RANGE AND HABITAT Leopard Seals occur widely in the Southern Ocean around Antarctica. They are most abundant in pack-ice and fast-ice habitats closer to the continent, especially along the Antarctic Peninsula. They regularly

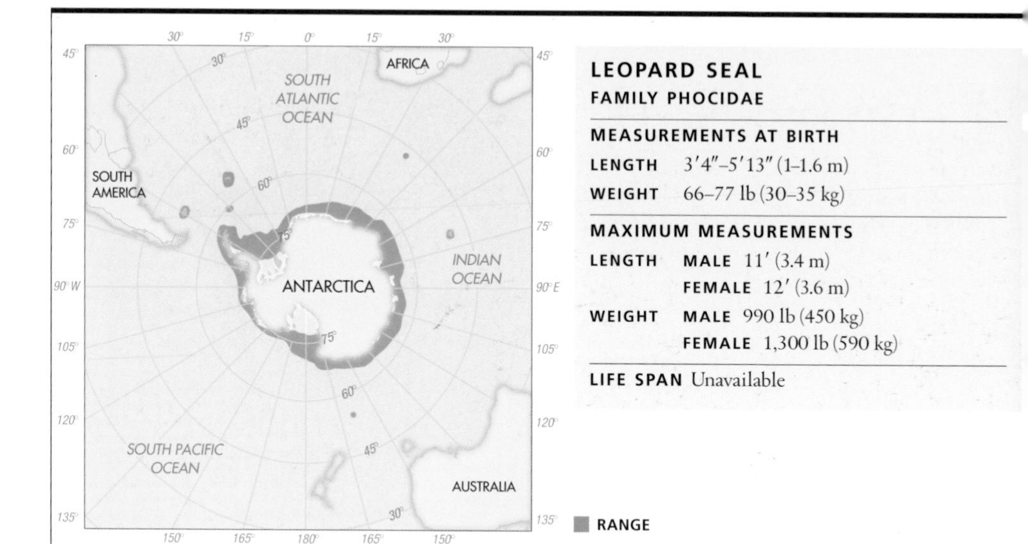

LEOPARD SEAL

FAMILY PHOCIDAE

MEASUREMENTS AT BIRTH

LENGTH	3′4″–5′13″ (1–1.6 m)
WEIGHT	66–77 lb (30–35 kg)

MAXIMUM MEASUREMENTS

LENGTH	**MALE**	11′ (3.4 m)
	FEMALE	12′ (3.6 m)
WEIGHT	**MALE**	990 lb (450 kg)
	FEMALE	1,300 lb (590 kg)

LIFE SPAN Unavailable

occur at South Georgia, the Falkland Islands, and Macquarie, South Orkney, South Shetland, and Kerguelen Islands, especially in September and October; they occasionally occur at Marion Island. Vagrants have been reported on mainland shores of South America, South Africa, Australia, and New Zealand. Their distribution overlaps substantially with that of the Crabeater Seal, a primary prey species in most areas.

SIMILAR SPECIES Crabeater, Weddell, and Ross Seals are found in many areas of the range of Leopard Seals. At close range, the Leopard Seal is easily distinguished by its massive head and slender body. Crabeater Seals have smaller heads and are spotted mostly on the sides around the foreflippers and hindflippers. Crabeater Seals move on the ice with snake-like, sinusoidal glides, whereas Leopard Seals make hunching movements, often keeping their heavy head on the ice. Ross Seals are substantially smaller, with a relatively small head, and their pelage is uniformly dark brown dorsally and light tan ventrally, with longitudinal streaks on the throat. Weddell Seals have robust dark bodies with scattered light blotches and small short heads.

Leopard Seals eat nearly anything that moves in the Southern Ocean, except perhaps cetaceans. At Seal Island off the Antarctic Peninsula, they kill and eat large numbers of Antarctic Fur Seal pups during the breeding season. Their teeth are massive and adapted for grabbing, ripping, and crushing.

The Leopard Seal has a distinctively massive head and gape. Its appearance is often described as reptilian.

BEHAVIOR Little is known about the social behavior of Leopard Seals. They usually appear to be solitary. Social interactions may be frequent, however, around large penguin colonies and between nursing mothers and pups and mating males and females. Leopard Seals are vocally active underwater, though they have mostly simple, low-frequency calls. Their haunting, underwater trill call is commonly heard near penguin colonies. They molt in late summer and early autumn.

REPRODUCTION Little is known about the reproductive biology of Leopard Seals. Females give birth on the sea ice and fast ice between September and January, with peak birthing in November and December. Mating occurs in December and early January, evidently in the water. During the breeding season, mature males apparently patrol waters where receptive females may aggregate. Nursing presumably lasts around four weeks. Females mate and abruptly wean their pups at about the same time.

FOOD AND FORAGING Leopard Seals are, along with Killer Whales, the key predators in the antarctic marine ecosystem. They eat other seals, particularly Crabeater Seals and Antarctic Fur Seals along the Antarctic Peninsula. They also hunt all species of penguin and a variety of fish and squid. Along the Antarctic Peninsula, krill is a key prey species. Indeed, these seals sometimes seem interested and able to eat anything that crosses their path; an adult platypus was found in the stomach of a Leopard Seal near Sydney, Australia. At Seal Island, off the northern tip of the Antarctic Peninsula, Leopard Seals regularly stalk, kill, and eat Antarctic Fur Seal pups, mostly between December and mid-February. This hunting accounts for the deaths of dozens of pups annually and appears to be limiting the population recovery of the species. Leopard Seals also prey upon chinstrap penguins throughout summer. They regularly migrate north in winter to South Georgia Island, where they eat large numbers of subadult Antarctic Fur Seals, juvenile Southern Elephant Seals, gentoo and macaroni penguins, and diving and cape petrels.

STATUS AND CONSERVATION There are an estimated 200,000 or more Leopard Seals throughout Antarctica. While there are no obvious threats to the vitality of the species, climatic warming will likely have some effect on distribution and local abundance, at least in regard to associated declines in the extent of stable fast ice and pack ice, where females give birth and nurse their pups. Leopard Seals are protected from harvest by the Convention for the Conservation of Antarctic Seals.

Cetaceans

The world's present-day cetacean fauna consists of more than 80 species. The suborder Odontoceti (odontocetes, or toothed cetaceans) includes the sperm whales, beaked whales, river dolphins, monodontids (Beluga and Narwhal), ocean dolphins, and porpoises. With 10 families, 40 genera, and at least 70 species, the odontocetes are a much more diverse group than the suborder Mysticeti (mysticetes, or baleen whales), which includes only four families, six genera, and at least 13 species. The two suborders are readily distinguished by two morphological features: Odontocetes have teeth and a single blowhole (nostril), while mysticetes have baleen and two blowholes. Another major difference is that odontocetes locate prey using echolocation, while the filter-feeding mysticetes do not echolocate (at least not in the strictest sense of the term).

The body forms of all cetaceans have a number of common features. All cetaceans are streamlined and elongated. They lack external hindlimbs, and the forelimbs have been drastically modified into flippers. These flippers vary in size and shape, from small and almost unnoticeable in the Sperm Whale to long and conspicuous in the Humpback Whale, or broad and paddle-like in the Killer Whale. The hand and finger bones are enclosed within a common skin surface, and cetaceans have no vestiges of nails.

Killer Whale, Johnstone Strait, British Columbia

The caudal peduncle, or tail stock, is muscular and varies in thickness. It is deep and "keeled" in the Sperm Whale, Dall's Porpoise, and some dolphins, but tapered and narrow in right whales and right whale dolphins. Propulsion comes from the backward thrust of the tail flukes, a pair of horizontal fins at the end of the tail. Cetaceans sometimes lift their flukes above the surface as they begin a dive, an action that is often called "fluking." They may also slap the water surface with their flukes, sometimes repeatedly, a behavior known as "lobtailing."

FLIPPER SHAPES *The flippers of cetaceans, which contain the bones of the arm and hand, come in many shapes and sizes. The flipper of some of the larger whale species may wave conspicuously in the air as the animal rolls onto its side near the water surface.*

Odontocetes have teeth, while mysticetes have baleen. This feature unequivocally differentiates the two cetacean suborders.

LEFT: *Common Bottlenose Dolphin*
RIGHT: *Gray Whale*

Most cetaceans have a dorsal fin situated at or behind the center of the back. It can be little more than a nubbin, as in the Blue Whale and the Ganges River Dolphin, thick and prominent, as in the pilot whales, or very tall and erect, as in the Killer Whale. Some species, like the Bowhead Whale, the Beluga, and the right whale dolphins, have no dorsal fin at all.

Cetacean skin is generally smooth and hairless, although sparse sensory bristles (vibrissae) can be found on the head surfaces of some species. There are no external ear flaps. In fact, it can be difficult to locate a cetacean's ear as the only external manifestation is a pinhole on the side

DORSAL FINS *The size, shape, and placement of a cetacean's dorsal fin help differentiate it from other species.*

TAIL SHAPES *The posterior margin, or trailing edge, of a cetacean's tail flukes differs markedly among species or groups. Nicks, scars, and pigmentation features on the flukes can sometimes be used to identify individuals.*

of the head behind the eye. The mammary glands, genitalia, and excretory organs are concealed within the body in order to reduce drag and avoid heat loss. Cetaceans have the same array of internal organs as other mammals. A key adaptation, however, is the layer of fatty tissue, or blubber, found immediately beneath the skin. The thickness of the blubber varies greatly among species, from only a couple of inches in some dolphins to more than 2 feet (60 cm) in the Bowhead Whale. Blubber serves the dual function of insulating the internal organs and storing energy.

Cetaceans exhibit a wide range of social organization and behavior. Some species appear to be largely solitary or only mildly social, while others are gregarious. Some are active and demonstrative at the surface, while others are more cryptic, showing little of themselves as they surface to breathe. Among the terms adopted from the lexicon of early whalers to describe cetaceans are: "spyhopping" (when an animal raises its head straight out of the water), "blow" (the column of warm air expelled from the lungs as a whale surfaces), "breaching" (when a cetacean jumps or leaps clear of the water surface), and "pod" (a group of closely associated individuals).

All toothed cetaceans have a single blowhole, while baleen whales have two.

LEFT: *Common Bottlenose Dolphin*
RIGHT: *Fin Whale*

Baleen Whales

The mysticetes, or baleen whales, are a suborder of the cetaceans and are classified into four families that currently comprise 13 tentatively recognized species. There is much debate about the precise number of species, and genetic analyses will undoubtedly continue to influence revisions of mysticete taxonomy.

Members of the family Balaenopteridae, also called rorquals, are characterized by their sleek body form and pleats on the underside of the mouth (rorqual is a corruption of a Danish word meaning "tubed" or "pleated" whale). This family includes the Blue, Fin, Sei, Bryde's, Minke, Antarctic Minke, and Humpback Whales, with the latter being alone in its own genus. The Gray Whale is alone in the family Eschrichtiidae, although this division has lately been challenged. The family Balaenidae includes the Bowhead Whale as well as the three species of right whale. All balaenids are robust animals with no dorsal fin, no ventral pleats, and very long, narrow baleen. The poorly studied Pygmy Right Whale of the Southern Hemisphere is the sole species in the family Neobalaenidae.

Mysticetes evolved their unique feature, baleen, in order to take advantage of the most abundant food sources in the ocean: small schooling fish and even smaller zooplankton. Baleen is an elaborate filtration system in the mouth that serves to filter prey from large volumes of seawater. Baleen consists of several hundred individual plates that hang

Gray Whale, Mexico

down from the upper jaw in two racks, one on each side of the mouth (the suborder's scientific name, Mysticeti, comes from the Greek word for "moustache"). The inner surface of the baleen features a dense mat of hair, which serves as a strainer. In plankton-feeding species such as the right whales, the hair is silky, forming a fine mesh for filtering copepods and other tiny zooplankton. In species such as the Humpback Whale that subsist on krill and fish, the inner hair is much coarser.

All mysticetes feed by engulfing large volumes of water and prey, then expelling the water out through the baleen while trapping prey organisms on the baleen's inner surface. Some species feed in discrete events, taking prey a mouthful at a time; these are known as "gulpers." Gulpers include all of the rorquals, which are able to greatly increase the capacity of their mouths during feeding by expanding their ventral pleats; contraction of the pleats then helps force water from the mouth. The balaenids (Bowhead Whale and right whales) are "skimmers," grazing through patches of zooplankton with their mouths open and continuously filtering prey as they swim. Sei Whales are unique in that they engage in both feeding styles. Gray Whales filter organisms mostly from bottom sediment.

Mysticetes tend to be large animals, ranging in maximum length from 21 feet (6.4 m) for the Pygmy Right Whale to well over 100 feet (30 m)

A Blue Whale mother and calf off the Pacific coast of Mexico.

in the case of the Blue Whale, whose mass greatly exceeds that of the largest dinosaurs. Large size may be an adaptation relating to food storage or heat conservation in cold water, but the evolution of this spectacular characteristic is not well understood.

The baleen whales are widely distributed, and several species are found in all major oceans. Bowheads are exclusively a Northern Hemisphere species, while Pygmy Right Whales are confined to the Southern Hemisphere. Gray Whales, which inhabited the North Atlantic as recently as the 1700s, are today found only in the North Pacific. Although the distribution of some mysticetes is frequently coastal, most species are capable of extensive transoceanic movements.

The popular idea that all baleen whales migrate annually from high-latitude summer feeding grounds to breed and calve in warm tropical waters is not correct. Certainly, such extensive migrations are well known in some species, such as the Humpback, Blue, and Gray Whales,

LEFT: *A Dwarf Minke Whale on the Great Barrier Reef off eastern Australia, where watching these diminutive baleen whales swim in clear waters has become popular in recent years.*

RIGHT: *An exceptional photograph of the little-known Pygmy Right Whale in its subantarctic ocean habitat, south of New Zealand's Chatham Islands. The strongly arched mouthline is a feature shared with the right whales.*

but others show little evidence of predictable seasonal movements. Bowhead Whales remain in high latitudes year-round, while some populations of Bryde's Whales may never leave the tropics.

With the sole exception of the Pygmy Right Whale, all of the baleen whales were subject to massive exploitation by the whaling industry. More than 2 million whales were killed by modern mechanized whaling in the 20th century, while right, Gray, and Bowhead whales were already greatly depleted by 1900. Although today many species appear to be recovering well , some remain highly vulnerable to extinction. Among the most endangered are the North Atlantic and North Pacific Right Whales, the North Atlantic populations of the Bowhead Whale, the western population of the Gray Whale, and probably most populations of the Blue Whale.

An intimate view of the eye and "eyebrow" callosity of a Southern Right Whale in Patagonia, Argentina.

Baleen Whales

Gray Whale *page 204*
Bryde's Whale
page 222
Bowhead Whale
page 198
Northern Right Whales
page 190
Southern Right Whale
page 194
Pygmy Right Whale *page 202*

Northern Right Whales

North Atlantic Right Whale ***Eubalaena glacialis*** **Müller, 1776**
North Pacific Right Whale ***Eubalaena japonica*** **Lacépède, 1818**

The northern right whales were the first of the great whales to be regularly hunted by commercial whalers. Slow, easy to catch, and rich in oil and baleen, they were considered the "right" whales to kill. All populations today remain highly endangered. Although some authorities consider the three oceanic populations of right whales—North Atlantic *(Eubalaena glacialis),* North Pacific *(E. japonica),* and Southern Hemisphere (see page 194)—to be a single species, recent genetic data indicate that they are evolving separately and have not interbred for millennia. The genus name means "true whale," while the specific names *glacialis* and *japonica* mean "ice" and "Japan," respectively. Northern right whales are also known as Black Right Whales and as Biscayan Right Whales and Nordkapers, after the Bay of Biscay and the North Cape of Norway, where they were once common.

DESCRIPTION The large robust northern right whales are immediately recognizable among the great whales by their lack of a dorsal fin. They are black with varying amounts of white on the underside. Right whales have a very broad back and paddle-shaped flippers. The head is huge and the mouthline strongly arched. The baleen plates number 220 to 260 per side and are very long (to 9 ft/2.7 m), narrow, and black, with a fine inner fringe. The head and rostrum are covered with callosities—raised roughened patches of skin—that usually appear white or cream-colored because of massive infestations of cyamids (called whale lice). These callosities occur on the whales in approximately the same places as facial hair on human males. The flukes are triangular and very broad, with smooth margins, and are usually raised during a dive. Northern right whales have a distinctive, V-shaped blow when seen from the front or from behind. Females are somewhat longer than males.

RANGE AND HABITAT Northern right whales occur mostly in coastal and shelf waters, but they have also been found offshore. Their present range is much diminished relative to pre-whaling days. The North Atlantic Right Whale ranges largely from Nova Scotia south to the southeastern United States, occupying waters off Cape Cod in winter and spring, the Bay of Fundy and

- LARGE AND ROTUND, WITH BROAD BACK AND NO DORSAL FIN
- CALLOSITIES ON HEAD
- STRONGLY ARCHED MOUTHLINE, WITH VERY LONG BALEEN
- WIDE, BLACK, TRIANGULAR FLUKES, WITH SMOOTH MARGINS
- RAISE FLUKES DURING A DIVE
- V-SHAPED BLOW
- OCCUR IN NORTH ATLANTIC AND NORTH PACIFIC OCEANS

the Scotian Shelf in summer, and calving grounds off the coasts of Florida and Georgia in winter. The whereabouts of much of the population during autumn and winter is unknown. Individuals are occasionally reported from Newfoundland and the Gulf of St. Lawrence, and the species is now very rare in European waters, where once right whales were found from Norway and Iceland to the British Isles, France, and Spain. The North Pacific Right Whale occurs from the Sea of Okhotsk and the Kuril Islands east through the Bering Sea and the Aleutians to the Gulf of Alaska. It has also been seen in waters off southern Japan in winter, where calving may take place. Calving grounds for the eastern North Pacific population are unknown. There have been occasional sightings of northern right whales off the west coast of the United States and Baja California, but these areas are not known to have been a significant part of their historical range.

SIMILAR SPECIES Northern right whales rarely overlap with Bowhead Whales in high latitudes of both the Atlantic and Pacific Oceans. Bowheads

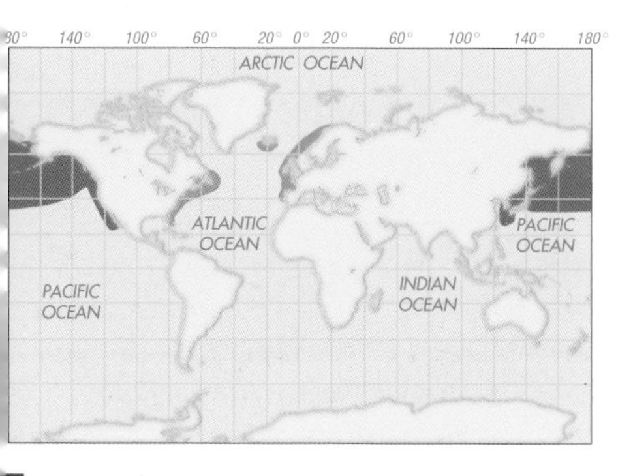

RANGE OF NORTH PACIFIC RIGHT WHALE
RANGE OF NORTH ATLANTIC RIGHT WHALE

NORTHERN RIGHT WHALES
FAMILY BALAENIDAE

MEASUREMENTS AT BIRTH
LENGTH 13–15′ (4–4.6 m)
WEIGHT 2,000 lb (910 kg)

MAXIMUM MEASUREMENTS
LENGTH Longest recorded 56′ (17 m), likely to 60′ (18.3 m); male about 3–5′ (1–1.5 m) shorter than female
WEIGHT About 200,000 lb (90,000 kg)

LIFE SPAN
Known to exceed 70 years

lack callosities on the head, and they have a distinctive white chin patch and a pronounced hump near the blowhole. All other large whales are readily distinguishable from northern right whales by the presence of a dorsal fin or dorsal hump.

BEHAVIOR Like other mysticetes, northern right whales occur singly or in unstable groups that are mostly short-lived and small. Courtship groups of three to as many as 30 whales are known. These usually involve multiple males jostling for position around a single female, with much rolling and churning evident. Aggregations of as many as 150 whales may occur in prime feeding habitats, such as the Bay of Fundy for the North Atlantic population. Northern right whales frequently breach and lobtail. They are occasionally curious enough to approach boats.

REPRODUCTION Females usually give birth to a single calf once every three to five years. Calves are born in winter after a gestation period of approximately 12 months. They are usually weaned and independent from their mothers toward the end of their first year, although separations as early as six to eight months old have been documented. The mating system of northern right whales is heavily based on sperm competition; at one ton (900 kg), male right whales have the largest testes in the animal kingdom. Females will often mate with more than

ABOVE: *Northern right whale breaching.* **RIGHT:** *The head of a North Atlantic Right Whale. The yellowish bumps are callosities, which can be used to identify individuals.*

The heads of four northern right whales in a courtship group. Such groups consist largely of multiple males around a central female.

one male in succession, or even with two simultaneously. Sexual behavior is observed year-round; however, since calving occurs only in winter after a 12-month gestation, copulation in seasons other than winter is presumed to be social in nature.

FOOD AND FORAGING Northern right whales are exclusively plankton feeders, subsisting on copepods, notably *Calanus finmarchicus,* and occasionally krill. They feed by skimming, moving through the water with mouth agape and constantly filtering prey through their finely meshed baleen. While northern right whales usually forage alone, they are occasionally seen feeding in groups.

STATUS AND CONSERVATION Northern right whales are among the world's most critically endangered mammals, and both species are protected throughout their range. In the North Atlantic, commercial whaling began with the Basques in the 11th century around the Bay of Biscay and spread to Labrador by 1530 and to New England by the 1600s. In the late 19th and early 20th centuries, a burst of Norwegian whaling all but extirpated the remaining eastern North Atlantic population. Today there are only about 300 North Atlantic Right Whales remaining, largely confined to waters off the United States and southern Canada. The species is significantly affected by mortalities from collisions with ships and entanglement in fishing gear, and it appears to be in decline. In the North Pacific, intensive whaling began in the Gulf of Alaska in 1835, and by 1900 the North Pacific Right Whale was already rare throughout most of its range. The eastern population was further reduced by illegal Soviet whaling during the 1960s and may presently number fewer than 100 animals. No calves have been seen in the eastern North Pacific for more than a century. In the western North Pacific, a few hundred northern right whales may remain, but no reliable estimates of population size are available.

Southern Right Whale

Eubalaena australis
Desmoulins, 1822

Although confined to the Southern Hemisphere, the Southern Right Whale is virtually identical in general appearance to its northern counterparts, the North Pacific and North Atlantic Right Whales. Nonetheless, most scientists have considered it a separate species, and this separation has recently been supported by molecular genetic data. The genus name *Eubalaena* means "true whale," while *australis* translates to "southern." Southern Right Whales were the first mysticete subjects of a long-term study based on individual identification, begun in 1970 by Roger Payne at Peninsula Valdés, Argentina.

DESCRIPTION With some minor exceptions, the Southern Right Whale is identical in appearance to the northern right whales. The body is black with white ventral patches and a large, rotund shape, and the broad back lacks a dorsal fin, making the species immediately recognizable within its range. The massive head has a narrow rostrum and a strongly arched mouthline. Callosities cover the head, occurring in approximately the same places that facial hair grows on human males. The Southern Right Whale has more prominent callosities on the lips than the northern rights. The callosities are black, but they usually appear white or cream-colored due to massive infestations of whale lice. There are 220 to 260 black baleen plates per side, which are long (to 9 ft/2.7 m) and narrow, with a very fine inner fringe. The flippers are large and paddle-shaped. The flukes are broad and triangular, with smooth margins, and are usually raised during a dive. The Southern Right Whale has a distinctive blow that appears V-shaped from in front or behind. Females are slightly longer than males.

RANGE AND HABITAT Like most baleen whales of the Southern Hemisphere, Southern Right Whales have a circumpolar distribution. They spend part of the year (the austral winter and spring) in inshore waters off the coasts of South America, South Africa, and Australasia, and are also known to occur off various remote offshore islands, including Crozet, Kerguelen, and Tristan da Cunha. Like the northern right whales, Southern Rights are highly migratory. In the southern spring (October to December) they

- LARGE AND ROTUND, WITH BROAD BACK AND NO DORSAL FIN
- CALLOSITIES ON HEAD
- STRONGLY ARCHED MOUTHLINE, VERY LONG BALEEN
- WIDE, BLACK, TRIANGULAR FLUKES WITH SMOOTH MARGINS
- RAISES FLUKES WHEN DIVING
- V-SHAPED BLOW
- CIRCUMPOLAR IN SOUTHERN HEMISPHERE

move south to spend the austral summer feeding in waters around Antarctica or elsewhere in the Southern Ocean. There appears to be relatively little interchange between populations that winter off South America, southern Africa, and Australasia, although this is not well understood.

SIMILAR SPECIES The Southern Right Whale is the only large whale in the Southern Hemisphere that lacks a dorsal fin. It cannot be easily confused with any other whale species.

BEHAVIOR The social structure of the Southern Right Whale appears to be identical to that of the northern rights, consisting of small, unstable groups and occasional large, unassociated feeding aggregations. Active groups of several males and one female clearly associated with sexual behavior often form at the ocean surface. Off Argentina, Southern Rights have been observed "sailing," a behavior in which they raise their flukes in the air at a 90° angle to the wind and "sail" downwind, often returning to their starting

180° 140° 100° 60° 20° 0° 20° 60° 100° 140° 180°
ARCTIC OCEAN
ATLANTIC OCEAN
PACIFIC OCEAN
PACIFIC OCEAN
INDIAN OCEAN

RANGE

SOUTHERN RIGHT WHALE

FAMILY BALAENIDAE

MEASUREMENTS AT BIRTH

LENGTH 13–15′ (4–4.6 m)

WEIGHT 2,000 lb (910 kg)

MAXIMUM MEASUREMENTS

LENGTH About 56′ (17 m); male 3–5′ (1–1.5 m) shorter than female

WEIGHT About 200,000 lb (90,000 kg)

LIFE SPAN

Probably similar to that of northern rights, whose life expectancy is known to exceed 70 years

point to repeat the behavior. It is uncertain whether this serves some function or whether the whales are simply playing. Lobtailing and breaching are other common activities.

REPRODUCTION As in northern rights, female Southern Right Whales usually give birth to a single calf every three to five years. Calving occurs in winter in warmer waters, such as off the coasts of southern Australia or South Africa. The gestation period is approximately 12 months. Calves are usually weaned and independent from their mothers toward the end of their natal year, although separations as early as six to eight months old have been documented. The mating system of right whales is heavily based upon sperm competition; at one ton (900 kg), a male right whale's testes are the largest in the animal kingdom. Females will often mate with more than one male in succession, or even with two

ABOVE LEFT: *The broad, smooth-edged, triangular tail of a Southern Right Whale.* **ABOVE RIGHT:** *Patches of roughened skin, called callosities, cover the head of this Southern Right Whale, off Argentina. The white underside is to the right.*

BELOW: *A Southern Right Whale breaches off the coast of Argentina, exposing its white underside and its paddle-shaped flippers.*

A Southern Right Whale feeds with its mouth partly open, showing the baleen.

simultaneously. Sexual behavior is observed year-round, but since calving occurs only in winter after a 12-month gestation, copulation during any other season than winter is presumed to be social in nature.

FOOD AND FORAGING Southern Right Whales feed largely on copepods and occasionally krill. They skim feed, moving through prey patches with mouth agape and continuously filtering prey through their long, finely meshed baleen.

STATUS AND CONSERVATION As with northern rights, the Southern Right Whale was exploited mainly prior to the modern era, with most catches occurring in the 19th century. Analyses of import records for whale oil and baleen suggest that American whalers alone killed some 60,000 Southern Rights in the 19th century. Illegal catches by the Soviets between 1951 and 1970 included a take of 3,212 Southern Rights, only four of which were reported to the International Whaling Commission. Despite this extensive exploitation, Southern Right Whales appear to be recovering well throughout much of their range. Four populations—in western Australia, eastern South America, and the eastern and western coasts of southern Africa (South Africa, Mozambique, and Namibia)—have been the focus of long-term research; all have shown high annual rates of increase in recent years. The species also appears to be slowly recovering in Brazil and Chile. The status of the Southern Right Whale around remote islands is largely unknown, with the exception of Campbell Island and the Auckland Islands, where recent surveys found substantial numbers of whales. The Southern Right Whale is protected throughout its range; while reliable estimates of abundance are not available, it is clear that the entire population consists of at least several thousand animals.

Bowhead Whale

Balaena mysticetus
Linnaeus, 1758

MATURE MALE

Of all large whales, the Bowhead is the most adapted to life in cold water, with a layer of blubber up to 1½ feet (50 cm) thick and a huge head that it uses to break through thick ice. Closely associated with sea ice through much of the year, the Bowhead Whale is found throughout arctic and subarctic areas in the Northern Hemisphere. Whalers hunted this species extensively until the early 20th century. The scientific name translates to "whale" (from the Latin words *balaena* and *cetus*) and "mustached" (from the Greek *mustakos*), referring to the very long baleen. The Bowhead is also known as the Greenland Right Whale.

DESCRIPTION The Bowhead Whale is large and very robust, with a huge head that in adults is fully one-third of its body length. The body is black, with a white chin patch that often has a line of black spots. The mouthline is strongly arched, and the rostrum very narrow. Baleen plates, numbering 230 to 360 on either side of the mouth, are black, narrow, and up to 14 feet (4.3 m) long. There is a peaked ridge, or "crown," before the blowholes and a notable depression behind them, particularly in adults. Bowheads have no dorsal fin and broad, triangular flukes with smooth margins, which they often raise during deep dives. Their blow is V-shaped when seen from the front or from behind.

RANGE AND HABITAT Bowhead Whales have a circumpolar distribution in high latitudes in the Northern Hemisphere. They are closely associated with ice for much of the year, wintering at the southern limit of the pack ice or in polynyas (large, semi-stable open areas of water within the ice), then moving northward as the sea ice breaks up and recedes during spring. A reverse movement occurs as ice cover spreads southward in autumn. There are five recognized populations of Bowheads. The largest winters in the Bering Sea and migrates northward into the Beaufort and Chukchi Seas in the spring. A second population summers along the western and perhaps northern portion of the Sea of Okhotsk, notably around the Shantar Islands; its wintering ground is largely unknown, but it is likely that most remain in the Sea of Okhotsk year-round. Three other populations occur in the Atlantic: in Davis Strait and Baffin Bay, Hudson Bay and Foxe Basin, and the area of Spitsbergen Island and the Barents Sea.

- LARGE AND ROTUND, WITH BROAD BACK AND NO DORSAL FIN
- ALL BLACK EXCEPT FOR WHITE CHIN PATCH
- HEAD ONE-THIRD BODY LENGTH, WITH BOWED MOUTHLINE
- PROMINENT "CROWN" AT BLOWHOLES, WITH DEPRESSION BEHIND
- OFTEN RAISES FLUKES WHILE DIVING
- V-SHAPED BLOW
- CIRCUMPOLAR IN HIGH LATITUDES OF NORTHERN HEMISPHERE

JUVENILE

SIMILAR SPECIES The North Atlantic and North Pacific Right Whales, the only whales that might be confused with the Bowhead, are easy to distinguish by the callosities on their heads. Unlike Bowheads, northern right whales are frequently white or marbled underneath, and their baleen, while sometimes similar in length, is never longer than 9 feet (2.7 m). They occur rarely in the extreme southern portion of the Bowhead's range and are unlikely to be associated with ice.

BEHAVIOR Bowhead Whales show little stability in their social organization beyond the mother-calf pair bond. Most other associations between individuals last only for hours or at most a few days. However, given that Bowhead vocalizations can be easily heard over several miles, the

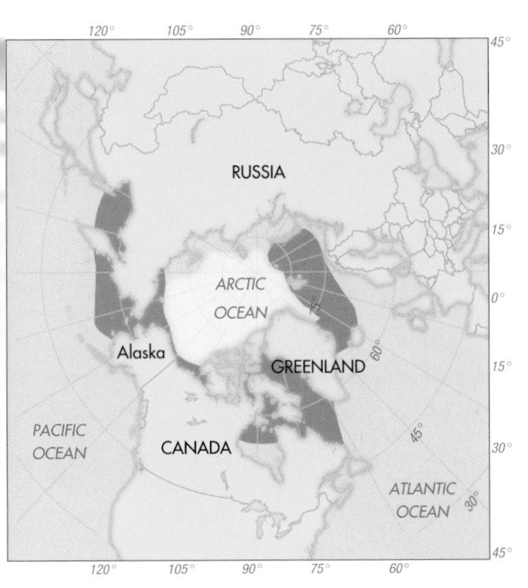

■ RANGE

BOWHEAD WHALE
FAMILY BALAENIDAE

MEASUREMENTS AT BIRTH
LENGTH 13–15′ (4–4.5 m)
WEIGHT 2,000 lb (900 kg)

MAXIMUM MEASUREMENTS
LENGTH 65′ (19.8 m)
WEIGHT About 200,000 lb (90,000 kg)

LIFE SPAN
Recent research suggests that this species may live considerably longer than 100 years

existence of some loose herd structure at times is possible. It appears likely that some Bowhead sounds function as primitive echolocation, as vocalizing Bowheads have been observed to alter their course around icebergs and other obstructions well before they would have been able to detect them visually. Bowheads are adapted for traveling long distances under ice. Their massive heads can reportedly break through ice up to 6 feet (1.8 m) thick. Both the migration and the distribution of Bowheads during the summer feeding season appear to be somewhat segregated by age and sex. Mothers and calves are generally the last to migrate in spring, and juveniles and adults often feed in different regions. Breaching and lobtailing are commonly observed in this species, although the function of these behaviors is unclear. Virtually nothing is known about the behavior of Bowheads during late fall and winter, when ice conditions and arctic darkness make observations impossible.

REPRODUCTION Females give birth every three to four years. The gestation period has never been

OPPOSITE TOP: *A Bowhead dives, showing its broad triangular tail.* **OPPOSITE BOTTOM:** *Two Bowhead Whales surface next to ice floes. The prominent white chin patch is an identifying feature of these whales.* **LEFT:** *A Bowhead raises its head above the surface in the open water of an ice lead.*

confirmed, but the best data suggest it lasts 13 to 14 months, with most calves born during the spring migration north. Weaning probably occurs when calves are 9 to 12 months old. Most conceptions are thought to occur in late winter or early spring, although mating behavior has been observed at other times of the year. Due to the male's unusually large testes, the mating system of the Bowhead Whale is thought to be based in part on sperm competition, involving a female mating with multiple males. Good evidence exists that, like Humpbacks, Bowhead males produce songs that may serve to advertise for females. These vocalizations are heard primarily in spring.

FOOD AND FORAGING Like right whales, Bowhead Whales are skim feeders; however, their diet is much more varied. Their primary prey are copepods and krill, and they also eat a wide variety of other invertebrates. More than 60 prey species have been identified in the stomachs of Bowheads killed by the Inuit hunt in Alaska. Bowheads are usually solitary while foraging, although they occasionally echelon feed together.

STATUS AND CONSERVATION Like the right whales, the Bowhead was the target of intensive whaling in the pre-modern era. Whaling for Bowheads began in the North Atlantic in the 16th century, with thousands of animals killed in waters from Spitsbergen Island to Labrador. The Bering-Chukchi-Beaufort population was first hunted in the mid-19th century, and the Sea of Okhotsk population was exploited shortly thereafter. Of the five populations recognized today, all but one remain highly endangered. The exception is the Bering-Chukchi-Beaufort population, estimated at more than 8,000 animals and steadily increasing despite continued hunting by Inuit. The Spitsbergen population is believed to be close to extinction, while the populations in Hudson Bay–Foxe Basin and Davis Strait–Baffin Bay may number a few hundred animals. The size of the Okhotsk Sea population is unknown but is probably at most a few hundred due to exploitation by the Soviet Union that continued into the 1960s. With the exception of the strictly managed Inuit hunt in Alaska, Bowheads are protected throughout their range.

Pygmy Right Whale

Caperea marginata
(Gray, 1846)

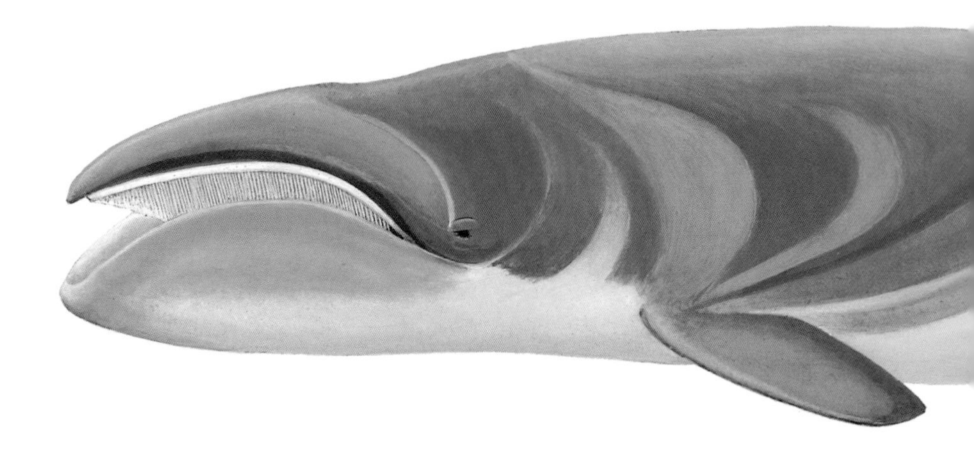

The Pygmy Right Whale is the smallest of the baleen whales and one of the least-known cetaceans. The species has rarely been observed at sea, and scientists have examined only a few dozen stranded specimens. The genus name means "wrinkle" in Latin, referring to the wrinkled appearance of the ear bone. The specific name, meaning "enclosed with a border," refers to the dark border around the baleen plates of some individuals.

DESCRIPTION The Pygmy Right Whale is relatively small and sleek, with a falcate dorsal fin. The body is dark gray dorsally and white ventrally. Many individuals have two chevrons across the back and sides, one above the flippers and one midway down the body. There are approximately 230 yellowish baleen plates on each side of the mouth; in some individuals, these have a dark border. The inner fringe of the baleen is fine and silky. The flippers are dark-colored and narrow, with rounded tips. The flukes are broad and distinctly notched in the middle. The thin, usually indistinct blow is often not visible at all. The blowholes do not appear simultaneously with the dorsal fin at the surface. Females appear to be a few inches longer than males.

RANGE AND HABITAT Pygmy Right Whales are found only in the Southern Hemisphere and presumably have a circumpolar distribution, having been recorded from Australia, South Africa, South America, and various locations in the Southern Ocean. Never observed south of the Antarctic Convergence, they seem to occur only in temperate waters. Details of their seasonal movements or possible migrations remain largely unknown. They have been recorded year-round off Tasmania, and there is some evidence that they move inshore during the spring and summer.

SIMILAR SPECIES Throughout its range, the Pygmy Right Whale is most likely to be confused with the Dwarf and Antarctic Minke Whales, and if neither the jaw nor the flippers are visible, it is difficult to reliably distinguish between these species. Dwarf Minkes often have a white flipper band, or patch, and both minke species lack the arched mouthline of Pygmy Rights.

- BOWED MOUTHLINE
- FALCATE DORSAL FIN
- TWIN CHEVRONS ON BACK AND SIDES OF AT LEAST SOME ANIMALS
- INDISTINCT BLOW
- CIRCUMPOLAR IN TEMPERATE WATERS OF SOUTHERN HEMISPHERE

BEHAVIOR, REPRODUCTION, AND FORAGING The anatomy of the Pygmy Right Whale is notable for the very wide, overlapping ribs and extensive muscle mass, suggesting that it is a strong swimmer. This theory is supported by observations of one animal's rapid acceleration when it was released after being temporarily restrained. Presumably the characteristic relates to predator avoidance. Virtually nothing is known about the social structure or feeding and mating behaviors of the Pygmy Right Whale. Rare observations of the species at sea are usually of solitary animals or pairs, although there are occasional records of large aggregations. Pygmy Right Whales appear to feed largely on copepods, and krill have also been found in the stomachs of stranded animals. Gestation is estimated (from insufficient data) to last about a year.

STATUS AND CONSERVATION There is no information on the abundance, population structure, or status of this species. However, the Pygmy Right Whale has rarely been hunted, and there is no reason to believe it is endangered.

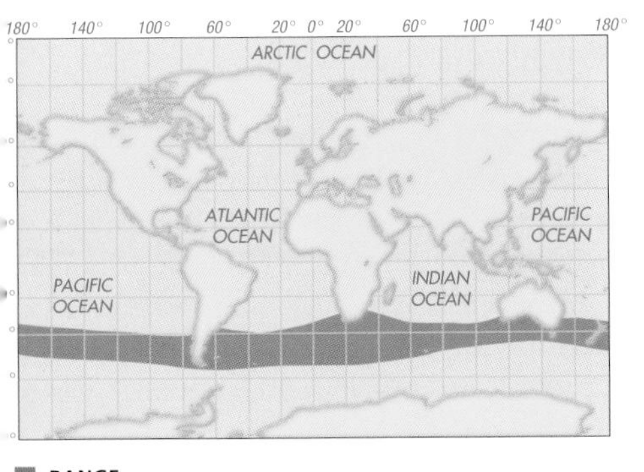

PYGMY RIGHT WHALE
FAMILY NEOBALAENIDAE

MEASUREMENTS AT BIRTH
LENGTH About 5′3″–7′3″ (1.6–2.2 m)
WEIGHT Unavailable

MAXIMUM MEASUREMENTS
LENGTH 21′ (6.4 m); male is a few inches shorter than female
WEIGHT Unavailable

LIFE SPAN Unavailable

Gray Whale

Eschrichtius robustus
(Lilljeborg, 1861)

The Gray Whale makes one of the longest annual migrations of any mammal, traveling some 5,000 miles (8,000 km) from its northern summer feeding grounds to winter calving areas in warm water. Only Humpbacks make similarly extensive migrations. The Gray Whale is the sole species in the family Eschrichtiidae, although recent genetic evidence has suggested that it is closely related to the balaenopterids (the rorquals). Whalers called the Gray Whale the "devil fish" for its ferocity, yet today the species is known for its curious approaches to boats, and in some areas it is the focus of a major whale-watching industry. Once found throughout the Northern Hemisphere, the Gray Whale became extinct in the Atlantic a few hundred years ago and now occurs only in the North Pacific. There are no austral populations, and no evidence exists that Grays ever lived south of the equator. The family and genus names refer to the Danish zoology professor Daniel Eschricht; *robustus* means "strong" or "oaken" in Latin.

DESCRIPTION The Gray Whale has a large body with a mottled gray coloration (calves are darker than adults). The head is narrow and triangular when seen from above. The mouth appears slightly arched and contains 130 to 180 creamy or yellowish baleen plates per side. The baleen is relatively short (to about 10 in/25 cm long), with coarse bristles on the inner fringe. Instead of the ventral pleats of rorqual whales, the Gray Whale has two to five deep longitudinal creases along the underside of the head. Gray Whales have no dorsal fin, but rather a hump followed by 6 to 12 bumps, or "knuckles," along the top of the caudal peduncle. Barnacles and whale lice grow in various places on the body and are usually most obvious on the head. The broad mottled flukes are frequently raised during a deep dive. The Gray Whale's blow can be either columnar or bushy in shape.

RANGE AND HABITAT Gray Whales occur most frequently in shallow coastal waters. Today two populations of Gray Whales exist in the eastern and western North Pacific. The eastern population (often called the California stock) migrates from summer feeding grounds in the Bering, Chukchi, and western Beaufort Seas to its winter breeding and calving areas off the coast of Baja California. Much of this migration is coastal and

- MOTTLED GRAY BODY
- DORSAL HUMP, WITH BUMPS ALONG TOP OF CAUDAL PEDUNCLE
- MANY BARNACLES AND WHALE LICE, NOTABLY ON HEAD
- FREQUENTLY RAISES FLUKES WHEN DIVING
- LOW, BUSHY, HEART-SHAPED BLOW
- FOUND ONLY IN NORTH PACIFIC

can be easily observed between November and March. In winter, concentrations of Grays, notably females with newborn calves, utilize lagoons on the Baja California peninsula. Although most eastern Grays migrate in spring to the Bering Sea and farther north, some summer in lower latitudes off British Columbia, Alaska, and California. In the western North Pacific, a small remnant population (known as the Korean stock) spends the summer in the Sea of Okhotsk, where a feeding ground has recently been identified off northeastern Sakhalin Island. The migratory destination of this population in winter is unknown, but most of these whales probably pass through the Sea of Japan and may breed and calve in tropical waters off southern China. The historical range of the now-extinct North Atlantic population is unclear but appears to have included coastal waters of Europe and Iceland, as well as the eastern coast of North America.

SIMILAR SPECIES The Humpback Whale, which occurs throughout the North Pacific, is black rather than gray, with a dorsal fin (although in a few animals this is so small as to be almost

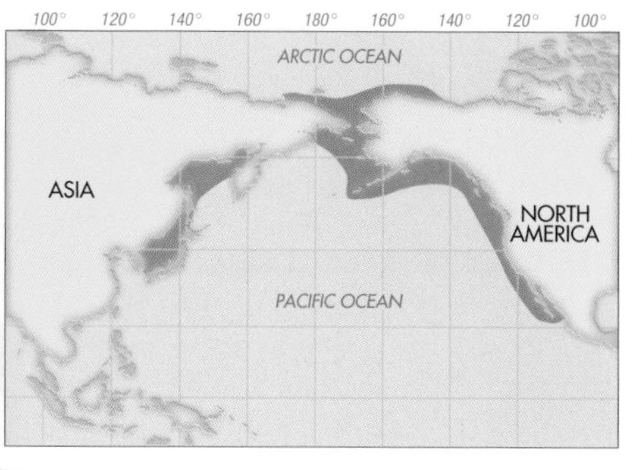

RANGE

GRAY WHALE
FAMILY ESCHRICHTIIDAE

MEASUREMENTS AT BIRTH
LENGTH 16′ (4.9 m)
WEIGHT 1,500 lb (680 kg)

MAXIMUM MEASUREMENTS
LENGTH 49′ (15 m)
WEIGHT 80,000 lb (35,000 kg)

LIFE SPAN
More than 40 years

absent) and very long flippers. The Sperm Whale occurs throughout the Gray Whale's range. It lacks the Gray's mottled coloration and abundant barnacles and whale lice, and is less likely to inhabitat coastal waters. The ranges of both the North Pacific Right Whale and the Bowhead may overlap with the Gray in high-latitude feeding areas; however, both these species are black and more rotund, with a strongly bowed mouthline, a smooth back, and no knuckles on the dorsal ridge.

BEHAVIOR Gray Whales are typical of mysticetes in that they do not form lasting associations. They frequently travel alone or in small unstable groups, although large aggregations can occur on both feeding and breeding grounds. Migrating Gray Whales move steadily in one direction, breathing and diving in predictable patterns. Radio tracking studies have shown that swim speeds increase as the northbound whales approach their summer feeding grounds. Breaching is relatively common in this species, although its function is unclear. The Gray Whale also regularly spyhops (raises its head out of the water) and can exhibit considerable curiosity toward boats.

REPRODUCTION Both breeding and calving are strongly seasonal. Females calve at intervals of two or three years. The gestation period is estimated to last 12 to 13 months, with some evidence that there is no fetal growth during the last month of pregnancy. Calves are born in winter and become independent by seven to nine months old, prior to the fall migration south. Most mating occurs in warm waters during winter, at which time groups consisting of a single female and multiple males are often observed. These groups, in combination with the relatively large size of the male testes, strongly suggest that sperm competition is an important feature of the Gray Whale's breeding behavior.

FOOD AND FORAGING The Gray Whale's principal prey are benthic amphipods, which they filter from bottom sediment in shallow shelf or coastal waters. Foraging whales often leave long trails of mud in their wake on the ocean bottom. Grays also feed on midwater prey, probably more than has usually been assumed.

STATUS AND CONSERVATION Gray Whales were heavily exploited in the North Pacific beginning in the 19th century, and the status of two extant populations differs greatly. The eastern population has been a model of stock recovery since its protection in 1937 and is now so large that it was removed from the U.S.

A Gray Whale calf swims along, staying close to its mother.

government's list of endangered species in 1994. The present population is estimated at 26,000 animals. The apparent starvation of many eastern Gray Whales in the late 1990s suggests that the population has met or exceeded the "carrying capacity" of this environment. The population is the target of a small aboriginal hunt off the Chukotka Peninsula in Russia. The protected western North Pacific population is one of the most critically endangered whale stocks in the world, with perhaps only 100 animals remaining. Considered extinct as recently as the 1970s, subsequent surveys found the remnant population, which feeds in the Sea of Okhotsk. The western Gray Whale was hunted commercially by Korea as late as 1966, and surviving animals continue to be threatened by entanglement in fishing gear, oil and gas exploration on their feeding grounds off Sakhalin Island, and even occasional exploitation by Japanese fishermen. If western Grays rely on coastal areas for calving (as perhaps they do in southern China), they may also be in danger from development and other threats. The North Atlantic population appears to have been extant as recently as the 17th century. The cause of its extinction is a mystery; whether early unrecorded coastal whaling was wholly responsible for or merely hastened the extinction of an already declining population remains unknown.

OPPOSITE: *A school of Gray Whales in the warm waters off Baja California, Mexico.*
LEFT: *The tail of a Gray Whale off Mexico shows the prominent scarring that is common in this species.*

Humpback Whale

Megaptera novaeangliae
(Borowski, 1781)

Perhaps the most familiar of the great whales, the Humpback Whale is best known for its acrobatic displays and its haunting songs. The migration of this species from summer feeding areas in higher latitudes to tropical mating and calving grounds is the longest of any mammal, with some whales making a round-trip journey of 10,000 miles (16,000 km). The Humpback Whale is the best studied of all the large whales, due largely to the fact that individuals can be easily identified by their natural markings; several populations have been under long-term study since the 1970s. The Humpback's long flippers, the longest of any cetacean, make the species virtually unmistakable at close range. The flippers also give this whale its scientific name, which translates to "big wing of New England"; the geographic reference is to the location where the first specimen was described.

DESCRIPTION The Humpback Whale has a large robust body. The head and lower jaw have a variable number of rounded protuberances, called tubercles. The baleen plates are all or mostly black, with 270 to 400 per side. Ventral pleats (14 to 22 in number) extend from the tip of the lower jaw to the umbilicus; these are fewer and wider than in other rorquals. The long narrow flippers are approximately one-third the length of the body. The dorsal fin is highly variable, from almost absent to high and falcate, and is often scarred. The dorsal fin is often located on a hump, which is particularly noticeable when the whale arches its back during a dive. The trailing margin of the flukes is prominently serrated. While diving, Humpbacks often raise their flukes, and their back often appears humped (giving the species its common name). The blow is often lower, rounder, and more bushy than those of other rorquals (although the blows of large Humpbacks can be similar to those of Fin Whales). Females are slightly larger than males.

The body is black above and black, white, or mottled below. The flippers are usually white ventrally; they have a variable dorsal surface that is usually mostly white on whales in the North Atlantic and Southern Hemisphere and mostly black on whales in the North Pacific. The flukes are black above and have a highly variable pattern below, from all white to all black, that is individually distinctive.

- EXTREMELY LONG FLIPPERS (ONE-THIRD BODY LENGTH)
- TUBERCLES ON HEAD AND LOWER JAW
- BODY BLACK ABOVE; WHITE, BLACK, OR MOTTLED BELOW
- FLUKES SERRATED ALONG TRAILING EDGE, WITH VARIABLE PATTERN ON UNDERSIDE
- FREQUENTLY RAISES FLUKES DURING DIVES
- BLOW OFTEN LOW, ROUNDED, AND BUSHY
- OFTEN USES BUBBLE NETS OR BUBBLE CLOUDS TO FEED
- FOUND IN ALL MAJOR OCEANS

NORTH PACIFIC

RANGE AND HABITAT Humpback Whales are found worldwide in all major oceans. They occur primarily in coastal and continental shelf waters, although they are also known to feed around some seamounts, and migrating whales often pass through deep waters. The Humpback is highly migratory. It feeds during summer in mid- and high latitudes and mates and calves during

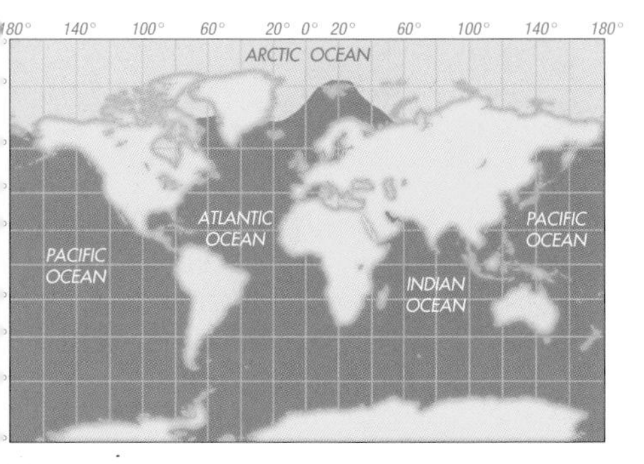

RANGE

HUMPBACK WHALE
FAMILY BALAENOPTERIDAE

MEASUREMENTS AT BIRTH
LENGTH 13–15′ (4–4.6 m)
WEIGHT About 1,500 lb (680 kg)

MAXIMUM MEASUREMENTS
LENGTH About 52–56′ (16–17 m); male slightly smaller than female
WEIGHT 90,000 lb (40,000 kg)

LIFE SPAN
Probably at least 50 years

NORTH ATLANTIC AND SOUTHERN HEMISPHERE

winter in tropical or subtropical waters, often concentrated around islands or reef systems. The Humpback population in the Arabian Sea appears unique in that it remains in tropical waters year-round.

SIMILAR SPECIES At close range, the Humpback Whale is difficult to mistake for any other species, because of its long flippers and the presence of tubercles on the head. The Humpback's range overlaps greatly with right, Fin, and Sei Whales. The three species of right whale lack a dorsal fin. Fin and Sei Whales are larger and sleeker, with dark gray bodies, and they rarely lift their flukes when diving. Gray Whales, which occur in the North Pacific, are gray, lack a dorsal fin, and have bumps along the top of the caudal peduncle.

BEHAVIOR Humpbacks exhibit marked seasonal contrasts in their social organization and behavior, but are typically observed alone or in small unstable groups in both summer and winter. Groups change frequently, and long-term associations, while known, are rare. In summer, whales may forage together for short periods, often cooperating to feed. Whales do not feed in winter, but subsist on fat reserves stored in the blubber. Dive times are typically three to five minutes in summer (occasionally much longer), but longer dives lasting 15 to 20, rarely to 40, minutes are the rule on winter breeding grounds. Humpbacks are known for their acrobatic behaviors, including breaching, lobtailing, and flipper slapping. The purpose of these behaviors is not always clear, but they all have different functions depending on the social or behavioral context.

REPRODUCTION The Humpback Whale's breeding behavior is largely confined to winter in low latitudes, where males sing long complex songs (also sometimes recorded in high latitudes), presumably to attract females and perhaps also to establish separation from or dominance with other males. Songs are population-specific and change over time. Males compete aggressively for females with lunges, tail slashes, charges, and blocks. Such competitive groups consist of two to more than 20 males around a single female and can last for hours, with frequent changes in group composition. The calving interval is usually two or three years. Most births occur in the tropics in midwinter. After a gestation period of about a year, females give birth to a single calf. Calves remain with their mothers for one year, rarely for two.

Individual Humpback Whales are easy to identify by the unique pattern of markings on the underside of the tail. The four whales shown here provide good examples of the variation in these patterns.

FOOD AND FORAGING Humpbacks feed on krill and a variety of small schooling fish, notably herring, capelin, and sandlance. They feed alone or cooperatively, lunging into schools of prey. Many populations use a bubble-feeding technique, blowing "nets" or "clouds" of bubbles to concentrate and trap prey (mostly fish). Bubble nets consist of a series of small bubble columns blown in a circle or spiral that enclose a central space. By contrast, a bubble cloud is a very large burst of bubbles with no open water in the middle.

STATUS AND CONSERVATION Humpbacks were hunted extensively by commercial whalers in the 20th century, during which time they were reduced to perhaps 10 percent of their original numbers worldwide. Approximately 200,000 Humpbacks were killed in the Southern Hemisphere between 1904 and 1983, and the species bore the brunt of illegal catches by the Soviet Union. Despite this, most Humpback populations appear to be recovering well from exploitation. The North Atlantic population is estimated at 11,600 animals, and the North Pacific at 6,000 to 8,000. The number of Humpbacks in the Southern Hemisphere is probably at least 17,000. Strong population growth rates have been reported in many areas. Today Humpbacks are hunted in small numbers by isolated aboriginal fisheries, and many more die from entanglement in fishing gear.

OPPOSITE: *A Humpback Whale in southeastern Alaska almost clears the water in a spectacular breach, revealing its long flippers and the ventral pleats on its underside.*

RIGHT: *Mouths wide open, a large group of feeding Humpback Whales lunges through the surface off Alaska.*

Minke Whale

Balaenoptera acutorostrata
Lacépède, 1804

Minke whales as a group are the smallest of the rorquals; of all baleen whales, only the Pygmy Right Whale is smaller. The Minke Whale has perhaps the most complex population structure of any whale, with evidence of considerable segregation by sex, age, and reproductive condition. The taxonomy of the Minke Whale is unclear. Currently scientists recognize two subspecies of the so-called "common" Minke Whale: the North Atlantic Minke *(Balaenoptera acutorostrata acutorostrata)* and the North Pacific Minke *(B. a. scammoni)*. Some consider a dwarf form, which occurs in the Southern Hemisphere, to be a third subspecies, but as yet this form has not been assigned a scientific name and is referred to simply as *Balaenoptera acutorostrata* subspecies (see page 216). Finally, there is the Antarctic Minke Whale, which was recently recognized as a separate species (see page 218). The Minke was supposedly derisively named after a novice whaler named Meincke, who called out a sighting of the whale at a time when it was considered too small to hunt. The genus name means "winged whale"; *acutorostrata* translates to "sharp snout" and refers to the Minke's distinctively shaped snout. Other names for the Minke Whale include Little Piked Whale and Lesser Rorqual.

DESCRIPTION The Minke Whale is small and sleek, with a sharply pointed head that looks V-shaped when seen from above. A sharp longitudinal ridge runs along the top of the rostrum, and there are 230 to 360 short baleen plates on each side of the mouth that are mostly cream-colored or white. The 50 to 70 ventral pleats terminate just behind the flippers, which are narrow and have pointed tips. Minkes have a falcate dorsal fin that appears simultaneously with the blowholes when they surface. Minkes arch their body prior to a dive but do not raise their flukes above the ocean surface. Their blow is usually not visible but may appear indistinctly in some individuals. Females are slightly longer than males.

The body is black or dark gray above, often with a gray chevron crossing the back behind the head, and white underneath. A white band across the flippers is diagnostic of the species.

RANGE AND HABITAT Minke Whales are among the most widely distributed of all the baleen

- SMALL SIZE
- SHARPLY POINTED SNOUT, SHARP RIDGE ON ROSTRUM
- WHITE BAND ON FLIPPERS
- DOES NOT RAISE FLUKES WHEN DIVING
- BLOW USUALLY NOT VISIBLE
- DISTRIBUTED THROUGHOUT NORTHERN HEMISPHERE

whales. They occur in the North Atlantic and North Pacific, from tropical to polar waters. Generally, they inhabit warmer waters during winter and travel north to colder regions in summer, with some animals migrating as far as the ice edge. Minkes are frequently observed in coastal or shelf waters; however, acoustic recordings in the North Atlantic indicate that during their winter migration many animals pass through deep water east and northeast of the Lesser Antilles. (The range of the Dwarf Minke is described on page 216).

SIMILAR SPECIES Small Bryde's or Sei Whales might be confused with large Minkes in warm-water areas where their distributions overlap; both of these whales, however, lack the Minke's white flipper band and have a more falcate dorsal fin and a rounded head, and Bryde's Whale has three rostral ridges to the Minke's one. Fin Whale calves, which might be difficult to distinguish from adult Minkes, lack the white flipper band and have a rounded head with a white lower right jaw. Blainville's Beaked Whales, widespread in

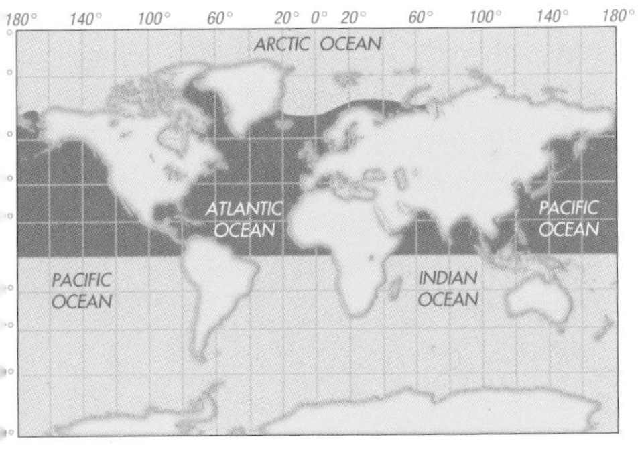

RANGE

MINKE WHALE
FAMILY BALAENOPTERIDAE

MEASUREMENTS AT BIRTH
LENGTH 8′2″–9′2″ (2.5–2.8 m)
WEIGHT About 710 lb (320 kg)

MAXIMUM MEASUREMENTS
LENGTH **MALE** 32′ (9.8 m)
FEMALE 35′ (10.7 m)
WEIGHT Probably 20,000 lb (9,200 kg)

LIFE SPAN
Uncertain; perhaps to 50 years

A Minke Whale lunge feeds in Haro Strait, British Columbia.

temperate and tropical waters of all oceans, have a very similar profile but lack the pointed head and white flipper band. Humpback Whales, found throughout the Minke Whale's range, have much longer flippers and tubercles on the head, and typically raise their flukes when diving.

BEHAVIOR As is typical of the baleen whales, Minke Whales are usually seen either alone or in small groups, although large aggregations sometimes occur in feeding areas. Minke populations are often segregated by sex, age, or reproductive condition, a complex population structure that implies an equally intricate social structure for the species. In Puget Sound, Washington, individual Minkes have adjoining ranges, a phenomenon that may indicate territoriality and that is unique among the baleen whales. Known for their curiosity, Minkes often approach boats. Predation by Killer Whales is probably a significant source of mortality; scientists have witnessed a number of attacks.

REPRODUCTION Very little is known about the reproductive and breeding habits of Minke Whales. There is good evidence that many females give birth annually, although this is unconfirmed. Calving is thought to occur in winter after a gestation of approximately 10 months. Acoustic detections and a few mother and calf sightings suggest that calving grounds may be in tropical waters, including the Caribbean and off Brazil. Calves appear to attain independence at about six months old; the

extreme rarity of mother-calf pair sightings anywhere in high latitudes suggests that females wean their calves prior to their arrival at the summer feeding grounds. Mating is presumed to occur in winter.

FOOD AND FORAGING Minke Whales feed on a variety of small schooling fish, including herring, capelin, and sandlance. Krill are also a major part of their diet, particularly in the Southern Hemisphere, where they are the Minke Whale's principal prey.

STATUS AND CONSERVATION Once considered too small to hunt, Minke Whales became the primary target of commercial whalers following overexploitation of the large whales in the 20th century. In the last hundred years, more than 100,000 Minkes have been killed in the Southern Hemisphere (many of these Antarctic Minkes), and thousands more were caught in the North Atlantic and North Pacific. Hunting for Minkes continues today, by Norway in the northeastern North Atlantic and by Japan in the North Pacific and the Antarctic. Although Minke Whales are widely believed to be abundant in most areas, a lack of reliable information on their population structure raises concerns about continued exploitation. International trade in the species is currently banned.

OPPOSITE: *Breaching is occasionally observed in Minke Whales, although it is less common than in some other species.*

ABOVE: *A Minke Whale surfaces in the Gulf of St. Lawrence, showing its sleek back and falcate dorsal fin. Minkes rarely have a visible blow.*

Dwarf Minke Whale

***Balaenoptera acutorostrata* subspecies**

Although the Dwarf Minke Whale has no recognized taxonomic status, it is widely accepted to be a Southern Hemisphere subspecies of the Minke Whale. It is sufficiently different in appearance to warrant separate treatment here. As its name suggests, the Dwarf Minke is significantly smaller at all ages than the other minke whales.

DESCRIPTION The Dwarf Minke Whale is generally similar in form to other minkes, with a small sleek body and a sharply pointed head that has a single rostral ridge. The baleen plates number 200 to 300 plates per side and are all or mostly white, with a narrow black border on some plates. The falcate dorsal fin appears at the surface simultaneously with the blowholes. Dwarf Minkes do not raise their flukes during a deep dive and rarely have a visible blow.

The body is dark gray or black above and white or cream-colored below. Unlike other minkes, the Dwarf Minke has a striking white flipper patch that often extends into a white shoulder patch immediately above the flipper.

RANGE AND HABITAT The Dwarf Minke Whale is found only in the Southern Hemisphere, primarily in lower and middle latitudes, although it has been recorded as far south as 65°S. It appears to have a circumpolar distribution, with sightings reported from Australia, South America, and South Africa.

SIMILAR SPECIES The Antarctic Minke Whale is larger, with a dorsal fin set farther back on the body, asymmetrically colored baleen plates, a longer rostrum, and no white flipper patch. Small Bryde's and Sei Whales, which may be confused with large Dwarf Minkes, may have a more falcate dorsal fin, a less pointed head, and usually a visible blow, and Bryde's Whales have three rostral ridges rather than one. The rare Pygmy Right Whale is distinguishable at close range by its sharply bowed mouthline. The ranges of all four species overlap greatly with the Dwarf Minke's.

BEHAVIOR, REPRODUCTION, AND FORAGING Little is known about the social system of the Dwarf Minke Whale; however it is likely similar to that of other baleen whales, including ephemeral associations and small group size. Scientists do not know if Dwarf Minkes segregate by

- SMALL AND SLEEK, WITH FALCATE DORSAL FIN AND SHARPLY POINTED HEAD
- BRIGHT WHITE PATCH ON FLIPPERS, OFTEN EXTENDING TO SHOULDER PATCH
- BALEEN ALL WHITE OR WHITE WITH NARROW BLACK BORDER
- DOES NOT RAISE FLUKES WHEN DIVING
- USUALLY NO VISIBLE BLOW
- CIRCUMPOLAR IN SOUTHERN HEMISPHERE

sex or age, as is true for common and Antarctic Minkes, or if they have any other significantly different behaviors. Breeding appears to be seasonal, with calves born in winter. Scant evidence suggests a gestation of about 10 months, with females giving birth every one or two years. Dwarf Minkes feed primarily on krill and schooling fish. They feed by lunging into schools of prey.

STATUS AND CONSERVATION There is not enough information to reliably assess the conservation status of the Dwarf Minke Whale. It is generally assumed to be abundant throughout its range. Dwarf Minkes have been hunted by commercial whalers; however, confusion with the Antarctic Minke Whale has complicated a reliable accounting of the number killed.

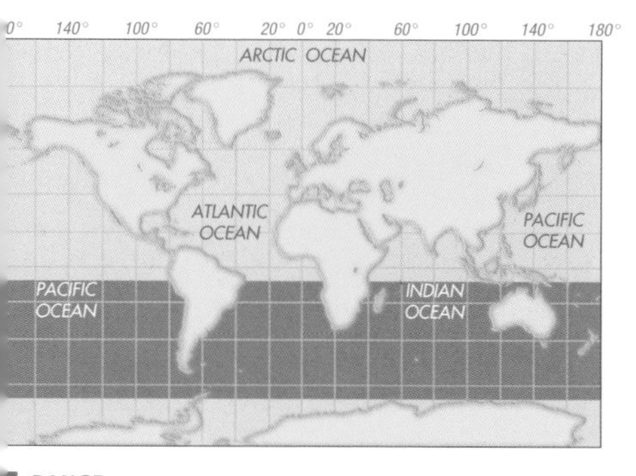

RANGE

DWARF MINKE WHALE
FAMILY BALAENOPTERIDAE

MEASUREMENTS AT BIRTH
LENGTH Probably about 6′7″ (2 m)
WEIGHT Unavailable

MAXIMUM MEASUREMENTS
LENGTH 26′ (7.8 m)
WEIGHT Probably about 14,000 lb (6,400 kg)

LIFE SPAN Unavailable

Antarctic Minke Whale

Balaenoptera bonaerensis
Burmeister, 1867

The Antarctic and common Minke Whales differ significantly in many external and skeletal features. Genetic analyses have shown that the two species have been separated for thousands of years; indeed, they are genetically closer to the Sei and Bryde's Whales than they are to each other. Even so, the Antarctic Minke Whale was only recently assigned separate species status. As its name suggests, the Antarctic Minke spends much of the year in waters around Antarctica, migrating to lower latitudes in winter. The specific name, *bonaerensis,* means "Buenos Aires," which is where the first described specimen was found. *Balaenoptera* translates to "winged whale."

DESCRIPTION Like the common Minke, the Antarctic Minke Whale is small and sleek, and it has a sharply pointed head, a single longitudinal ridge on the rostrum, and a falcate dorsal fin. Both species are dark gray above and white below. There are 200 to 300 baleen plates on each side of the mouth. The baleen of Antarctic Minkes has an asymmetrical coloration, with more anterior white plates on the right side than on the left; the remaining baleen is dark gray. The flippers are light gray, sometimes with a distinct band. Antarctic Minkes have 22 to 38 ventral pleats that extend almost to the umbilicus. Their blow is usually not visible, and they do not raise their flukes when diving.

RANGE AND HABITAT The Antarctic Minke Whale appears to have a circumpolar distribution in the Southern Hemisphere. As befits its name, it summers largely in waters surrounding Antarctica, from the ice edge to areas far from land. Its winter grounds appear to range between about 7°S and 35°S, and it is possible that some whales occasionally reach or even cross the equator. As with the common Minke, the migratory and seasonal movements of the Antarctic Minke Whale are complex and poorly understood. Scientific analyses have failed to clearly identify discrete populations of Antarctic Minkes, possibly because different populations occupy the same areas at different times. Matters are further complicated by the presence of the similar-looking Dwarf Minke Whale, which overlaps with the Antarctic Minke. Studies off South

- SMALL BODY, POINTED HEAD
- ASYMMETRICAL BALEEN COLORATION
- DOES NOT RAISE FLUKES WHEN DIVING
- BLOW USUALLY NOT VISIBLE
- CIRCUMPOLAR IN SOUTHERN HEMISPHERE

Africa suggest that the Antarctic Minke occurs there later in the year (after June) and has a more offshore distribution than the Dwarf Minke.

SIMILAR SPECIES The range of the Antarctic Minke overlaps greatly with those of several other baleen whales. The Dwarf Minke is significantly smaller than the Antarctic Minke and lacks the asymmetrical coloration of the baleen plates. It has a white band across the flippers that often extends to a white shoulder patch, a shorter rostrum, and a dorsal fin set farther forward on the back. Small Bryde's and Sei Whales could be confused with large Antarctic Minkes; both of these species have a more falcate dorsal fin and rounded head, and Bryde's Whale has three rostral ridges rather than one. Fin Whales are much larger than Antarctic Minkes, although Fin Whale calves may cause confusion; they have a more rounded head and a white lower right jaw. Humpback Whales have much longer flippers and tubercles on the head; they typically raise their flukes when diving. The rare Pygmy Right Whale is distinguishable at close range by its sharply bowed mouthline. Antarctic Minkes might overlap with some beaked whales in deep

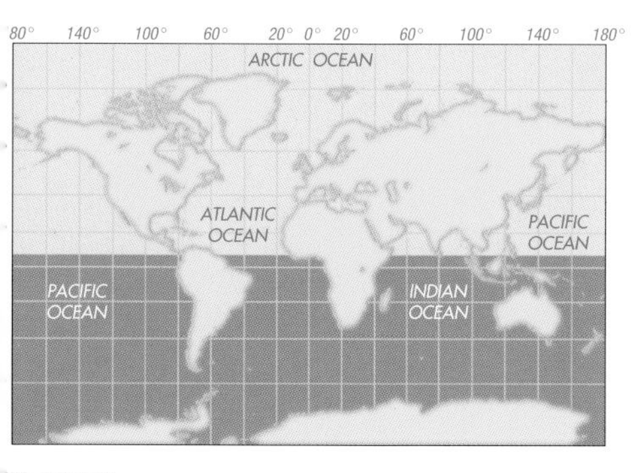

RANGE

ANTARCTIC MINKE WHALE

FAMILY BALAENOPTERIDAE

MEASUREMENTS AT BIRTH

LENGTH About 9′2″ (2.8 m)

WEIGHT 990 lb (450 kg)

MAXIMUM MEASUREMENTS

LENGTH 35′ (10.7 m)

WEIGHT Probably about 20,000 lb (9,100 kg)

LIFE SPAN Unavailable

water, although the austral distributions of these two species are unclear. The heads of the beaked whales are characterized by a pronounced melon, a very narrow rostrum, and a curved mouthline.

BEHAVIOR Very little is known about the social structure or behavior of the Antarctic Minke Whale. As is true for other baleen whales, they frequently travel alone or in small groups and sometimes gather in large feeding aggregations. There is good evidence that populations are segregated by age, sex, or reproductive condition, even during migrations. Antarctic Minkes appear briefly at the surface. They occasionally approach boats, apparently out of curiosity. These whales are the frequent targets of predation by Killer Whales.

REPRODUCTION Little is known about the mating system of the Antarctic Minke Whale. Breeding is seasonal, and calving and mating take place in winter or spring (timing may vary among populations). Females are thought to give birth to a single calf every one or two years, after a gestation period of about 10 months. Calves are probably weaned by six months old.

OPPOSITE: *An Antarctic Minke Whale spyhops in Paradise Bay, exposing its triangular head.*

ABOVE LEFT: *An Antarctic Minke Whale surfaces close to an iceberg in the Antarctic.*

ABOVE RIGHT: *Two Antarctic Minke Whales move rapidly through the water, exposing their sharply falcate dorsal fins.*

FOOD AND FORAGING Antarctic Minkes feed largely on krill, although they may sometimes take small schooling fish. Like most balaenopterids, they are gulp feeders, lunging into schools of prey.

STATUS AND CONSERVATION More than 100,000 minke whales were killed in the Southern Hemisphere during the 20th century, most of which were probably Antarctic Minkes. Japan continues to kill several hundred minkes a year, mainly Antarctics, for the purposes of scientific research, the value of which is much debated. Estimates of abundance for this species vary, although it is generally believed that there are several hundred thousand. International trade in minke whale products is banned under CITES, but there has been considerable pressure from whaling nations to lift this restriction.

Bryde's Whale

Balaenoptera edeni
Anderson, 1879

Bryde's Whales are unique among mysticetes in that at least some populations do not migrate. Instead they spend much or all of their lives in warm tropical waters. The species is currently the subject of considerable taxonomic debate. A pygmy form differs significantly in size from the regular form and will likely be assigned separate species status in the near future. Indeed, the regular and pygmy forms each appear to be genetically more closely related to the Sei Whale than they are to each other. Furthermore, there is good evidence that the regular form occurs in two distinct types—inshore and offshore—that may differ in their diet and reproductive cycles. Bryde's Whale is named for Johan Bryde, a Norwegian who initiated a whaling operation in South Africa. The specific name refers to Ashley Eden, the British Chief Commissioner for Burma who provided the type specimen described by Anderson. This species is often confused with Sei Whales; in the past, Bryde's Whales caught by whalers were often mistakenly recorded as Seis.

DESCRIPTION Bryde's Whale is a moderately large and very sleek whale. The body is dark gray above and gray or white, sometimes with a pinkish tinge, below. Three longitudinal ridges on top of the rostrum are diagnostic of the species (although in some animals, the longitudinal ridges are reportedly indistinct or absent). There are 250 to 370 slate-gray baleen plates that reportedly differ in length and breadth between the inshore, offshore, and pygmy forms. The dorsal fin is tall, extremely falcate, and often noticeably ragged on the trailing edge. There are 40 to 50 ventral pleats that extend from the chin to the navel. The pygmy form is considerably smaller, and some animals may have white or cream-colored lower jaws. Bryde's Whale frequently arches its back when diving but does not raise its flukes above the ocean surface. The blow can be either columnar or bushy.

RANGE AND HABITAT Bryde's Whales are found worldwide in tropical to temperate waters, rarely above latitudes of about 35°. They inhabit areas where there is unusually high productivity, including the Caribbean Sea, areas off equatorial and southern Africa, the Arabian Sea, the Indian Ocean, waters off Australasia, and equatorial and

- DARK GRAY BODY
- THREE LONGITUDINAL ROSTRAL RIDGES (IN MOST INDIVIDUALS)
- PROMINENT, FALCATE DORSAL FIN
- DORSAL FIN USUALLY VISIBLE SIMULTANEOUSLY WITH BLOWHOLES ON SURFACING
- DOES NOT RAISE FLUKES WHEN DIVING
- FOUND WORLDWIDE IN TROPICAL AND TEMPERATE WATERS BELOW 35°

temperate Pacific waters. The pygmy form is known from the eastern Indian Ocean, Australasia, and the western Pacific; its occurrence elsewhere is unclear. Off South Africa and perhaps elsewhere the inshore form occurs closer to the coast than the offshore form and may not migrate at all. Offshore Bryde's may make limited north–south seasonal movements.

SIMILAR SPECIES Bryde's Whale overlaps with several other species in tropical or temperate waters to about 40°N and 40°S and is most easily confused with Sei and Fin Whales. Sei Whales frequently occur with Bryde's Whales in lower latitudes; they are usually larger, with only one rostral ridge (rather than three) and are less likely to arch the back during a dive. Fin Whales have a white right lower jaw (in Bryde's Whales, both jaws are dark gray, although they may both be white in the pygmy form) and a more variably shaped dorsal fin that appears at the surface after the blowholes. Minke Whales are much smaller and rarely have a visible blow. The Humpback Whale is darker and stockier, usually with a less prominent and less falcate dorsal fin, and frequently raises its flukes while diving.

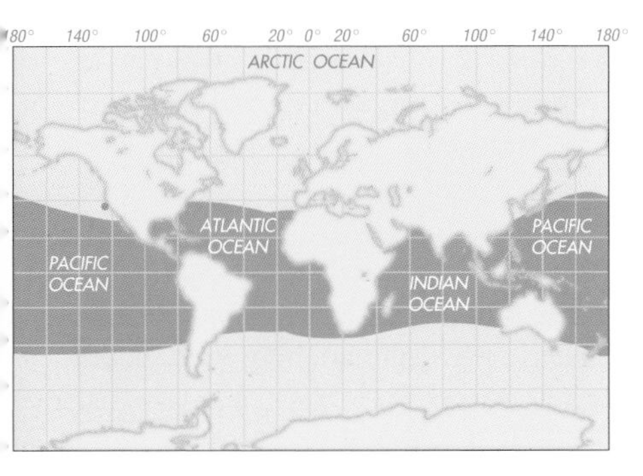

RANGE
EXTRALIMITAL RECORD

BRYDE'S WHALE
FAMILY BALAENOPTERIDAE

MEASUREMENTS AT BIRTH

LENGTH About 13′ (4 m)

WEIGHT Possibly about 1,500 lb (680 kg)

MAXIMUM MEASUREMENTS

LENGTH 51′ (15.6 m)

WEIGHT 90,000 lb (40,000 kg)

LIFE SPAN Unavailable

BEHAVIOR Very little is known about the social organization and behavior of Bryde's Whales. Like other mysticetes, they usually travel alone or in small, ephemeral groups, although large aggregations may occur in feeding areas. Bryde's Whales occasionally breach.

REPRODUCTION The mating system of Bryde's Whale is entirely unknown. The inshore form off South Africa and perhaps elsewhere is unique among mysticetes in that it lacks a distinct breeding season and apparently mates and calves year-round. Presumably these South African whales are able to forage year-round in their tropical habitat (rather than fasting in winter, as many other species do), so that there is no need to time pregnancy to coincide with seasonal peaks in prey availability. Both the offshore and pygmy forms of Bryde's Whale appear to breed seasonally, in late autumn or winter. Females usually give birth every two years, following a gestation period of about one year. Calves are probably weaned at about six months of age, but this information remains unconfirmed.

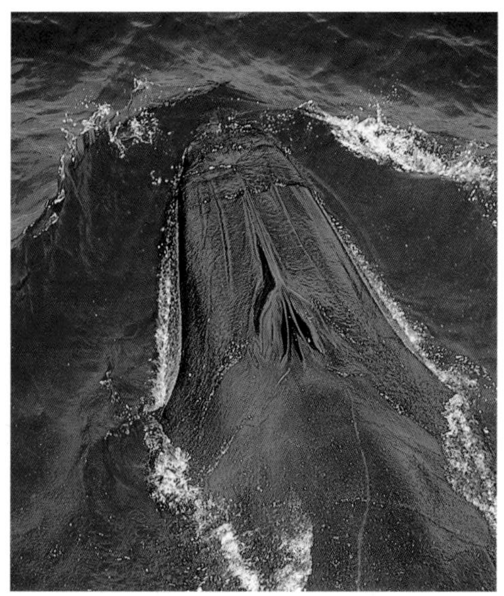

OPPOSITE: *Underwater view of a Bryde's Whale.* **ABOVE:** *Bryde's Whale has a sleek back and a falcate dorsal fin.* **RIGHT:** *Head of a Bryde's Whale viewed from above. The three rostral ridges are characteristic of this species.*

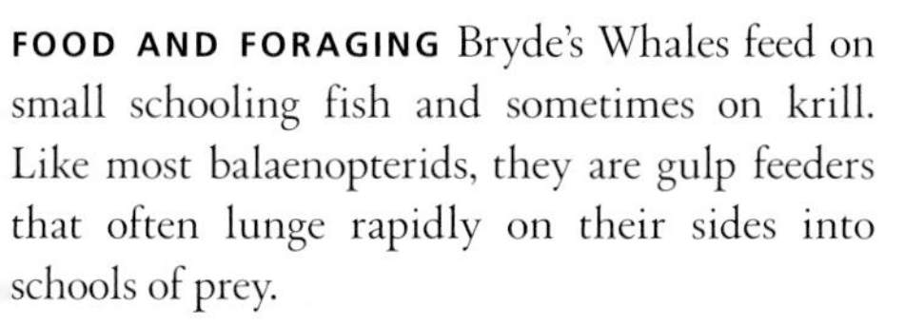

FOOD AND FORAGING Bryde's Whales feed on small schooling fish and sometimes on krill. Like most balaenopterids, they are gulp feeders that often lunge rapidly on their sides into schools of prey.

STATUS AND CONSERVATION Bryde's Whales have never been exploited to the same extent as other large whales, and there is no concern that they are at risk of extinction. Fewer than 8,000 of these animals were killed in the Southern Hemisphere in the 20th century, although it is probable that many earlier catches of Bryde's were erroneously recorded as Sei Whales. In recent years, Japanese whalers have taken Bryde's Whales in the North Pacific. There are no reliable estimates of abundance for any population of Bryde's Whale.

Sei Whale

Balaenoptera borealis
Lesson, 1828

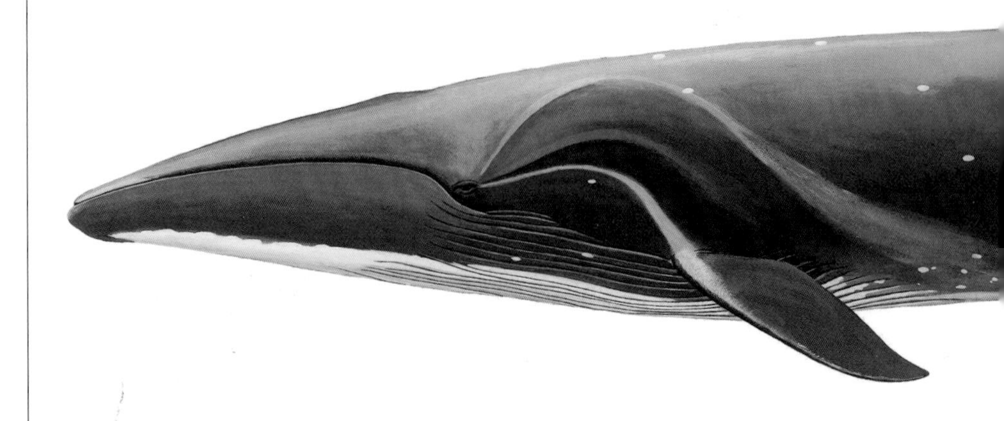

The sleek, fast, and enigmatic Sei Whale differs significantly in distribution and behavior from other rorquals. The species is notable for its often unpredictable distribution, characterized by major influxes into an area followed by absence from it for years or even decades. The common name (pronounced "sigh") derives from the Norwegian word for pollack (coalfish), because the appearance of pollack off the Norwegian coast sometimes coincides with the arrival of Sei Whales. The genus name means "winged whale," while *borealis* translates to "northern." Together with Fin Whales, Seis are arguably the fastest of the great whales. Sei Whales feed on the broadest range of prey of any balaenopterid, from fish and krill to copepods, and they have varied feeding behaviors that reflect their catholic diet. The Sei Whale is also known (although not commonly) as Rudolphi's Rorqual, after an early 19th century German naturalist, D. K. A. Rudolphi, who gave an early description of the species.

DESCRIPTION The Sei Whale has a large, sleek body that is dark gray dorsally and often white or cream-colored on the underside. Oval scars often cover the body, presumably caused by bites from cookie-cutter sharks. The baleen plates, numbering 300 to 410 on each side of the mouth, are dark gray or black, with a fine, white inner fringe. A single longitudinal ridge forms the midline on top of the Sei Whale's head, distinguishing it from the very similar Bryde's Whale, which has three rostral ridges. The dorsal fin is large, prominent, and usually very falcate. There are 30 to 60 ventral pleats, which extend to a point well forward of the umbilicus. Sei Whales have a tall, columnar blow. While diving, they do not arch the back as much as some other whales. Females are slightly longer than males.

RANGE AND HABITAT The Sei Whale occurs worldwide from subtropical or tropical waters to high latitudes and inhabits both shelf and oceanic waters. Its distribution is poorly understood. Scientists have proposed various population divisions within each major oceanic region (North Atlantic, North Pacific, and Southern Hemisphere); however, the validity of these designations is questionable, as too little is known about the movements of individuals between areas. Seis are known worldwide for their unpre-

- SINGLE ROSTRAL RIDGE
- DARK GRAY BODY, OFTEN WITH OVAL-SHAPED SCARS
- PROMINENT, FALCATE DORSAL FIN, USUALLY VISIBLE WHEN BLOWHOLES SURFACE
- DARK GRAY OR BLACK BALEEN, WITH FINE WHITE INNER FRINGE
- DOES NOT RAISE FLUKES WHEN DIVING
- RARELY ARCHES BACK WHEN BEGINNING A DEEP DIVE
- WORLDWIDE FROM SUBTROPICAL OR TROPICAL WATERS TO HIGH LATITUDES

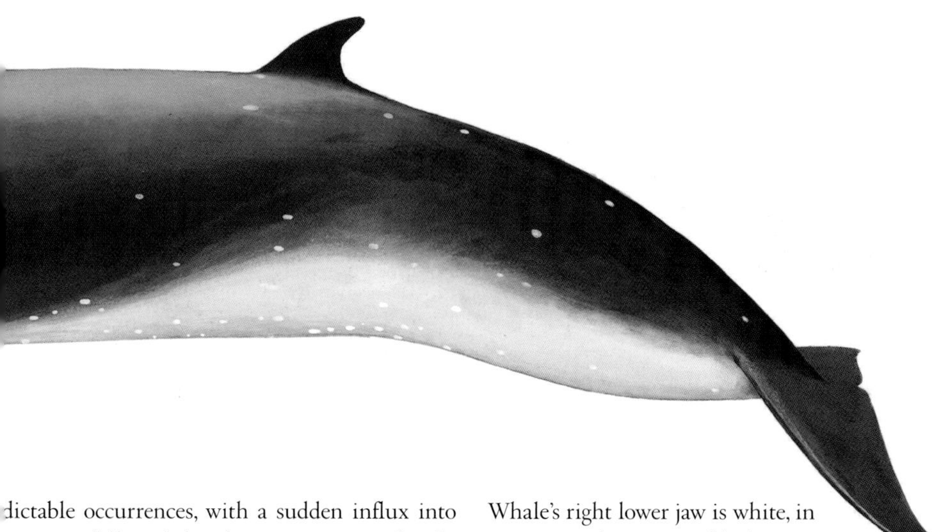

dictable occurrences, with a sudden influx into an area followed by disappearance and subsequent absence for years or even decades. It is generally accepted that many or most Sei Whales undertake seasonal movements between high latitudes in summer and tropical waters in winter. The pattern and extent of these migrations remains unclear.

SIMILAR SPECIES Bryde's Whale, which occurs with the Sei Whale in lower latitudes, can be reliably distinguished only by its three rostral ridges, as opposed to the Sei's single ridge. The Fin Whale's right lower jaw is white, in contrast to the Sei's two black jaws. Unlike the Sei Whale, the Fin Whale surfaces with its dorsal fin appearing after (not with) the blowholes, and arches its back ("rounds out") during a deep dive. The Minke Whale is much smaller and rarely has a visible blow. The Blue Whale is larger, with blue-gray mottling on the body, a much smaller dorsal fin, a broad head, and a larger splashguard in front of the blowholes. The Humpback Whale is darker and stockier, usually lacks the Sei Whale's prominently falcate dorsal fin, and frequently raises its flukes while diving.

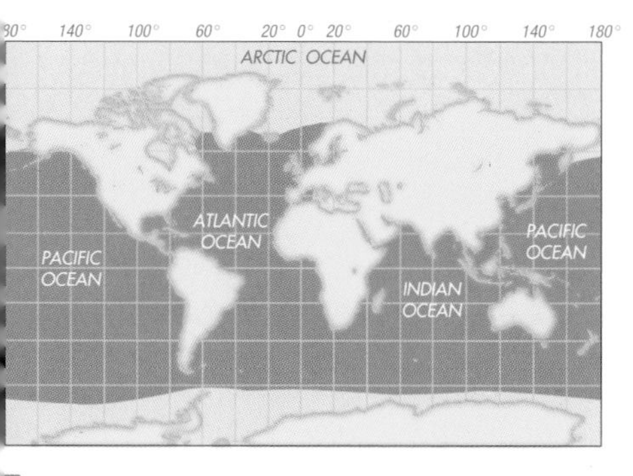

RANGE

SEI WHALE
FAMILY BALAENOPTERIDAE

MEASUREMENTS AT BIRTH

LENGTH 15′ (4.5 m)
WEIGHT About 1,500 lb (680 kg)

MAXIMUM MEASUREMENTS

LENGTH 64′ (19.5 m); male slightly shorter than female
WEIGHT 100,000 lb (45,000 kg)

LIFE SPAN

Probably more than 50 years

LEFT: *The tall, columnar blow of a surfacing Sei Whale.*

ABOVE: *A lateral view of a Sei Whale reveals its falcate dorsal fin.*

BEHAVIOR Very little is known about the social behavior of Sei Whales. Like other baleen whales, their groups are usually small and associations between individuals ephemeral. Sei Whales are usually seen traveling alone or in small groups, although large unstable pods have been recorded in some areas, and aggregations of unassociated individuals can occur on feeding grounds. Together with Fin Whales, Seis are probably the fastest of the large whales, capable of swimming at up to 25 knots for short distances.

REPRODUCTION Calves are born in winter, presumably in tropical waters, after a gestation period of 11 to 12 months. They separate from their mothers at six to eight months old. Females usually give birth at intervals of two or three years. Nothing is known about the Sei Whale's mating system, other than that breeding is seasonal, occurring in winter. There are unconfirmed reports from whaling operations that Seis occasionally hybridize with Fin Whales.

OPPOSITE: *This underwater view of a Sei Whale shows the pitted scars that are typical of this species and probably come from the bites of cookie-cutter sharks.*
LEFT: *Two Sei Whales underwater. Sei Whales have a single rostral ridge on the head.*

FOOD AND FORAGING The Sei Whale feeds on small fish, squid, krill, and smaller zooplankton, notably copepods. It is the only mysticete that feeds both by gulping (taking one mouthful of water and prey at a time, like a Humpback) and skimming (swimming with the mouth open and constantly filtering food, like a right whale). The Sei Whale uses gulping when feeding on fish or krill, and it skims for copepods. The very fine hairs on the inside of the Sei Whale's baleen (relative to that of other balaenopterids) make up the smaller mesh needed to efficiently filter copepods.

STATUS AND CONSERVATION Like most rorquals, Sei Whales lay beyond the reach of whalers until the introduction of faster whaling boats. They were then hunted worldwide, notably in Antarctica, where 200,000 were killed during the 20th century. An accurate count of the number of Sei Whales killed is impossible since for many years members of this species were confused with Bryde's Whales. (Seis were also sometimes even recorded simply as "finners," an old term meaning either Fin Whale or any undifferentiated rorqual.) Their status today is uncertain; however, they are believed to be reasonably abundant in the North Atlantic and North Pacific, while Southern Ocean populations remain greatly depleted. Sei Whales are protected internationally as an endangered species, although this status is contested for some northern populations.

Fin Whale

Balaenoptera physalus
(Linnaeus, 1758)

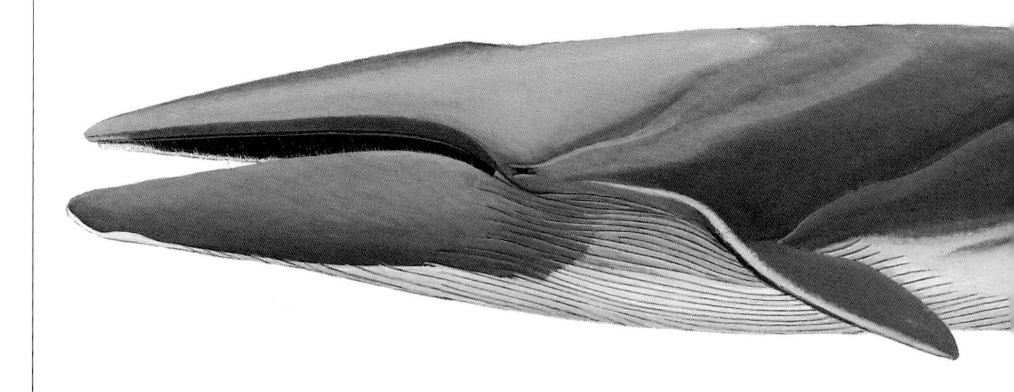

Known as the "greyhound of the oceans," the Fin Whale is a sleek fast animal whose huge size is exceeded only by that of the Blue Whale. The asymmetrical pigmentation of the Fin Whale's lower jaw is unique among cetaceans and has never been satisfactorily explained. *Physalus* has been variously translated as "bellows" (possibly referring to the whale's ventral pleats or its prominent blow), "rorqual whale," or even "a kind of toad that puffs itself up," which presumably refers to the inflation of the ventral pouch during feeding. Alternate common names for this species include Finback, Common Rorqual, and Razorback; the latter name reflects the sharp dorsal ridge on the Fin Whale's caudal peduncle.

DESCRIPTION The Fin Whale is very large and sleek, with a sharp, variably shaped dorsal fin that is often pointed or falcate. Behind the dorsal fin, the upper edge of the caudal peduncle is characterized by a sharp ridge. There are 260 to 480 baleen plates on each side of the mouth. The baleen is black or olive green, except for the white or cream-colored front third on the right side. The ventral pleats extend at least to the umbilicus and number 50 to 100. Fin Whales rarely raise their flukes while diving. They have a tall, columnar blow. Females are slightly longer than males.

The Fin Whale's body is dark gray above and white or cream-colored below, and the flukes are bordered with gray underneath. The unique asymmetrical pigmentation of the lower jaw—black on the left, white on the right—is diagnostic at close range. Most individuals have swirls (called the "blaze") on the right side of the head and a V-shaped chevron across the back behind the head. The patterns of these markings, together with the shape of the dorsal fin, are often used to identify individuals.

RANGE AND HABITAT A cosmopolitan species, the Fin Whale occurs in all major oceans, usually in temperate to polar latitudes and less commonly in the tropics. Although Fin Whales tend to concentrate in coastal and shelf waters, they can also be found in the deep ocean. Fin Whales also occur in the Mediterranean Sea. Their seasonal movements appear to be complex. Fin Whales can be found over a broad latitudinal range throughout the year, which may indicate that the species does not migrate. However, there

- RIGHT LOWER JAW WHITE
- RIGHT FRONT BALEEN PLATES WHITE OR CREAM-COLORED
- CHEVRON PATTERN BEHIND HEAD OFTEN VISIBLE, ESPECIALLY FROM ABOVE
- RARELY RAISES FLUKES WHILE DIVING
- BLOWHOLES SURFACE BRIEFLY BEFORE DORSAL FIN EMERGES
- OCCURS IN ALL MAJOR OCEANS

The Fin Whale's jaw is black on the left side and white on the right.

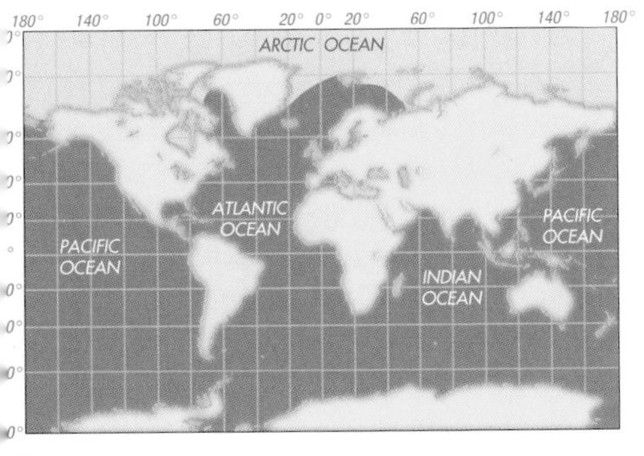

RANGE

FIN WHALE

FAMILY BALAENOPTERIDAE

MEASUREMENTS AT BIRTH

LENGTH 19′6″–21′ (6–6.5 m)

WEIGHT 4,000–6,000 lb (1,800–2,700 kg)

MAXIMUM MEASUREMENTS

LENGTH *Northern Hemisphere:* 79′ (24 m); *Southern Hemisphere:* 89′ (27.1 m); male up to 6′7″ (2 m) shorter than female

WEIGHT 260,000 lb (120,000 kg)

LIFE SPAN

At least 80 years

is evidence that some populations may shift in winter to occupy the summer habitats of others. Genetic data indicate that the Mediterranean Sea population is distinct from that of the North Atlantic.

SIMILAR SPECIES This whale is most easily confused with Sei and Bryde's Whales. Both species lack the Fin Whale's diagnostic white right lower jaw, and both have a highly falcate dorsal fin that is set farther forward on the back; in addition, Sei and Bryde's Whales surface with their blowholes and dorsal fin appearing simultaneously, whereas the Fin Whale's blowholes surface first. Though similar in size to the Fin Whale, the Blue Whale has distinctive blue-gray mottling on the body, a much smaller dorsal fin, a broad, U-shaped head, and a larger splashguard in front of the blowholes. The Humpback Whale is darker, smaller, and stockier than the Fin Whale, with very long flippers; it frequently raises its flukes while diving. All these species have far-ranging distributions similar to the Fin Whale's.

BEHAVIOR The social system of the Fin Whale is typical of the mysticetes, in that stable groups appear to be rare. These whales are usually found traveling alone or in small groups, although large unstable pods have been recorded in some areas, and large unassociated aggregations can occur on feeding grounds. Loose segregation of Fin Whales by age class or reproductive state appears to occur in some regions. Like Blue Whales, both male and female Fin Whales make very loud, low-frequency vocalizations that can travel over hundreds of miles in deep water. Whether these are used to communicate over such distances is unknown. Fin Whales are among the fastest of all the large whales, capable of bursts of speed up to 25 knots. These whales sometimes associate with Blue Whales, and occasional interspecific mating apparently occurs.

REPRODUCTION No distinct breeding or calving grounds have been identified for the Fin Whale, and scientists know very little about the species' mating system. Females usually give birth every two or three years to a single calf; while twins have occasionally been recorded in utero, there is no evidence that any survive. Calving occurs in

OPPOSITE TOP: *A rare view of a Fin Whale's flukes, in the Gulf of California. Fin Whales do not usually raise their tails when diving.* **LEFT:** *Right-side view of a Fin Whale, showing the powerful columnar blow. The white of the right lower jaw is also visible through the water.*

winter, after a gestation period of 11 to 12 months. Calves stay with their mothers for six to eight months. Breeding is seasonal in winter. Chasing has occasionally been observed among pairs of Fin Whales or groups of three in late autumn on the feeding grounds, a behavior that may be related to courtship. Repetitive, low-frequency vocalizations recorded from Fin Whales have been interpreted as male breeding displays similar to Humpback Whale songs.

FOOD AND FORAGING The Fin Whale feeds on krill and various small schooling fish, notably herring, capelin, and sandlance. Fin Whales feed by lunging, mouth agape, into prey schools; some of these lunges may involve bursts of speed of up to 25 knots.

STATUS AND CONSERVATION Until the invention of steam power and the explosive harpoon, Fin Whales were too fast for whalers to catch. However, they now have the dubious distinction of having been hunted in larger numbers than any other whale species during the 20th century, with 725,000 killed in the Southern Hemisphere alone. The status of the Fin Whale today is uncertain. They are believed to be generally abundant in the North Atlantic and North Pacific. They may be making a slow recovery off Antarctica, although no reliable estimate exists for these once huge but now heavily depleted populations. Fin Whales are protected as an endangered species internationally, a status that has recently been challenged for some Northern Hemisphere populations.

OPPOSITE BOTTOM: *Two Fin Whales crash through the water on their sides in a feeding lunge.* **RIGHT:** *A remarkable underwater view of a feeding Fin Whale in the Gulf of California. The whale's underside has expanded as the mouth fills with water, and the baleen is clearly visible in the mouth.*

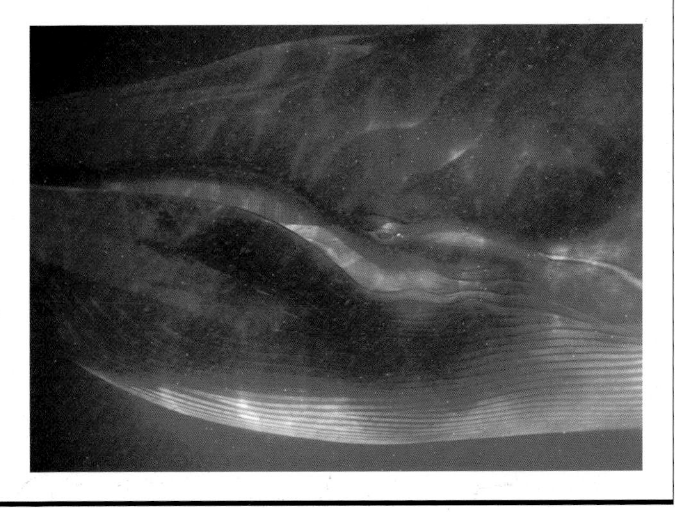

Blue Whale

Balaenoptera musculus
(Linnaeus, 1758)

Blue Whales are the largest animals ever to live in our planet's history. The largest reliably recorded Blues, killed by whalers in the Southern Hemisphere prior to World War I, measured more than 100 feet (30 m) in length and probably weighed nearly 200 tons (180,000 kg). The genus name means "winged whale"; *musculus* could mean "muscular" or "mouse," the latter definition possibly a joke on the part of Linnaeus, the Swedish taxonomist who named the species. The subspecies Pygmy Blue *(Balaenoptera musculus brevicauda)* is generally recognized as significantly different from the "true" Blue Whale and may merit separate species status. *Brevicauda* translates to "short tail," a reference to the Pygmy Blue's short caudal peduncle. True Blue Whales are tentatively divided into *B. m. musculus* in the Northern Hemisphere and *B. m. intermedia* in the Southern Ocean. Blues are also known as "sulphur-bottoms," a reference to the yellowish sheen some individuals acquire from diatoms (algae) on their skin.

DESCRIPTION The Blue Whale has a huge body, which is a mottled, blue-gray color that appears aquamarine underwater. The mottling pattern is highly variable and can be used to identify individuals. Diatoms sometimes coat the body and are most visible as a yellowish sheen on the whale's lighter underside. The head is U-shaped when seen from above and flat in profile, with a large splashguard that rises before the two large blowholes. There are 260 to 400 baleen plates per side that are black with a coarse inner fringe. The ventral pleats number 55 to 68 and extend at least as far as the umbilicus. The tiny dorsal fin is set far back on the body and appears well after

BLUE WHALE
FAMILY BALAENOPTERIDAE

MEASUREMENTS AT BIRTH

LENGTH	23–26′ (7–7.9 m)
WEIGHT	About 6,000–7,900 lb (2,700–3,600 kg)

MAXIMUM MEASUREMENTS

LENGTH	*Northern Hemisphere:* 98′ (29.8 m); *Southern Hemisphere:* 110′ (33.3 m); male 5–10′ (1.5–3 m) shorter than female
WEIGHT	400,000 lb (180,000 kg)

LIFE SPAN
Likely more than 70 years

- HUGE SIZE
- FLAT, U-SHAPED HEAD WITH LARGE SPLASHGUARD BEFORE BLOWHOLES
- MOTTLED, BLUE-GRAY COLORATION
- VERY SMALL DORSAL FIN, SET FAR BACK ON BODY
- OFTEN RAISES FLUKES DURING A DIVE
- TALL COLUMNAR BLOW
- OCCURS WORLDWIDE, MOST ABUNDANT IN EASTERN NORTH PACIFIC

the blowholes when the whale surfaces. The triangular flukes are often raised during a dive. The Blue Whale's blow is very tall and columnar. Females are somewhat larger than males. The Pygmy Blue has shorter, broader baleen, and a shorter caudal peduncle; however, reliably distinguishing between Pygmy Blue and true Blue Whales is likely to be difficult for any but the most experienced observers.

RANGE AND HABITAT The true Blue Whale is a wide-ranging species that occurs in all oceans and inhabits coastal, shelf, and oceanic waters. Pygmy Blues probably occur largely in the Southern Hemisphere. They have been reported from Chile east to Australia, from the tropics to 60°S (occasionally farther south), and appear to be particularly common in the southern Indian Ocean. In Antarctica, true Blue Whales feed farther south than Pygmy

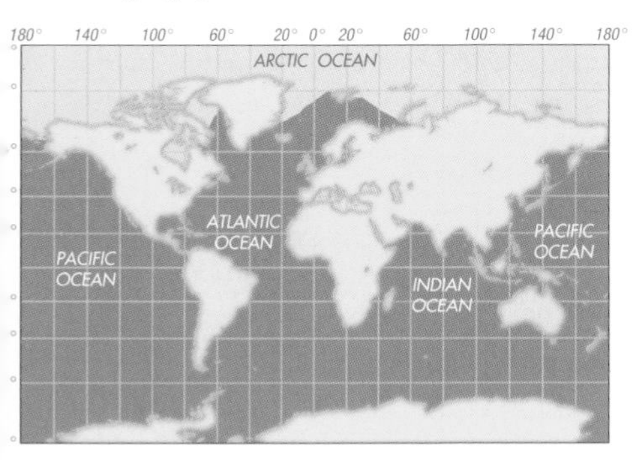

Blues, penetrating above 60°S to the ice edge. The seasonal movements of Blue Whales are complex and not well understood. Much of the population migrates to low latitudes during winter, sometimes to productive areas where feeding continues.

SIMILAR SPECIES The Fin Whale, whose distribution largely overlaps that of the Blue Whale, is similar in size and also produces a very tall blow. Fin Whales lack the Blue's broad rounded head and its mottled bluish coloration, their dorsal fin is usually larger, and their lower right jaw is white. Fin Whales do not raise their flukes during a dive.

BEHAVIOR Very little is known about the social structure of the Blue Whale. Like other mysticetes, Blue Whales tend to travel alone or in small, short-lived groups. They occasionally travel in association with Fin Whales. Blue Whales possess the loudest voice in the animal kingdom, emitting low-frequency sounds that can travel literally hundreds of miles in deep water. Researchers do not know, however, whether the whales use their powerful vocalizations to communicate over great distances. It has been speculated that some of the sounds serve to image underwater features such as seamounts or islands for the purpose of long-distance navigation and orientation.

REPRODUCTION Females usually give birth to a single calf every two or three years. Calving takes place in winter, probably in tropical or subtropical waters, after a gestation period of approximately 11 months. Nursing calves gain 200 pounds (90 kg) a day and separate from their mothers at six to eight months old. Nothing is known about the mating system of the Blue Whale, other than that breeding occurs seasonally, in winter. Blue Whales are known to occasionally hybridize with Fin Whales, and unlikely as it would seem given the considerable differences in size and morphology between the two species, there is one well-documented report of a Humpback–Blue Whale hybrid from the South Pacific.

FOOD AND FORAGING Blue Whales feed almost exclusively on krill and are occasionally reported taking pelagic crabs. They feed by lunging into prey schools. The largest individuals probably consume six tons (5,500 kg) of krill a day. It has been suggested that, because of their enormous size, Blues require larger prey patches than other balaenopterids.

STATUS AND CONSERVATION Beginning around 1900, Blue Whales were hunted extensively worldwide, most notably in the Southern Hemisphere, where 360,000 were killed during the 20th century. Almost all populations were drastically reduced in size, some by perhaps 99 percent of their original numbers. The only population of Blue Whales that may be thriving today is the one that summers off California, estimated at 2,000 animals. Other North Pacific populations may number only a few hundred whales, and the population that once used Japanese coastal waters has been extirpated. Perhaps a few hundred to a thousand Blue Whales remain in the North Atlantic. In Antarctica, the once-huge population of true Blues are gone, leaving perhaps a few hundred animals scattered through circumpolar waters. Off South Georgia Island, hunting of Blue Whales continued until 1936, with more than 39,000 whales caught, apparently resulting in the extirpation of that population. The status of the Pygmy Blue Whale is less clear, but probably all populations are depleted; more than 10,000 Pygmy Blues were killed illegally by the Soviet Union in the 1960s. Blue Whales are now protected worldwide.

OPPOSITE TOP: *A Blue Whale calf breaches off the coast of Baja California, Mexico.* **OPPOSITE BOTTOM:** *A Blue Whale displays its beautifully curved flukes as it dives.* **BELOW:** *An exceptional photo of a Blue Whale as it completes a feeding lunge. The ventral pleats have expanded during the lunge, greatly increasing the capacity of the whale's mouth.*

Sperm Whales

This section treats the three species in the families Physeteridae and Kogiidae as a single group. Sperm whales get their name from an unusual shared characteristic, spermaceti (literally meaning "sperm of the whale"), a liquid wax that fills the spermaceti organ, or "case," in the head of these whales; this organ's function is uncertain. Spermaceti was commercially valuable in the 18th and 19th centuries because of its use in making candles, and later as a base for cosmetics. Although some authorities classify the Pygmy and Dwarf Sperm Whales as a subfamily within the family Physeteridae, the current trend is to recognize a separate family, Kogiidae, with one genus, *Kogia.*

FAMILY PHYSETERIDAE Herman Melville's classic tale, *Moby Dick,* helped make the Sperm Whale an enduring archetype. With its huge boxcar-like head, narrow undershot lower jaw, and massive conical teeth, this whale dominated the iconography of 19th-century Yankee whaling. Once subjected to intensive hunting, the Sperm Whale is now regarded as a generally benign and vulnerable creature.

The family Physeteridae included as many as 20 genera during the Pliocene and Miocene epochs (between about 2 and 24 million years ago). It has since dwindled to a single genus, *Physeter,* with one species. Sperm Whales are among the easier cetaceans to identify at sea. They can be

Sperm Whale

encountered almost anywhere in the world's oceans and seas, but especially along the edges of continental and island shelves. Their distinctive blow can be discerned at a considerable distance. They usually lift their flukes as they dive, providing a good distant signal of their presence.

FAMILY KOGIIDAE The Pygmy and Dwarf Sperm Whales are among the smallest species to be called "whales." They are superficially almost identical, although the Pygmy is larger. Cryptic animals, the kogiids do not form large groups or draw attention to themselves with aerial acrobatics, bow riding, or splashing at the water's surface. Their dolphin-like dorsal fin makes it easy to mistake them for more familiar species when glimpsed only briefly. The relatively high frequency of strandings in some areas suggests that kogiids are more numerous than supposed.

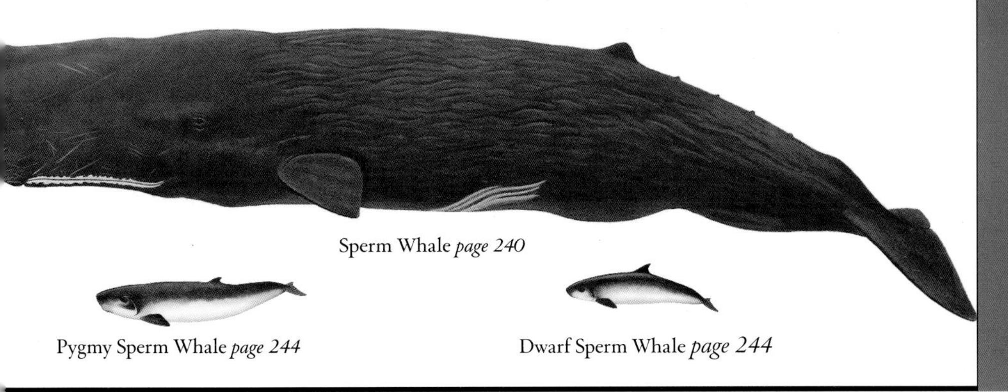

Sperm Whale *page 240*

Pygmy Sperm Whale *page 244*

Dwarf Sperm Whale *page 244*

Sperm Whale

Physeter macrocephalus
Linnaeus, 1758

The Sperm Whale, the largest of the toothed whales, is an animal of extremes. It dives deeper, and for longer, than any other whale, with the possible exception of the bottlenose whales. Among the cetaceans, it displays the greatest difference in size between males and females. Its distribution is truly cosmopolitan, stretching from the Arctic to Antarctica and including most large gulfs and seas. Sperm Whales probably were, and still remain, the most abundant of the large whales, despite having been prime targets of the whaling industry in the past. Herman Melville's classic novel *Moby Dick* was informed by knowledge that came from hunting and butchering Sperm Whales. In the last few decades, innovative and nonlethal research has begun to reveal the Sperm Whale's life in nature. The specific name *macrocephalus* is derived from the Greek for "big head."

DESCRIPTION A disproportionately large head, especially in males, dominates the Sperm Whale's body. The skin posterior to the head region is often wrinkled, in contrast to the smooth taut skin of other whales. The lower jaw is narrow and rod-like, with the mouth underslung and barely visible when viewed from the side. The blowhole is set forward on the head and skewed strongly to the left. There are 2 to 10 short deep grooves on the throat. The dorsal fin is low, thick, and rounded or obtuse. A roughened, callous-like patch occurs on the dorsal fin of most females (about 75%) and some immature males (about 30%), but never on adult males. The ridge along the back behind the fin is lined with bumps, and the caudal peduncle is thick. The flippers are short and broad with rounded tips, and the flukes are triangular with a straight trailing edge. Males are roughly one-third longer than females and twice as heavy. There are 20 to 26 pairs of large, conical teeth along the lower jaw that fit into sockets in the upper jaw.

Sperm Whales are an even dark gray but may appear brown in bright sunlight. The upper lips and lingual portion of the lower jaw are white, and there are often irregular whitish blotches on the belly and flanks.

RANGE AND HABITAT Sperm Whales inhabit ice-free marine waters worldwide from the equator to the edges of the polar pack ice. They tend to occur in highest densities in canyon waters, near

- SOMEWHAT BUSHY BLOW ANGLES FORWARD AND LEFT FROM FRONT OF HEAD
- OFTEN LIES STATIONARY AT OCEAN SURFACE
- USUALLY ENCOUNTERED OFFSHORE IN DEEP WATER
- TRIANGULAR FLUKES LIFTED HIGH AT START OF DIVE, PRECEDED BY THICK CAUDAL PEDUNCLE
- SEEN ALONE (USUALLY ADULT MALE) OR IN GROUPS, OFTEN IN ECHELON
- WORLDWIDE DISTRIBUTION

the edges of banks, and over continental slopes. While Sperm Whales are usually seen in deep, offshore waters, they can occur near the shore where the continental or island shelf is narrow and the water deep. There is a marked difference in migratory behavior between adult males and females. Only adult males move into high latitudes for feeding, while all age classes and both sexes range throughout tropical and temperate seas. Groups of females and immatures generally remain for periods of at least a decade within areas about 600 miles (1,000 km) wide, while adult males range over much larger distances, especially latitudinally. Movements by males across and between ocean basins have resulted in a surprisingly high degree of genetic uniformity among Sperm Whales worldwide.

SIMILAR SPECIES At a distance, the Sperm Whale can be confused with any other large whale that has a conspicuous bushy blow, especially Humpback, Gray, and right whales; however, in Sperm Whales the blow originates

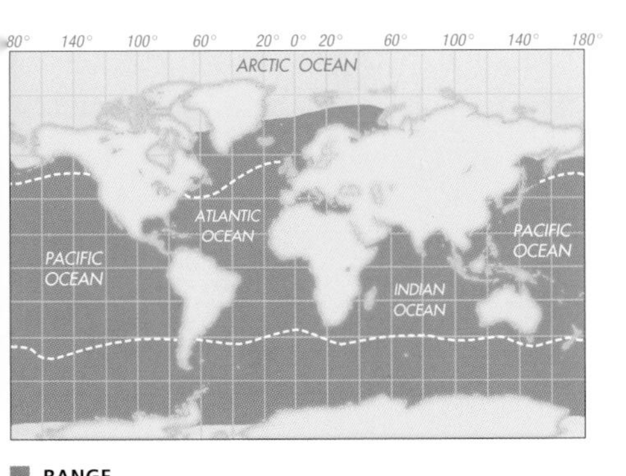

SPERM WHALE
FAMILY PHYSETERIDAE

MEASUREMENTS AT BIRTH

LENGTH	13′ (4 m)
WEIGHT	About 2,200 lb (1,000 kg)

MAXIMUM MEASUREMENTS

LENGTH	MALE	60′ (18.3 m)
	FEMALE	More than 36′ (11 m)
WEIGHT	MALE	120,000 lb (57,000 kg)
	FEMALE	55,000 lb (24,000 kg)

LIFE SPAN

At least 60–70 years

from the front of the head, whereas in other large whales it emerges from much farther back on the head. Humpback Whales are almost as cosmopolitan as Sperm Whales but are generally more coastal. They are readily distinguished by the knobs on their head, their very long, often white flippers, and the serrated rear margin on their flukes. Gray Whales occur in the North Pacific but generally closer to shore than Sperm Whales. They have a generally lighter, mottled coloration and are often encrusted with barnacles, and their flukes are strongly convex along the rear margin. Right whales, present (or once present) throughout most of the Sperm Whale's range in temperate to subpolar latitudes, are relatively easy to distinguish up close by their head callosities, smooth black back with no dorsal fin, broad flippers, and flukes that are concave along the rear margin. The bottlenose whales and Baird's and Arnoux's Beaked Whales are most common in the temperate and polar portions of the Sperm Whale's range and may cause some confusion, particularly with female and juvenile Sperm Whales; however, these whales have triangular or falcate dorsal fins and prominent beaks.

BEHAVIOR Stable, long-term groups of females form the core units of Sperm Whale society. These groups consist of up to about a dozen adult females accompanied by their female and young male offspring. Males begin leaving the family groups at about six years old to join "bachelor schools." The cohesion among individuals within a bachelor school declines as the males age, and males are essentially solitary during their breeding prime and in old age. Because adult male Sperm Whales can have heavy scarring on their heads and occasional broken jaws and teeth, scientists presume that serious fighting may take place between them. Sperm Whales are acoustically active, presumably depending on echolocation as well as passive listening to navigate and find food in the dark ocean depths. They

ABOVE LEFT: *In an uncharacteristic pose for a Sperm Whale, this one from the Gulf of California, Mexico, opens its mouth to reveal the pink palate and white lips. Sperm Whales have teeth only in the narrow lower jaw.*

ABOVE RIGHT: *Sperm Whales usually lift their flukes high above the surface as they begin a deep, or "terminal," dive. Researchers use the frayed margins of the flukes, with their distinctive patterns of nicks and notches, for identification of individuals.*

OPPOSITE: *This dorsal view of a Sperm Whale in the Gulf of California shows the blowhole, located far to the left on the head. The smooth skin of the head gives way to a more wrinkled texture back toward the dorsal fin.*

LEFT: *Set against the rugged New Zealand coastline, this male Sperm Whale exhibits the forward-angled blow typical of the species.*

typically dive to depths of hundreds or thousands of feet for as long as 40 minutes, although they are capable of diving a mile or deeper for more than two hours. Between dives, Sperm Whales spend long periods (typically up to 10 minutes) "rafting" at the ocean surface. Their average swimming speed is 2 to 3 knots, perhaps up to 7 knots, and they are reputed to swim as fast as 12 knots when being chased. Entire schools of Sperm Whales occasionally strand on beaches.

REPRODUCTION Because of slow maturation and a long-term parental investment in offspring, Sperm Whale populations grow very slowly. As a consequence, the species is unable to recover quickly from overexploitation. Prime-aged females give birth only every four to six years, and the calving interval is even longer for older females. The gestation period lasts well over a year, possibly as long as 18 months. Females nurse their calves for at least two years and sometimes much longer; milk has been found in the stomachs of 13-year-old males and 7½-year-old females. Breeding grounds are located in tropical and subtropical waters. Although some mating activity takes place from midwinter to midsummer, most conceptions occur in spring.

FOOD AND FORAGING Sperm Whales are thought to forage mainly on or near the ocean bottom. Their main food items are large and medium-size squid; one individual's stomach contained a giant squid 40 feet (12 m) long and weighing 440 pounds (200 kg). Sperm Whales also consume octopuses and demersal rays, sharks, and various bony fish. While foraging, they often ingest stones, sand, sponges, and other nonfood items. Sperm Whales are thought to feed regularly throughout the year, consuming about 3 to 3½ percent of their weight each day.

STATUS AND CONSERVATION Whaling caused major reductions in Sperm Whale populations worldwide. Sperm Whales were hunted mainly for their body oil and for spermaceti. Their body oil is an exceptional lubricant used, for example, in submarines. Spermaceti, a semiliquid waxy oil found in the Sperm Whale's head, was used in the 18th and 19th centuries to make candles that illuminated streets and homes. Byproducts of the hunt included ambergris (a waxy substance used as a perfume fixative), which is occasionally found in the whales' lower intestines, and tooth ivory, used in carving. Today the principal threats to Sperm Whales are entanglement in fishing gear and collisions with ships. Selective hunting of mature males by whalers may have caused reduced pregnancy rates and increased the vulnerability of groups of females and young to predation by Killer Whales. There are still many tens of thousands of Sperm Whales in the world's oceans. With continued protection from large-scale commercial whaling, there is reason to hope that populations will recover.

Pygmy and Dwarf Sperm Whales

Pygmy Sperm Whale, *Kogia breviceps* (Blainville, 1838)
Dwarf Sperm Whale, *Kogia sima* (Owen, 1866)

Very similar in appearance, the Pygmy and Dwarf Sperm Whales were considered a single species until as recently as 1966, when a scientist at the Smithsonian Institution in Washington, D.C., published a definitive account distinguishing the two whales and establishing the existence of the Dwarf species. Pygmy and Dwarf Sperm Whales both have a spermaceti organ, a feature they share with the Sperm Whale. Unlike the Sperm Whale, the Pygmy and Dwarf Sperm Whales also have a sac in the lower intestine filled with about 13 quarts (12 liters) of a syrupy, dark reddish-brown liquid, which they expel into the water when startled. The liquid creates a dense cloud that may deter a predator or conceal the whale's escape. In ancient Japan, these whales were called *uki-kujira,* or "floating whales," because of their tendency to lie motionless at the ocean surface. In parts of the Lesser Antilles, they are known as "rat porpoises," possibly because of their underslung lower jaw and sharp teeth or their furtive behavior. Because Pygmy and Dwarf Sperm Whales have a superficial resemblance to sharks, many of their strandings have been misreported. The genus name *Kogia* is of uncertain origin. It may be a variant of the English term "codger," or it may be intended to honor a Turk named Cogia Effendi, who observed cetaceans in the Mediterranean Sea.

DESCRIPTION The Pygmy and Dwarf Sperm Whales are very similar in appearance, with only a few significant differences. Both species have a small but extremely robust body that tapers rapidly from the dorsal fin to the flukes. The head appears conical when seen from above or below. The head shape changes with age, becoming blunter and more squarish on older individuals. The Dwarf Sperm Whale may have several short longitudinal grooves on the throat. The falcate dorsal fin is short, small, and positioned behind the midpoint of the back on the Pygmy Sperm Whale, while it is more prominent and perhaps set somewhat farther forward on the Dwarf species. The small flippers appear to be positioned unusually far forward on the body. The Pygmy has 12 to 16 (occasionally 10–11) pairs of teeth in the lower jaw that are thin, back-curved, very sharp, and without enamel. The Dwarf has 7 to 12 (occasionally 13) pairs of teeth in the lower jaw and sometimes as many as three pairs of teeth in the upper jaw.

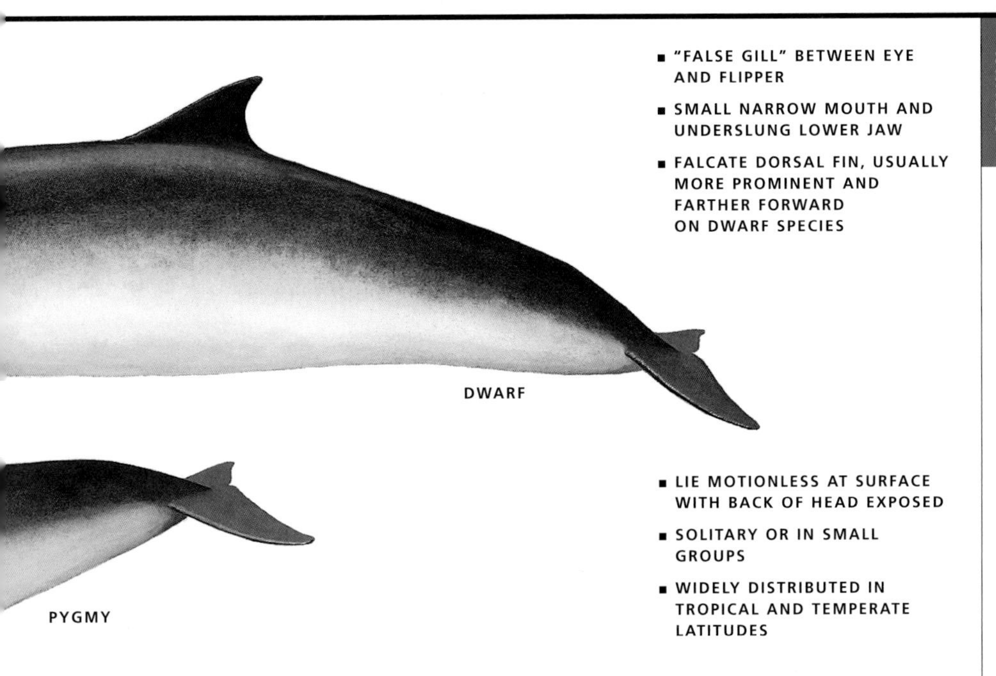

- "FALSE GILL" BETWEEN EYE AND FLIPPER
- SMALL NARROW MOUTH AND UNDERSLUNG LOWER JAW
- FALCATE DORSAL FIN, USUALLY MORE PROMINENT AND FARTHER FORWARD ON DWARF SPECIES

- LIE MOTIONLESS AT SURFACE WITH BACK OF HEAD EXPOSED
- SOLITARY OR IN SMALL GROUPS
- WIDELY DISTRIBUTED IN TROPICAL AND TEMPERATE LATITUDES

The body coloration is a drab, bluish-steel gray, shading to dull white or pinkish on the belly. A light, bracket-shaped marking, often called a "false gill" because of its resemblance to a fish's gill cover, is present between the eye and flipper. There are often circular scars around the mouth caused by squid bites, and tooth rakes elsewhere on the body caused either by interspecific fighting or predation by sharks.

RANGE AND HABITAT These whales occur in tropical and temperate latitudes worldwide. Judging by stomach contents and rare sightings at sea, Pygmy Sperm Whales are thought to reside primarily seaward of the continental shelf. Dwarf Sperm Whales are somewhat more coastal, inclined to inhabit shelf-edge and slope waters. There is no evidence that either species migrates. Pygmy Sperm Whales have been recorded in the

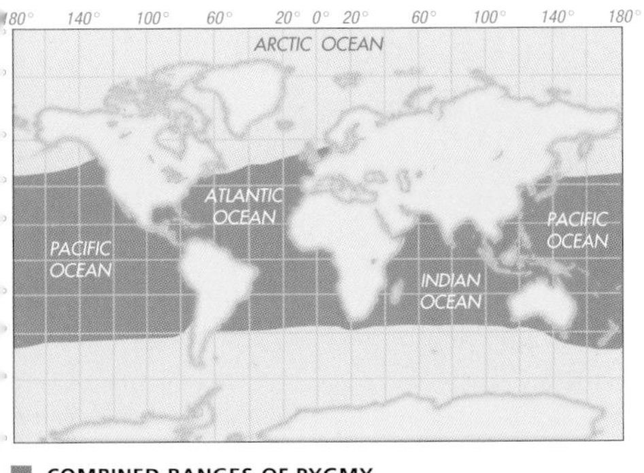

COMBINED RANGES OF PYGMY AND DWARF SPERM WHALES

PYGMY AND DWARF SPERM WHALES
FAMILY KOGIIDAE

MEASUREMENTS AT BIRTH

LENGTH *Dwarf:* 3′4″ (1 m)
Pygmy: 3′11″ (1.2 m)

WEIGHT *Dwarf:* Unavailable
Pygmy: 51–55 lb (23–25 kg)

MAXIMUM MEASUREMENTS

LENGTH *Dwarf:* 8′10″ (2.7 m)
Pygmy: 11′6″ (3.5 m)

WEIGHT *Dwarf:* More than 460 lb (210 kg)
Pygmy: 900 lb (410 kg)

LIFE SPAN Unavailable

Atlantic from Nova Scotia, the Azores, and the Netherlands south to Uruguay and Cape Province, South Africa; and in the Pacific from Honshū (Japan), Hawaii, and northern Washington state south to the Tasman Sea, the Juan Fernández Islands, and Chile. Strandings or other records of Dwarf Sperm Whales in the Atlantic range from Virginia and the Mediterranean Sea south to southern Brazil and the southern tip of Africa, and in the Pacific from Honshū (Japan) and British Columbia to the Mariana Islands, New Zealand, and Chile; the species is known to occur along the rim of the Indian Ocean from southern Africa to Oman, Indonesia, and southern Australia.

SIMILAR SPECIES It is extremely difficult to distinguish between the Pygmy and Dwarf Sperm Whales in the field. The Dwarf Sperm Whale is somewhat smaller (by about 1½ feet/45 cm), with a proportionally smaller head that in adults is more squarish; the dorsal fin is usually more prominent and erect, and may be set slightly farther forward on the body. To accurately distinguish between these two species, a close examination of the whole animal may be necessary, including body measurements and tooth counts.

BEHAVIOR Very little is known about the social organization of Pygmy and Dwarf Sperm Whales; however, they are not considered gregarious as they are usually observed alone or in small groups of up to six (Pygmy) or ten (Dwarf). Their movements are relatively slow, and they do not often engage in aerial displays. They roll slowly, exposing the back and dorsal fin, and dive without lifting their flukes. Their blow is generally faint or invisible. Because of their tendency to raft motionless at the surface, it is often possible to approach them closely before they startle and dive; this tendency could also make them especially vulnerable to shark attacks and collisions with ships. One Pygmy Sperm Whale found stranded in northern California had been attacked by a white shark. Echolocation-type clicks have been recorded from live stranded animals.

REPRODUCTION The gestation period for Pygmy and Dwarf Sperm Whales is thought to be about

OPPOSITE: *A rare sighting of a Dwarf Sperm Whale breaching in the Caribbean Sea, displaying the light, bracket-shaped mark on the side of its head just before the flipper.* **ABOVE:** *Pygmy and Dwarf Sperm Whales often lie motionless at the surface and appear startled when a vessel approaches closely. The prominent, slightly falcate dorsal fin is a characteristic feature of the Dwarf Sperm Whale.*

a year. It is not unusual for females to be pregnant while still lactating. The mating and calving season for Pygmy Sperm Whales spans about seven months, from autumn through spring, while Dwarf Sperm Whales, at least off southern Africa, appear to give birth mainly from December to March (summer). Weaning probably occurs when calves are about a year old.

FOOD AND FORAGING Pygmy and Dwarf Sperm Whales prey on cephalopods (including squid), crustaceans (especially shrimp and crabs), and fish. Some of their prey items are known to migrate vertically over a 24-hour period, so the whales could forage in deeper water during the day or farther up in the water column at night. At least some foraging is thought to take place near the ocean bottom; this is based on the head structure of these whales, including an underslung lower jaw, as well as the presence of benthic fish and crabs in the stomachs of some individuals.

STATUS AND CONSERVATION Pygmy and Dwarf Sperm Whales have never been the targets of large-scale commercial whaling, although some 19th-century sperm whalers harpooned them casually. More recently, shore-based whalers have hunted them in Indonesia, the Lesser Antilles, and Japan. They sometimes become entangled in fishing gear, although the scale of mortality from this is not known to be great enough to threaten their populations. A number of stranded whales were found to have gut blockages caused by the ingestion of plastic bags, and this may constitute a conservation concern. Neither species is considered abundant; however, the incidence of strandings in some areas, such as Florida and South Africa, is frequent enough to suggest that they are fairly common in local waters. Shipboard surveys off California produced estimates of about 3,000 Pygmy Sperm Whales, plus another 900 whales that were either Pygmy or Dwarf Sperm Whales. Vessel surveys have produced estimates of about 11,000 Dwarf Sperm Whales in the eastern tropical Pacific and a few hundred in the northern Gulf of Mexico. The somewhat more nearshore distribution of the Dwarf Sperm Whale could make it more vulnerable to threats from fishing and other human activities.

Beaked Whales

The beaked whales, or ziphiids, include all species in the family Ziphiidae (although whales in the genera *Hyperoodon* and *Berardius* are often referred to as "bottlenose whales.") They are so difficult to observe that several species have never been seen alive, and only a few have been studied in any detail. Since the 1980s, rapidly developing techniques of molecular analysis, and efforts by field scientists to obtain biopsies and to sample stranded or bycaught specimens, have resulted in a remarkable surge in knowledge about beaked whale systematics and biology. The taxonomy of this group is tentative and uncertain; new species are regularly described and old species frequently revised. At latest count, there were at least 20 species of beaked whales in six genera.

In addition to having a beak (or "rostrum") of variable length, all beaked whales have a pair of grooves on the throat that converge anteriorly. Beaked whales also have "flipper pockets," slight depressions in the body wall into which the flippers can be tucked, perhaps to reduce drag. The flukes are proportionally large and usually lack a median notch. In many respects, the key to distinguishing the various beaked whales from one another (apart from molecular differences) lies in the size and shape of the teeth and their position in the lower jaw. Most species have retained only a single pair of teeth, which develop and erupt from

Arnoux's Beaked Whale

the gums only in adult males (females and juveniles are effectively toothless). In several species, the teeth of adult males are exposed outside the closed mouth and can be considered tusks. Judging from the long linear scars that often crisscross the bodies of adult males, it seems that males use these tusks in aggressive encounters with other males.

In general, beaked whales are deep divers that spend relatively little time near the surface. They are found mostly in water at least 1,000 feet (300 m) deep. The diet of most species appears to be dominated by mesopelagic or bathypelagic squid and fish.

GENUS *ZIPHIUS* AND GENUS *TASMACETUS* These two genera each include only one species. The single species in the genus *Ziphius,* Cuvier's Beaked Whale, has a nearly cosmopolitan distribution, occurring in tropical to cold temperate waters of all oceans except the Arctic. Sometimes known as the Goose-beaked Whale due to the profile of its head, this whale's rostrum is shorter and less well defined externally than those of the other ziphiids. The two erupted teeth at the tip of the lower jaw of adult males project somewhat forward and are often colonized by stalked barnacles. The sole member of *Tasmacetus,* Shepherd's Beaked Whale, is the only ziphiid with a full set of functional teeth in the upper and lower jaws. The species was discovered as recently as 1933 and is still known mainly from stranded specimens found in various parts of the cold temperate Southern Hemisphere.

GENUS *BERARDIUS* The two species in this genus, Baird's and Arnoux's Beaked Whales, are distinguished by their large body size, steep (bluff) melon, and long beak. They are exceptional in having two pairs, rather than a single pair, of teeth, set at the front of the lower jaw. The anterior pair are large and triangular, while the posterior pair are smaller and peg-like. The teeth erupt in adults of both sexes, with the anterior pair jutting conspicuously from the tip of the jaw even when the mouth is closed. While Baird's occurs only in deep waters of the temperate and subarctic North Pacific, Arnoux's appears to be circumpolar in cool waters of the Southern Hemisphere.

GENUS *INDOPACETUS* AND GENUS *HYPEROODON* There is longstanding uncertainty surrounding the identity of Longman's Beaked Whale, the single species in the genus *Indopacetus.* It was known for many decades only on the basis of two weathered skulls—one found in eastern Australia and the other on the east coast of Africa. From the 1960s to the 1990s, scientists on cruises in the tropical Pacific and Indian Oceans reported occasional observations of moderately large beaked whales that most closely resembled the Southern and Northern

Spade-toothed Whale

Mesoplodon traversii

The Spade-toothed Whale, recently known as Bahamonde's Beaked Whale, is one of the "newest" cetaceans. A partial skull found on Robinson Crusoe Island in the Juan Fernández archipelago, Chile, led a team of South American researchers to describe a new species in 1995, calling it Bahamonde's Beaked Whale *(Mesoplodon bahamondi).* Subsequently, a New Zealand team of beaked whale specialists discovered that a specimen found on White Island, New Zealand, was genetically indistinguishable from the Juan Fernández animal and also identical to a much older tooth and jaw from the Chatham Islands, described in 1874 as *Dolichodon traversii,* and later erroneously considered to be a Strap-toothed Whale. At the time this field guide was being written, the South American and New Zealand scientists were still in the process of redescribing and renaming the species. Since the species has never been identified in the flesh, its external appearance can only be inferred. It may be similar in form to Andrews' Beaked Whale, with which it is thought to be closely related because of similarities in skull characteristics. The Spade-toothed Whale may reach body lengths of about 16 to 18 feet (5–5.5 m). Nothing is known about its behavior, reproduction, abundance, or conservation status. However, it seems unlikely that the species is either widespread or abundant, given the paucity of records of its occurrence.

Beaked whales are notoriously difficult to observe and identify, much less photograph. These exceptional images reflect the hard work and persistence of field biologists, combined with a dose of good luck.

Bottlenose Whales of the genus *Hyperoodon*. However, Bob Pitman, an American scientist who has observed these and other ziphiids many times, recently argued convincingly that these are, in fact, Longman's Beaked Whales. The Southern Bottlenose Whale is generally viewed as a southern circumpolar inhabitant of cool temperate and antarctic waters, while the Northern Bottlenose is confined to similarly high latitudes of the North Atlantic. The bottlenose whales usually have one pair of teeth at the tip of the lower jaw that erupt only in adult males.

GENUS *MESOPLODON* There are at least 14 species in the genus *Mesoplodon*. Mesoplodont teeth vary greatly in size and shape among the species; their placement also varies, ranging from the tip of the jaw in some species to well back along it in others. Mesoplodonts range in size from less than 12 feet (3.7 m) long, in the Pygmy Beaked Whale, to more than 20 feet (6 m), in the Strap-toothed Whale. Their color patterns can be complex; however, given the scarcity of observations of living specimens, the details of coloration are unknown for several species. Some have white areas that stand out from the generally dark body pigmentation (for example, a white or partly white beak). All mesoplodonts have a falcate or triangular dorsal fin, usually small or moderate in size. Their melons range from convex or slightly bulbous to low and flattened.

Beaked Whales

Cuvier's Beaked Whale *page 254*

Northern Bottlenose Whale *page 268*

Pygmy Beaked Whale *page 284*

Shepherd's Beaked Whale *page 264*

Hector's Beaked Whale *page 274*

Gervais' Beaked Whale *page 278*

True's Beaked Whale *page 276*

Gray's Beaked Whale *page 282*

Strap-toothed Whale *page 292*

Sowerby's Beaked Whale *page 280*

Stejneger's Beaked Whale *page 296*

Baird's Beaked Whale
page 260
Southern Bottlenose Whale
page 272
Longman's Beaked Whale page 266
Arnoux's Beaked Whale page 258
Hubbs' Beaked Whale page 288
Ginkgo-toothed
Beaked Whale
page 290
Blainville's Beaked Whale page 294
Andrews' Beaked Whale page 286

Cuvier's Beaked Whale

Ziphius cavirostris
G. Cuvier, 1823

Cuvier's Beaked Whale has the widest distribution of all the beaked whales. Until the last decade or two, almost everything known about this species came from examinations of stranded animals. Many survey results still have an "unidentified" category that combines mesoplodonts and Cuvier's Beaked Whale because they are so difficult to distinguish at sea, but experienced cetacean observers are becoming more confident of their ability to identify Cuvier's. As a result, scientists are now getting a better understanding of the species' distribution, abundance, and behavior. This whale is also widely known as the Goose-beaked Whale, because of the supposed goose-like profile of its head. The specific name *cavirostris* is from the Latin *cavus* for "hollow" or "concave," and refers to the prenarial basin, a concavity in the skull just ahead of the nasal bones that is unique to male Cuvier's Beaked Whales.

DESCRIPTION Cuvier's Beaked Whale has a rotund body shape. Because its beak is shorter than those of other ziphiids and its melon somewhat bulbous and bluff, the body seems almost truncated anteriorly. The melon slopes steeply but smoothly onto the short thick beak. The mouthline is short and has a broadly U-shaped profile. The dorsal fin is relatively small, falcate, and set about two-thirds of the way back on the body. The small, narrow flippers fit into "flipper pockets," slight depressions in the body wall that are also typical of the mesoplodonts. In mature males, two teeth erupt from the tip of the lower jaw and point somewhat forward as well as upward. They are exposed even when the mouth is closed and are sometimes colonized by stalked barnacles.

In general, the body is dark gray to reddish brown, with some countershading. Adult males have a white head, and the whiteness can continue onto the dorsal neck region. In females, the lighter head coloration is less pronounced, although the beak and throat can be almost white. There tends to be a dark area around the eyes, with light, crescent-shaped or chevron-like markings fore and aft. White oval scars caused by bites from cookie-cutter sharks or possibly lampreys are typically present, especially on the belly and sides, and can give a mottled appearance. Linear scarring is also evident on adult males, probably caused by the teeth of other

- ROBUST BODY
- BLUFF MELON, SLOPING SMOOTHLY ONTO SHORT BEAK
- WHITE HEAD OF ADULT MALE
- COSMOPOLITAN DISTRIBUTION

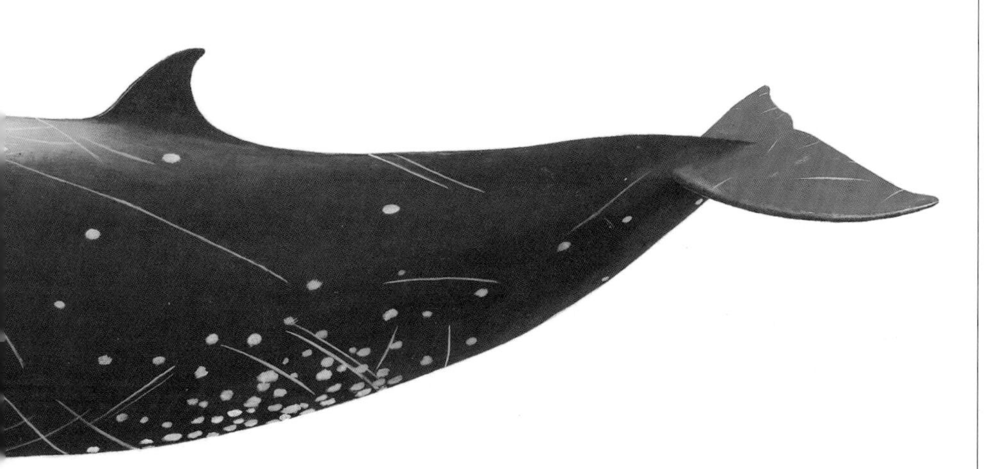

males. Calves are generally dark above and lighter below.

RANGE AND HABITAT Cuvier's Beaked Whale has a cosmopolitan distribution in deep offshore, tropical to cool temperate marine waters. Strandings have occurred in the Atlantic as far north as Massachusetts and the Shetland Islands and as far south as Tierra del Fuego and southern Africa, and in the Indo-Pacific from the Aleutians to southern Australia, New Zealand, and the Chatham Islands. This is the most common ziphiid in the Gulf of Mexico and probably also in the Caribbean and Mediterranean Seas. Sightings are seasonal in the southern Bay of Biscay, one of the only areas where Cuvier's have been studied. They are found there primarily in late winter and spring, between February and May, and are highly concentrated spatially, occurring mainly in a narrow latitudinal band (43°30'N–45°30'N) in water 5,000 to 11,500 feet (1,500–3,500 m) deep. These whales seem to have very specific habitat requirements—slope waters with steep depth gradients.

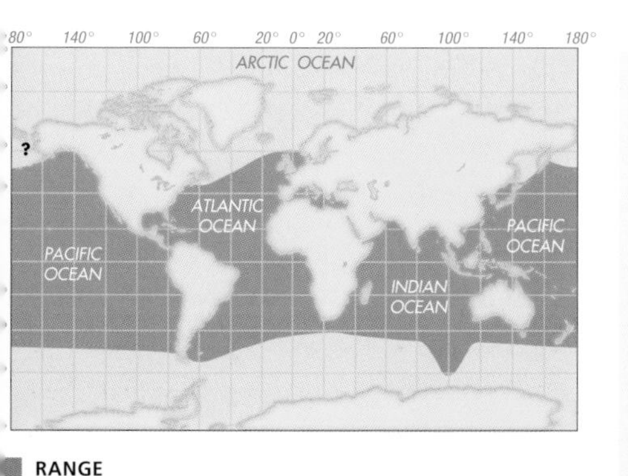

RANGE
? POSSIBLE RANGE

CUVIER'S BEAKED WHALE
FAMILY ZIPHIIDAE

MEASUREMENTS AT BIRTH
LENGTH 8′10″ (2.7 m)
WEIGHT About 550–660 lb (250–300 kg)

MAXIMUM MEASUREMENTS
LENGTH About 23′ (7 m)
WEIGHT **MALE** At least 5,700 lb (2,600 kg)
FEMALE About 6,600 lb (3,000 kg)

LIFE SPAN
Probably at least 40 years, possibly more than 60

SIMILAR SPECIES The bottlenose whales present particular identification difficulties in temperate latitudes of the North Atlantic Ocean and the Southern Hemisphere. Adult bottlenose whales are at least several feet longer, and males in particular have a steeper, more bulbous melon than Cuvier's Beaked Whales. The mysterious Indo-Pacific "bottlenose" whale, tentatively identified as Longman's Beaked Whale, may also cause confusion in warmer latitudes. In addition, various mesoplodonts occur throughout the range of Cuvier's Beaked Whale. These generally have longer, better-defined beaks, and in some species, the teeth of adult males erupt posterior to the tip of the lower jaw. Detailed views of living animals are needed to distinguish Cuvier's Beaked Whale from young and female individuals of many species of beaked whales.

BEHAVIOR Cuvier's Beaked Whales are usually observed alone or in small groups of up to about seven animals. As is true for mesoplodonts, their blow is low and diffuse, and they are fairly cryptic at the surface. On a few occasions they have been seen breaching. They can dive for a half hour or longer. On the supposition that these whales respond to the approach of a vessel by diving, the best chance for detecting them comes from looking astern. Worldwide, there are more strandings of Cuvier's Beaked Whales than of any other ziphiid. On several occasions, entire groups appear to have come ashore, most or all of them still alive.

REPRODUCTION Almost nothing is known about reproduction in this species.

An extraordinary view of the face of a Cuvier's Beaked Whale, off Cape Hatteras, North Carolina, shows the complex color pattern and distinctive head shape. The exposed teeth at the tip of the lower jaw identify this individual as an adult male.

FOOD AND FORAGING Based on studies of their stomach contents, Cuvier's Beaked Whales seem to eat mostly squid, although they also eat fish and possibly crustaceans. Most of their prey are open-ocean species that occur well below the surface, including on or near the seafloor in deep waters. Because they lack functional teeth, Cuvier's Beaked Whales presumably capture most of their prey by suction, as do most or all other ziphiids.

STATUS AND CONSERVATION Because it is difficult both to detect and to identify this species, there are few estimates of its abundance. Surveys off the west coast of the United States indicate a population in the low thousands there. The very wide distribution and common occurrence of Cuvier's Beaked Whales relative to other beaked whales could mask their vulnerability to human activities. They have been hunted in Japan, the Lesser Antilles, Indonesia, and possibly Taiwan. They are also entangled regularly in drift gillnets set in deep water. It is strongly suspected that these whales are vulnerable to particular types of noise, a fact that could jeopardize their survival in areas of intense industrial, military, or perhaps even recreational activity. Mass strandings in the Mediterranean Sea and the Bahamas have been linked to military exercises involving powerful underwater sound transmissions.

LEFT: *Linear scarring on the body of this Cuvier's Beaked Whale is probably the result of wounds inflicted by the teeth of companions or competitors.* **RIGHT:** *The blow of Cuvier's Beaked Whale is low and diffuse, and its beak is short compared with those of most other beaked whales.*

Arnoux's Beaked Whale

Berardius arnuxii
Duvernoy, 1851

This large beaked whale, whose distribution is circumpolar in the Southern Hemisphere, is so similar to Baird's Beaked Whale, which lives in the North Pacific, that many scientists are skeptical about their status as separate species. The two forms are believed to be indistinguishable at sea except by virtue of their occurrence in different hemispheres. Based on the measurements of a few specimens found stranded or trapped in the ice, Arnoux's Beaked Whales appear not to grow as large as Baird's Beaked Whales.

DESCRIPTION Arnoux's Beaked Whale has a long, almost tubular body with a small dorsal fin set far behind the center of the back. The fin is usually blunt or rounded at the peak and has a somewhat falcate rear margin. The flippers have rounded tips. The melon is prominent and bulbous, sloping smoothly to the long beak, and the mouthline is long and curves upward at the back. The lower jaw extends past the upper jaw, so that the two large triangular teeth at the beak tip are exposed even when the mouth is closed. Two smaller teeth erupt behind the larger teeth but are concealed inside the mouth. Both males and females have erupted teeth.

Most of the body is slate gray to light brown, with the head slightly lighter in color. With age, these whales acquire numerous white linear scars on the melon, back, and sides. Older animals sometimes also have scars on the belly. Overall, this scarring can give the skin a marbled appearance.

RANGE AND HABITAT Arnoux's Beaked Whale probably has a circumpolar distribution in the Southern Hemisphere, ranging from the Antarctic Continent and ice edge, at about 78°S, northward to about 34°S. Most records are from south of 40°S. Individuals have sometimes become trapped by ice and forced to overwinter in the Antarctic.

SIMILAR SPECIES The Southern Bottlenose Whale occurs throughout most of Arnoux's Beaked Whale's range; as the next-largest beaked whale in the Southern Hemisphere, it is probably the species most likely to cause confusion in identification. Southern Bottlenose Whales have a bluffer melon that sometimes overhangs the beak, a shorter beak with no exposed teeth at the tip, and a more erect and falcate dorsal fin, set farther forward on the body.

- LONG GRAY BODY WITH MANY LINEAR SCARS
- BULBOUS MELON AND LONG BEAK
- TWO PAIRS OF TEETH AT TIP OF LOWER JAW, ANTERIOR PAIR VISIBLE WHEN MOUTH IS CLOSED
- SMALL DORSAL FIN, SET FAR BACK ON BODY
- FOUND MAINLY IN ANTARCTIC AND SUBANTARCTIC WATERS

BEHAVIOR, REPRODUCTION, AND FORAGING Arnoux's Beaked Whales are generally shy and difficult to observe, much less to identify, at sea. Although considered gregarious, they are usually seen in close-knit groups of 10 or fewer, with occasional aggregations of a few dozen. On one occasion in the Antarctic, researchers followed a school of about 80 whales for several hours. The whales eventually split into groups of 8 to 15 and dispersed into the loose pack ice. Most of the dives that were timed by the researchers lasted 15 to 25 minutes; one group dived for longer than an hour and traveled more than 4 miles (6 km) underwater before resurfacing. Arnoux's Beaked Whales are presumed to have a diet similar to that of Baird's Beaked Whales, which is dominated by squid and deep-sea fish. Essentially nothing is known about reproduction in this species.

STATUS AND CONSERVATION Arnoux's Beaked Whale is sighted far less frequently than the Southern Bottlenose Whale, but there are no good abundance estimates. It has not been exploited and is presumably in no immediate danger.

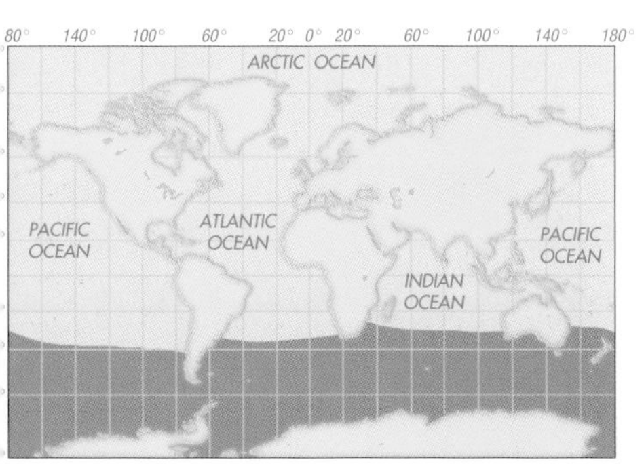

RANGE

ARNOUX'S BEAKED WHALE

FAMILY ZIPHIIDAE

MEASUREMENTS AT BIRTH

LENGTH About 13′ (4 m), possibly longer

WEIGHT Unavailable

MAXIMUM MEASUREMENTS

LENGTH **MALE** Less than 32′ (9.8 m)
FEMALE 32′ (9.8 m)

WEIGHT Unavailable

LIFE SPAN Unavailable

Baird's Beaked Whale

Berardius bairdii
Stejneger, 1883

Baird's Beaked Whale, sometimes called the Giant Bottlenose Whale, is the largest of the beaked whales. In waters off Japan, its large size and occurrence within reach of shore stations have made this species a target of commercial whaling. Particularly in Chiba Prefecture, Baird's is highly esteemed for its meat, which is eaten fresh or dried. The Japanese name for Baird's Beaked Whale is *tsuchi-kujira, tsuchi* referring to an old-fashioned wooden hammer shaped like a bottle, and *kujira* meaning "whale." Like other beaked whales, Baird's lives in deep water and is a challenging research subject. Thus far, studies of the species have consisted mainly of dissections of animals killed and brought to land in Japan for processing and subsequent sale in local markets.

DESCRIPTION Baird's Beaked Whale has a long, almost tubular body, with a maximum girth typically 50 to 60 percent of its length. The dorsal fin, set far behind midback, is small and has a straight or slightly falcate rear margin and usually a somewhat rounded tip. The flippers are rounded on the ends and not tapered. The melon is prominent and bulbous (but not bluff, as in the bottlenose whales), and slopes smoothly onto a long beak. The mouthline is long and sinuous, curving upward at the back. Because the lower jaw extends farther forward than the upper jaw, the two large triangular teeth at the tip of the lower jaw are visible when the mouth is closed. These teeth are occasionally colonized by stalked barnacles, but they nevertheless flash brightly in sunlight and are usually noticeable. Two smaller teeth, concealed inside the closed mouth, erupt behind the larger pair. Both males and females have erupted teeth.

The body is evenly colored, ranging from slate gray to black. There are often cloudy patches on the throat, near the umbilicus, and in the genital region. As these whales age, they acquire numerous white linear scars on the melon, back, and sides, and sometimes on the belly in older animals. This scarring can give the skin a marbled appearance. Patches of diatoms sometimes give the skin a brown or greenish-brown sheen.

RANGE AND HABITAT Baird's Beaked Whale is a deepwater species that occurs near shore only in areas where the continental shelf is very narrow. It occurs in highest densities in the vicinity of submarine escarpments and seamounts and over

- LONG GRAY BODY WITH MANY LINEAR SCARS
- BULBOUS MELON SLOPING TO LONG BEAK
- TWO PAIRS OF TEETH AT BEAK TIP, LARGER ANTERIOR PAIR VISIBLE WHEN MOUTH IS CLOSED
- SMALL DORSAL FIN SET FAR BACK ON BODY
- LIMITED TO COOL WATERS OF NORTH PACIFIC

continental slopes. Baird's distribution is thought to be continuous across the North Pacific, north of about 35°N. In the western North Pacific, it is present year-round in the Sea of Japan, along the southern Kuril Islands, and in the Sea of Okhotsk; it generally does not occur south of 34°N or in the Yellow and East China Seas. During winter and spring in the Sea of Okhotsk, these whales have been observed in the drift ice, sometimes even in narrow cracks of open water. In the eastern North Pacific, Baird's ranges from the southern half of the Bering Sea, along the Aleutian Islands, and south normally to about 28°N off Baja California, although it is known to enter the Gulf of California occasionally.

SIMILAR SPECIES The ranges of several mesoplodonts overlap with that of Baird's Beaked Whale, most notably Stejneger's and Hubbs' Beaked Whales. Both species are considerably smaller, with a less bulbous melon, a relatively shorter beak, and teeth positioned well behind

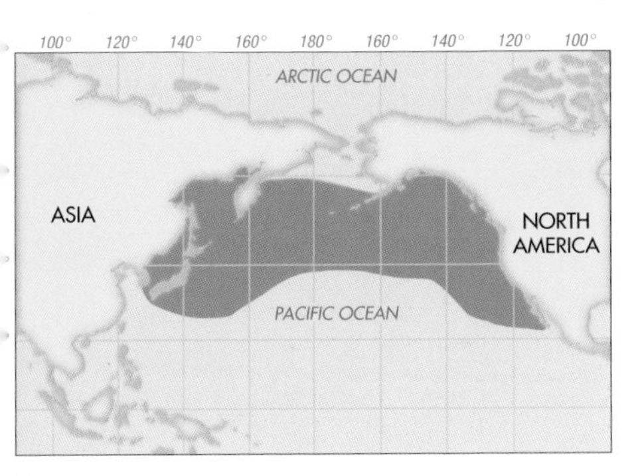

RANGE

BAIRD'S BEAKED WHALE
FAMILY ZIPHIIDAE

MEASUREMENTS AT BIRTH
LENGTH 15′ (4.5–4.6 m)
WEIGHT Unavailable

MAXIMUM MEASUREMENTS
LENGTH **MALE** About 39′ (12 m)
FEMALE 42′ (12.8 m)
WEIGHT **MALE** More than 20,000 lb (10,000 kg)
FEMALE More than 24,000 lb (11,000 kg)

LIFE SPAN
MALE 84 years
FEMALE 54 years

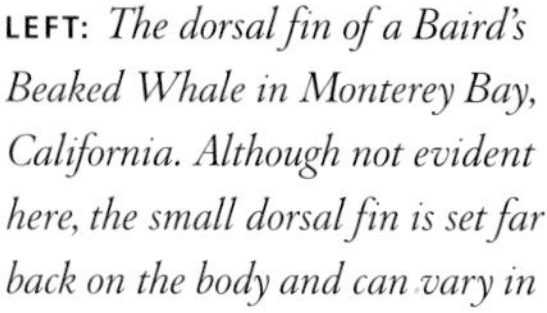

LEFT: *The dorsal fin of a Baird's Beaked Whale in Monterey Bay, California. Although not evident here, the small dorsal fin is set far back on the body and can vary in shape from triangular to slightly falcate.*

RIGHT: *An exceptional view of a Baird's Beaked Whale breaching. Its long beak, long body, small flippers, and narrow caudal peduncle are all clearly evident. The rear margin of the flukes lacks a median notch, as is typical in beaked whales.*

the tip of the lower jaw and exposed only in adult males. Cuvier's Beaked Whale also occurs throughout the range of Baird's Beaked Whale; it has a short beak (shorter even than the mesoplodonts), and adult males have a light-colored head. In the lowest latitudes of its range, Baird's may be confused with "tropical bottlenose whales" (see Longman's Beaked Whale). These whales have a proportionally larger, taller, more pointed dorsal fin and a bluffer melon.

BEHAVIOR Baird's Beaked Whales usually travel in close-knit groups of up to 10 individuals, with a maximum size of about 30. When chased, members of the group remain together, surfacing and diving synchronously. In spite of the fact that Baird's has been hunted for a very long time, the composition of pods has not been well studied. Scientists are puzzled by the fact that more than two-thirds of the whaling catch of these whales are males, even though the females' larger size should make them a more desirable target for whalers. Mass strandings are rare for this species. However, on one occasion in Mexico, a group of seven animals, including four mature males and two mature females, stranded together.

Baird's Beaked Whales tend to be shy of vessels. They sometimes breach, spyhop, and slap the water's surface with their flippers or flukes. They spend most of their time, however, far below the surface on deep dives. They typically remain submerged for a half hour, sometimes an hour or longer, and usually spend only about five minutes at the surface between dives. While feeding, they routinely dive to depths of more than 3,300 feet (1,000 m). These exceptional diving capabilities, however, apparently do not keep them out of reach of predators. Their remains have been found in the stomachs of several Killer Whales taken off Japan, and some of the linear scarring on their bodies has the characteristic spacing of Killer Whale teeth.

REPRODUCTION Much uncertainty exists about reproduction in Baird's Beaked Whale. Studies of animals killed in the Japanese coastal hunt indicate that gestation could last either 10 months or 17 months. Females probably do not give birth more than once every three years. This species is exceptional among cetaceans in that males mature faster and appear to live considerably longer than females, so that at any one time there are substantially more sexually mature males than mature females in the population. Old males may play an important social role. Some researchers have speculated that males help care for weaned calves, thereby allowing females to give birth more often.

FOOD AND FORAGING Baird's Beaked Whales feed on squid and deep-sea fish. Most feeding occurs on or near the seafloor at depths of 3,300 to 10,000 feet (1,000–3,000 m). An important summer and autumn feeding ground of Baird's Beaked Whales in waters off Japan is apparently influenced in some way by the cold subsurface Oyashio Current, which may cause upwelling and thus high productivity.

STATUS AND CONSERVATION There is no estimate for the overall population of Baird's Beaked Whales, but there are probably no more than a few tens of thousands. Only about 400 are thought to occur off the west coast of the United States (not including those in Alaskan waters). Surveys conducted in Japan produced estimates of about 6,000 whales there, two-thirds of which occurred off the Pacific coast. In recent years, 54 Baird's Beaked Whales have been taken annually in Japan under a national quota. Baird's Beaked Whales were exploited much more extensively in the past. Nearly 4,000 were reportedly killed in Japan between 1948 and 1986 (with a peak catch of well over 300 reported in 1952). In addition, Soviet whalers reported killing 176 of these whales between 1933 and 1974, and whalers off North America killed about 60 between 1912 and 1966. A few animals also continue to die each year in fishing gear, especially drift gillnets set in deep water. However, the species is not considered threatened.

A close-knit group of Baird's Beaked Whales off Guadalupe Island, Mexico. The scarring on the animals' backs is typical of adults, probably caused by the teeth of fellow Baird's Beaked Whales.

Shepherd's Beaked Whale

Tasmacetus shepherdi
Oliver, 1937

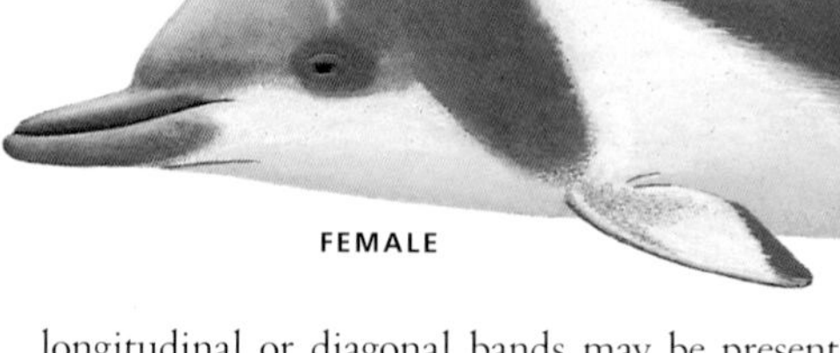

Shepherd's Beaked Whale was discovered only in 1933 and remains one of the least-known cetaceans. Virtually everything scientists know about it comes from a few stranded carcasses. Its appearance and behavior at sea have not been reliably documented. Shepherd's Beaked Whale is unique among the beaked whales in having numerous teeth in both the upper and lower jaws.

DESCRIPTION Shepherd's Beaked Whale has a cigar-shaped body, similar to that of mesoplodonts. However, its melon is much more bluff, more like that of Cuvier's Beaked Whale. The long, well-demarcated beak narrows to a pointed tip. The dorsal fin is about 12 to 14 inches (30–35 cm) high, falcate, and positioned about two-thirds of the way back from the beak tip. The flippers are fairly short and tapered. There are two large teeth at the tip of the lower jaw and 17 to 27 pairs of smaller teeth lining both the upper and lower jaws.

Shepherd's Beaked Whale has a dark back and lighter sides and belly. The front part of the head may also be noticeably lighter. Light areas continuous with the belly coloration intrude onto the sides above and forward of the flippers. Light longitudinal or diagonal bands may be present elsewhere on the sides and flanks. The dorsal fin, flippers, and flukes are apparently all dark.

RANGE AND HABITAT Shepherd's Beaked Whale is considered an offshore, oceanic species. Its distribution, as inferred from stranding records, is probably circumpolar in cold temperate waters of the Southern Hemisphere. It has not been reported north of 30°S. Strandings have been documented in Peninsula Valdés and Tierra del Fuego, Argentina; Tristan da Cunha Island; South Africa; South Australia; New Zealand and the Chatham Islands; and the Juan Fernández Islands off Chile. The majority of records have come from New Zealand.

SIMILAR SPECIES Other ziphiids that occur in southern cold temperate waters may cause confusion. Cuvier's Beaked Whale has a much shorter and less sharply demarcated beak. Both Arnoux's Beaked Whale and the Southern Bottlenose Whale are larger (especially Arnoux's), with thicker, sturdier beaks that are not tapered at the tip, and sometimes with bluffer melons. Mesoplodonts that probably overlap with Shepherd's

- RELATIVELY BLUFF MELON
- WELL-DEMARCATED, LONG, ALMOST POINTED BEAK
- DORSAL FIN SET FAR BACK ON BODY
- LIMITED TO COLD TEMPERATE SOUTHERN HEMISPHERE IN DEEP OFFSHORE WATERS

Beaked Whale include Gray's, Andrews', and Hector's Beaked Whales and the Strap-toothed Whale. Most mesoplodonts are smaller and have less bulbous melons.

BEHAVIOR, REPRODUCTION, AND FORAGING Nothing is known for certain about this species' social organization, behavior, or reproduction. However, the scarcity of at-sea observations suggests that it occurs only in small groups that are difficult to detect. Alternatively, it may be extremely rare, avoid vessels, and dive deeply, spending little time at the surface. Unlike the other ziphiids, Shepherd's Beaked Whale apparently eats fish at least as often as it eats squid. The stomach of a stranded adult female in Argentina contained only bottom-dwelling fish.

STATUS AND CONSERVATION Nothing is known about the species' abundance, and no meaningful evaluation of its conservation status is possible. There is no evidence of deliberate exploitation or incidental mortality in fishing gear.

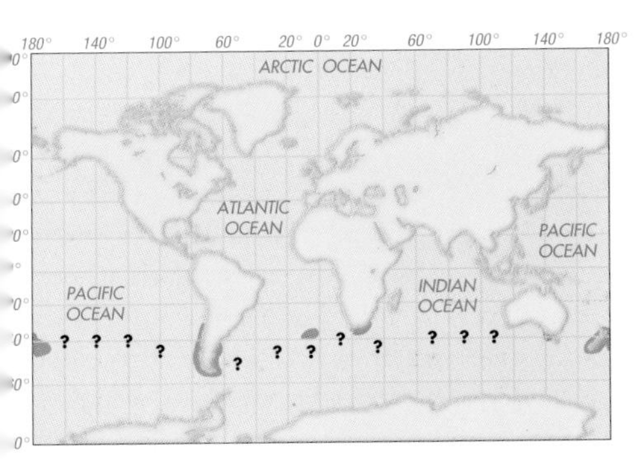

SHEPHERD'S BEAKED WHALE

FAMILY ZIPHIIDAE

MEASUREMENTS AT BIRTH
LENGTH Possibly about 10′ (3 m)
WEIGHT Unavailable

MAXIMUM MEASUREMENTS
LENGTH 19′6″–23′(6–7 m)
WEIGHT Unavailable

LIFE SPAN Unavailable

Longman's Beaked Whale

Indopacetus pacificus
(Longman, 1926)

This species represents one of cetology's greatest long-standing mysteries. It was described by H. A. Longman (as *Mesoplodon pacificus*) on the basis of a single skull and jaw found on a Queensland, Australia, beach in 1882. In the 1960s, Joseph Curtis Moore, a beaked whale expert from the Field Museum in Chicago, assigned it to a new genus, *Indopacetus.* Moore included *"Indo-"* in the genus name because he had just learned of a second specimen, a skull found on the coast of Somalia, Africa. (The genus name indicates that these whales occur in both the Indian and Pacific Oceans.) Some authorities, however, place the species in the genus *Mesoplodon,* pending a thorough review and revision of the group. A mysterious cetacean that has been seen in tropical Indo-Pacific waters and often called the "tropical bottlenose whale" may in fact be Longman's Beaked Whale.

DESCRIPTION There are no confirmed observations of a living Longman's Beaked Whale. However, some field researchers are convinced that the "bottlenose whales" seen and photographed in the tropical Pacific and Indian Oceans, sometimes tentatively identified as a species of *Hyperoodon,* are Longman's Beaked Whales. These animals look very much like Southern Bottlenose Whales. They are about 23 to 26 feet (7–8 m) long, with a falcate pointed dorsal fin, a bulbous (sometimes overhanging) melon, and a moderately long beak. The skulls currently providing the basis for recognizing the existence of Longman's Beaked Whale indicate that there are only two teeth, both near the tip of the lower jaw. Presumably these would erupt through the gum only in adult males. Erupted teeth have not been observed in sightings of "tropical bottlenose whales."

The color pattern of "tropical bottlenose whales" seems to be variable but is dominated by tan to grayish-brown tones. Younger individuals often have a pale melon and white sides.

RANGE AND HABITAT Longman's Beaked Whale occurs in tropical waters of the Indian and Pacific Oceans and is almost certainly a pelagic species. The reported "tropical bottlenose whales" have been seen mainly in waters warmer than 79°F (26°C). One observation in the Gulf of Mexico suggests that the species may also occur in the tropical Atlantic, but this has yet to be confirmed.

LONGMAN'S BEAKED WHALE

FAMILY ZIPHIIDAE

No reliable measurements or estimations of life span are yet available for this species.

SIMILAR SPECIES Here it is assumed that the animals reported as "tropical bottlenose whales" are Longman's Beaked Whales. Although several mesoplodonts overlap Longman's in distribution, they should be easy to distinguish because of their smaller body size, proportionally smaller dorsal fin, lack of visible blow (usually), and relatively flat or at least less bulbous melon. Southern Bottlenose Whales, which would be very hard to distinguish from these animals, are not known to move north of about 30°N. However, recent discoveries of new specimens of Longman's Beaked Whales from South Africa suggest that there may be some range overlap in the southern Indian Ocean. Cuvier's Beaked Whale also occurs in the tropics but has a comparatively shorter beak and sloping melon.

BEHAVIOR, REPRODUCTION, AND FORAGING Reported group sizes of "tropical bottlenose whales" have been larger than those reported for other beaked whales in the tropical Indo-Pacific, ranging into several tens and occasionally as high as 100, with an average of roughly 15 to 20. Pods appear close-knit and cohesive. Dives last up to 18 to 25 minutes. These animals are seen occasionally in association with Short-finned Pilot Whales, and less often with bottlenose dolphins. They have been observed breaching, and their low, bushy blow is visible. Nothing is known about reproduction or feeding in Longman's Beaked Whale.

STATUS AND CONSERVATION Longman's Beaked Whale is thought to be rare. There are no reports of incidental mortality or direct hunting.

■ DISTRIBUTION APPARENTLY RESTRICTED TO TROPICAL PACIFIC AND INDIAN OCEANS

★ STRANDINGS

? POSSIBLE RANGE

Northern Bottlenose Whale

Hyperoodon ampullatus
(Forster, 1770)

These deep-diving whales are the best-studied of the beaked whales, in part because of their long history of commercial exploitation. Several 19th-century British and Norwegian commercial whalers were literate naturalists, and they provided detailed descriptions of the whales they hunted. The whaling industry also made large numbers of carcasses available for scientific study. During the last two decades, with whaling for this species prohibited, a group of scientists based at Dalhousie University in Nova Scotia has carried out a singularly intensive, long-term study of living bottlenose whales. Their work provides a classic example of the contrast between what can be learned by studying whales alive in their natural habitat versus what can be learned by hunting them down and dissecting their remains. The genus name *Hyperoodon* is a misnomer, referring to what were mistakenly considered small teeth on the palate but are actually nothing more than bony bumps. The specific name, derived from the Latin *ampulla* for "bottle" or "flask," refers to the shape of the beak.

DESCRIPTION The Northern Bottlenose Whale is a long but rotund animal with a pronounced beak and a bluff melon. The dorsal fin is up to about 12 inches (30 cm) high and is situated well behind the middle of the back. It is falcate and usually pointed at the tip. The flippers are small and rounded at the tips. The appearance of the head distinguishes adult males. It is white and flattened in front, while that of females is gray and bulbous. (That of subadult males is intermediate.) Males grow about 3 feet (1 m) longer than females. There are two teeth at the tip of the lower jaw. They erupt only in adult males and are oriented slightly forward. Stalked barnacles, of a species also common on ships and driftwood, sometimes attach to the teeth.

Adult coloration is countershaded: The back is mid- to dark gray and the belly lighter. There are white or yellowish oval scars on the body that become more extensive with age, especially on the belly and sides. The only patterning is on the heads of adults: Males have a white patch on the front of the melon that can extend back to the eyes; females have a white neck collar behind the blowhole.

RANGE AND HABITAT The Northern Bottlenose Whale is found only in the North Atlantic

- LARGEST BEAKED WHALE IN NORTH ATLANTIC
- BLUFF MELON AND PROMINENT BEAK
- WHITE MELON ON ADULT MALE
- WHITISH NECK COLLAR ON ADULT FEMALE
- FOUND ONLY IN MIDDLE TO HIGH LATITUDES OF NORTH ATLANTIC

Ocean, occurring mainly in cold temperate to subarctic waters. It inhabits Davis Strait, the Labrador Sea, and the Scotian Shelf in the west, and the Greenland and Barents Seas in the east; it generally is not common in the Gulf of St. Lawrence, Hudson Bay, or the North and Baltic Seas. It is rarely seen farther south than Nova Scotia, the Azores, or the Iberian Peninsula. References to "bottlenose whales" in the northern North Pacific are, in most instances, Baird's Beaked Whales. The Northern Bottlenose Whale prefers deep water (more than 3,300 ft/1,000 m) and sometimes occurs in or near pack ice. Its movement patterns are complicated and do not seem to follow the "usual" north–south, summer–winter cycle of other well-studied larger cetaceans. It is seen year-round in and near the Gully, a submarine canyon off Nova Scotia. These animals may constitute a separate "resident" population. Bottlenose whales are seasonally present in the southern Bay of Biscay (May–August), and there are also large numbers in much higher latitudes of the Barents and Greenland Seas at this time. Again, this may mean that there is a "resident" low-latitude population in the eastern Atlantic.

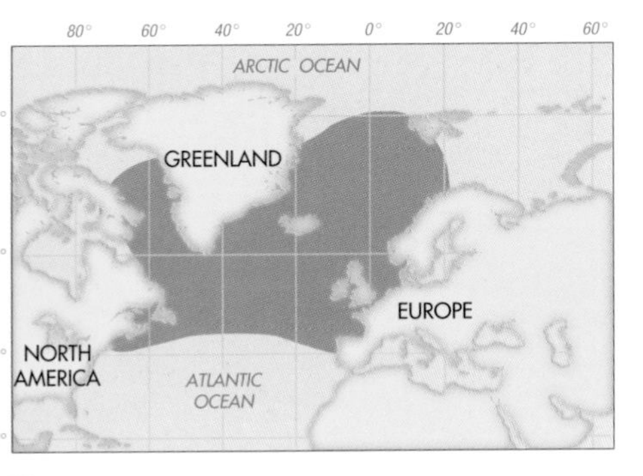

RANGE

NORTHERN BOTTLENOSE WHALE
FAMILY ZIPHIIDAE

MEASUREMENTS AT BIRTH
LENGTH 10′–11′6″ (3–3.5 m)
WEIGHT Unavailable

MAXIMUM MEASUREMENTS
LENGTH **MALE** 32′ (9.8 m)
FEMALE 29′ (8.7 m)
WEIGHT 13,000–17,000 lb (5,800–7,500 kg)

LIFE SPAN
At least 37 years

SIMILAR SPECIES The two other beaked whales most likely to be encountered in much of the range of the Northern Bottlenose, especially in the more temperate portions, are Cuvier's and Sowerby's Beaked Whales, both of which are considerably smaller. Adult male Cuvier's also have a white head, but for both sexes the much less bluff melon and shorter, less well-defined beak should make Cuvier's distinguishable at close range. Sowerby's Beaked Whale has a much flatter melon than the Northern Bottlenose Whale, and the adult male's tusk-like teeth are exposed on the sides of the closed mouth.

BEHAVIOR Northern Bottlenose Whales occur in groups averaging about four individuals. Few groups contain more than 10. Occasionally, many such groups are within sight at the same time. There is some segregation, as groups can consist exclusively of mature and maturing males, females, and young, or a combination that includes one or several mature males. An unusual feature of Northern Bottlenose Whales is their tendency to approach and swim around a stationary or slow-moving vessel. Their behavior at the surface ranges from periods of quiescence to erratic bursts of rapid swimming in different

LEFT: *An adult Northern Bottlenose Whale is accompanied by a much darker-colored calf in the Gully, a protected marine area off Nova Scotia.*

RIGHT: *A Northern Bottlenose Whale lobtailing. Beaked whales usually have no notch between the flukes.*

OPPOSITE: *This young Northern Bottlenose Whale has the bluff melon and well-defined beak typical of the species. Also, the light coloration above the eye and on the melon anticipate the more sharply defined adult color pattern.*
LEFT: *Northern Bottlenose Whales viewed from the crow's nest of a research vessel in the Gully, off Nova Scotia, where a study led by Hal Whitehead at Dalhousie University has provided a wealth of information on this species.*

directions. Most dives last less than 10 minutes, but about 10 percent last more than an hour. Whalers claimed that Northern Bottlenose Whales sometimes remained submerged for up to two hours after being harpooned. Whales may dive routinely to depths of 2,600 feet (800 m) or more, and maximum dive depths are at least 5,000 feet (1,500 m).

REPRODUCTION Most births of Northern Bottlenose Whales appear to take place in spring or early summer (April–June), although newborn-size whales are seen in the Gully as late as August. Gestation and lactation each last at least a year, the latter perhaps much longer. Although whaling data suggest that females give birth every other year, it is likely that the average calving interval is more than two years.

FOOD AND FORAGING The diving capabilities of Northern Bottlenose Whales allow them to forage on the bottom in very deep water. A particular species of squid, *Gonatus fabricii*, forms the bulk of their diet in all areas where stomach contents have been examined. They also eat herring, various deep-sea fish, shrimp, and even sea cucumbers and sea stars.

STATUS AND CONSERVATION Bottlenose whales have spermaceti (a waxy substance used in cosmetics and other items) in their heads, and it, together with their body oil, made them worthy targets of 19th- and early 20th-century whalers. The combined recorded catch by the British and Norwegian fleets in the last quarter of the 19th century exceeded 22,000. This figure does not include the many killed or seriously injured whales that were lost. Norway continued hunting Northern Bottlenose Whales through 1973, when the British market for whale meat (used as pet food) was legally closed. There is no doubt that whaling reduced abundance in some areas, especially in the northeastern Atlantic, but the extent of the depletion was hotly debated. All populations have now been protected for more than 25 years, and substantial recovery may have occurred. A survey estimate in the late 1980s indicated that there were at least 5,000 Northern Bottlenose Whales in waters around Iceland and the Faeroe Islands. About 230 whales spend some of their time in the Gully. Although these whales are not known to be killed in large numbers in fishing gear, they may be threatened by disturbance from oil and gas development near the Gully.

Southern Bottlenose Whale

Hyperoodon planifrons
Flower, 1882

The Southern Bottlenose Whale is less well known than its close relative, the Northern Bottlenose Whale. It has never been hunted commercially, nor has it been the subject of any sort of dedicated field study. Consequently, most of what is known about it comes from stranded specimens or opportunistic observations made by scientists engaged in research on other species. Sightings of "tropical bottlenose whales" in the tropical Pacific and Indian Oceans have been a source of confusion and speculation among field researchers for several decades. Although these animals look very much like Southern Bottlenose Whales, observers who have seen both types are convinced that they belong to a different species, possibly the elusive Longman's Beaked Whale.

DESCRIPTION The Southern Bottlenose Whale has a long but robust body. The beak is well demarcated from the melon and may appear stubbier on adult males than on females or juveniles. The melon of adults is bulbous to bluff. The dorsal fin is falcate and prominent, situated well behind midbody. There is a single pair of teeth at the tip of the lower jaw that erupt only in adult males; they are visible outside the closed mouth. The body is generally tan or grayish brown. In some individuals, the melon is pale and clearly demarcated from the darker color of the back. The pale area extends behind the blowhole. The back and flanks are dark, and the belly may be noticeably lighter than the sides. Adult males can appear almost white at sea.

RANGE AND HABITAT The Southern Bottlenose Whale is circumpolar in the Southern Hemisphere, from the Antarctic north to at least Valparaiso (Chile), North Island (New Zealand), and New South Wales (Australia) in the South Pacific; to Western Australia and about 31°S in the western Indian Ocean; and to Cape Province (South Africa) and Rio Grande do Sul (Brazil) in the South Atlantic. It apparently is most common south of 30°S, especially between 58°S and 62°S in the Atlantic and eastern Indian Ocean sectors, and is seen as far south as 73°S in the Ross Sea. In summer, the highest densities of these whales are within 60 nautical miles of the ice edge. The seasonality of sightings off KwaZulu-Natal, South Africa, suggests that the whales become common there in summer but are absent in winter.

- BULBOUS OR BLUFF MELON AND WELL-DEFINED BEAK
- PALE MELON ON SOME INDIVIDUALS
- SOUTHERN CIRCUMPOLAR DISTRIBUTION

SOUTHERN BOTTLENOSE WHALE

FAMILY ZIPHIIDAE

MEASUREMENTS AT BIRTH

LENGTH Unavailable

WEIGHT Unavailable

MAXIMUM MEASUREMENTS

LENGTH **MALE** 23′ (6.9 m)
FEMALE 25′ (7.5 m)

WEIGHT Unavailable

LIFE SPAN Unavailable

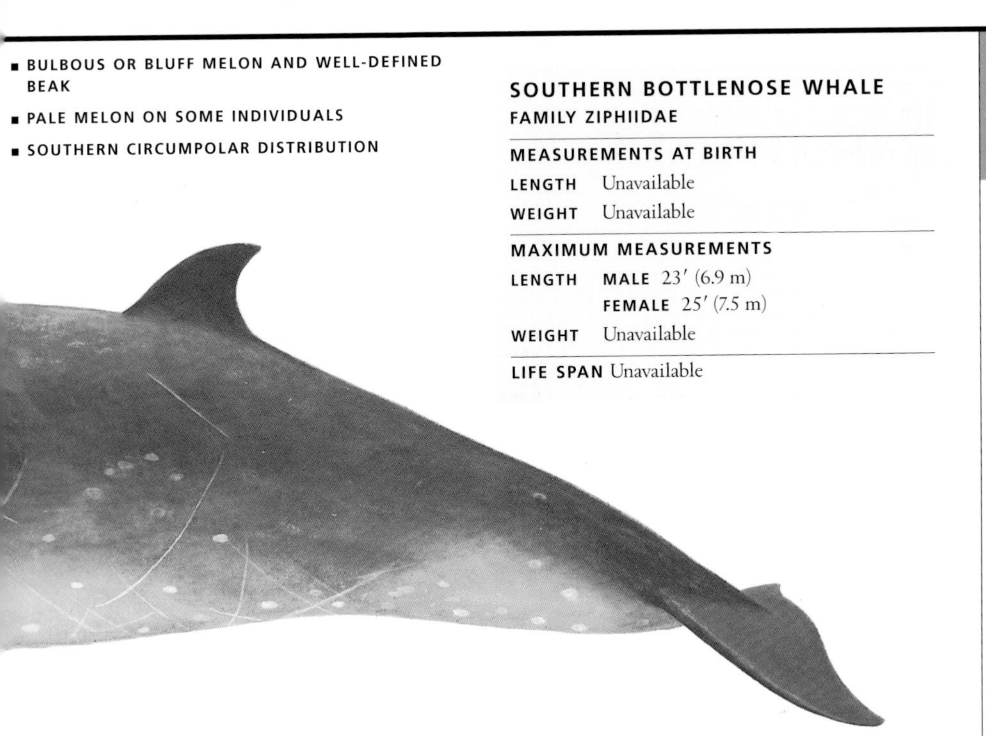

SIMILAR SPECIES The Southern Bottlenose Whale is most likely to be confused with Arnoux's Beaked Whale, which overlaps its entire range. Arnoux's is somewhat larger and has a longer beak, a less bulbous melon, and a smaller dorsal fin. Cuvier's Beaked Whale, which occurs at least occasionally within the range of the Southern Bottlenose, has a much less bulbous or bluff melon and an abbreviated beak.

BEHAVIOR, REPRODUCTION, AND FORAGING Southern Bottlenose Whales occur in small groups, usually of fewer than 10 individuals. These whales breach occasionally, and their bushy blows are readily visible in most conditions. Almost nothing is known about reproduction in the Southern Bottlenose Whale. Based on a few observations of newborns, it is assumed that calving off South Africa takes place in spring or early summer. Southern Bottlenose Whales are squid specialists. Some of the antarctic and subantarctic squid included in their diet are also prey of Sperm Whales, but Southern Bottlenose Whales tend to consume smaller size classes of these species.

ARCTIC OCEAN
ATLANTIC OCEAN
PACIFIC OCEAN
INDIAN OCEAN
PACIFIC OCEAN

RANGE

STATUS AND CONSERVATION Although poorly known, this species is thought to be the most abundant beaked whale in the Antarctic, based on the frequency of sightings. It has not been exploited commercially and is not regularly killed incidentally in fisheries. There are thought to be about 500,000 Southern Bottlenose Whales south of the Antarctic Convergence in summer (January).

Hector's Beaked Whale

Mesoplodon hectori
(Gray, 1871)

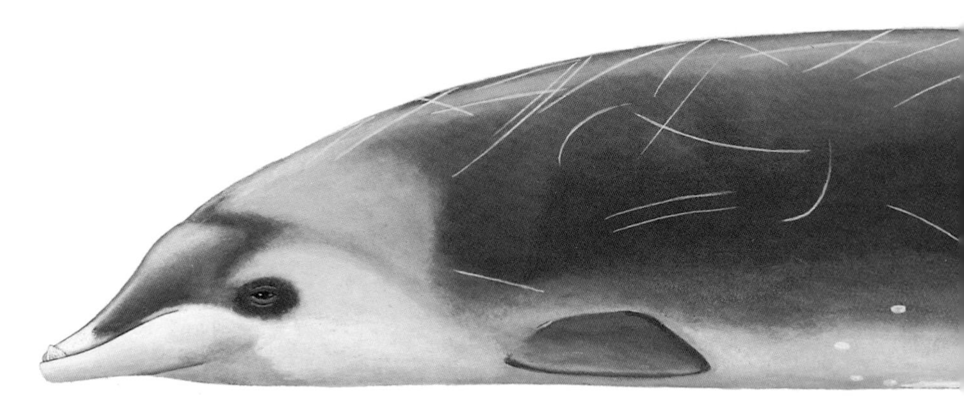

Hector's Beaked Whale is one of two cetacean species (the other being Hector's Dolphin) named after Sir James Hector, founding curator of the colonial museum in Wellington, New Zealand. Hector's Beaked Whale was long viewed as occurring only in the Southern Hemisphere, but during the 1970s several strandings and sightings were reported from southern California. Although based on these records much of the literature now states that the species occurs there as well, recent analyses of the DNA of the stranded animals from California suggest that they belonged to a different, as yet undescribed, species.

DESCRIPTION The small Hector's Beaked Whale has the typical mesoplodont body form, with a small head, long trunk, and short tail. The dorsal fin is small, triangular, and slightly falcate. The erupted teeth of adult males, positioned about an inch behind the tip of the lower jaw, are large and prominent. Their exposed portion is shaped roughly like an isosceles triangle.

Published descriptions of color pattern in this species are confounded by the fact that many of the details are based on specimens from California that now appear to represent a different species.

RANGE AND HABITAT Hector's Beaked Whale is found in cool temperate waters of the Southern Hemisphere, with strandings from the east and west coasts of South America, the Falkland Islands, Cape Province in South Africa, Tasmania, and North and South Islands of New Zealand. A high proportion of the documented strandings of this species have been in New Zealand.

SIMILAR SPECIES The species most likely to cause confusion are the other mesoplodonts of the cool temperate Southern Hemisphere, including Andrews', Gray's, Blainville's, the Ginkgo-toothed, and True's Beaked Whales and the Strap-toothed Whale. Of these, adult male Blainville's Beaked and Strap-toothed Whales are probably the easiest to distinguish, the former by their highly arched lower jaw and the latter by their considerably larger size and distinctive, strap-like teeth. The other species are less likely to be distinguishable at sea, although the exposed teeth in the lower jaw of adult males are farther back along the jaw in all except True's.

BEHAVIOR, REPRODUCTION, AND FORAGING Almost nothing is known about the social

- SMALL BODY AND SMALL TRIANGULAR DORSAL FIN
- PROBABLY LIMITED TO COOL TEMPERATE WATERS OF SOUTHERN HEMISPHERE

organization, behavior, or reproduction of this species. Like other mesoplodonts, Hector's Beaked Whale is presumably a squid specialist.

STATUS AND CONSERVATION Judging by the frequency of strandings, these whales may be more common in the South Pacific around New Zealand than elsewhere. However, the lack of reliable identifications at sea makes it impossible to evaluate the conservation status of Hector's Beaked Whale. It has not been hunted, nor is it known to experience significant incidental mortality in fishing gear.

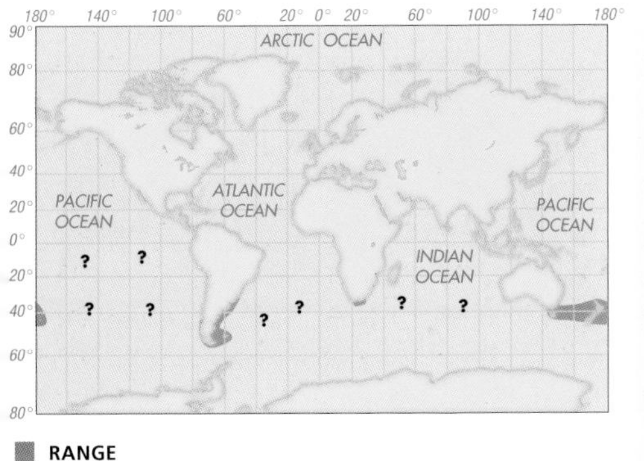

HECTOR'S BEAKED WHALE

FAMILY ZIPHIIDAE

MEASUREMENTS AT BIRTH

LENGTH Less than 6′3″ (1.9 m)

WEIGHT Unavailable

MAXIMUM MEASUREMENTS

LENGTH **MALE** At least 14′ (4.2 m)
FEMALE Unavailable (A reported 14′6″/4.4-m specimen was misidentified)

WEIGHT Unavailable

LIFE SPAN Unavailable

True's Beaked Whale

Mesoplodon mirus
True, 1913

True's Beaked Whale, once thought to occur only in the North Atlantic, is now recognized as having one or more populations in the Southern Hemisphere. Because the species appears to be absent from tropical waters, there may yet prove to be a basis for establishing separate subspecies of True's Beaked Whale. The whale's common name refers to Frederick W. True, a curator at the United States National Museum (now the Smithsonian Institution), who described the species on the basis of an animal that stranded in North Carolina.

DESCRIPTION True's Beaked Whale is notably rotund in the middle and tapered toward both ends. The dorsal fin is small and can be either triangular or falcate, and the spinal area behind the fin forms a sharp dorsal ridge. The melon is well rounded, almost bulbous, and slopes fairly steeply onto the short beak. There can be a distinct crease in the area behind the blowhole. The teeth of adult males erupt slightly at the tip of the lower jaw, which extends just beyond the tip of the upper jaw.

In the North Atlantic, the body is generally medium to brownish gray on the back and lighter ventrally. The lips and dorsal fin are darker, and there is a dark oval area around each eye. A narrow dark blaze may extend along the back from the head to the dorsal fin. There is a white patch in the genital area. In the Southern Hemisphere, a freshly stranded adult female was mostly bluish black, with an extensive white zone that encompassed the dorsal fin and most of the caudal peduncle, a white genital patch, and a light gray jaw and throat with black speckling.

RANGE AND HABITAT In the Northern Hemisphere, True's Beaked Whale inhabits warm temperate waters of the North Atlantic. Strandings have occurred as far north as Nova Scotia and Ireland and as far south as Florida, the Bahamas, and the Canary Islands. Its distribution in the Southern Hemisphere is known only from strandings in South Africa and Western Australia. There appears to be a distributional hiatus in the tropical belt of the Atlantic Ocean, and there is no evidence that True's occurs in the tropical Pacific, North Pacific, or northern Indian Oceans.

SIMILAR SPECIES Cuvier's and Blainville's Beaked Whales occur throughout the entire range of

- WELL-ROUNDED MELON SLOPING ONTO RELATIVELY SHORT BEAK
- DARK-COLORED DORSAL FIN, STRONGLY CONTRASTING WITH LIGHTER BACK (NORTH ATLANTIC)
- OCCURS PRIMARILY IN WARM TEMPERATE NORTH ATLANTIC AND OFF SOUTH AFRICA AND WESTERN AUSTRALIA

TRUE'S BEAKED WHALE

FAMILY ZIPHIIDAE

MEASUREMENTS AT BIRTH

LENGTH 7′3″ (2.2 m)

WEIGHT Unavailable

MAXIMUM MEASUREMENTS

LENGTH 17′6″ (5.3 m)

WEIGHT **MALE** At least 2,200 lb (1,020 kg)
FEMALE About 3,000 lb (1,400 kg)

LIFE SPAN Unavailable

True's Beaked Whale. Cuvier's is larger and has a much shorter, less distinct beak, and Blainville's has a flatter melon and a strongly curved (female) or prominently arched (male) mouthline.

BEHAVIOR, REPRODUCTION, AND FORAGING In an unusual sighting that almost certainly involved this species, three whales were observed moving together in deep water (about 600 fathoms) along the shoreward side of the continental shelf edge off Hatteras Inlet, North Carolina. Each surfacing lasted 10 to 12 seconds, with the animal's beak breaking the surface first, followed by the head, exposed to below eye level. The blow was a low column of vapor about as high as the head was long. After blowing, the animal would arch its back and slip beneath the surface without lifting the flukes. One of the three whales was smaller, and the observer speculated that the group consisted of two adult females and a large calf or juvenile. Little is known about reproduction in this species. A stranded female accompanied by an 11-foot (3.4 m) calf was both pregnant and lactating. True's Beaked Whales are presumed to eat mainly squid.

STATUS AND CONSERVATION True's Beaked Whale has rarely been observed at sea, and scientists have little data on its abundance. The lack of live sightings could indicate that the species is naturally rare, or it could result from the fact that True's is difficult to observe or lives mainly in areas that are infrequently or never surveyed. It is not known to have been hunted regularly anywhere in its range, nor is it known to be taken frequently as a bycatch in any fishery.

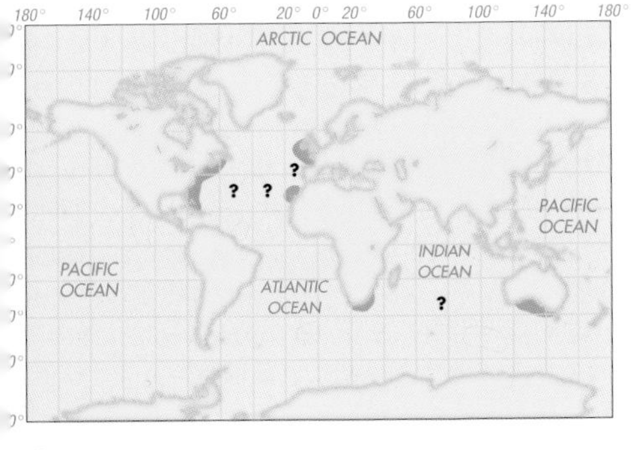

■ RANGE
? POSSIBLE RANGE

Gervais' Beaked Whale

Mesoplodon europaeus
(Gervais, 1855)

This Atlantic species has never been positively identified alive at sea, yet it seems to strand on North American shores more frequently than any other mesoplodont. In the early literature, it is referred to as *Mesoplodon gervaisi.* For many years the only record from the eastern North Atlantic was that of the type specimen found floating dead in the English Channel in 1840. In fact, most researchers used to call this whale the Gulf Stream or Antillean Beaked Whale on the supposition that the carcass from the English Channel had been carried by the Gulf Stream far east of the species' normal range. Recent strandings in Ireland, the Canary Islands, western Africa, and even Ascension Island in the tropical mid-Atlantic offer a more complete picture of the distribution of Gervais' Beaked Whale.

DESCRIPTION Gervais' Beaked Whale is a small to medium-size mesoplodont with a slender, elongated, laterally compressed body. The head is small and tapered, with a narrow beak and essentially straight mouthline. The erupted teeth of adult males are positioned well behind the tip of the lower jaw and are inconspicuous.

Judging from fresh stranded specimens, the color pattern is subtle. The body is generally dark gray dorsally and medium gray ventrally. At least some adult females have a white ano-genital patch. Females also may have light smudges extending up onto the sides of the head from the throat, with the area around the eye remaining dark. Juveniles have a white belly that darkens with age.

RANGE AND HABITAT Gervais' Beaked Whales appear to be found only in the tropical and warm temperate Atlantic Ocean. They strand relatively frequently in the western North Atlantic (especially along Florida's coast), the Gulf of Mexico, and the Caribbean Sea.

SIMILAR SPECIES The mesoplodonts with ranges known to overlap that of Gervais' Beaked Whale include Blainville's, True's, and to a lesser extent Sowerby's Beaked Whales. Blainville's and True's probably occur throughout the entire range of Gervais', while Sowerby's overlaps mostly in the northern portion. Blainville's Beaked Whale is distinguished by the adult

■ MOST LIKELY TO BE SEEN IN TROPICAL TO WARM TEMPERATE ATLANTIC OCEAN, ESPECIALLY IN WEST

GERVAIS' BEAKED WHALE

FAMILY ZIPHIIDAE

MEASUREMENTS AT BIRTH

LENGTH 6′11″ (2.1 m)

WEIGHT Unavailable

MAXIMUM MEASUREMENTS

LENGTH **MALE** At least 14′9″ (4.5 m)
FEMALE At least 17′ (5.2 m)

WEIGHT At least 2,600 lb (1,200 kg)

LIFE SPAN

Unknown, but more than 48 growth layer groups were found in a sectioned tooth of one specimen

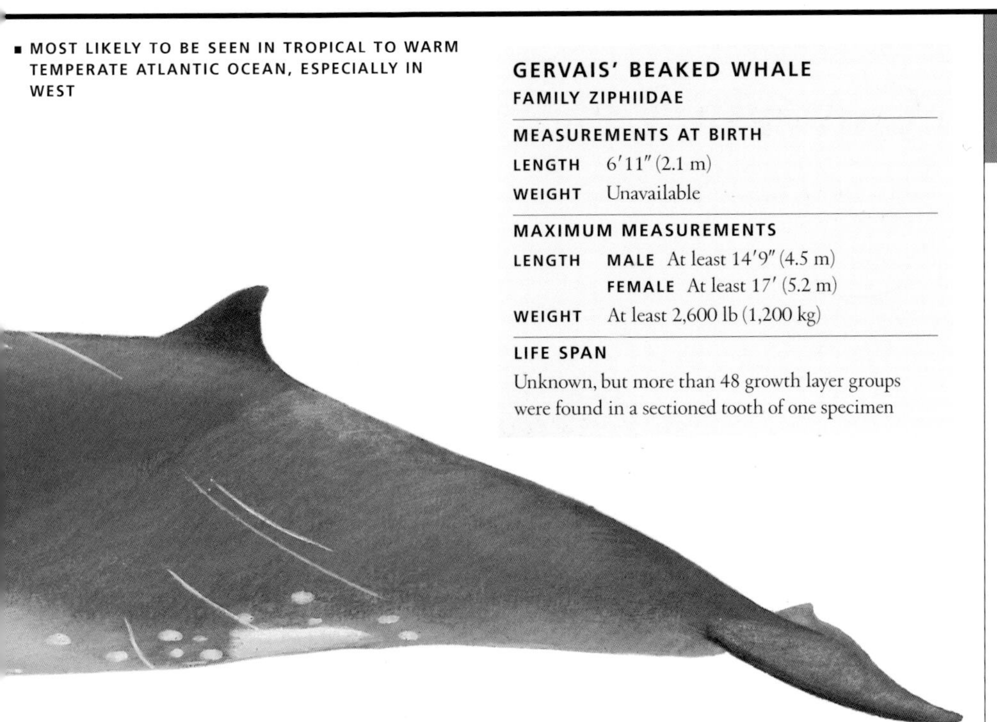

male's highly arched lower jaw, with the crown of the tooth on each side erupting at the apex. Female and juvenile Blainville's, as well as Sowerby's and True's Beaked Whales of both sexes, may be almost impossible to distinguish. Cuvier's Beaked Whale is relatively common in much of Gervais' Beaked Whale's range; it has a higher, steeper melon and a considerably shorter beak, and adult males have a white head and two exposed teeth that jut forward from the tip of the lower jaw.

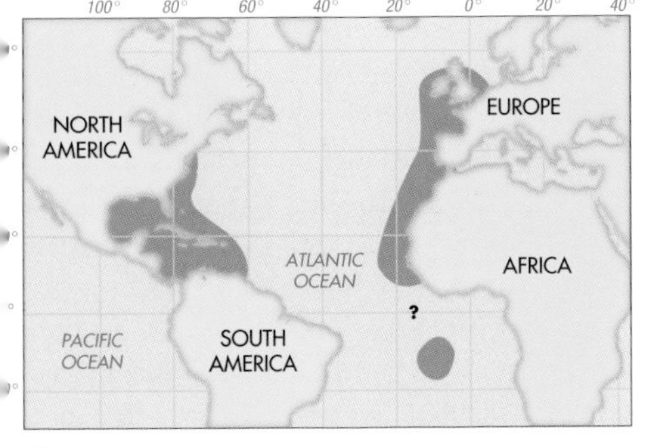

■ RANGE ? POSSIBLE RANGE

BEHAVIOR, REPRODUCTION, AND FORAGING Group size is generally small, but there is no reliable information on social organization or behavior for this species, and essentially nothing is known about its reproductive parameters. Gervais' Beaked Whale is presumed to eat mostly squid.

STATUS AND CONSERVATION Gervais' Beaked Whale is probably fairly rare. However, it is the most common mesoplodont to strand along the Atlantic coast of the United States. It is taken only occasionally in fishing gear and is not known to be hunted anywhere.

Sowerby's Beaked Whale

Mesoplodon bidens
(Sowerby, 1804)

This mesoplodont is found only in the North Atlantic and is sometimes referred to as the North Atlantic or North Sea Beaked Whale. This was the first living ziphiid to be described, and the specific name (from the Latin *bi-*, for "two," and *dens*, for "teeth") refers to the fact that these animals have only two teeth, which was considered to be a distinctive feature at the time. We now know that all mesoplodonts, as well as some other beaked whale species, share this characteristic.

DESCRIPTION Sowerby's Beaked Whale has a typical mesoplodont body shape, with a moderately long beak and a small falcate dorsal fin set behind the middle of the back. Its melon is convex but not bulbous. The two teeth in the lower jaw of adult males erupt at a position well behind the tip of the jaw. The long gray beak is often exposed at the start of a surfacing, and the erupted teeth of adult males may be easily seen.

The color pattern is nondescript, basically consisting of a dark gray dorsal surface and a lighter gray ventral surface. The back appears slate gray at sea. Long, linear, grayish or whitish scars and spots are present on the sides (and sometimes the back), especially of adult males.

RANGE AND HABITAT Sowerby's Beaked Whale occurs in cool and warm temperate waters of the North Atlantic, from at least as far north as Labrador in the west and 71°30'N in the Norwegian Sea in the east. Its normal southern limit is probably between about 33°N and 41°N, or roughly Nantucket, Massachusetts, in the west and Madeira in the east. A single stranding in the Gulf of Mexico is considered to have been outside the normal range of the species. Water depths where sightings have occurred range from 650 to 5,000 feet (200–1,500 m).

SIMILAR SPECIES True's, Gervais', and Blainville's Beaked Whales are found within the normal range of Sowerby's. Females and young males of the four species cannot be reliably differentiated at sea, while adult males potentially can be distinguished by the appearance and position of the exposed teeth. The teeth of True's are at the tip of the jaw, and those of Blainville's are on a prominently raised platform toward the rear of the jaw. The teeth of Gervais' are in about the same position as those of Sowerby's but are only slightly visible. Cuvier's Beaked Whale is present throughout most of the range of

- LONG, UNIFORMLY GRAY BEAK
- EXPOSED TEETH OF ADULT MALE WELL BEHIND TIP OF LOWER JAW
- OCCURS ONLY IN NORTH ATLANTIC, MAINLY NORTH OF ABOUT 30°N

SOWERBY'S BEAKED WHALE

FAMILY ZIPHIIDAE

MEASUREMENTS AT BIRTH

LENGTH 7′10″ (2.4 m)

WEIGHT About 370 lb (170 kg)

MAXIMUM MEASUREMENTS

LENGTH **MALE** 18′ (5.5 m)
FEMALE At least 16′6″ (5 m)

WEIGHT 2,200–2,900 lb (1,000–1,300 kg)

LIFE SPAN Unavailable

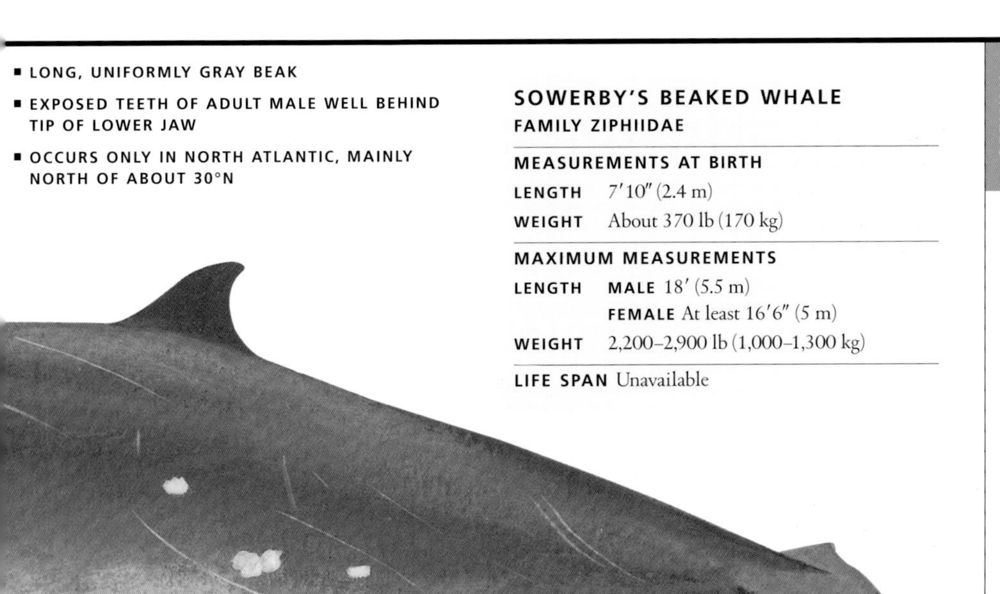

Sowerby's, but it has a much shorter beak, and the beak and adjoining head region of adults is noticeably lighter than the rest of the body. Northern Bottlenose Whales also occur throughout much of Sowerby's range but are considerably larger and have a more prominent melon.

BEHAVIOR, REPRODUCTION, AND FORAGING

Sowerby's Beaked Whales can occur in groups of up to 8 to 10 closely associated individuals, including adult males and females with calves. They are known to dive for at least 28 minutes. They strand occasionally in groups; the largest mass stranding to have been documented involved six animals. Squid are presumed to be important in the diet of Sowerby's Beaked Whale, but parts of fish thought to be Atlantic cod have been found in the stomachs of stranded individuals. Nothing is known about reproduction in this species.

STATUS AND CONSERVATION There are no abundance estimates of Sowerby's Beaked Whales. Although Norwegian whalers occasionally killed them off Iceland and in the Barents Sea in the past, these whales are not known to be hunted anywhere today. Some deaths are caused by entanglement in fishing gear, but the scale of this mortality is not thought to be high enough to represent a major threat to the species.

RANGE ? POSSIBLE RANGE

Gray's Beaked Whale

Mesoplodon grayi
von Haast, 1876

Gray's Beaked Whale may be the most gregarious mesoplodont. It strands relatively frequently (especially in New Zealand), sometimes en masse, and is one of the better-known members of its genus. This whale is sometimes called the Scamperdown Whale, supposedly after the locality of an early stranding in New Zealand. The common name refers to John Edward Gray, a zoologist at the British Museum. Gray's Beaked Whale occurs primarily in the Southern Ocean.

DESCRIPTION Gray's Beaked Whale is a slender, medium-size mesoplodont with a small, falcate dorsal fin. It has a notably long, pointed beak, with a relatively straight mouthline. The melon bulges slightly just in front of the blowhole and slopes smoothly onto the beak. This species is unique among the mesoplodonts in that adults of both sexes often have a regularly spaced row of 17 to 22 small teeth in each side of the upper jaw toward the back of the mouth, beginning at approximately the same position along the jaw as the single pair of teeth in the lower jaw. The teeth of the upper jaw normally protrude only a few millimeters past the gum. In adult males, the emergent portion of the two lower-jaw teeth is triangular in shape. These teeth are positioned roughly halfway along the mouthline and tilt slightly forward.

The color pattern is generally dark dorsally and lighter ventrally. Adult females have a light gray lower jaw and may be slightly lighter dorsally than adult males, with white areas around the umbilicus and in the genital region. Adults of both sexes have a white beak. Adult males are extensively scarred with long linear and small, round, white markings.

RANGE AND HABITAT Gray's Beaked Whale is widely distributed in temperate waters of the Southern Hemisphere, mainly between 30°S and 45°S. Strandings have been documented from both the east and west coasts of South America, the Falkland Islands, South Africa, Australia, and New Zealand, including the Chatham Islands. There have been confirmed sightings far offshore to the south and east of Madagascar and in Antarctic regions. A single stranding in the Netherlands has always been considered extralimital.

- SLENDER BODY
- LONG NARROW BEAK, WHITE IN ADULTS
- STRAIGHT MOUTHLINE
- OCCURS ALMOST EXCLUSIVELY IN TEMPERATE WATERS OF SOUTHERN HEMISPHERE

GRAY'S BEAKED WHALE

FAMILY ZIPHIIDAE

MEASUREMENTS AT BIRTH

LENGTH About 6′11″–7′3″ (2.1–2.2 m)

WEIGHT Unavailable

MAXIMUM MEASUREMENTS

LENGTH **MALE** 19′ (5.7 m)
FEMALE At least 17′6″ (5.3 m)

WEIGHT About 2,400 lb (1,100 kg)

LIFE SPAN Unavailable

SIMILAR SPECIES Gray's Beaked Whale is most likely to be confused with the other Southern Hemisphere temperate-region mesoplodonts, primarily Andrews', Hector's, Blainville's, the Ginkgo-toothed, and True's Beaked Whales and the Strap-toothed Whale. The beaks of all these species are shorter and stouter than that of Gray's. Blainville's has a raised area at the rear of the lower jaw, and the Strap-toothed has strap-like teeth that curve over the upper jaw, as well as a distinctive color pattern. The tusks of adult male Hector's and True's Beaked Whales erupt near the tip of the lower jaw. Adult male Blainville's can be distinguished by the high arch of the jawline. Those of the adult male Andrews' and Ginkgo-toothed are also somewhat arched.

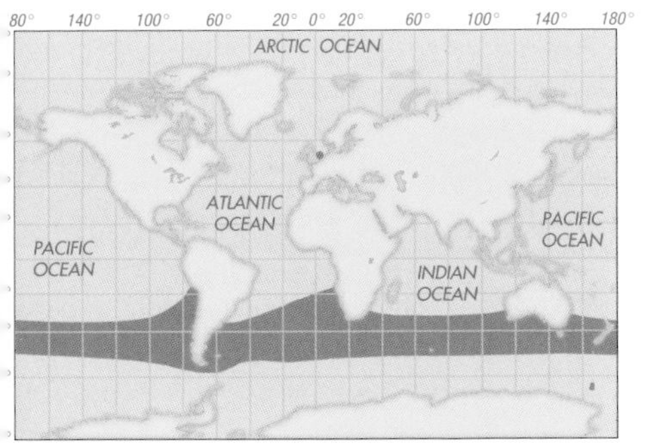

BEHAVIOR, REPRODUCTION, AND FORAGING An unusual aspect of Gray's Beaked Whale is that it occasionally mass strands. One such stranding of more than 20 individuals at the Chatham Islands indicated a much larger school size than is normally reported for mesoplodonts. On another occasion, at the Chatham Islands, eight whales stranded together. Groups of five or six animals have been encountered in the Indian Ocean, southeast of Madagascar. Some authors have speculated that the upper-jaw teeth of Gray's Beaked Whale function in helping the animal grasp its prey. Nevertheless, these whales presumably specialize in preying on squid and likely capture their prey by suction. Little is known about reproduction in this species.

STATUS AND CONSERVATION Based on the high number of strandings, together with the large size of stranded groups, Gray's Beaked Whale may be relatively abundant. There is no indication that these whales have been hunted deliberately or that they entangle often in fishing gear.

Pygmy Beaked Whale

Mesoplodon peruvianus
Reyes, Mead, and Van Waerebeek, 1991

This recently described species was at first thought to be limited to the eastern South Pacific, where scientists would occasionally encounter specimens while examining the Peruvian catch of small cetaceans. However, in 1990 two specimens—one a rotten carcass, the other a nearly complete skeleton—were discovered farther north in Bahía de la Paz, at the southern tip of Baja California, while another was found stranded near Kaikoura, New Zealand. The common name is based on the fact that this is the smallest known mesoplodont. This species is often called the Lesser Beaked Whale.

DESCRIPTION The Pygmy Beaked Whale has a typical spindle-shaped mesoplodont form, with its maximum girth between the flippers and the dorsal fin. The dorsal fin is small and triangular, slightly falcate, and positioned about two-thirds of the way back on the body. The caudal peduncle is unusually thick or deep, and the flukes unusually broad (about one-fourth of the body length). The melon bulges slightly ahead of the blowhole, then slopes onto a relatively short beak. The mouthline curves up toward the back. Two teeth in the lower jaw are positioned well behind the tip of the beak and probably protrude only slightly outside the gum tissue in adult males.

Adults are generally dark gray on the back and sides and much lighter ventrally, especially on the lower jaw and throat and on the belly behind the umbilicus. The dorsal fin, flippers, and flukes are dark gray.

RANGE AND HABITAT The Pygmy Beaked Whale is known to occur in waters off Peru and Chile between roughly 11°S and 29°15'S, in the southwestern Gulf of California, off the central west coast of Mexico, and off southern California. A group of three individuals was sighted in offshore waters at about 11°N, 98°W, and tentative sightings have been made in Bahía de Banderas, Mexico. It is likely that the species' range includes much of the eastern tropical Pacific. The New Zealand stranding suggests that the species occurs in the western South Pacific as well.

SIMILAR SPECIES Most other beaked whales that occur in the eastern tropical Pacific are appreciably larger; however, juveniles of various mesoplodonts could be mistaken for Pygmy Beaked Whales.

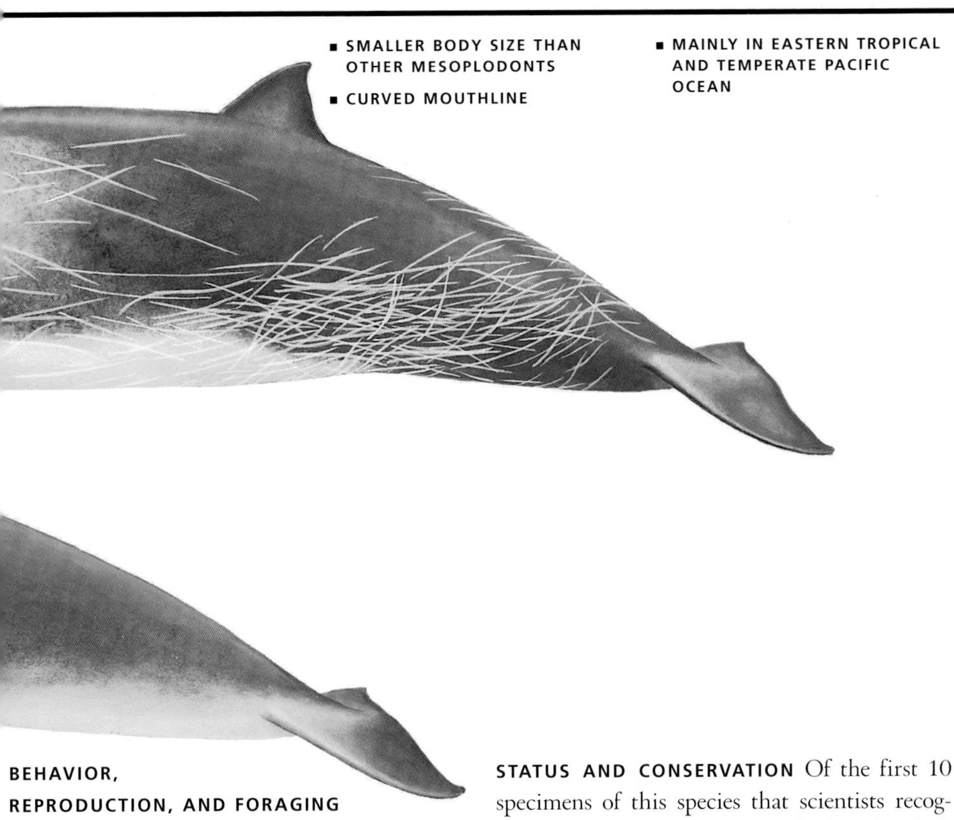

BEHAVIOR, REPRODUCTION, AND FORAGING Almost nothing is known about social organization, behavior, or reproduction in the Pygmy Beaked Whale. A small group of these whales seen far offshore to the west of Central America consisted of two adults and a calf. The stomach contents of an immature male consisted entirely of fish remains, suggesting that the Pygmy Beaked Whale preys on organisms other than, or in addition to, squid.

STATUS AND CONSERVATION Of the first 10 specimens of this species that scientists recognized and examined, at least six had died in drift gillnets set for sharks off Peru during the 1980s. Because the scientists were sampling only a relatively small proportion of the total cetacean kill in Peru, the overall bycatch rate could have been (and may continue to be) substantial. Nothing is known about the abundance of the Pygmy Beaked Whale, and until more is learned, it is impossible to judge whether the species is depleted or secure.

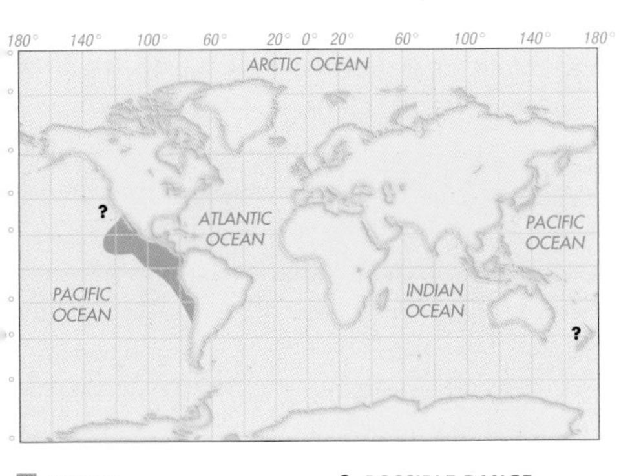

■ RANGE ? POSSIBLE RANGE

PYGMY BEAKED WHALE
FAMILY ZIPHIIDAE

MEASUREMENTS AT BIRTH
LENGTH 5′3″ (1.6 m)
WEIGHT Unavailable

MAXIMUM MEASUREMENTS
LENGTH 13′ (3.9 m)
WEIGHT Unavailable

LIFE SPAN Unavailable

Andrews' Beaked Whale

Mesoplodon bowdoini
Andrews, 1908

This poorly known mesoplodont probably has a southern circumpolar distribution north of the Antarctic Convergence. There are scarcely more than 35 records of its occurrence, all of them consisting of stranded carcasses found mostly in New Zealand and Australia. Roy Chapman Andrews, a curator at the American Museum of Natural History in New York, named the species after George S. Bowdoin, a trustee and donor to the museum. New Zealand scientist Alan Baker recently published a detailed study of Andrews' Beaked Whale's morphology, distribution, and status.

DESCRIPTION Andrews' Beaked Whale has a robust body, a low melon, and a short thick beak. The massive lower jaw rises halfway along to form a prominent arch, with the large exposed teeth of adult males erupting at the apex. The dorsal fin, situated somewhat behind the middle of the back, is small, triangular, and blunt-tipped. The crowns of the teeth may extend above the level of the rostrum.

Andrews' Beaked Whale is generally dark gray to black. Adult males have a grayish dorsal "saddle" from immediately behind the blowhole extending back to the dorsal fin. Up to half of the beak is white or light gray, and some individuals have a light patch on each side of the head in front of the eyes. Females are thought to have less white on the beak, a slate gray back, and grayish-white flanks and belly. The bodies of adult males are marked by linear scars presumably made by the teeth of other males. Both males and females have at least a few oval scars from bites by cookie-cutter sharks.

RANGE AND HABITAT Andrews' Beaked Whale is known only from the Southern Hemisphere, where it may be circumpolar in temperate latitudes (generally between 32°S and 54°30'S). Strandings have been documented in southern Australia, New Zealand (including Campbell Island), Macquarie Island, the Falkland Islands, and Tristan da Cunha. Published records attributed to this species from the Kerguelen Islands in the southern Indian Ocean and from Peru were misidentified specimens of other *Mesoplodon* species.

SIMILAR SPECIES Among the mesoplodonts known to overlap in range with Andrews' Beaked

■ WHITE FRONT PORTION OF ADULT MALE'S BEAK

■ KNOWN DISTRIBUTION LIMITED TO TEMPERATE LATITUDES OF SOUTHERN HEMISPHERE

Whale, the most likely to be confused with it are probably the Strap-toothed Whale and Gray's, Hector's, Blainville's, and the Ginkgo-toothed Beaked Whales. Detailed examination of specimens, in some cases perhaps even genetic analyses, are usually required to confirm the identity of females and young males. However, it may be possible to distinguish adult males by the position and shape of their teeth.

BEHAVIOR, REPRODUCTION, AND FORAGING As there are no confirmed sightings at sea and no records of mass strandings, scientists know almost nothing about this species' social organization and behavior or its reproductive and foraging habits. The calving season in New Zealand waters appears to be during summer and autumn.

STATUS AND CONSERVATION Nothing is known about the abundance of Andrews' Beaked Whale. It is not known to be hunted anywhere, nor is there evidence of large-scale incidental mortality in fisheries.

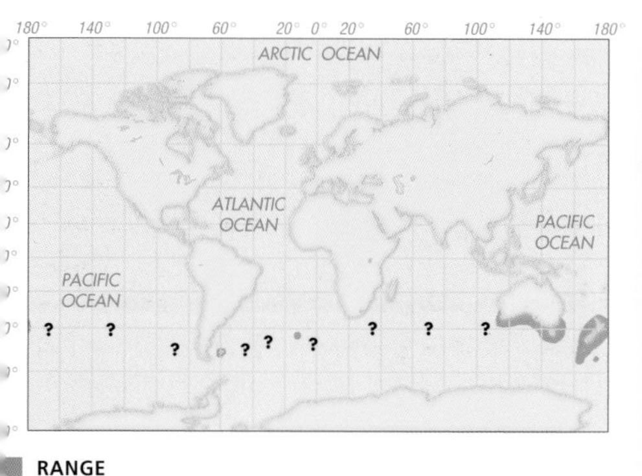

RANGE

? POSSIBLE RANGE

ANDREWS' BEAKED WHALE

FAMILY ZIPHIIDAE

MEASUREMENTS AT BIRTH

LENGTH 7′3″ (2.2 m)

WEIGHT Unavailable

MAXIMUM MEASUREMENTS

LENGTH **MALE** At least 15′ (4.5 m)

FEMALE 16′ (4.9 m)

WEIGHT Unavailable

LIFE SPAN Unavailable

Hubbs' Beaked Whale

Mesoplodon carlhubbsi
Moore, 1963

This mesoplodont is a North Pacific endemic. Carl Hubbs, an influential American ichthyologist, published a description of a whale found thrashing in the surf near his office at Scripps Institution of Oceanography in La Jolla, California, in 1945. He concluded that it was a specimen of Andrews' Beaked Whale. However, Joseph Curtis Moore, a mammalogist at Chicago's Field Museum and an expert in beaked whales, reassigned the specimen to a new species, which he named *carlhubbsi* in honor of his colleague.

DESCRIPTION The body of Hubbs' Beaked Whale is rotund in the midsection and tapered at both ends. The slightly falcate dorsal fin, placed well behind midbody, is consistently 8½ to 9 inches (22–23 cm) high in adults. The melon is moderately bulbous. The mouthline is prominently arched in the back half. Adult males have two partially exposed teeth in the lower jaw that can project to the top of the beak or higher when the mouth is closed.

Adult males are generally dark gray to black, with white areas on the upper jaw, the front half of the lower jaw, and the melon. The large white area on the melon, variable in size, is reminiscent of a cap, or "beanie." Adult males are extensively scarred on the flanks and belly. Females and juveniles are countershaded, grading from medium gray dorsally to white ventrally; their beak is noticeably lighter than the rest of the body, although not brilliant white as in adult males.

RANGE AND HABITAT Hubbs' Beaked Whales are known only from temperate waters of the North Pacific. In the western Pacific, the species appears to be restricted to the northeastern coast of Honshū, Japan, where the cold south-flowing Oyashio Current meets the warm north-flowing Kuroshio Current at about 38°N. In the eastern Pacific, the northern limit of these whales is thought to be approximately Prince Rupert, British Columbia (54°N), and they occur at least as far south as San Diego, California (33°N).

SIMILAR SPECIES Hubbs' Beaked Whale may overlap with Stejneger's Beaked Whale in the northern portions of its range. The adult male Stejneger's has a much flatter melon and an all-dark head (lacking Hubbs' white "beanie" and beak), and the exposed portions of its tusks are

- WHITE "BEANIE" ON MELON OF ADULT MALE
- WHITE BEAK OF ADULT MALE
- PROMINENT ARCH IN BACK HALF OF MOUTHLINE
- TUSKS ON ADULT MALE
- LIMITED TO TEMPERATE NORTH PACIFIC

HUBBS' BEAKED WHALE

FAMILY ZIPHIIDAE

MEASUREMENTS AT BIRTH	
LENGTH	8′2″ (2.5 m) or less
WEIGHT	Unavailable
MAXIMUM MEASUREMENTS	
LENGTH	18′ (5.4 m)
WEIGHT	About 3,300 lb (1,500 kg)
LIFE SPAN	Unavailable

much larger. The Ginkgo-toothed Beaked Whale overlaps extensively with Hubbs' Beaked Whale but lacks the brilliant white beak and "beanie"; also, adult males have barely exposed tusks and are usually much less scarred. Cuvier's and Baird's Beaked Whales occur throughout the range of Hubbs' Beaked Whale. Both species have teeth only at the tip of the lower jaw. Cuvier's, especially the adult male, has a similarly pale head but a much shorter beak. Baird's is much larger, with a long beak.

BEHAVIOR, REPRODUCTION, AND FORAGING An interesting feature of this species is the extensive scarring acquired over time by adult males. Although several other species of beaked whales share this feature, adult male Hubbs' Beaked Whales are typically covered with white linear scars. The scars appear to be from the teeth of other males and presumably signify aggressive interactions, possibly related to striving for social dominance and reproductive success. They appear as long, single or double scratch marks. Hubbs' Beaked Whales prey mainly on mesopelagic squid and fish. Little is known about reproduction in this species. Calving is thought to take place mainly in summer.

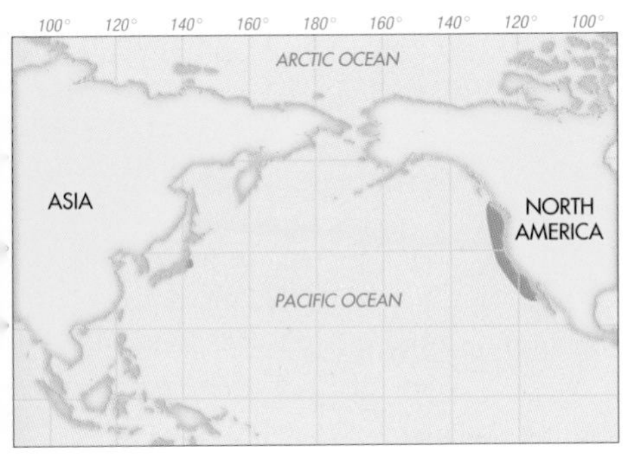

RANGE

STATUS AND CONSERVATION Nothing is known about the abundance of Hubbs' Beaked Whales. Japanese fishermen hunting for Cuvier's Beaked Whales and other small cetaceans have killed them occasionally. They are killed often enough in drift gillnets off California, and probably elsewhere, to raise concerns about at least local depletion. Their range seems to be highly specific to particular oceanographic conditions, and this could make them vulnerable to changes in climate.

Ginkgo-toothed Beaked Whale

Mesoplodon ginkgodens
Nishiwaki and Kamiya, 1958

The Japanese scientists who described and named this species noted that the teeth of mature males were shaped like the leaves of the ginkgo tree, with a central point and nearly symmetrical, curving sides. The Ginkgo-toothed Beaked Whale was formally described only in 1958, although Japanese shore-based whalers had been taking small numbers of this and other mesoplodonts (most notably Hubbs' Beaked Whale) in a hunt directed primarily at Cuvier's Beaked Whales. While scientists have examined a few specimens killed by whalers, most of what is known about the Ginkgo-toothed Beaked Whale comes from examinations of stranded carcasses.

DESCRIPTION The Ginkgo-toothed Beaked Whale has a robust body. The dorsal fin is typical of the mesoplodonts: small, often falcate, and positioned about two-thirds of the way back from the beak tip. The melon is well rounded and slopes fairly steeply onto the moderate-length beak. In adult males, the mouthline curves sharply upward about halfway back from the beak tip. The large teeth of adult males are mostly covered by gum tissue and the lips, and do not project above the plane of the upper-jaw surface.

Based on descriptions of stranded animals (less than 20 fresh specimens have been examined), adult males are basically dark gray, with lighter areas on the front half of the beak. Females are lighter overall, with a medium gray back and light gray belly. Adult males usually have many white spots, 1¼ to 1½ inches (3–4 cm) in diameter, on the ventral surface of the caudal peduncle. They seem to lack the linear scars typical of many adult male mesoplodonts. With further observations of live animals, the Ginkgo-toothed Beaked Whale may prove to have a different and more complex color pattern.

RANGE AND HABITAT The Ginkgo-toothed Beaked Whale occurs in tropical and warm temperate waters of the Indo-Pacific. Strandings have been documented from as far north as the Pacific coast of Honshū (Japan) and southern California to as far south as the Galápagos Islands and New South Wales, Australia. In the Indian Ocean, there are records from Sri Lanka, the Maldives, and the Strait of Malacca.

SIMILAR SPECIES Several mesoplodonts share some or much of the Ginkgo-toothed Beaked

- UPWARD CURVE IN REAR HALF OF MOUTHLINE OF ADULT MALE
- LIMITED MAINLY TO TROPICAL AND WARM TEMPERATE INDO-PACIFIC WATERS

GINKGO-TOOTHED BEAKED WHALE

FAMILY ZIPHIIDAE

MEASUREMENTS AT BIRTH

LENGTH 7′10″ (2.4 m) or less

WEIGHT Unavailable

MAXIMUM MEASUREMENTS

LENGTH **MALE** 16′ (4.8 m)
FEMALE 16′ (4.9 m)

WEIGHT Unavailable

LIFE SPAN Unavailable

Whale's range, including Hubbs', Blainville's, and possibly True's. Hubbs' is distinguished by the adult male's white melon and beak. The adult male Blainville's has an extremely high arch in the lower jaw. The adult male True's has an essentially straight mouthline and therefore lacks the prominent upward curve in the rear portion of the lower jaw typical of the male Ginkgo-toothed. The presumed range of Longman's Beaked Whale overlaps much of the Ginkgo-toothed Beaked Whale's range. Assuming that the tropical "bottlenose" whale is Longman's, it is considerably larger, with a bluff melon that is often cream-colored, a short to medium-length beak, and a straight mouthline.

BEHAVIOR, REPRODUCTION, AND FORAGING Little is known about social organization or behavior in this species. Because adult males are generally devoid of linear scars, it has been suggested that the Ginkgo-toothed Beaked Whale differs from other mesoplodonts in that it does not engage in male-to-male combat. The Ginkgo-toothed Beaked Whale's diet is not well known but presumably consists mainly of mesopelagic squid and possibly fish. Little is known about reproduction in this species.

STATUS AND CONSERVATION Whalers and fishermen in Japan and Taiwan have harpooned these whales opportunistically. Ginkgo-toothed Beaked Whales are also susceptible to entanglement in gillnets, especially drift gillnets set in deep water. There are no reliable survey data to provide a basis for evaluating abundance or trends for this poorly known species.

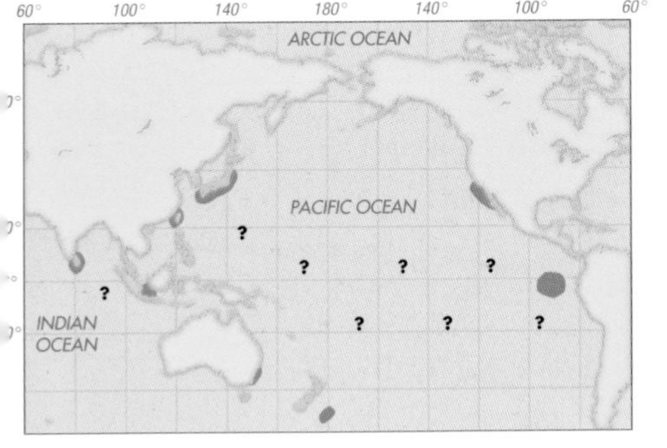

RANGE

? POSSIBLE RANGE

Strap-toothed Whale

Mesoplodon layardii
(Gray, 1865)

This species gets its common name from the unique dentition of adult males. Two teeth in the lower jaw grow to wrap over the upper jaw, crossing and pressing tightly against it. As a result, adult male Strap-toothed Whales can barely open their mouths. Scientists consider it likely that the teeth are used primarily in fighting among Strap-toothed Whales. The "strap teeth" may also limit the size of squid that the animal can suck into its mouth. The Strap-toothed Whale is also known as Layard's Beaked Whale, after E. L. Layard, a curator of the South African Museum during the 19th century. Layard prepared drawings of a skull and sent them to John Edward Gray, the prolific British taxonomist who described the species.

DESCRIPTION The Strap-toothed Whale is the largest and one of the more boldly marked mesoplodonts. Its overall form is like that of other mesoplodonts—a small head, large thorax and abdomen, and short tail. The dorsal fin is small, falcate, and set about two-thirds of the way back on the body. The melon is moderately bulbous and slopes steeply downward before merging with the beak. The beak is long and slender, with a straight mouthline. The strap-like, laterally compressed teeth erupt in adult males well behind the tip of the lower jaw. They are oriented so that they slant back toward the melon at a 45° angle, and grow to wrap over and press tightly against the upper jaw. The exposed portions of the teeth commonly host large stalked barnacles.

On adults, at least the front half of the beak, the entire throat, and a band behind the eye are white. The dorsal surface from just ahead of the blowhole back to almost midbody is light gray to white, reminiscent of a cape. There is also a fairly large white or light gray area in the ano-genital region and a small white spot at the base of each flipper. The rest of the body is black. Young Strap-toothed Whales are evenly colored with countershading.

RANGE AND HABITAT Based on the many stranding localities and a few at-sea observations, it seems clear that the Strap-toothed Whale is widely distributed, and probably circumpolar, in cool temperate waters of the Southern Hemisphere.

- WHITE THROAT AND FRONT HALF OF BEAK
- WHITE BAND BEHIND EYE, JOINS LIGHT DORSAL COLORATION
- DARK "MASK" ON FACE
- STRAP-LIKE TEETH OF ADULT MALE CURVE OVER UPPER JAW
- OCCURS ONLY IN COOL TEMPERATE SOUTHERN HEMISPHERE

STRAP-TOOTHED WHALE

FAMILY ZIPHIIDAE

MEASUREMENTS AT BIRTH

LENGTH 7′3″–7′10″ (2.2–2.4 m)

WEIGHT Unavailable

MAXIMUM MEASUREMENTS

LENGTH **MALE** 19′6″ (5.9 m)
FEMALE 20′ (6.2 m)

WEIGHT Probably at least 2,200–2,900 lb (1,000–1,300 kg)

LIFE SPAN Unavailable

SIMILAR SPECIES The beaked whales with ranges that overlap that of the Strap-toothed Whale include primarily Arnoux's, Andrews', Gray's, Hector's, Blainville's, the Ginkgo-toothed, True's, and Cuvier's Beaked Whales and the Southern Bottlenose Whale. Both Arnoux's Beaked and the Southern Bottlenose are larger, even though the Strap-toothed Whale is the largest mesoplodont. Particular difficulty could arise in distinguishing Andrews' and Gray's, as much or all of the beak, at least in adult males, is white. None of these other species, of course, has teeth that in any way resemble those of the adult male Strap-toothed Whale.

RANGE

BEHAVIOR, REPRODUCTION, AND FORAGING Nothing is known about the social organization and behavior of this species. Like other ziphiids, Strap-toothed Whales are squid eaters. Adult males, however, are considerably restricted by their teeth in the extent to which they can open their mouths. In one comparison, the gapes of an adult female and an immature male measured 2½ inches (6.5 cm), while the gapes of two adult males measured only 1¼ to 1½ inches (3.2–4.0 cm). With such small gapes, adult males can eat only relatively small, slender squid, generally weighing less than 3½ ounces (100 g) and having mantle lengths less than 6¼ inches (16 cm).

STATUS AND CONSERVATION Strap-toothed Whales have not been hunted regularly anywhere in their range, nor are they known to be killed often in fishing gear. In view of their wide distribution they may be more secure than some other beaked whales with more limited distributions.

Blainville's Beaked Whale

Mesoplodon densirostris
(Blainville, 1817)

Blainville's Beaked Whale appears to be the most widely distributed mesoplodont. It is the only member of its genus that has proven amenable to at-sea observational research. Groups are regularly observed in at least three areas: off the Waianae coast of Oahu, Hawaii; off Moorea in the Society Islands of the South Pacific; and in the northeastern Bahamas. In the last of these areas, a photo-identification study is under way. The species is sometimes called the Dense-beaked Whale. The Latin name derives from *densus* for "thick" or "dense" and *rostrum* for "beak."

DESCRIPTION Blainville's Beaked Whale has a deep and robust body that is laterally compressed, especially in the tail region. The dorsal fin is small, triangular or falcate, and set about two-thirds of the way back on the body. The beak is moderately long, and the melon comparatively small and flat. The mouthline is distinctive, with an abrupt, rising step at midlength. The small pointed crown of a massive tooth erupts from each side of the lower jaw in adult males, approximately as they reach maturity. These teeth tilt forward markedly and rise above the upper jaw. Clusters of single-stalked barnacles frequently attach to exposed portions of the teeth.

The color pattern is subtle and countershaded. The dorsal and lateral surfaces are dark bluish gray, changing abruptly to light gray ventrally. The head can be brownish, shading to light gray on the edges of the upper lip and on the lower jaw. There are often many white oval scars, especially in the genital area, possibly caused by cookie-cutter shark bites. Some long white "scratch" scars are usually present on adult males.

RANGE AND HABITAT Blainville's Beaked Whale is widespread in tropical and warm temperate waters of all oceans. Incursions into the higher latitudes are probably related to warm current systems, such as the Gulf Stream in the North Atlantic and the Agulhas Current off southern Africa. Strandings have been reported north to Nova Scotia, Iceland, and the British Isles in the North Atlantic and Japan and central California in the North Pacific; and south to Rio Grande do Sul (Brazil) and South Africa in the South Atlantic and central Chile, Tasmania, and northern New Zealand in the South Pacific.

- STRONGLY ARCHED REAR HALF OF LOWER JAW IN ADULT MALE
- FORWARD-TILTING ERUPTED TEETH IN ADULT MALE
- WIDESPREAD IN TROPICAL AND WARM TEMPERATE WATERS

BLAINVILLE'S BEAKED WHALE

FAMILY ZIPHIIDAE

MEASUREMENTS AT BIRTH

LENGTH 6′7″ (2 m)

WEIGHT 132 lb (60 kg)

MAXIMUM MEASUREMENTS

LENGTH **MALE** At least 14′6″ (4.4 m)
FEMALE At least 15′ (4.6 m)

WEIGHT **MALE** More than 1,800 lb (800 kg)
FEMALE At least 2,200 lb (1,000 kg)

LIFE SPAN Unavailable

There is no evidence of seasonal movements or migrations. Blainville's Beaked Whales are regularly sighted in slope waters roughly 1,600 to 3,300 feet (500–1,000 m) deep with even deeper gullies nearby.

SIMILAR SPECIES The only other mesoplodonts with a similar distinct arch at the rear of the lower jaw are Stejneger's, Hubbs', and the Ginkgo-toothed Beaked Whales, and to a lesser extent Andrews' Beaked Whale, but in none of these species is the arch so massive and prominent as it is in the adult male Blainville's. In addition, adult male Hubbs' can be distinguished by their white beak and their distinctive white patch on the melon ("beanie").

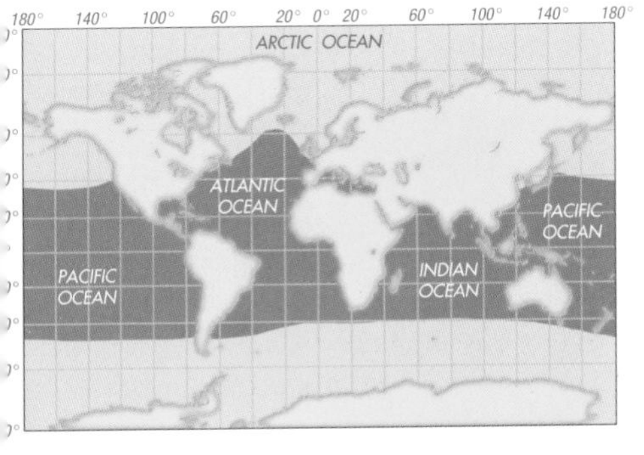

RANGE

BEHAVIOR, REPRODUCTION, AND FORAGING Blainville's Beaked Whale occurs in small groups of three to seven individuals. It usually surfaces slowly, with little splashing. Dives can last at least 22 minutes. This whale feeds on squid and small fish. Little is known about its reproduction.

STATUS AND CONSERVATION Blainville's Beaked Whales have been taken occasionally in hunts for small cetaceans but have not been regular targets of hunting anywhere. Several individuals that stranded in the Bahamas in March 2000 are thought to have been victims of acoustic trauma; naval military exercises had taken place just prior to the strandings.

Stejneger's Beaked Whale

Mesoplodon stejnegeri
True, 1885

This species was described in 1885 from a single skull found on Bering Island by Leonhard Stejneger, a curator at the U. S. National Museum (now the Smithsonian Institution). Thereafter, almost 100 years passed with nothing known about this whale beyond the fact of its existence. Since the late 1970s, there has been a flurry of interest in Stejneger's Beaked Whale as museum scientists have accumulated numerous stranded specimens from Japan and the Aleutian Islands. Most notably, in 1994 a group of four adult females stranded in fresh condition on Adak Island in the Aleutians. This event provided an unprecedented opportunity to study the species' biology and external appearance. Stejneger's Beaked Whale is occasionally called the Bering Sea Beaked Whale. Although it certainly has the most northerly distribution of the North Pacific mesoplodonts, its range encompasses much more than just the Bering Sea. The exposed portions of this species' tusk-like teeth are larger than those of any other mesoplodont except the Strap-toothed Whale, and it is for this reason that Stejneger's Beaked Whale has sometimes been called the Saber-toothed Whale.

DESCRIPTION Stejneger's Beaked Whale has a long body, tapered at both ends as in other mesoplodonts. The melon is notably flat or depressed in comparison to most other mesoplodonts. It slopes smoothly onto the rather pointed, medium-length beak. The falcate dorsal fin is situated well behind midback. The mouthline curves smoothly upward posteriorly, forming a prominent arch. In adult males, the flattened teeth erupt immediately ahead of the apex of this arch, tilting forward and inward.

The overall color of both males and females is uniformly gray to black. The upper surface of the head is dark brownish gray (forming a "cranial cap"); this darkness extends downward over the eyes, giving the animal an almost helmeted appearance. Apparently the overall coloration darkens with age, so that in old animals both the mottled undersides and the contrast between the dark "cranial cap" and "flipper pockets" give way to an even darker gray. The undersides of the flukes (at least of females) have a starburst pattern of light gray or white markings radiating from the posterior margin; these markings intensify with age. White, lightly pigmented, circular or

- DARK "CRANIAL CAP"
- PROMINENT ARCH IN BACK HALF OF MOUTHLINE
- LARGE, EXPOSED, TUSK-LIKE TEETH ON ADULT MALES
- DISTRIBUTION IN SUBARCTIC AND COOL TEMPERATE NORTH PACIFIC

FLUKES

oval-shaped scars, typical of cookie-cutter shark bites, are usually present, concentrated on the posterior half of the body. The density of these scars increases dramatically with age. Animals in the Sea of Japan, where cookie-cutter sharks are not known to occur, seem not to have such scars. Females have few linear scars or tooth rakes, while adult males are covered with them.

RANGE AND HABITAT Stejneger's Beaked Whale occurs in subarctic and cool temperate waters of the North Pacific and southwestern Bering Sea, including the Seas of Japan and Okhotsk. Strandings have been recorded as far south as Miyagi Prefecture, Japan, and Monterey, California. There is a strong peak in strandings in the Sea of Japan in winter and spring, suggesting that these whales may migrate. The presence of cookie-cutter shark scars on animals from the Aleutian Islands implies that they range south into warm temperate waters, where these sharks are most abundant. Stejneger's Beaked Whales are observed primarily in deep slope waters (about 2,500–5,000 ft/750–1,500 m deep).

SIMILAR SPECIES Mesoplodonts observed in Alaskan waters will almost certainly be Stejneger's Beaked Whales, as no other *Mesoplodon*

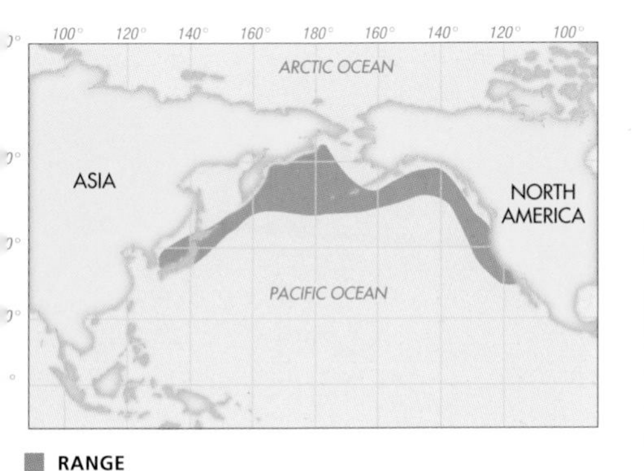

RANGE

STEJNEGER'S BEAKED WHALE
FAMILY ZIPHIIDAE

MEASUREMENTS AT BIRTH
LENGTH 6′11″–7′7″ (2.1–2.3 m)
WEIGHT Unavailable

MAXIMUM MEASUREMENTS
LENGTH **MALE** At least 17′6″ (5.25 m)
FEMALE At least 18′ (5.5 m)
WEIGHT Unavailable

LIFE SPAN Unavailable

species is known to occur there. Hubbs' Beaked Whales are likely to be confused with Stejneger's in lower latitudes, but Hubbs' have a much more prominent bulge ahead of the blowhole, and adults (especially males) have a light-colored or white melon. Adult males also have a white beak, and their tusks are substantially less exposed than those of Stejneger's. Cuvier's and Baird's Beaked Whales overlap the entire distribution of Stejneger's Beaked Whale. Both have a more prominent melon than Stejneger's, and their teeth are situated at the tip of the lower jaw rather than on a raised area well behind the tip. Cuvier's Beaked Whales have a shorter beak than Stejneger's, and adult Cuvier's, especially males, have paler dorsal coloration. Baird's Beaked Whale is considerably larger and has a longer beak than Stejneger's.

BEHAVIOR Stejneger's Beaked Whales occur in groups ranging from three or four to about 15 animals. The whales are tightly bunched when traveling near the surface, suggesting that groups are cohesive. A stranding of four adult females together suggests that there may be some degree of segregation by sex or age. Examinations of the bones of adult males have shown healed fractures, especially in the lower jaws. This evidence, together with the extensive body scarring, suggests that they fight with one another, presumably for access to females.

REPRODUCTION Apparently, some Stejneger's calves are born in spring (March–May) off central Honshū. By assessing the lengths of fetuses found in stranded females in Alaska, scientists projected that births would have taken place in approximately April to May and September.

FOOD AND FORAGING The stomach contents of Stejneger's Beaked Whales are dominated by meso- and bathypelagic squid from two families, Gonatidae and Cranchiidae. These squid do not make daily vertical migrations, so the whales clearly forage at considerable depths. Their suction mode of feeding causes them to ingest debris occasionally, including plastic bags and twine.

STATUS AND CONSERVATION Stejneger's Beaked Whales were formerly hunted in Japan along with Cuvier's Beaked Whales. They have been taken occasionally in the drift gillnet fishery for swordfish and sharks off the west coast of North America. In the absence of any reliable information on abundance, it is impossible to judge whether the species is in trouble in any part of its range.

River Dolphins

The river dolphins are a disparate group of four species that have been classified in a number of different ways. Scientists have sometimes assigned them to one family, but the current convention is to view them as belonging to four separate families, each containing a single genus. The river dolphins share a few primitive cranial features, and all possess long narrow jaws, but they are otherwise quite distinct from each other. Only three of the four species actually live in rivers.

FAMILY PLATANISTIDAE The Ganges and Indus River Dolphins, commonly called *susu* and *bhulan,* respectively, are considered to comprise a single species, although much of the recent literature refers to them as separate species. These small gray animals are virtually blind; they rely almost entirely on echolocation, and probably passive listening and touch, to navigate and forage in dark river waters of the Indian Subcontinent. Throughout most of their range, which includes the Indus, Ganges, Brahmaputra, Meghna, Karnaphuli, and Sangu River systems, they are the only cetaceans present in upstream waters. Ganges and Indus River Dolphins roll quickly at the surface, exposing their low, triangular dorsal hump, and sometimes also their long, forceps-like beak. Their habitat has been severely degraded by water development projects, and these dolphins are endangered throughout their range.

Amazon River Dolphin

FAMILY INIIDAE The Amazon River Dolphin, generally called *boto* in Brazil, occurs in both the Orinoco and Amazon River basins. Throughout most of its range, it co-occurs with the smaller Tucuxi, which is generally more active at the surface and less solitary. The Amazon River Dolphin has a long beak, small eyes, and a long fleshy dorsal keel instead of a dorsal fin. Often called the Pink Dolphin, it is dark gray when young and lightens with age, with some adults becoming uniformly pink or blotched with pink. The flippers are large and paddle-like, the flukes broad and triangular. Amazon River Dolphins have extremely flexible bodies and are able to twist and bend more than most dolphins. They can be bold and curious toward boats and swimmers, but also cryptic, as they sometimes surface to breathe without arching their back. This species remains relatively common in much of its range.

FAMILY LIPOTIDAE The Yangtze River Dolphin, also known by its Chinese name, *baiji,* lives only in the main channel of the Yangtze River of China, which it shares with a freshwater population of Finless Porpoises. It has a long slim beak and a triangular dorsal fin with a blunt peak. Its eyes are small but not nearly as reduced as those of the Ganges and Indus River Dolphins. Although generally light gray dorsally and white ventrally, the color pattern is complex, with white stripes or brush

Because they live in freshwater environments and compete with humans for food and water, river dolphins, like these two Amazon River Dolphins, are exceptionally vulnerable.

strokes sweeping up onto the sides of the face and the caudal peduncle. This dolphin is the world's most critically endangered cetacean.

FAMILY PONTOPORIIDAE The fourth species, the Franciscana or La Plata Dolphin, is a river dolphin in name only. It inhabits coastal marine waters from northern Argentina to southern Brazil, with highest densities in areas influenced by continental runoff. Like the other river dolphins, the Franciscana has a long narrow beak and a plump body. Its flippers are large, broad, and blunt at the tips. The dorsal fin rises at a low angle from the back and is straight or slightly undercut on the rear margin, with a blunt peak. This small gray dolphin frequently entangles in gillnets, causing concern that its numbers have been greatly reduced over much of its range since the 1940s.

Ganges and Indus River Dolphins

Platanista gangetica
(Roxburgh, 1801)

The Ganges and Indus River Dolphins inhabit several large river systems of the Indian subcontinent. They are among the most endangered cetaceans in the world and among the most intriguing. They are functionally blind. In any event, underwater vision would be useless in much of their extremely turbid habitat. Although they must surface in the normal manner, with the blowhole uppermost, they swim mostly on their side, nodding their head and sweeping the water column with an almost constant barrage of echolocation clicks. Scientists long recognized the animals in the Indus River basin as a distinct species, *Platanista minor;* however, it was recently decided that more analyses, including genetic comparisons, are necessary for a proper diagnosis. The two populations are currently regarded as subspecies: the Indus River Dolphin *(Platanista gangetica minor)* and the Ganges River Dolphin *(P. g. gangetica).* In the literature and locally, the Ganges subspecies is sometimes referred to by its Hindi name, *susu* (*shushuk* in Bangladesh), while the Indus subspecies is referred to as *bhulan,* from the Punjabi and Sindhi.

DESCRIPTION Indus and Ganges River Dolphins have plump, uniformly gray bodies. The head is small, with a rounded melon bisected by a longitudinal ridge. The beak is strikingly long and narrow and is thickened at the tip. The eyes are the size of pinholes and are, in fact, often harder to detect than the ear openings. There is a small, triangular dorsal prominence, or hump, no more than one or two inches high, positioned behind the middle of the back. The flippers are large and broadly splayed, with a scalloped outer margin. There are 26 to 39 pairs of teeth in the upper jaw and 26 to 35 pairs in the lower jaw. The lower teeth are much longer than the upper ones, and toward the front of the jaw the teeth interlock; in young animals, the front teeth overlap the sides of the jaw like fangs and are visible when the mouth is closed.

RANGE AND HABITAT The Indus River Dolphin occurred historically throughout the Indus River basin, including its tributaries, the Sutlej, Ravi, Chenab, and Jhelum Rivers; however, its current range is only about a fifth of what it was in the 19th century. It is now essentially confined to

- LONG, FORCEPS-LIKE BEAK
- LOW, SOMEWHAT TRIANGULAR DORSAL HUMP
- UNPATTERNED GRAY COLORATION
- OCCUR IN INDIAN SUBCONTINENT, IN FRESH AND POSSIBLY BRACKISH WATERS

the main channel of the Indus River in Pakistan, downstream of Chashma Barrage. Its highest densities are in Sind Province, between Guddu and Sukkur Barrages, and the next highest densities are in Punjab Province, between Taunsa and Guddu Barrages.

The Ganges River Dolphin occurs in the Ganges, Brahmaputra, Meghna, Karnaphuli, and Sangu River systems. Most of its distribution is in India and Bangladesh. Relatively high densities have been documented recently in the Vikramshila Gangetic Dolphin Sanctuary in India and in the lower Sangu River in southern Bangladesh. There are also at least hundreds in the Sundarbans of India and Bangladesh (a swampy region formed by the delta of the Ganges and Brahmaputra Rivers). The Ganges subspecies has been essentially extirpated from Nepal, except for about 20 animals in the Karnali River, in the western part of the country.

SIMILAR SPECIES The Irrawaddy Dolphin overlaps with the Ganges River Dolphin in the Sundarbans; it has a dorsal fin and a rounded head with no beak. There is no danger of confusing the Indus River Dolphin with any other

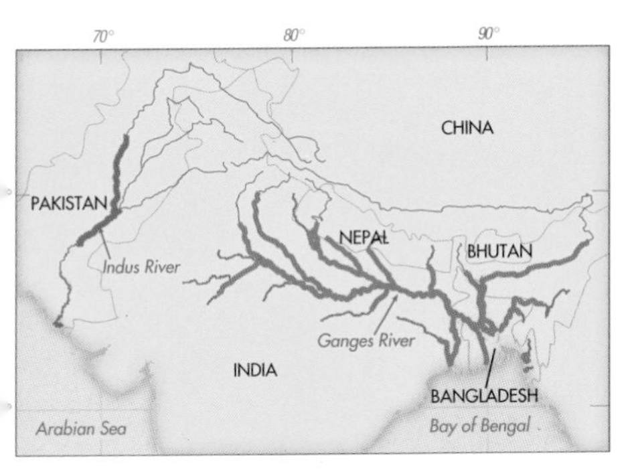

RANGE OF INDUS RIVER DOLPHIN
RANGE OF GANGES RIVER DOLPHIN

GANGES AND INDUS RIVER DOLPHINS
FAMILY PLATANISTIDAE

MEASUREMENTS AT BIRTH
LENGTH 28–35″ (70–90 cm)
WEIGHT Unavailable

MAXIMUM MEASUREMENTS
LENGTH **MALE** 6′11″ (2.12 m)
FEMALE 8′3″ (2.52 m)
WEIGHT 185 lb (84 kg)

LIFE SPAN Unavailable

LEFT: *Ganges and Indus River Dolphins do not often lift their tails out of the water as they dive. In fact, their surfacing is usually quick and inconspicuous. In this instance, the animal has provided a good look at the low triangular dorsal hump and the deeply concave flukes.*
RIGHT: *Ganges and Indus River Dolphins live in close proximity to people, in areas such as the Karnali River in Nepal. They share their immediate environment with livestock, ferries, houseboats, and bathers. Sometimes when they surface to breathe, the long, narrow beak comes into view before the animal quickly arches the back and disappears.*

species because it is the only cetacean that presently inhabits the Indus River. Historically, the Finless Porpoise, Hump-backed Dolphin, and one or both of the bottlenose dolphins may have overlapped with the Indus subspecies in the Indus Delta.

BEHAVIOR Very little is known about the social organization of Ganges and Indus River Dolphins. They frequently appear to be solitary; however, loosely associated groups of a few dozen are occasionally seen in areas of the river where currents are interrupted. These dolphins are almost always active, which is not surprising, given their constantly moving environment. When surfacing, they sometimes leap clear of the water, and their long beak is often visible as they roll for a breath. Dives average between about 30 and 90 seconds in length but can last for several minutes.

REPRODUCTION Young dolphins appear to begin taking solid food within a month or two after birth, and they are almost certainly weaned well before one year of age. However, reproductive parameters of these dolphins are highly uncertain, and estimates scattered throughout the literature should be viewed with skepticism. It has proven impossible to obtain reliable samples of dead dolphins, and observations of living animals are hampered by the extreme difficulty of distinguishing individuals, for example, through photo-identification. Also, environmental conditions during the monsoon season and the socioeconomic and political situation in most areas where the dolphins are available for study at other seasons preclude the kind of sustained fieldwork necessary for obtaining good life-history data.

FOOD AND FORAGING As these river dolphins are known to root in bottom sediments, many of their prey items must live on or near the river bottoms. Young Ganges and Indus River Dolphins seem to eat mainly shrimp and small fish, and adults also sometimes take fairly large fish. Pairs or small groups have been observed diving in the same area and surfacing synchronously, leading to suggestions that they sometimes forage cooperatively.

STATUS AND CONSERVATION Although these river dolphins still have a fairly extensive range, their distribution has contracted, and their numbers have declined dramatically in many areas. For the past 30 years it was thought that Indus River Dolphins numbered only in the hundreds. However, a range-wide survey in 2001 revealed that there were still at least 1,000 of these dolphins. Ganges River Dolphins are more abundant and occupy a much wider range, but there is no good overall abundance estimate for them. Entanglement in fishing gear represents a major threat to these dolphins. Locally, dolphin oil is valued as a liniment, a remedy for livestock ailments, and an aphrodisiac, and fishermen in parts of India and Bangladesh use dolphin oil and minced dolphin meat to lure catfish to their boats. Consequently, there is a strong incentive to obtain dolphin carcasses, whether "incidentally" in nets or deliberately by harpooning. Ganges and Indus River Dolphins are also adversely affected by threats to their habitat, including the discharge of contaminants into rivers and the extraction of water for agricultural, industrial, and household uses. In addition, irrigation and hydroelectric dams have fragmented the populations, especially in the Indus River basin. In Pakistan, dolphins sometimes become marooned in irrigation canals during floods and are unable to return to the river. A rescue program for these animals is carried out by the Pakistan office of the World Wide Fund for Nature.

Amazon River Dolphin

Inia geoffrensis
(Blainville, 1817)

This freshwater cetacean is unquestionably the most abundant of the world's river dolphins, found throughout the Amazon and Orinoco River basins of South America. During the late 1950s and 1960s, the Amazon River Dolphin was sought after by oceanariums in the United States, Japan, Europe, and South America. More than 70 were imported into the United States between 1956 and 1966, only one of which survived to the mid-1980s. As a result of its former presence in oceanariums, the Amazon River Dolphin is more familiar to the public than the other river dolphins. *Boto* (incorrectly spelled *boutu* in some of the older literature) is the dolphin's Brazilian name. It is also known as *tonina,* primarily in Venezuela, and as *bufeo* in other Spanish-speaking countries, such as Peru, Ecuador, and Bolivia. The Amazon River Dolphin is often called the Pink Dolphin.

DESCRIPTION The Amazon River Dolphin's appearance is truly bizarre and unlike that of any other cetacean. The body is long and unusually flexible. The head is small, with a long beak. The melon changes shape according to the animal's disposition and can be bulging or lumpy. The long mouthline, angling up toward the beady eyes, is suggestive of a false, untrustworthy smile. (Indeed, folklore about the Amazon River Dolphin indicates that local people regard it with respectful dread rather than with the reverence they feel for the Tucuxi.) The dorsal fin is low, with an exceptionally long base and a rounded peak, forming a triangular lobe; it is located approximately at midback. The flippers are extraordinarily long and broad, with scalloped rear margins. The flukes are broad and triangular, and often ragged along the rear margins. The beak and the leading edge of the flippers are often abraded or scarred, especially on old individuals. There are 24 to 35 pairs of teeth in both the upper and lower jaws. The rearmost 8 to 10 pairs on each jaw have a flange, or cusp, on the outside and a depression on the inside; these molar-like "cheek teeth" presumably enhance this dolphin's ability to crush hard-bodied prey such as armored catfish.

The coloration varies from solid gray to brilliant pink. Young Amazon River Dolphins are uniformly gray. As they age, the flanks and ventral surfaces become white, pink, or pink-tinged. On old animals, the back, too, can become pink or blotched with pink, although the blowhole

- BULBOUS LUMPY MELON
- VERY LONG BEAK
- LONG, LOW DORSAL FIN (OR RIDGE)
- WHITISH OR PINK COLORATION ON ADULT
- LIMITED TO AMAZON AND ORINOCO RIVER BASINS

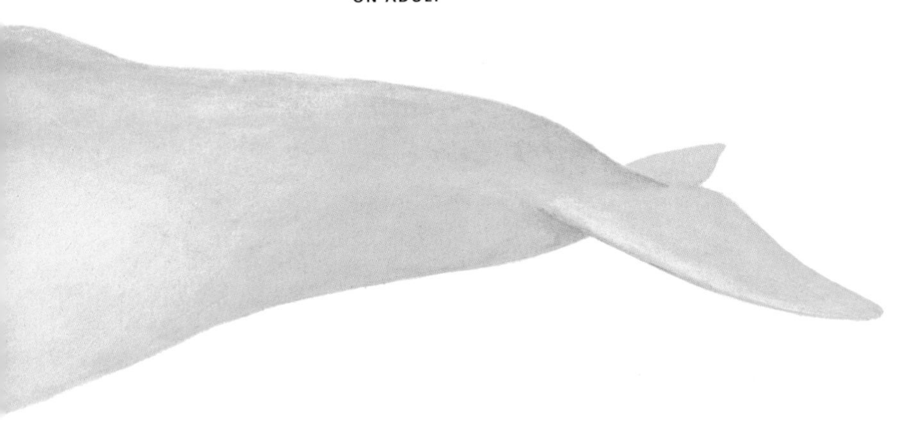

region and the middle of the back tend to remain at least partially gray. The pink skin is caused by blood flow through the capillaries and fades quickly after the animal dies. The pink coloration is also affected by activity and water clarity; pink Amazon River Dolphins occur mainly in turbid waters, and generally become pinker when aroused.

RANGE AND HABITAT There are three geographically separate subspecies: one in the Orinoco River basin *(Inia geoffrensis humboldtiana)*, one in the Amazon River basin, but not including the upper Madeira River *(I. g. geoffrensis)*, and one in the Madeira River system between Pôrto Velho and Guajará-Mirim *(I. g. boliviensis)*. They occur in the river mouths, and their upstream limits are defined only by the presence of impassable rapids or waterfalls. Amazon River Dolphins may enter connecting rivers, lakes, streams, and side channels of any size, and even flooded forests. Confluences, where two or three large rivers converge or where a tributary joins a main channel, are the species' preferred habitat.

80° 70° 60° 50° 40° 10° 0° 10° 20°
Orinoco R.
Amazon R.
ATLANTIC OCEAN
PACIFIC OCEAN
SOUTH AMERICA

RANGE

AMAZON RIVER DOLPHIN
FAMILY INIIDAE

MEASUREMENTS AT BIRTH

LENGTH	30–32″ (76–82 cm)
WEIGHT	15–18 lb (7–8 kg)

MAXIMUM MEASUREMENTS

LENGTH	MALE	8′2″ (2.5 m)
	FEMALE	7′3″ (2.2 m)
WEIGHT	MALE	400 lb (180 kg)
	FEMALE	260 lb (120 kg)

LIFE SPAN

MALE 36 years
FEMALE At least 21 years

SIMILAR SPECIES The Tucuxi, the only cetacean that shares the Amazon River Dolphin's habitat, occurs in most of the river dolphin's Amazon range and at least the lower reaches and delta of the Orinoco River system. Tucuxis usually occur in larger groups and are smaller and quicker, with a more typically dolphin-like dorsal fin and a comparatively short beak. At the water's surface, Tucuxis tend to roll quickly, providing a clear view of the dorsal fin; they also frequently leap clear of the water in an arc.

BEHAVIOR Amazon River Dolphins are solitary or only mildly social, rarely occurring in groups of more than three or four individuals. However, individuals and small groups often form larger temporary aggregations, sometimes to feed around concentrations of fish and sometimes in curious approaches to humans who are boating, fishing, or swimming. They often follow boats, as though anticipating some kind of benefit, such as discarded or injured fish. Amazon River Dolphins can appear to be sluggish swimmers. They typically surface in two ways. At times, they "sneak" along the water's surface, maintaining a horizontal position and exposing only the upper

ABOVE: *The Amazon River Dolphin is sometimes called the Pink Dolphin, because its coloration is often a diffuse combination of gray and pink. This profile view of an Amazon River Dolphin in a Venezuelan river shows the low, triangular dorsal fin, the bulbous melon, and the long beak typical of this freshwater species.*

BELOW: *These captive Amazon River Dolphins, mother and calf, exhibit the long beak, large flippers, and "puffy" face characteristic of their species.*

Amazon River Dolphins can easily maneuver along the flooded rainforest floor as well as in the rivers and lakes of South America. Their long beak is useful for digging and probing, and their neck is exceptionally flexible.

melon and blowhole. At other times, they arch and roll, with the head and often the entire beak surfacing first, followed by the dorsal fin and much of the back. While all dolphins chuff as they exhale, Amazon River Dolphins often take this to an extreme, making a very loud snort or sneezing sound.

REPRODUCTION In the Brazilian Amazon, calving takes place mainly in May, June, and July, as water levels reach their peak and begin to decline. Presumably, lactating females are better able to meet their energy needs at this time, as fish are forced by the receding waters to leave the flooded forests and concentrate in the rivers. In an Amazon River Dolphin population studied in an Orinoco tributary, however, calves began appearing only near the end of the low-water period. As water levels rose, their numbers increased, and they were never seen during periods of falling water levels. It is possible that the timing of reproductive events differs across the Amazon River Dolphin's extensive, varied, and highly dynamic range. The gestation period lasts 10 to 11 months. Females, at least occasionally, are simultaneously lactating and pregnant.

FOOD AND FORAGING Amazon River Dolphins are versatile foragers, known to consume more than 50 species of fish from 19 different families. The fish prey are usually 2 to 30 inches (5–80 cm) long (averaging about 8 in/20 cm). Amazon River Dolphins also at least occasionally take mollusks, crustaceans, and small turtles. Their diet varies seasonally in relation to water levels. Generally, Amazon River Dolphins can be more selective during periods of low water levels when fish are constrained to the main river channels and the deeper areas of lakes. During periods of high water, fish are able to disperse much more widely, and they become more difficult to find and capture.

STATUS AND CONSERVATION Amazon River Dolphins are presently not endangered. They continue to occupy most of their historic range and number in at least the tens of thousands. However, many of the traditional beliefs that kept people from killing these dolphins are rapidly disappearing. Some fishermen now kill or mutilate river dolphins they discover in their nets. Others shoot or poison them to protect fishing gear and reduce what they perceive as competition for valuable fish resources. Like other river cetaceans, Amazon River Dolphins are vulnerable to the degradation of water quality caused by poor land-use practices, such as deforestation, unmanaged or poorly managed disposal of mining and agricultural wastes, and discharge of untreated sewage and chemical contaminants. They also suffer from the effects of water development projects. While there are relatively few dams in the main channels of the Amazon and Orinoco compared with rivers in southern Asia, many schemes for regulating the flow of these rivers are either already planned or in the early stages of development.

Yangtze River Dolphin

Lipotes vexillifer
Miller, 1918

Western scientists were unaware of the Yangtze River Dolphin's existence until the early 1900s. By the end of the 20th century, however, this freshwater dolphin endemic to China had come to be regarded as the world's most endangered cetacean. The Yangtze River Dolphin is often called by its Chinese name, *baiji* (pronounced byé-gee). The scientific name has interesting origins. The genus name has nothing to do with lipids, or fats, as one might suppose. Rather, it comes from the Greek word *leipo,* meaning "one who is left behind." When Gerrit S. Miller, Jr., a curator at the Smithsonian Institution, formally described the species in 1918, he apparently intended to convey the idea that the Yangtze River Dolphin was an evolutionary relic, left to evolve in isolation in its freshwater Yangtze River environment. The specific name comes from the Latin word *vexillum,* meaning "banner," and the suffix *-fer,* "to bear"; Miller had been incorrectly informed that the river dolphin's Chinese name meant "flag bearer," supposedly referring to the appearance of the dorsal fin. Some of the older literature refers to the Yangtze River Dolphin as the White Flag Dolphin.

DESCRIPTION The Yangtze River Dolphin has a robust body with a small head and a long narrow beak that is slightly upturned at the tip. The mouthline curves up slightly at the corners. The melon is marked by a bulbous protrusion, giving the head a bluff appearance in profile. The eyes are small and inconspicuous, set unusually high on the sides of the head. The dorsal fin, set behind midback, is low, blunt-peaked, and triangular. The flippers are broad, with rounded tips. Yangtze River Dolphins have 30 to 34 pairs of teeth in the upper jaw and 32 to 36 pairs in the lower jaw.

The dorsal surfaces are light bluish gray, and the ventral surfaces are white. The whiteness on the throat intrudes onto the sides of the head and neck in irregular stripes that look like brush strokes. Similar intrusions of white are present on the sides of the caudal peduncle. The facial color pattern is variable and may be used to distinguish individuals, particularly in combination with scars and nicks on the dorsal fin. The difficulties of finding, approaching, and photographing these dolphins, however, severely limit the feasibility of photo-identification as a research tool.

- LOW, TRIANGULAR DORSAL FIN
- LONG NARROW BEAK AND BLUFF MELON
- GRAY DORSAL COLORATION WITH STROKES OF WHITE ON FACE AND TAIL
- DISTRIBUTION LIMITED TO YANGTZE RIVER, CHINA

RANGE AND HABITAT The Yangtze River Dolphin is endemic to China, with its current distribution confined to the middle and lower reaches of the Yangtze River's main channel. Historically, it ranged somewhat farther upriver and also inhabited the Yangtze's two large lake systems, Dongting and Poyang, and apparently the Fuchun River to the south of the Yangtze. Like other river dolphins, Yangtze River Dolphins prefer countercurrent eddies that form in the meandering, low-gradient portions of the river where tributary confluences, mud banks, or sandbars interrupt the water flow.

115°
120°
Yangtze River
CHINA
East China Sea

■ RANGE

SIMILAR SPECIES The Finless Porpoise, the only other cetacean that regularly inhabits the Yangtze River basin, occurs throughout the Yangtze River Dolphin's range. It is smaller than the river dolphin and generally a darker shade of gray, with a blunt head and no dorsal fin. The Indo-Pacific Hump-backed Dolphin is similar in size and appearance to the Yangtze River Dolphin but has a more falcate dorsal fin. It has been known to wander as far as 90 miles (150 km) up the Yangtze River, and so may occur in the extreme lower end of the Yangtze River Dolphin's range. (In Hong Kong, the Hump-backed Dolphin is often called

YANGTZE RIVER DOLPHIN

FAMILY LIPOTIDAE

MEASUREMENTS AT BIRTH

LENGTH 35–37″ (90–95 cm)

WEIGHT Unavailable

MAXIMUM MEASUREMENTS

LENGTH **MALE** 7′6″ (2.29 m)
FEMALE 8′4″ (2.53 m)

WEIGHT **MALE** 290 lb (130 kg)
FEMALE 370 lb (170 kg)

LIFE SPAN Unavailable

Observations of a captive dolphin, seen here at the Wuhan Institute of Hydrobiology, China, have provided almost everything that is known about the behavior of Yangtze River Dolphins, and there is little prospect of learning more about their life in the wild. Recent surveys of the species' range have resulted in only a few sightings, and it is feared that the Yangtze River Dolphin is in serious danger of extinction.

the Chinese White Dolphin, a name that can be easily confused with the Yangtze River Dolphin's antiquated name, White Flag Dolphin.)

BEHAVIOR Nothing is known about the social relations of these river dolphins. Yangtze River Dolphins typically occur in small groups of two to six. Although group size can be as large as 15 or 16 individuals, Yangtze River Dolphins have not been seen in groups of 10 or more for many years. These dolphins are strong swimmers, able to move at least short distances against the Yangtze's strong current. One group of three dolphins tracked by scientists traveled about 60 miles (100 km) upriver in three days. Another Yangtze River Dolphin, identified by a distinctive notch in its dorsal fin, was photographed 125 miles (200 km) farther upriver than it had been seen the year before.

REPRODUCTION Little is known about reproduction in this species. Scientists believe that Yangtze River Dolphins mate in April or May and give birth mainly in February and March, suggesting a gestation period of about 10 or 11 months.

FOOD AND FORAGING Judging by the stomach contents of a few individuals and by observations of captive dolphins, Yangtze River Dolphins prey on fish up to about 3½ inches (9 cm) long. Given the size of their mouth and throat, they probably avoid eating fish that weigh more than about 9 ounces (250 g). However, one captive adult male regularly eats fish weighing 10 to 18 ounces (300–500 g). Yangtze River Dolphins swallow their prey headfirst, presumably to keep the sharp spines on the fish from snagging on their throat or esophagus.

STATUS AND CONSERVATION While there is great uncertainty about how many Yangtze River Dolphins exist, there is no doubt that their numbers are critically low, probably only in the tens. Their rapid decline in the last few decades may be attributed to a combination of factors: accidental mortality in fisheries (especially from "rolling hooks" and electrofishing), vessel collisions, underwater blasting, habitat degradation caused by intensive ship traffic, the damming of tributary streams and lakes, reductions in prey populations, and pollution. For more than a decade, the Chinese government has been attempting to capture Yangtze River Dolphins and relocate them into a so-called "seminatural reserve," where they can be better protected. This policy has failed: Thus far only one animal, an adult female, has been caught and relocated, and she was found dead, entangled in netting, less than six months after being released into the reserve. A male, Qi Qi (pronounced chee-chee), rescued in 1980 and rehabilitated after being badly injured by fishhooks, remains the only Yangtze River Dolphin in captivity.

ABOVE: *The Yangtze River Dolphin has small eyes, a rounded melon, and puffy "cheeks."* **RIGHT:** *Qi Qi, a male Yangtze River Dolphin that has been in captivity in Wuhan since 1980, exhibits a subtle color pattern, with gray tones dorsally and white undersides; the colors blend irregularly on the sides.*

Franciscana

Pontoporia blainvillei
(Gervais and d'Orbigny, 1844)

- EXTREMELY LONG BEAK
- LOW, ROUNDED DORSAL FIN
- NONDESCRIPT GRAY COLORATION
- LOW SURFACING PROFILE AND BRIEF APPEARANCE AT SURFACE
- RESTRICTED TO TEMPERATE COASTAL WATERS OF EASTERN SOUTH AMERICA

This small, cryptic dolphin lives exclusively in estuarine and coastal marine waters of eastern South America. Franciscanas are difficult to observe, and little is known about them. However, given the species' alarmingly high rate of mortality from entanglement in gillnets over the last few decades, scientists fear it is in serious trouble. The Franciscana is also called La Plata Dolphin, a name that was applied to the species because the first described specimen was collected in the Río de la Plata estuary in Uruguay. The international scientific community has since adopted the name Franciscana, its local name in Argentina and Uruguay. In Brazil, the Franciscana is known as *toninha.*

DESCRIPTION The Franciscana is one of the smallest cetaceans. It has a plump body with a somewhat bulging melon that gives it a bluff appearance in profile. The beak is extremely long and slender. The relative length of the beak increases with age, from less than 10 percent of the total body length in young animals to as much as 15 percent in older animals. The mouthline is long and straight. The dorsal fin, set somewhat behind midback, is moderate in size with a rounded tip. It rises at a shallow angle on the leading edge and has a straight trailing edge. The flippers are large and broad, with rounded tips. There are 53 to 58 pairs of small teeth in the upper jaw and 51 to 56 pairs in the lower jaw.

Franciscanas are basically gray overall, with a lighter ventral coloration. Individuals in the southern parts of the species' range are said to appear more brown than gray.

RANGE AND HABITAT Franciscanas live in coastal waters from approximately Espírito Santo state, Brazil (18°25'S), south to San Matías Gulf, Argentina (41°10'S). Differences in body size, parasites, and genetic characteristics suggest that there are at least two distinct populations. The animals in south-central Brazil are smaller than those in southern Brazil and Uruguay. Franciscanas usually remain within about 30 nautical miles of shore and in waters less than about 100 feet (30 m) deep, although they are occasionally found in areas to about 200 feet (60 m) deep. Franciscanas occur in highest densities in areas with large volumes of continental runoff, such as Lagoa dos Patos (in Brazil) and Río de la Plata.

SIMILAR SPECIES Burmeister's Porpoise overlaps with the Franciscana throughout most of its range, from southern Brazil (about 29°S) southward. Burmeister's is generally larger and darker, with no beak and a more tapered, pointed dorsal fin. Its relatively small size, low surfacing profile, and gray coloration could cause confusion in brief sightings. The Tucuxi overlaps with the Franciscana along the coast of Brazil to about 28°S; its general appearance may be similar, apart from the much shorter beak and more pointed and slightly falcate dorsal fin.

BEHAVIOR Franciscanas occur in relatively small groups, ranging from a few to about 15 individuals. They are not known to engage in aerial displays, and they make little commotion when surfacing, making them difficult to detect at sea. While remains of Franciscanas have been found in the stomachs of large sharks, scientists do not know whether these represent cases of scavenging or active predation.

REPRODUCTION Franciscanas reach sexual maturity at an early age, and most females give birth by four or five years old. The gestation period is estimated to be 10½ to 11 months. Lactation probably lasts at least nine months, although calves begin taking some solid food at about three months old. Females are sometimes simultaneously lactating and pregnant.

FOOD AND FORAGING Franciscanas are known to eat at least 24 species of fish, primarily small bottom-dwellers such as juvenile croakers and weakfish. They also consume shrimp and squid. Their diet varies greatly by area, season, and age.

STATUS AND CONSERVATION The nearshore fisheries in Brazil, Uruguay, and Argentina have developed rapidly since World War II, resulting in increased incidental entanglement of Franciscanas in gillnets. Until recently, most local studies of Franciscanas have consisted only of surveying beaches to find and examine stranded carcasses. Researchers have used the stranding rates to estimate trends in the bycatch. This approach is not very satisfactory. Among its shortcomings is the fact that the proportion of by-caught dolphins that end up on the beach is unknown, and therefore so is the total incidental kill. The best way to monitor the bycatch is by placing observers on fishing vessels, but in the small-scale fisheries that affect Franciscanas, such programs are difficult to implement. South American scientists recently carried out the first successful aerial surveys of Franciscanas. They estimate that there are about 40,000 dolphins in the coastal waters of the Brazilian state Río Grande do Sul and Uruguay.

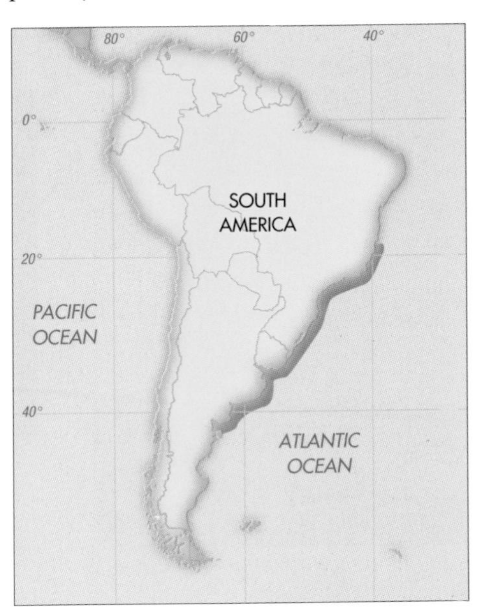

FRANCISCANA
FAMILY PONTOPORIIDAE

MEASUREMENTS AT BIRTH
LENGTH 30–31″ (75–80 cm)
WEIGHT 17–19 lb (7.3–8.5 kg)

MAXIMUM MEASUREMENTS
LENGTH **MALE** 5′2″ (1.58 m)
FEMALE 5′9″ (1.74 m)
WEIGHT **MALE** 95 lb (43 kg)
FEMALE At least 115 lb (52 kg)

LIFE SPAN
MALE 18–20 years
FEMALE 15 years

■ RANGE

Beluga and Narwhal

The Beluga and the Narwhal are the only two living species in the family Monodontidae. Although there has never been much doubt about the close systematic relationship between the two species, some taxonomists have assigned the Beluga to a separate family or subfamily, sometimes linking it with the Irrawaddy Dolphin, which is superficially similar. Several authorities have even placed the Beluga and the Narwhal in the family Delphinidae, the ocean dolphins.

The monodontids are medium-size whales with almost no beak, broad flippers, and no dorsal fin. The flukes of adults in both species are strongly convex along the rear margin. The family name, from the Greek *monos,* meaning "one," and *odon,* meaning "tooth," is something of a misnomer. The Beluga, far from being single-toothed, has eight or nine pairs of peg-like teeth in each jaw. Narwhals have no functional teeth inside the mouth, but a long, spiraled tooth erupts through the left side of

Narwhals

Beluga

the upper jaw of adult males, extending forward and slightly downward to a length of more than 6 feet (2 m).

These whales are, along with the Bowhead Whale, the only cetaceans that live year-round in the Arctic. They sometimes occur in close proximity, but in general, Narwhals live in deep sounds and channels and seem more adapted to heavy pack-ice conditions. Belugas frequent shallow estuaries in summer, and some populations spend much of the year in ice-free waters or in areas with relatively light ice cover. Both species are deep divers. Belugas and Narwhals continue to be hunted in many northern communities for their skin, a delicacy among natives of the Arctic, and in the case of the Narwhal for its valuable tusk ivory.

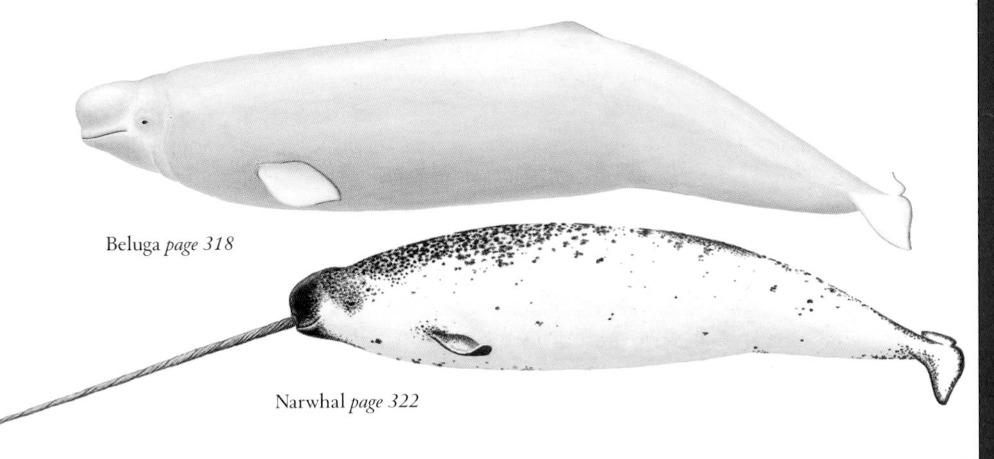

Beluga *page 318*

Narwhal *page 322*

Beluga

Delphinapterus leucas
(Pallas, 1776)

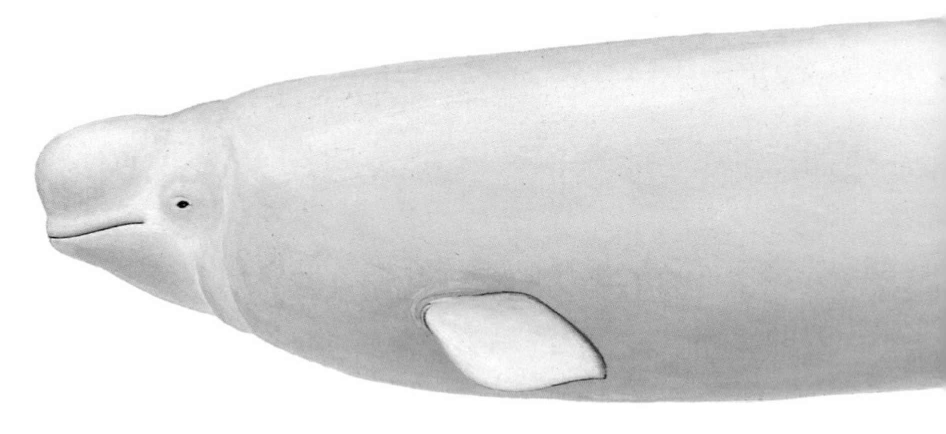

Known by some early whalers as "sea canaries" because of their loquacious natures, these whales are abundant and widespread in the Arctic and Subarctic. For many centuries, Belugas, also known as White Whales, have been a staple of arctic societies, providing food, fuel oil, and even soft durable leather. They were among the first cetaceans to be brought into captivity. Their resilience and adaptability, stunning appearance, engaging disposition, and trainability have made them popular performers in oceanariums. Several areas where Belugas congregate have become whale-watching meccas, most notably eastern Canada's lower St. Lawrence River and the Churchill River estuary in western Hudson Bay. Over the past 15 years, there has been a flurry of research on the species, much of it involving satellite telemetry. These studies have shown that the Beluga has impressive diving abilities and is even more ice-adapted and abundant than was previously believed.

DESCRIPTION The Beluga has a rounded midsection that tapers toward the head and tail. Its torso is markedly rotund when the animal is well fed. The head is unlike that of any other cetacean, with a bulging melon that one researcher described as feeling like a balloon filled with warm lard. A Beluga is able to change the shape of its melon at will, presumably by moving air around in various sinuses. The neck is unusually mobile because the cervical vertebrae are not fused, and Belugas readily turn or nod their heads. The beak is short and broad, with a cleft upper lip and a labile mouth that can be puckered. The belly and sides may be lumpy, with folds and creases of fat. There is no dorsal fin, but there is a narrow ridge on the back where a dorsal fin would otherwise be. The broad flippers are upcurled at the tips in large males. The flukes become increasingly ornate as the animal ages, and those of mature adults are strongly convex on the rear margin. There are eight to nine pairs of peg-like teeth in both the upper and lower jaws, sometimes worn down to the gum in older adults.

Young Belugas are evenly gray. They lighten as they age and eventually become completely white except for dark pigment on the dorsal ridge and along the edges of the flukes and flippers. The white skin of adults can sometimes have a yellowish cast when they begin congregating in estuaries in summer, but this disappears after they molt.

- WHITE ADULT COLORATION
- ROUNDED, MALLEABLE MELON AND FLEXIBLE NECK
- SHORT BROAD BEAK WITH CLEFT UPPER LIP
- BROAD FLIPPERS AND ORNATELY SHAPED FLUKES
- LACK OF DORSAL FIN
- OCCURS ONLY IN HIGH LATITUDES OF NORTHERN HEMISPHERE

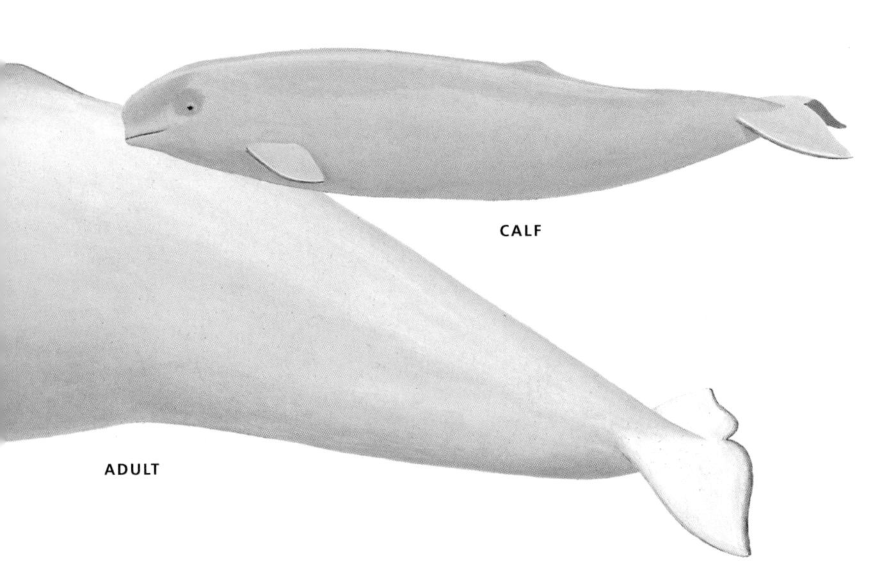

RANGE AND HABITAT Belugas have an essentially circumpolar distribution in the Northern Hemisphere, centered mainly between 50°N and 80°N. Nearly 30 stocks are provisionally recognized for management purposes. Stocks are defined primarily on the basis of summering grounds, most of which are centered on estuaries where the animals molt. Belugas exhibit a high degree of philopatry, or loyalty to a site, and individuals (females in particular) tend to return, year after year, to the estuary visited by their mother in the year of their birth. In fall, Belugas are driven away from bays and estuaries by ice, and they winter primarily in polynyas, near the edges of pack ice, or in areas of shifting, unconsolidated ice. They appear to be equally at home in shallow river mouths, where they may become stranded between tides, and in deep submarine

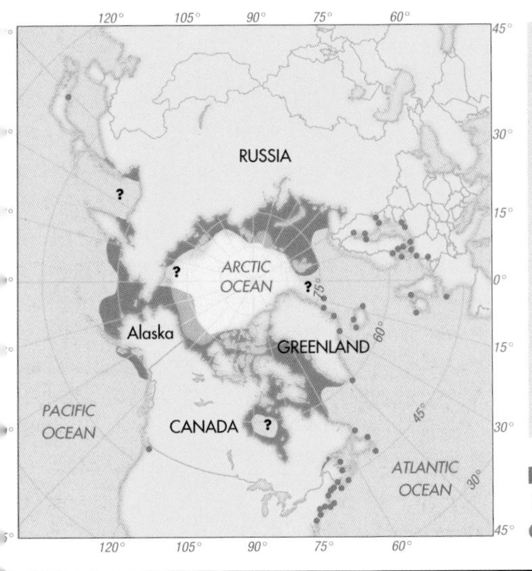

BELUGA

FAMILY MONODONTIDAE

MEASUREMENTS AT BIRTH

LENGTH 4′11″–5′3″ (1.5–1.6 m)

WEIGHT 176–220 lb (80–100 kg)

MAXIMUM MEASUREMENTS

LENGTH **MALE** 14–16′ (4.2–4.9 m)
FEMALE 13–14′ (3.9–4.3 m)

WEIGHT **MALE** 2,400–3,500 lb (1,100–1,600 kg)
FEMALE 1,500–2,600 lb (700–1,200 kg)

LIFE SPAN

At least 25 years, possibly more than 50

■ RANGE
? POSSIBLE RANGE
● EXTRALIMITAL RECORDS

trenches, where they dive to depths in excess of 2,600 feet (800 m).

SIMILAR SPECIES The Narwhal is the species most likely to be confused with the Beluga, but mainly in latitudes north of about 65° N. Adult male Narwhals usually have a spiraled tusk jutting forward from the upper lip, making them reasonably easy to distinguish, and the mottled or spotted skin of adult Narwhals is in contrast to the even gray or white of Belugas. In the Arctic and Subarctic at times, particularly from an aerial perspective, the silvery flashes from a shoal of Harp Seals may superficially resemble a pod of young Belugas rolling at the surface. The tails of seals, however, move from side to side rather than vertically, and Harp Seals tend to be quicker, more active, and inclined to remain at or just below the surface. Whitecaps, small bits of floating ice, and even seabirds can be difficult to distinguish from Belugas at a distance. One experienced researcher describes a Beluga at ½ to 1¼ miles (1–2 km) away as a white spot that appears, grows, shrinks, and disappears, remaining in view for about three seconds.

BEHAVIOR Belugas are highly social, occurring in close-knit pods, often of the same sex and age class. Groups of large males, numbering several hundred, are observed in summer, as are smaller groups consisting of mothers and their dependent calves. Aggregations of Belugas in estuaries can build to thousands of animals when undisturbed by hunting. Belugas have a diverse vocal repertoire that encompasses trills, squawks, bell-like sounds, sharp reports (possibly caused by jaw clapping), and a sound like that made by rusty gate hinges. Bill Schevill, a pioneer in the field of cetacean bioacoustics, described their "high-pitched resonant whistles and squeals, varied with ticking and clucking sounds slightly reminiscent of a string orchestra tuning up, as well as

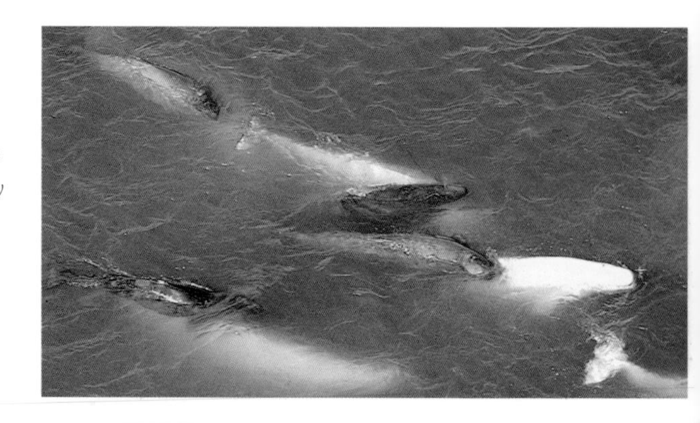

ABOVE: *The Beluga's melon is bulbous and malleable. This animal's short, broad beak is well demarcated from the melon. Its skin appears to be in transition from gray to white as occurs as Belugas approach maturity.*
RIGHT: *The all-gray Beluga calves are easily distinguishable from the essentially all-white adults.*

A closely spaced pod of five adult Belugas moves along the coast of Alaska in pack ice. Adaptation to living in an icy environment has allowed the Beluga to disperse throughout most of the Arctic and Subarctic.

mewing and occasional chirps." Sometimes their calls reminded him of a crowd of children shouting in the distance. The most serious hazards for wild Belugas, apart from human hunters, are Killer Whales and Polar Bears. The bears quickly converge on areas where Belugas are ice-entrapped, taking a heavy toll by swiping at the animals with their powerful paws and dragging them onto the ice.

REPRODUCTION The timing of reproductive events varies by region. In general, conception takes place in late winter or spring when the animals are least accessible for observation (late February to mid-April in Alaska; May in eastern Canada and West Greenland). Credible estimates of the gestation period range from somewhat less than a year to 14½ months. Young Belugas are nursed for two years and may continue to associate with their mothers for a considerable time thereafter. The calving interval probably averages three years.

FOOD AND FORAGING The diets of Belugas vary according to regional and seasonal prey availability. Stomach contents of individuals from various regions demonstrate that the species' overall diet includes a great variety of organisms: fish (from salmon to arctic cod to herring and capelin), cephalopods (squid and octopuses), crustaceans (shrimps and crabs), marine worms, and even large zooplankton. Many prey items are bottom-dwelling organisms. This probably explains why many dives (monitored with time-depth recorders) have a "square" profile, characterized by a steep and continuous descent and ascent, with a distinct bottom phase in between. The whales are almost certainly foraging near the seabed, at depths of at least 1,000 feet (300 m). The Beluga's puckered lips serve to create suction as the animal forages (and also enable Belugas to shoot streams of water at oceanarium spectators).

STATUS AND CONSERVATION Although there are well over 100,000 Belugas in the circumpolar Arctic today, their aggregate abundance was much greater in the past, before commercial hunting decimated some groups. Among the more robust populations today are those in the Beaufort Sea (40,000), the eastern High Arctic of Canada (28,000), western Hudson Bay (25,000), and the eastern Bering Sea (18,000). The whales in these four areas are hunted locally, but the removal rates are thought to be sustainable. In contrast, a number of other populations are in great peril and should not be, but are, still hunted. These include those in Cook Inlet, Ungava Bay, and some parts of southeastern Baffin Island and West Greenland. The animals in the St. Lawrence River have high contaminant burdens in their bodies and high cancer rates. Some formerly important Beluga estuaries are now infested with motorboats and hunters, rendering them unsuitable to support large aggregations of the whales. Hunt management is the most critical immediate imperative for Beluga conservation.

Narwhal

Monodon monoceros
Linnaeus, 1758

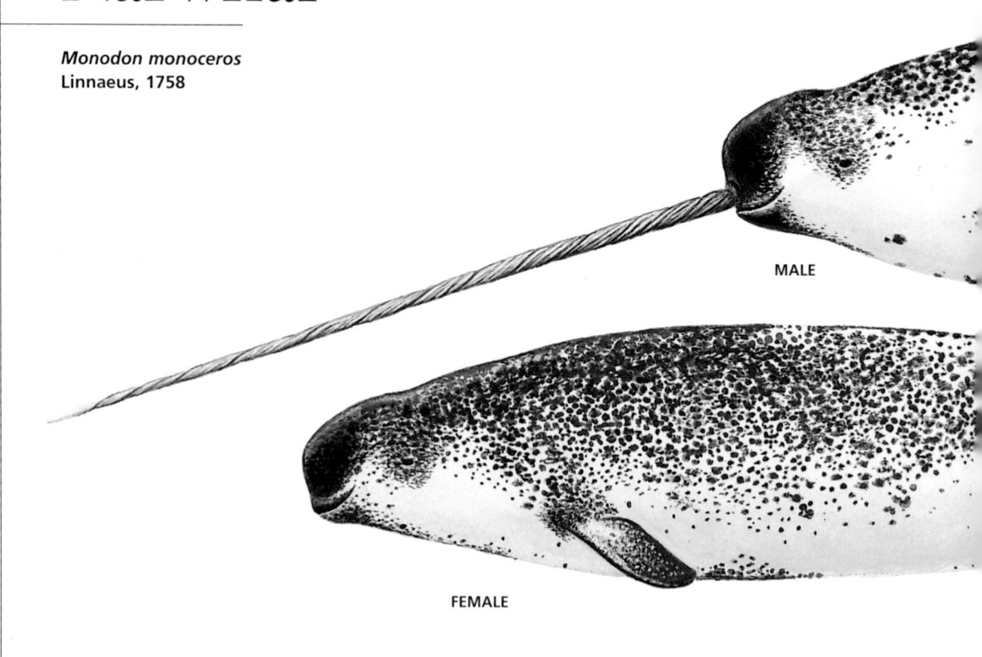

In the popular imagination, the Narwhal is inextricably linked with the fabled unicorn, and for good reason: The adult male's long, spiraled tusk came closer than anything else in nature to "proving" the unicorn's existence. In fact, from the Middle Ages onward, traders and chemists conspired to keep the existence of the Narwhal a secret, while selling its tusks as "unicorn horns" for immense profit. Although it is no longer linked to the mythical horned horse, the Narwhal is still a compelling creature because of the remoteness and harshness of its arctic environment as well as its unusual appearance. Thanks to the efforts of native local people and adventurous scientists equipped with powerful new high-tech tools, we are finally beginning to learn some of the details about the lives of Narwhals, even during the dark polar winter.

DESCRIPTION The Narwhal has a short rounded head with no beak. The melon is bluff, protruding somewhat forward of the small upturned mouth. The small flippers are broad but short. The flukes of adults are strongly convex on the rear margin; from an aerial perspective, they are reminiscent of butterfly wings. Like Belugas, Narwhals have no dorsal fin but rather a low fleshy ridge along the posterior half of the back. All Narwhals lack functional teeth inside the mouth, and most females remain essentially toothless throughout life. In males (and rarely females), the left of two upper-jaw teeth erupts through the lip at two or three years of age and keeps growing. The erupted portion of this tusk can be up to 9 feet (2.7 m) long, and the entire tusk can weigh more than 22 pounds (10 kg). In most cases, the surface of the tusk has a leftward spiral, but the axis is straight. Occasionally even the axis itself is twisted. The right tooth sometimes also erupts so that the animal is "double-tusked."

Adult Narwhals are spotted black and white on the back and upper sides. Old individuals can

- LONG TUSK ON ADULT MALES
- SMALL ROUNDED HEAD WITH NO BEAK
- NO DORSAL FIN
- SPOTTED, BLACK-AND-WHITE DORSAL COLORATION IN ADULT
- ARCTIC DISTRIBUTION

be almost completely white, with black areas limited to the center of the back, the top of the head, and the edges of the appendages. Newborn Narwhals are light gray but become almost black by the time they are weaned. Thereafter, they become mottled as white areas begin to appear on the belly and sides.

RANGE AND HABITAT Narwhals have a discontinuous arctic distribution. They are most abundant in deep waters that branch northward from the North Atlantic basin, especially Hudson Strait, northwestern Hudson Bay, Foxe Basin, Davis Strait, Baffin Bay, and Lancaster Sound. Another center of distribution is in the Greenland Sea, with small groups also occurring in parts of the northern Barents Sea. Their migrations are tuned to the formation and movement

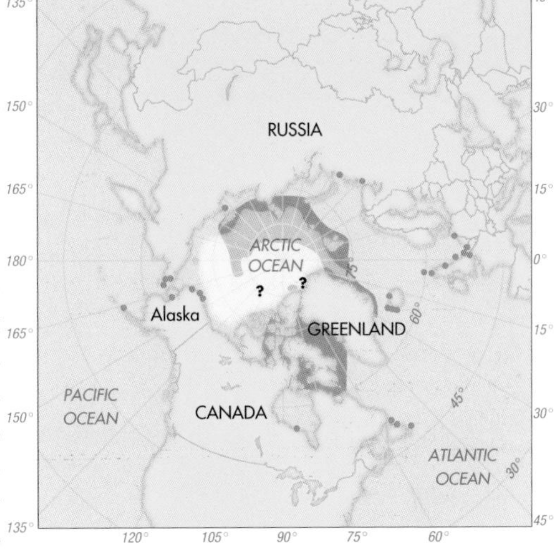

NARWHAL
FAMILY MONODONTIDAE

MEASUREMENTS AT BIRTH

LENGTH	5′3″ (1.6 m)
WEIGHT	176 lb (80 kg)

MAXIMUM MEASUREMENTS

LENGTH	**MALE**	15′6″ (4.7 m)
	FEMALE	14′ (4.2 m)
WEIGHT	**MALE**	3,500 lb (1,600 kg)
	FEMALE	2,200 lb (1,000 kg)

LIFE SPAN
At least 25 years, possibly 50

Narwhals occasionally lift their flukes as they dive. The ornately curved flukes are distinctive in both color and shape.

of sea ice. As the ice disintegrates and breaks up in spring, Narwhals follow the receding edge of the pack ice and use small cracks and melt holes to penetrate deep sounds and fjords as quickly as possible. They reside in these areas throughout the summer and early fall. As the ice cover re-forms, they head for offshore wintering areas where the ice is constantly in motion, allowing them to find breathing space between the floes.

SIMILAR SPECIES The Beluga is the only species that might be confused with the Narwhal, primarily with females and juveniles since the tusk of adult male Narwhals is so distinctive. Belugas are either solid gray or white, never black, mottled, or spotted. Both species are fairly gregarious, and usually at least a few individuals within a group have readily identifiable features. Belugas can occur in all areas inhabited by Narwhals, and occasionally the two species are seen together. However, they normally do not form mixed groups or schools; both species tend to form large single-species concentrations, particularly in summer.

BEHAVIOR Narwhals often form large aggregations of several hundred animals during summer. Such aggregations, however, consist of smaller, fairly close-knit groups of a few up to about 20 individuals. These groups are typically homogeneous, consisting of animals of the same sex or a single age class. In winter while distributed in the pack ice, Narwhals seem to be more scattered and solitary, perhaps owing to the patchiness of cracks and holes in the ice. The presence of scars and wounds in the head region, and the high incidence of broken tusks, suggest that adult males fight one another. Such fighting could play a role in establishing dominance and thus access to mating opportunities. While Narwhals have been seen apparently crossing tusks above the surface, there is no concrete evidence that they fence with them. Polar Bears are known to kill Narwhals that are trapped in small pools of open water, and Killer Whales prey on them in their inshore summering areas. Although they do not mass strand like pilot whales, Narwhals are subject to catastrophic mortality from entrapment by wind-driven or fast-forming ice. The frequency and scale of such mortality are especially high in the Disko Bay region of West Greenland.

REPRODUCTION Narwhals mate during late winter and spring (peaking in April), when the animals are generally inaccessible for observation. Gestation lasts about 15 months, and most calves are born in summer (July–August, peaking around the first of August) when the animals are in fjords. Lactation and nursing lasts for at least a year, so the calving interval is at least two years and probably averages three.

FOOD AND FORAGING Narwhals are deep divers. They forage in the entire water column, taking pelagic fish (especially arctic cod), squid, and shrimp, as well as bottom-dwelling species such as Greenland halibut. Dives can last as long

As they migrate toward their summering areas in deep arctic fjords, Narwhals take advantage of cracks and leads in the pack ice, crowding one another for breathing space. The two individuals in the foreground appear to be young males, their tusks projecting forward for only a foot or two, and their dark bodies only beginning to whiten.

as 20 minutes and reach depths of more than 3,300 feet (1,000 m). Narwhals apparently suck prey into their mouth and swallow it whole. They do not use the tusk to spear fish.

STATUS AND CONSERVATION Narwhals have long been hunted by native peoples for food, oil, and ivory. The skin (called "maktaq," variously spelled) is considered a delicacy. Commercial whalers hunted Narwhals but generally only on a casual basis, as Bowhead Whales were their preferred quarry in the Arctic. For a brief period in the early 20th century, the Hudson's Bay Company purchased Narwhal skins and tusks for export (the former to be used to make soft leather gloves). Tusks continue to be profitable export items, and maktaq has high commercial value in northern towns in both Canada and Greenland. Population estimates based on aerial surveys are about 35,000 Narwhals in Baffin Bay, 1,400 in Hudson Strait, and 300 in Scoresby Sound (East Greenland). These numbers were not corrected to account for submerged animals, and the true range-wide abundance may be greater than 50,000. The principal known threat to Narwhal populations is hunting, particularly since it is now facilitated by fast motorized boats and high-powered rifles.

This aerial view of four Narwhals, taken in the eastern Canadian Arctic, shows many of the species' distinctive features, including the long spiraled tusk, the small rounded head, the mottled, black-and-white coloration, and the absence of a dorsal fin. The low dorsal ridge appears as a dark line along the middle of the back of the older whiter animals.

Ocean Dolphins

Members of the Delphinidae, the largest cetacean family, are commonly known as "ocean dolphins," despite the fact that some are called "whales," and others are not solely ocean dwellers but also inhabit rivers and lakes. Most of the species are indeed "ocean dolphins," but they vary greatly in external appearance, size, dentition, food habits, social behavior, and habitat preferences.

A few features are common to most, if not all, delphinid species. Like other toothed whales, they have a single blowhole situated on the top of the head, slightly left of center. Their teeth, although variable in number and size, are generally conical and undifferentiated. Most have a dorsal fin at approximately the center of the back. The flukes are always cleaved on the rear margin by a pronounced notch. The length and width of the beak is highly variable among species and genera, with some having a well-defined beak and others no beak at all. Delphinid coloration can be boldly patterned in black and white, with or without intermediate gray or tan patches; all dark brown or black; evenly gray; spotted or speckled; and occasionally almost completely white.

The ocean dolphin family includes several subfamilies, 17 extant genera, and at least 33 species. Systematists have disagreed about how to classify this group, particularly with regard to subfamilies and genera. This guide follows the traditional scheme that includes six subfamilies.

Pacific White-sided Dolphin

SUBFAMILY CEPHALORHYNCHINAE This peculiar group of four small, coastal dolphin species in one genus is limited to specific areas within the temperate latitudes of the Southern Hemisphere, with offshore populations occurring in only one species. The *Cephalorhynchus* dolphins have only the hint of a beak, and their dorsal fins, while large, are rounded or triangular, quite unlike the falcate form typical of most other oceanic dolphins.

SUBFAMILY STENONINAE This subfamily, sometimes called Steninae, always includes at least two genera, *Steno* and *Sotalia,* and occasionally a third genus, *Sousa.* The genera *Steno* (Rough-toothed Dolphin) and *Sotalia* (Tucuxi) both have a single species. *Sousa* (hump-backed dolphins) has multiple forms, but its taxonomy is unresolved, and scientists have debated whether to recognize one, two, or three species.

SUBFAMILY DELPHININAE This large subfamily includes at least six genera and 17 species, all marine. As a group, they cover almost the entire globe and are found in habitats ranging from the tropics to cold subpolar waters, and from shallow coastal and inshore waters to the deep pelagic realm.

The familiar bottlenose dolphins (genus *Tursiops*) are very widely distributed along tropical and temperate coastlines. They have generally gray coloration, prominent beaks, and pointed, falcate dorsal fins. The genus *Stenella* holds the distinction of having by far the largest number of individuals. There are millions of spotted and Spinner dolphins, hundreds of thousands of Striped Dolphins, and probably at least tens of thousands of Clymene Dolphins. These are generally gregarious, tropical to warm temperate species.

The six currently recognized *Lagenorhynchus* species, or "lags," have a short thick beak that is sometimes hardly noticeable. Their dorsal fins are large and falcate, and their pigmentation tends to be complex. Their distributions are anti-tropical, meaning that different species occur in the different ocean basins, separated by a hiatus in warm tropical zones. The common dolphins (genus *Delphinus*) are similar in size, shape, behavior, and range to *Stenella* dolphins, while the almost cosmopolitan Risso's Dolphin (genus *Grampus*) is a true oddball, with its broad cleft melon and clown-like appearance.

SUBFAMILY LISSODELPHINAE The two right whale dolphins (genus *Lissodelphis*), found in the cool temperate North Pacific and in the temperate Southern Hemisphere, are the only delphinids that entirely lack a dorsal fin. Their bodies appear exceptionally streamlined, such that they are sometimes described as "eel-like." Both species are black and white, although the northern form is mostly black.

TOP LEFT: *Indo-Pacific Hump-backed Dolphin* **TOP RIGHT:** *Risso's Dolphin*
BOTTOM: *Southern Right Whale Dolphins*

LEFT: *Hector's Dolphin* **RIGHT:** *Long-finned Pilot Whale*

SUBFAMILY GLOBICEPHALINAE This is one of the more contentious subfamily groupings, in terms of classification. It always includes the two pilot whales (genus *Globicephala*) and usually the other so-called "blackfish" (genera *Orcinus, Pseudorca, Feresa,* and *Peponocephala*). However, *Orcinus* (the Killer Whale) is sometimes ascribed to a separate subfamily, Orcininae, while the Irrawaddy and Risso's Dolphins are sometimes included in the Globicephalinae (or sometimes the Orcininae). Members of this subfamily are called "whales," although not all are significantly larger than the larger dolphin species. All members of this subfamily lack a noticeable beak and have a prominent dorsal fin. Most have dark bodies with muted color patterns that are often not discernible except at close range (thus the whaler's term "blackfish").

As a group, the blackfish have a cosmopolitan distribution, with a strong presence in tropical and subtropical zones. They have a reputation for pugnacity. The Killer Whale is a well-known predator of other mammals, and the False and Pygmy Killer Whales and the pilot whales have also been observed to be aggressive toward other cetaceans. Most species have relatively few large, conical teeth, with the exception of the Melon-headed Whale, which has many small sharp teeth.

SUBFAMILY ORCAELLINAE The sole member of this subfamily, the Irrawaddy Dolphin, inhabits the Irrawaddy River in southern Asia, as well as several other large tropical river systems and marine coastal waters throughout much of southeastern Asia to northern Australia. Its lack of a beak, unpatterned light gray pigmentation, and small dorsal fin distinguish it.

Ocean Dolphins

Pantropical Spotted Dolphin page 366
Atlantic Spotted Dolphin page 370
Striped Dolphin page 380
Clymene Dolphin page 378
Long-beaked Common Dolphin page 392
Spinner Dolphin page 374
Short-beaked Common Dolphin page 388
Heaviside's Dolphin page 339
Rough-toothed Dolphin page 346
Chilean Dolphin page 336
Melon-headed Whale page 426
Hector's Dolphin page 342
Risso's Dolphin page 422
Commerson's Dolphin page 332
Tucuxi page 354
Short-finned Pilot Whale page 444
Long-finned Pilot Whale page 440
Atlantic Hump-backed Dolphin page 350
Indo-Pacific Hump-backed Dolphin page 350
Fraser's Dolphin page 384

Commerson's Dolphin

Cephalorhynchus commersonii
(Lacépède, 1804)

While passing through the Strait of Magellan during his round-the-world voyage in 1767, Philibert Commerson, a French physician and botanist, observed small dolphins with shiny white bodies and black extremities. His letter describing these animals found its way to Comte de Lacépède, a naturalist at the National Museum of Natural History in Paris, who formally introduced Commerson's Dolphin to the scientific literature. The Spanish name, *tonina overa*, means "black-and-white dolphin." The species was long thought to occur only off southern South America; however, in the 1950s scientists determined that the black-and-white dolphins seen around the Kerguelen Islands in the southern Indian Ocean were Commerson's Dolphins. This population is now regarded as an isolated, morphologically distinct (but as yet unnamed) subspecies. Commerson's Dolphin has been maintained successfully in captivity, and there have been several captive births. This handsome, active animal is a popular performer in oceanariums.

DESCRIPTION Commerson's Dolphin is among the smallest of the oceanic dolphins. Its girth can be as much as two-thirds of its body length, but the body tapers considerably toward the flukes. The head is conical and lacks a noticeable beak. The dorsal fin rises at a shallow angle and has a broadly rounded peak and a concave rear margin. The flippers are also rounded at the tip; the anterior edge of the left flipper, and less often the right flipper, can be serrated, giving a saw-toothed appearance. The Kerguelen subspecies is considerably larger than the South American form and also differs in that it has a more well-defined beak and tapered flippers. There are 28 to 34 pairs of teeth in the upper jaw and 26 to 35 pairs in the lower jaw.

The color pattern is striking and bold. The head is black except for a large white throat patch. The blackness sweeps downward from behind the blowhole in a broad diagonal band that engulfs the flippers. A wide black stripe extends along the spine from just in front of the dorsal fin to about the middle of the caudal peduncle. The dorsal fin, flukes, posterior half of the caudal peduncle, and area immediately around the genital slit are black. In adults of the South American subspecies, the rest of the body is pure white. In the Kerguelen subspecies and South American juveniles, the sides and the area on the back between the black zones are gray.

- SMALL BODY SIZE, WITH LOW-PROFILE, ROUNDED DORSAL FIN
- CONICAL HEAD WITH SMALL OR INCONSPICUOUS BEAK
- STRIKING BLACK-AND-WHITE, OR BLACK, WHITE, AND GRAY COLOR PATTERN
- RAPID MOVEMENTS AND READINESS TO BOW RIDE
- COASTAL DISTRIBUTION OFF SOUTHERN SOUTH AMERICA AND KERGUELEN ISLANDS

RANGE AND HABITAT Commerson's Dolphins inhabit cold temperate and subantarctic waters of southern South America and southward, and around the Kerguelen Islands. The largest population occurs in coastal and inshore waters of Patagonia, the Strait of Magellan, Tierra del Fuego, and the Falkland Islands. Some animals apparently wander south through the Drake Passage as far as the South Shetland Islands. Reports of their occurrence around South Georgia Island are unreliable. A small population inhabits coastal waters around the Kerguelen Islands, some 5,300 miles (8,500 km) from the South American population. The Kerguelen population presumably became established at around the end of the last glacial period, about 10,000 years ago, by animals that moved eastward from South America in the West Wind Drift.

Commerson's Dolphins inhabit shallow, nearshore waters, especially where the substrate is sandy or muddy and the tidal range is great. They typically move closer inshore with the tide

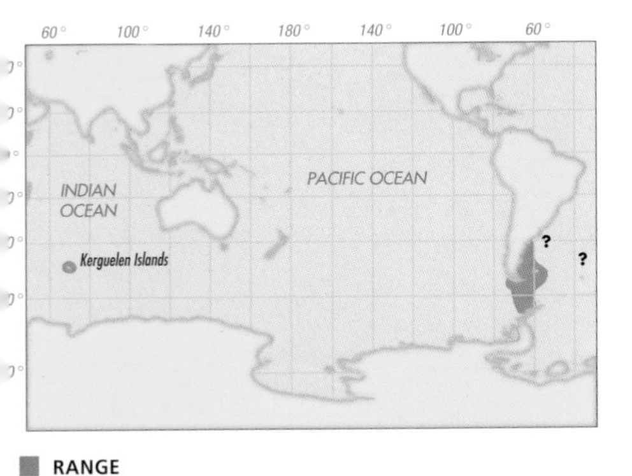

RANGE
? POSSIBLE RANGE

COMMERSON'S DOLPHIN
FAMILY DELPHINIDAE

MEASUREMENTS AT BIRTH
LENGTH 22–30″ (55–75 cm)
WEIGHT 10–17 lb (4.5–7.5 kg)

MAXIMUM MEASUREMENTS
LENGTH **MALE** *South America:* 4′9″ (1.44 m)
Kerguelen Islands: 5′6″ (1.67 m)
FEMALE *South America:* 4′10″ (1.47 m)
Kerguelen Islands: 5′9″ (1.74 m)
WEIGHT 190 lb (86 kg)

LIFE SPAN
At least 18 years

and are often seen swimming in the breakers. They appear to be at home in kelp beds and in the narrow passages of the Strait of Magellan, where current velocities can exceed 10 miles per hour (15 km/h). They are sometimes seen offshore, near the shelf edge in the Kerguelens and near Burdwood Bank off Argentina.

SIMILAR SPECIES The Chilean Dolphin overlaps with Commerson's in the Strait of Magellan and Beagle Channel; it is similar in size and shape but is white only on the undersides and lacks the conspicuous white area on the sides and back that is typical of the adult Commerson's. Burmeister's Porpoise inhabits the entire South American continental range of Commerson's, and the two species sometimes occur in close proximity to one another. Burmeister's is all dark on the back and sides and has a lower dorsal fin with a more pointed tip. The Spectacled Porpoise, which occurs along the eastern and southern coasts of South America and around the Falkland Islands, is white only on its lower half.

BEHAVIOR Commerson's Dolphins usually occur in groups of only two or three individuals, although more than 100 may sometimes congregate in the same area. They are known for their high-speed swimming and seem to delight in darting through groundswells, tidal bores, and breakers. Commerson's Dolphins often ride the bow waves or play in the wakes of high-speed vessels. While traveling, they can make racing, low-angle leaps; at other times, they may make explosive jumps high above the ocean surface. They habitually swim upside-down and spin underwater on their longitudinal axes.

Groups of Commerson's Dolphins, such as these near the Falkland Islands, can appear playful, sometimes catching rides and racing along the crests of large waves.

OPPOSITE TOP: *The stark contrast of black and white makes Commerson's Dolphin one of the most handsome cetaceans. These two animals swim near kelp in the clear green waters off the Falkland Islands.* **LEFT:** *A small group of Commerson's Dolphins caught in South American waters was taken to Sea World in San Diego in the 1980s. The calves were found to be more precocious than bottlenose dolphin calves, and they grew rapidly, one of them attaining 80% of its mother's size by its first birthday.*

REPRODUCTION Most births apparently take place between early spring and late summer (October–March), after a gestation period of about a year. It is uncertain how long calves are nursed in the wild. Captive-born calves begin mouthing food at two months of age and take fish by four months.

FOOD AND FORAGING Commerson's Dolphins are versatile in their feeding habits, and they rely on both pelagic and benthic prey. They feed mainly on mysid shrimp and small fish (such as silversides, sardines, and hake) but will take many other organisms, including squid, octopus, marine worms, tunicates, and even algae. They sometimes hunt cooperatively by encircling small schooling fish and taking turns darting through the concentrated school. They also drive small fish, such as sardines, against the shoreline and may become temporarily stranded in the process. In narrows or in areas with wide tidal flats, they might take advantage of the turbulence as it dislodges sessile prey and uncovers sand- or mud-dwelling animals. Commerson's Dolphins in captivity consume about 7 to 9 pounds (3–4 kg) of herring daily.

STATUS AND CONSERVATION The nearshore distribution of Commerson's Dolphin makes it vulnerable to incidental capture in gillnets and other fishing gear used in coastal waters. These dolphins are also killed at least occasionally in midwater trawl nets deployed for shrimp on the Argentine shelf. In the ancient past, people living along the coasts of Patagonia and Tierra del Fuego harpooned Commerson's Dolphins for their meat and oil. In the 1970s and 1980s, crab fishermen in southern Argentina and Chile hunted them for bait. This hunting is now illegal and has reportedly declined in recent years. There has been much concern about the status of Commerson's Dolphin, primarily in regard to the possibility that the crab bait fishery has reduced its abundance in the southern portions of its range. However, data for assessing the species' population status are generally inadequate, and survey results are difficult to interpret because its seasonal movements are largely unknown. While no acute or immediate threats to the species are known, mortality from entanglement in fishing gear is not recorded systematically and may be higher than reports indicate. The small population around the Kerguelen Islands is certainly isolated and therefore naturally vulnerable.

Chilean Dolphin

Cephalorhynchus eutropia
(Gray, 1846)

For many years, this dolphin was known as the Black Dolphin, an ill-chosen and unhelpful name given by scientists who had only observed either dead specimens that were darkened from exposure to air and sun, or animals seen in the water at a distance. Today the species is called the Chilean Dolphin in recognition of the fact that it apparently occurs only in the coastal waters of Chile. It remains one of the most poorly studied cetaceans.

DESCRIPTION The Chilean Dolphin has a chunky body with a girth up to about two-thirds of its length. The head is conical, and the mouthline is fairly long. Compared with other *Cephalorhynchus* species, the Chilean Dolphin's head is longer and slightly more pointed. There is no demarcation to define a beak and melon, and a slight groove is present on each side of the face. The large eyes are positioned just behind and above the corners of the mouth. The dorsal fin is moderate in size, with a smoothly rounded peak and an undercut, almost S-shaped rear margin. The flippers are also moderate in size and have rounded tips. Saw-toothed serrations sometimes occur along the leading edge of the flippers, a characteristic feature of species in this genus. There are 28 to 34 pairs of teeth in the upper jaw and 29 to 33 pairs in the lower jaw.

The throat and belly are white, and a small white patch is present in each axilla (but is not visible when the flippers are pressed against the body). The rest of the body is a complex and subtle pattern of gray tones. The center of the back, dorsal fin, flippers, and flukes are dark gray, and dark gray stripes connect the flippers and the corners of the mouth. The head and sides are lighter gray. An asymmetrical pair of crescent-shaped dark streaks sweep forward onto the rostrum from behind the blowhole. There is a dark ventral band between the flippers and a dark genital patch.

RANGE AND HABITAT The Chilean Dolphin is endemic to coastal waters of Chile, from near Valparaíso (33°S) south to Navarino Island near Cape Horn (55°15'S). It is relatively common in the many channels of Chile's convoluted coastline south from Chiloé Island, showing a particular preference for areas with rapid tidal flow, tide rips, and the sills at the mouths of fjords. North of Chiloé Island, it occurs along open coasts and in bays and river mouths. In at least some areas,

- SMALL SIZE
- ROUNDED DORSAL FIN
- SUBTLE PIGMENTATION DOMINATED BY SHADES OF GRAY ON BACK AND SIDES
- OCCURS IN INSHORE AND COASTAL WATERS OF SOUTHWESTERN SOUTH AMERICA

Chilean Dolphins are present year-round and therefore probably do not migrate.

SIMILAR SPECIES Commerson's Dolphin overlaps with the Chilean Dolphin in the Strait of Magellan and Beagle Channel, and the two species are similar in size and shape. A well-demarcated and conspicuous white area on the sides and back distinguishes Commerson's. Burmeister's Porpoise is present throughout the Chilean Dolphin's range. It is all dark, about the same size, and has no beak; however, Burmeister's has a slender dorsal fin that is positioned farther back on the body and has a much lower profile and a more

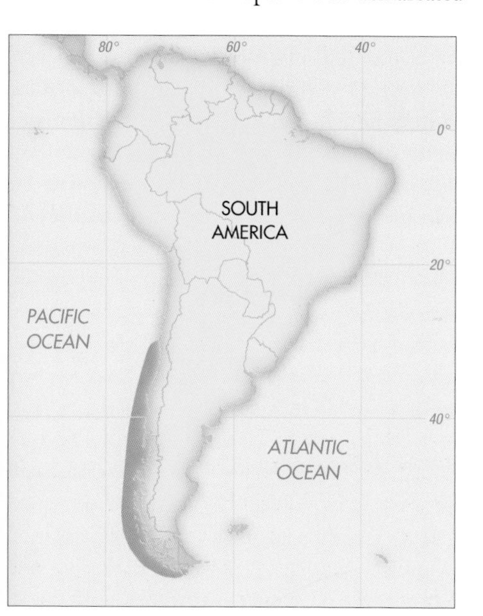

RANGE

CHILEAN DOLPHIN
FAMILY DELPHINIDAE

MEASUREMENTS AT BIRTH

LENGTH	Unavailable
WEIGHT	Unavailable

MAXIMUM MEASUREMENTS

LENGTH	MALE	At least 5′6″ (1.67 m)
	FEMALE	At least 5′5″ (1.65 m)
WEIGHT	MALE	At least 139 lb (63 kg)
	FEMALE	At least 126 lb (57 kg)

LIFE SPAN Unavailable

LEFT: *These two Chilean Dolphins, seen near Chiloé Island, exhibit the species' moderately large, rounded dorsal fin.*

RIGHT: *Chilean Dolphins tend to occur in small groups of only two or three, and they are shy of vessels. This view of a pair of dolphins near Chiloé Island suggests the contrast between the dark dorsal fin and the lighter gray sides.*

pointed peak. The Spectacled Porpoise also occurs along the southwestern coast of South America but may be distinguished by its blunter head and more upright dorsal fin.

BEHAVIOR Chilean Dolphins typically occur in groups of two or three, although occasionally groups may number 10 to 15, and in northern parts of their range there can be schools of 20 to 50 animals. Much larger aggregations were reported historically. Although they sometimes bow ride in the northern parts of their range, in the southern channels Chilean Dolphins tend to move away from and avoid boats. They are powerful and agile swimmers, able to navigate between breakers and the shore and within turbulent waters along rock walls.

REPRODUCTION Scientists know very little about the reproductive behavior of the Chilean Dolphin, as it is one of the least studied cetaceans.

FOOD AND FORAGING This species preys mainly on small schooling fish (such as sardines and anchovies), squid, and crustaceans. Chilean Dolphins observed near a salmon hatchery appeared to be playing with the fish and may have eaten young released salmon.

STATUS AND CONSERVATION Although regarded as rare, Chilean Dolphins are locally common in some parts of their range. They are taken often in coastal gillnets, and although most of this catch is incidental, the carcasses are used for bait or fish meal. Some shooting and harpooning also occurs, with the dolphins used for bait or human consumption. In addition to deaths from entanglement and hunting, Chilean Dolphins may suffer from encroachment on their habitat by salmon aquaculture facilities.

Heaviside's Dolphin

***Cephalorhynchus heavisidii* (Gray, 1828)**

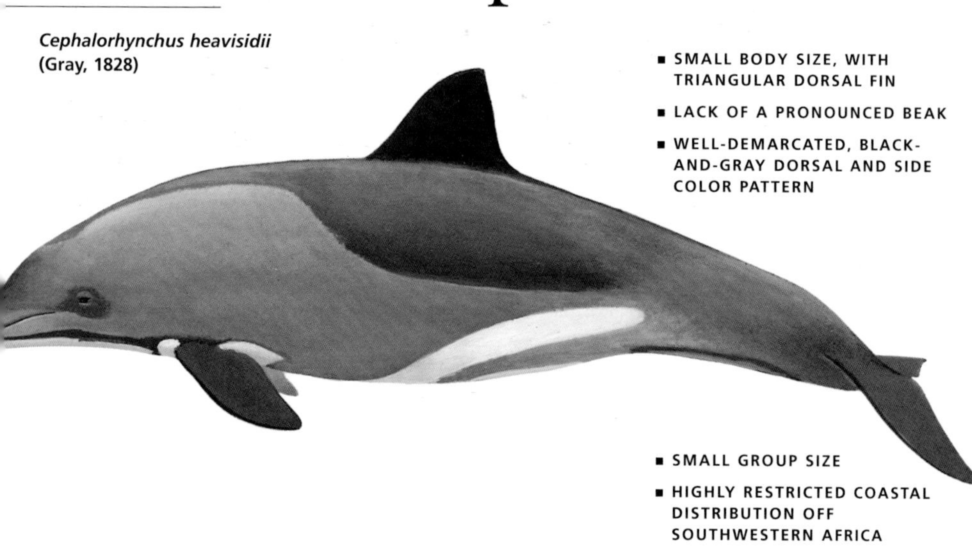

- SMALL BODY SIZE, WITH TRIANGULAR DORSAL FIN
- LACK OF A PRONOUNCED BEAK
- WELL-DEMARCATED, BLACK-AND-GRAY DORSAL AND SIDE COLOR PATTERN
- SMALL GROUP SIZE
- HIGHLY RESTRICTED COASTAL DISTRIBUTION OFF SOUTHWESTERN AFRICA

This small handsome dolphin, a southeastern Atlantic endemic, is not particularly well known. It is only during the last few decades that scientists in South Africa and Namibia have made progress in documenting its morphology, ecology, and behavior. Most of what is known about the species still comes from opportunistic research rather than intensive focused studies. Heaviside's Dolphin is named after Captain Haviside, who was responsible for transporting a specimen from Africa to England in 1827. However, its name was mistakenly spelled after Captain Heaviside, an eminent surgeon who collected anatomical specimens.

DESCRIPTION Heaviside's Dolphin is a stout animal. Its girth immediately anterior to the dorsal fin can be nearly two-thirds of the body length. The head is cone-shaped and lacks a pronounced beak. The dorsal fin, set at about midback, is moderately large and peculiarly shaped; it is roughly triangular, with the front margin longer than the rear margin. The front margin of the fin is slightly convex, and the rear margin is either straight or slightly concave. The small flippers are only moderately tapered and have blunt tips. As is true of other *Cephalorhynchus* species, there are irregular serrations on the leading edge of the flippers. There are 22 to 28 pairs of teeth in both the upper and lower jaws, fewer than in other members of this genus.

The color pattern is unusual and thus distinctive. A gray, cape-like area dominates the front third to half of the body. The dorsal surface from immediately in front of the dorsal fin to the flukes is bluish black. A dark strip along the dorsal midline connects a dark area around the blowhole with the dark back. There is also a dark zone around the eye. There are four sharply demarcated white ventral markings. The largest is trident-shaped, extending posteriorly from between the flippers and splitting into three branches at midbody; the two lateral branches flare up onto the flanks to form white markings that flash conspicuously when the animal arches above the sea surface. There is a white, diamond-shaped wedge on the front of the chest and a small white patch in each axilla. The dorsal fin is bluish black.

RANGE AND HABITAT Heaviside's Dolphin has an extremely limited range along the southwestern

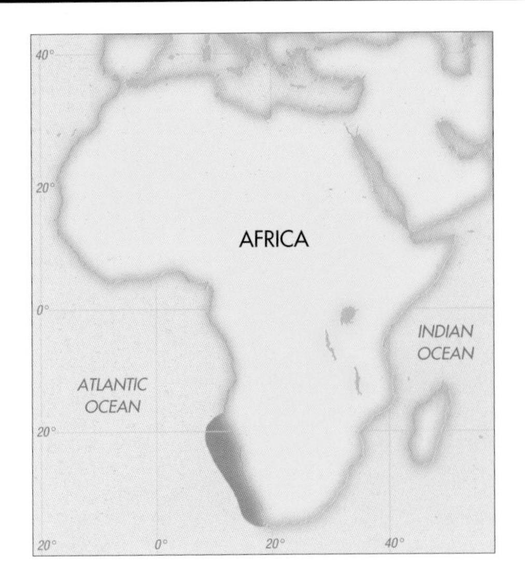

HEAVISIDE'S DOLPHIN
FAMILY DELPHINIDAE

MEASUREMENTS AT BIRTH
LENGTH About 31–33″ (80–85 cm)
WEIGHT About 20–22 lb (9–10 kg)

MAXIMUM MEASUREMENTS
LENGTH **MALE** 5′9″ (1.74 m)
FEMALE 5′6″ (1.68 m)
WEIGHT 165 lb (75 kg)

LIFE SPAN Unavailable

RANGE

coast of Africa, from near the Angola–Namibia border (at about 17°S) south to Cape Point, South Africa (near Cape Town). It occurs mainly close to shore in water less than 330 feet (100 m) deep but has been seen up to 45 nautical miles from shore in water as deep as 600 feet (180 m). Most sightings have occurred where surface water temperatures range from 48 to 60°F (9–15°C).

SIMILAR SPECIES Heaviside's Dolphin should not be confused with any other species within its limited range.

BEHAVIOR Small groups of 10 or fewer are typical of this species. The few occasions when 20 or 30 individuals are seen together may represent a temporary convergence of two or more groups. When swimming at a high speed, Heaviside's Dolphins often clear the surface at a low angle. At other times, such as while engaged in social interactions, they make more vertical jumps, and either arch to reenter the water headfirst with little splash or fall onto their side or back with a noisy splash. A leap with a headfirst reentry may end up as a somersault if the momentum causes

OPPOSITE: *Heaviside's Dolphins, seen here in Brittania Bay, South Africa, lack a pronounced beak. Dark areas around the eyes contrast with the lighter gray zones on the sides of the head and torso, and there is a dark line along the dorsal midline from the blowhole back to the dark dorsal saddle.*

ABOVE: *A leaping Heaviside's Dolphin in Brittania Bay displays the species' complex, sharply demarcated color pattern.*

the tail to continue past the vertical. While their reactions to vessels vary, Heaviside's Dolphins sometimes approach to bow ride or play in the wake.

REPRODUCTION Little is known about reproduction in Heaviside's Dolphin. Some, if not most, calving occurs in the summer (November–January).

FOOD AND FORAGING Heaviside's Dolphins eat various coastal fish and cephalopods. Many of their prey organisms, such as juvenile hake and octopus, occur on or near the ocean bottom, but they also take pelagic animals, such as squid and small gobies. Little is known about their feeding behavior.

STATUS AND CONSERVATION There are no estimates of abundance for Heaviside's Dolphin. However, the species is clearly vulnerable because of its restricted coastal distribution. At least a few animals are killed incidentally in gillnets, purse seines, beach seines, and midwater trawls annually. Some are illegally shot or harpooned, apparently for their meat.

Hector's Dolphin

Cephalorhynchus hectori
(van Bénéden, 1881)

Of the four members of its genus, Hector's Dolphin is the most intensively studied. Through the 1990s, researchers and government authorities in New Zealand brought worldwide attention to the species' precarious conservation status. At the same time, the growing popularity of nature tourism led to the development of dolphin-watching operations in a number of New Zealand's harbors. Few areas in the world can offer a more beautiful setting for this activity than Akaroa Harbor and adjacent Banks Peninsula, or a more striking animal to watch than Hector's Dolphin. Sir James Hector, the first curator of the colonial museum in Wellington, New Zealand, initially identified a dolphin shot in Cook Strait as a form of Dusky Dolphin. Hector sent it for confirmation to Belgian paleontologist P. J. van Bénéden, who recognized it as a new species and honored Hector by naming it after him.

DESCRIPTION Hector's Dolphin has a chunky body that tapers dramatically at the posterior third. The head is conical with no beak. The dorsal fin is uniquely shaped: low in profile with a broadly rounded peak and a strongly convex, undercut rear margin. The small flippers are rounded and only slightly tapered; they often have the saw-toothed serrations on the leading edge that are typical of members of this genus. There are 24 to 31 pairs of teeth in both the lower and upper jaws.

The color pattern is a complex combination of light gray, creamy white, and black, with sharp demarcations between the different zones. The back and sides are light gray, and the throat and belly are cream-colored, except for a dark band between the flippers and a dark gray genital patch. The appendages and adjacent areas are black. Broad black bands extend forward from the flippers to engulf the eyes, the sides of the face, and the tips of both jaws. A thin, crescent-shaped black band crosses the head just behind the blowhole. The cream-colored ventral area splits into three prongs behind the umbilicus, with a finger-like projection intruding onto each flank. A small white patch is present on each side immediately behind the flipper.

RANGE AND HABITAT Hector's Dolphin is endemic to coastal waters of New Zealand. There are two main populations, one centered off the

- SMALL BODY SIZE
- CONICAL HEAD WITH NO BEAK
- ROUNDED DORSAL FIN WITH UNDERCUT REAR MARGIN
- WELL-DEMARCATED, BLACK, WHITE, AND GRAY COLORATION
- DISTRIBUTION LIMITED TO COASTAL WATERS OF NEW ZEALAND

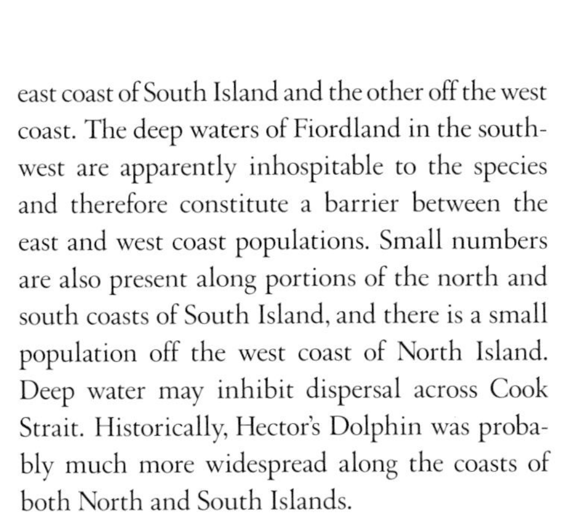

east coast of South Island and the other off the west coast. The deep waters of Fiordland in the southwest are apparently inhospitable to the species and therefore constitute a barrier between the east and west coast populations. Small numbers are also present along portions of the north and south coasts of South Island, and there is a small population off the west coast of North Island. Deep water may inhibit dispersal across Cook Strait. Historically, Hector's Dolphin was probably much more widespread along the coasts of both North and South Islands.

During summer, Hector's Dolphin occurs almost entirely within 6 miles (10 km) of shore, often just beyond the surf line or inside harbors. Its distribution shifts slightly offshore and becomes less concentrated during winter. Hector's Dolphin is considered nonmigratory.

SIMILAR SPECIES Hector's Dolphin is the only species in New Zealand coastal waters with a rounded dorsal fin. The Dusky, Short-beaked Common, and Common Bottlenose Dolphins all overlap with Hector's, but they are larger and have

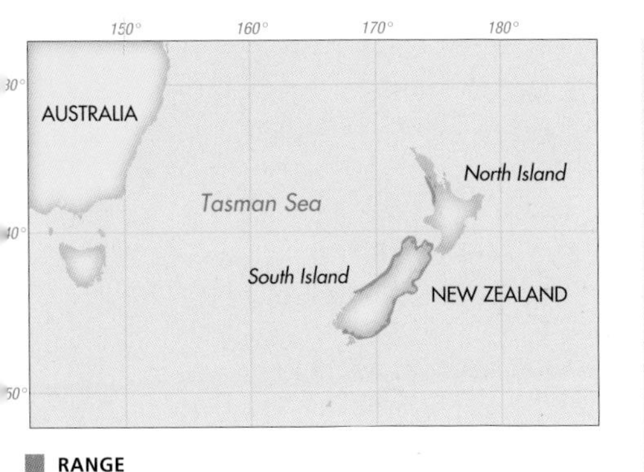

HECTOR'S DOLPHIN
FAMILY DELPHINIDAE

MEASUREMENTS AT BIRTH

LENGTH 24–30″ (60–75 cm)

WEIGHT Unavailable

MAXIMUM MEASUREMENTS

LENGTH **MALE** 4′6″ (1.38 m)
FEMALE 5′ (1.53 m)

WEIGHT **MALE** 117 lb (53 kg)
FEMALE 128 lb (58 kg)

LIFE SPAN

About 20 years

sickle-shaped dorsal fins. Of the three, only the Dusky Dolphin has a similarly shaped head without a well-defined beak.

BEHAVIOR Groups usually range in size from two to eight. Group membership, however, is somewhat fluid, with members from different groups mixing freely. Hector's Dolphins readily bow ride and are attracted to small boats traveling at speeds of less than 10 knots. They are playful and acrobatic, sometimes blowing bubbles, playing with seaweed, or surfing the waves. Their leaping has been classified into three categories: horizontal, vertical, and noisy. Their low-angle horizontal leaps are associated with traveling or keeping pace with a vessel. Vertical leaps are associated with social interactions and often involve two or more individuals leaping in close succession and reentering the water headfirst with very little splash. Noisy leaps involve landing on the back, side, or belly with a splash. In their shallow-water habitat, Hector's Dolphins rarely dive for longer than about a minute and a half. Large sharks, which are common in some parts of their range, are confirmed predators of these dolphins.

REPRODUCTION Calves are born mainly during spring and early summer, from early November to mid-February. The gestation period is uncertain but probably lasts about a year. Calves stay with their mothers for at least a year, often for two. They may start eating some solid foods before six months of age. The calving interval is at least two years.

ABOVE: *This breaching Hector's Dolphin, likely to land on its side with a noisy splash, shows the deeply concave rear margin of the flukes and the mainly white throat and belly.* **OPPOSITE:** *Hector's Dolphins can appear curious and playful at times. The individuals pictured here seem to be playing with kelp fronds.*

FOOD AND FORAGING Hector's Dolphins feed on many kinds of small fish and squid. Many of their prey are benthic or epibenthic species, such as flatfish, stargazers, and cod. However, they also

take midwater and surface species, such as mullet and sprat. Hector's Dolphins sometimes follow trawlers fishing for bottom species, perhaps to take advantage of injured or disoriented fish.

STATUS AND CONSERVATION Surveys conducted during the mid-1980s suggested a total population of only 3,000 to 4,000 Hector's Dolphins. Entanglement in gillnets has been a major threat to the species. According to one analysis, mortality in gillnets since 1970 has reduced the population by at least 50 percent and possibly by more than 65 percent. A sanctuary created in 1988 to restrict gillnetting activity off South Island around Banks Peninsula has reduced, but certainly not eliminated, this threat. Increasing amounts of high-speed vessel traffic in the harbors inhabited by Hector's Dolphins pose a new threat. Several young animals are known to have died recently after being struck by boat propellers. Pollution, port development, and aquaculture facilities are additional concerns. A recent survey covering the southern, eastern, and northern coasts of South Island produced an estimate of somewhat less than 2,000 Hector's Dolphins. The North Island population is feared to have fewer than 100 individuals, occupying a much smaller range than in the past.

These Hector's Dolphins, swimming abreast, exhibit their distinctive dorsal fin shape—a rounded lobe with an undercut rear margin.

Rough-toothed Dolphin

Steno bredanensis
(G. Cuvier *in* Lesson, 1828)

Rough-toothed Dolphins are found in the open ocean in tropical latitudes. They rarely occur close to land except around islands with steep drop-offs near shore. Oceanic animals generally acclimate less well to captivity than coastal species, yet oceanariums in Hawaii and Japan have had good success with Rough-toothed Dolphins. Several of the captives have gained renown as quick learners and creative performers. They have been used not only in shows, but also in at-sea studies, swimming untethered alongside a research vessel. At Sea Life Park in Hawaii, a Rough-toothed Dolphin mother and a Common Bottlenose Dolphin father produced a calf that lived for four years. The species' common name refers to the rough texture of the teeth, caused by many fine longitudinal grooves.

DESCRIPTION The Rough-toothed Dolphin is readily distinguished from other long-beaked oceanic dolphins by the shape of its head, which lacks a crease at the base of the melon. Rather, the "forehead" slopes smoothly from the blowhole onto the long narrow beak. The tall dorsal fin is erect and set at midback; it is usually moderately falcate. The flippers are relatively large. There are 19 to 26 pairs of teeth in the upper jaw and 19 to 28 pairs in the lower.

The color pattern of the Rough-toothed Dolphin is complex but muted. A dark dorsal cape narrows markedly between the blowhole and the dorsal fin, then widens at the dorsal fin and posteriorly. In dorsal view, the cape has a distinctive "neck" in front of the dorsal fin. The lighter gray sides give way to a white belly and throat. There is often considerable dark spotting or flecking, giving the sides, throat, and belly a mottled appearance. In addition, white and yellowish scars are usually present, caused either by fighting with other Rough-toothed Dolphins or by the bites of large squid and cookie-cutter sharks. The lips are white, but the upper surface of the beak is dark (except at the tip, which is often white). The eyes are darkly shaded. There is a wide eye-to-flipper stripe, but it does not contrast strongly with the ground color and is therefore not prominent.

RANGE AND HABITAT The Rough-toothed Dolphin is limited to deep tropical and warm temperate waters worldwide, including portions of the Gulf of Mexico, the Gulf of Aden, and the

- LONG BEAK WITH NO CREASE AT MELON
- LARGE FLIPPERS AND PROMINENT, TALL DORSAL FIN
- DISTINCTIVE DORSAL CAPE THAT NARROWS BETWEEN BLOWHOLE AND DORSAL FIN
- WHITE LIPS AND THROAT
- IRREGULAR SPOTTING AND BLOTCHING ON VENTRAL SURFACE
- OCCURS WORLDWIDE, MAINLY IN WARM DEEP WATERS

Mediterranean Sea. It typically occurs seaward of continental or island shelves. The species' distributional limits (determined from strandings and other records) are Virginia and the Netherlands in the North Atlantic; Rio Grande do Sul, Brazil, and about 32°S (far offshore of Africa) in the South Atlantic; KwaZulu-Natal and northwestern Australia in the Indian Ocean; the Pacific coast of central Honshū and Baja California in the North Pacific; and northern New Zealand and northern Chile in the South Pacific. The Rough-toothed Dolphin is a known year-round resident of many areas, such as Hawaii, the northern Gulf of Mexico, and the Mediterranean Sea. It is not known to be migratory.

SIMILAR SPECIES One or both bottlenose dolphin species occur throughout the Rough-toothed Dolphin's range, and the three can be confused at a distance because of the overall similarity in body size, dorsal fin appearance, and back coloration. At close range, however, the bottlenose dolphins' well-demarcated and shorter beaks should make them easily distinguishable.

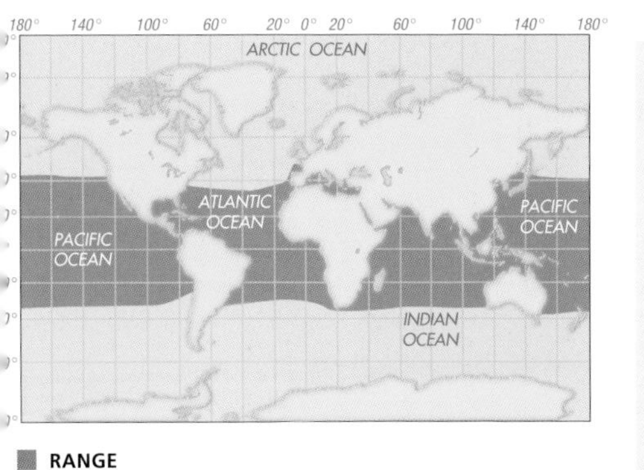

ROUGH-TOOTHED DOLPHIN

FAMILY DELPHINIDAE

MEASUREMENTS AT BIRTH

LENGTH 3′4″ (1 m)

WEIGHT Unavailable

MAXIMUM MEASUREMENTS

LENGTH **MALE** 8′8″ (2.65 m)
FEMALE 8′4″ (2.55 m)

WEIGHT **MALE** 350 lb (160 kg)
FEMALE Unavailable

LIFE SPAN

32 years

Also, their capes do not narrow ahead of the dorsal fin; instead, they often have a white or light gray blaze that intrudes into the cape from the gray overlay. The ranges of the Pantropical Spotted Dolphin and the Spinner Dolphin also overlap extensively with that of the Rough-toothed Dolphin. Like the bottlenose dolphins, both have a distinct crease separating the melon from the beak. The Pantropical Spotted Dolphin can be distinguished by its spotted pattern and wide dorsal cape. The Spinner Dolphin has a slimmer build and "cleaner" color pattern, with little of the scarring and blotching seen on the Rough-toothed.

ABOVE: *The Rough-toothed Dolphin's smoothly sloping forehead distinguishes this species from other ocean dolphins of similar size, most of which have a crease that divides the melon from the beak. The fairly tall, erect dorsal fin and the white tip on the beak, identifiable on these Rough-toothed Dolphins around Moorea, French Polynesia, are also distinguishing features.*

BELOW: *The narrow, dark gray dorsal cape of the Rough-toothed Dolphin shows up only in good lighting, while the white lips and lower jaw are generally more conspicuous.*

The Rough-toothed Dolphin has proven difficult to study in the wild. One of the few places in the world where it is encountered regularly is in waters near Moorea, French Polynesia, where this breaching individual was photographed.

BEHAVIOR Rough-toothed Dolphins typically form close-knit groups of 10 to 20 individuals. Groups are seldom larger than about 50 animals, but these dolphins sometimes form dispersed aggregations of several hundred. Rough-toothed Dolphins often occur with other species such as Short-finned Pilot Whales, Common Bottlenose Dolphins, Pantropical Spotted Dolphins, or Spinner Dolphins. Very hard to follow at sea, they are thought to be deep divers and can remain submerged for up to 15 minutes. Rough-toothed Dolphins occasionally strand in large groups. Of 62 individuals that came ashore on the Gulf coast of Florida in 1997, four were rehabilitated and released successfully.

REPRODUCTION Almost nothing is known about reproduction in the Rough-toothed Dolphin.

FOOD AND FORAGING The Rough-toothed Dolphin's diet generally consists of fish and cephalopods. It is known to take fish of various sizes, including fairly large ones. Dolphins off Hawaii have been observed "sharing" large (more than 11 lb/5 kg) mahi mahi, possibly caught cooperatively. Observations in captivity suggest that these dolphins behead and eviscerate fish, regardless of how small, before consuming them.

STATUS AND CONSERVATION The species is widespread and common in many tropical areas, but not nearly as abundant as some other pantropical dolphins, such as spotted, Spinner, Striped, common, bottlenose, and Risso's. Survey results suggest that there may be around 150,000 Rough-toothed Dolphins in the eastern tropical Pacific and 850 in the northern Gulf of Mexico. Relatively few are reportedly taken in fishing gear. Small numbers are taken by dolphin hunters in Japan, the Solomon Islands, western Africa, and the Lesser Antilles. Rough-toothed Dolphins are notorious, especially in Hawaii, for stealing bait and catch from fishing lines. This behavior makes them extremely unpopular with recreational and commercial fishermen.

Hump-backed Dolphins

Atlantic Hump-backed Dolphin, *Sousa teuszii* (Kükenthal, 1892)
Indo-Pacific Hump-backed Dolphin, *Sousa chinensis* (Osbeck, 1765)

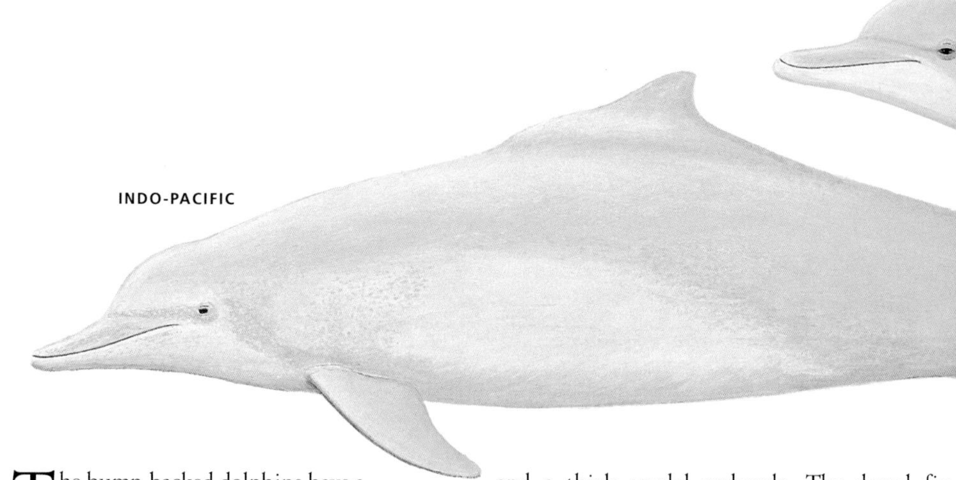

The hump-backed dolphins have a complex taxonomy that remains unresolved. They are presented together here because of the uncertainty of describing, classifying, and naming different taxa within the genus *Sousa*. At least two species have generally been recognized: the Atlantic Hump-backed Dolphin, endemic to the west coast of Africa, and the Indo-Pacific Hump-backed Dolphin of the coastal Indo-Pacific region. Some scientists argue that the Indo-Pacific form should be further split into two species: the Indian (or Plumbeous) Hump-backed Dolphin *(S. plumbea)*, occurring along the Indian Ocean coasts, and the Pacific Hump-backed Dolphin *(S. chinensis)*, which is found along the coasts of Southeast Asia, New Guinea, and northern Australia. All hump-backed dolphins live close to shore, usually centered in and near the mouths of large rivers. Their common name refers to the structure of the dorsal fin region on animals in the western part of the range. Hump-backed dolphins in China are commonly known as Chinese White Dolphins.

DESCRIPTION Hump-backed dolphins have a robust body with a relatively long, slender beak and a thick caudal peduncle. The dorsal fin and hump region varies considerably. Hump-backed dolphins in the Atlantic Ocean and in the western and northern Indian Ocean have a long thick hump at midback topped by a small, falcate dorsal fin. Those in the Pacific Ocean and the eastern Indian Ocean have a less-pronounced hump and in some cases no hump at all, although the base of the dorsal fin is unusually long; the fin itself is low and only somewhat falcate with a blunt peak. The hump occurs only in the *teuszii* and *plumbea* forms. The flippers are broad and tapered, with blunt tips. Atlantic Hump-backed Dolphins have 27 to 31 pairs of teeth in both the upper and lower jaws; hump-backed dolphins elsewhere have 30 to 38 pairs in each jaw.

The color pattern is highly variable, both within and between populations. Hump-backed dolphins off western Africa are slate gray dorsally and lighter gray ventrally. Animals off southern Africa are often described as "plumbeous" (lead-colored) or brownish gray dorsally, with lighter gray sides shading to off-white ventrally; those in the northern Indian Ocean are uniformly plumbeous or brownish gray. Hump-backed

- DISTINCTIVE HUMP AT BASE OF SMALL DORSAL FIN (MUCH LESS PRONOUNCED IN EAST)
- LONG SLENDER BEAK
- GRAY, PINK, WHITE, OR SPECKLED COLORATION
- SMALL GROUP SIZE
- OCCUR IN SHALLOW, NEARSHORE WATERS OF EASTERN ATLANTIC, WESTERN PACIFIC, AND INDIAN OCEANS

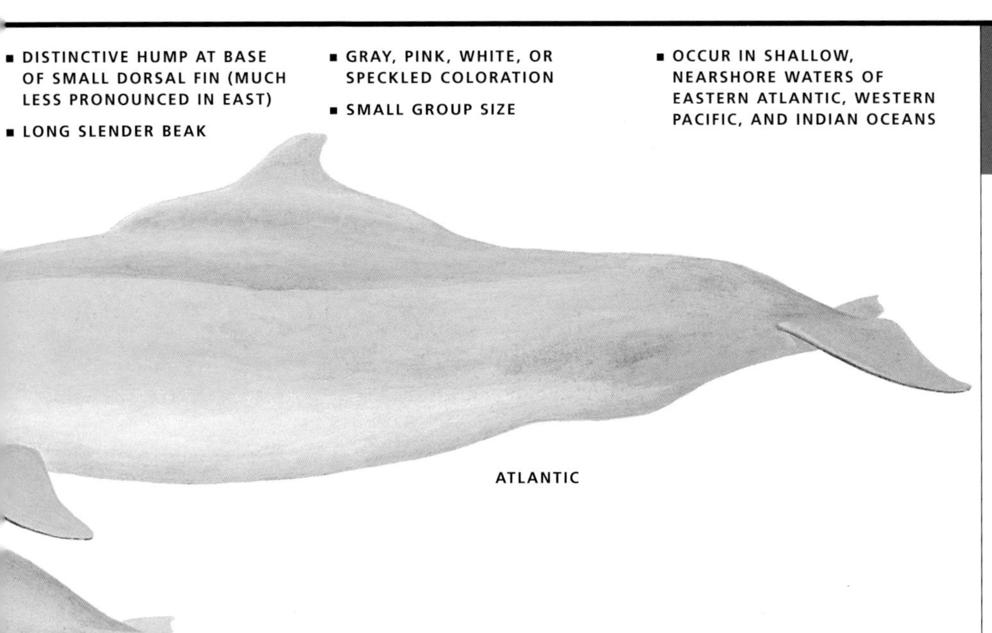

dolphins in the Indian Ocean can have a few dark spots on the belly (southern Africa) or heavier spotting (southern and southeastern Asia). In Australia, they are pale gray dorsally and off-white ventrally. Large adults in Chinese waters are often pure white or pink; the young are solid gray, becoming lighter and spotted with age. A sharp color line often separates the dorsal hump and fin from the rest of the back, especially on hump-backed dolphins in the Indian Ocean; in some animals, the hump and fin are white, in others just a different shade of gray.

RANGE AND HABITAT Hump-backed dolphins have a tropical and subtropical distribution, generally in waters warmer than 60°F (15°C). They occur in the eastern Atlantic Ocean from Dakhla Bay on the northwestern coast of Africa (at about 24°N) south to Cameroon, including the Bijagós Archipelago off Guinea-Bissau; coastally around the rim of the Indian Ocean from False Bay, South Africa, to the western coast of the Bay of Bengal, including Madagascar, the Red Sea, and the Persian Gulf; and in the western Pacific Ocean from the Gulf of Thailand and the Strait

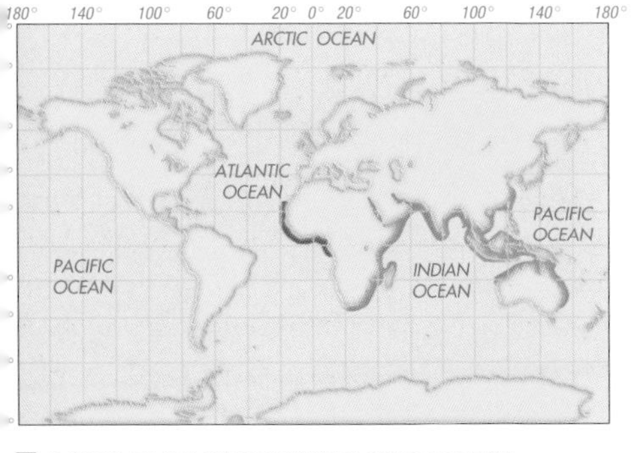

■ RANGE OF ATLANTIC HUMP-BACKED DOLPHIN
■ RANGE OF INDO-PACIFIC HUMP-BACKED DOLPHIN

HUMP-BACKED DOLPHINS

FAMILY DELPHINIDAE

MEASUREMENTS AT BIRTH

LENGTH 3′4″ (1 m)

WEIGHT *South Africa:* 31 lb (14 kg)

MAXIMUM MEASUREMENTS

LENGTH **MALE** *South Africa:* 9′2″ (2.79 m)
FEMALE *South Africa:* 8′2″ (2.49 m)

WEIGHT **MALE** 570 lb (260 kg)
FEMALE At least 370 lb (170 kg)

LIFE SPAN

At least 40 years

of Malacca south and east to Australia and north to the East China Sea. Hump-backed dolphins typically remain within a few miles of shore and are rarely seen in water deeper than 80 feet (25 m). Off South Africa, they typically occur over or near rocky reefs within 1,500 feet (400 m) of shore. On open coasts, they are often spotted just beyond the surf line. Their preferred habitat appears to be in or near bays and estuaries, and they can also be common in turbid channels within mangrove forests and deltaic sandbanks. In China they appear to occur exclusively in estuaries.

SIMILAR SPECIES Bottlenose dolphins occur throughout the range of hump-backed dolphins and are the species most likely to cause confusion;

ABOVE: *Adult Indo-Pacific Hump-backed Dolphins in the South China Sea can be almost entirely pink, in striking contrast to calves, which are dark gray.*

BELOW LEFT: *The coloration of hump-backed dolphins changes with age. While calves are generally dark gray, they become mottled or spotted as juveniles or young adults. Unlike this animal near Hong Kong, hump-backed dolphins off West Africa and in the western and northern Indian Ocean have a pronounced hump at the base of the dorsal fin.* **BELOW RIGHT:** *An Indo-Pacific Hump-backed Dolphin off the coast of Australia shows its flukes as it dives.*

they are usually darker and more evenly colored, with a larger, often strongly falcate dorsal fin (without a hump) and a shorter, thicker beak. Bottlenose dolphins tend to form larger, more active groups. Irrawaddy Dolphins, which overlap in range from the Bay of Bengal to Australia, are distinguished by a rounded head (with no beak) and a smaller dorsal fin.

Hump-backed dolphins are among the longer-beaked marine dolphins. Recent field studies in the South China Sea have resulted in a proliferation of photographs of the animals in that region.

BEHAVIOR Hump-backed dolphins occur in fluid groups of typically about three to eight animals. Aggregations of more than 20 seen off Hong Kong are usually associated with trawlers, which seem to function as magnets for the hungry dolphins. In comparison with bottlenose dolphins, with which they often co-occur in some areas, hump-backed dolphins are slower moving and less demonstrative at the surface and are therefore more cryptic. They can, however, be playful, lifting their heads clear of the water, breaching, and slapping the surface with their flippers or flukes. Hump-backed dolphins are shy of boats and do not bow ride. In Moreton Bay, Queensland, hump-backed dolphins mix with bottlenose dolphins in the vicinity of shrimp trawlers. In parts of their range, such as off South Africa, they live in shark-infested waters and are frequently subject to shark predation.

REPRODUCTION Little is known about reproduction in hump-backed dolphins. The gestation period is probably 10 to 12 months, and most calves are born in spring or summer. Based on observations in South Africa, the calving interval is thought to average three years. Calves may begin taking solid food at six months of age but may not be fully weaned until two years old or older.

FOOD AND FORAGING Hump-backed dolphins prey mainly upon species that live on or near the ocean bottom and are associated with reefs or the brackish waters of estuaries. These include small fish, squid, and octopus. In some areas, hump-backed dolphins appear to feed most actively during the rising tide. Off Hong Kong and Queensland, Australia, they regularly follow trawlers, feeding on escaped, injured, or discarded organisms.

STATUS AND CONSERVATION For most of the hump-backed dolphins' range, there are no good estimates of abundance. More than 1,000 inhabit the Pearl River delta and adjacent portions of China's coast, and close to 500 are present along South Africa's Eastern Cape. Because they tend to occur very near shore in small, largely isolated populations, hump-backed dolphins are exceptionally vulnerable. Throughout their range, they are threatened by entanglement in fishing gear, especially gillnets and trawls, and in Australia and South Africa they also become entangled in the anti-shark nets set along beaches to protect swimmers.

Hump-backed dolphins are taken deliberately for food in some areas (such as Mozambique, possibly Madagascar, and parts of western Africa). There is concern that intensive fishing in areas such as Dakhla Bay has depleted the dolphins' prey resources. Levels of pollutants found in the tissues of hump-backed dolphins in South Africa and Hong Kong are alarmingly high. Heavy vessel traffic degrades their environment and also poses the risk of injury or death from collisions. A massive die-off of marine wildlife, including at least 140 hump-backed dolphins, occurred in the western Persian Gulf in 1986; the cause was never determined.

Tucuxi

Sotalia fluviatilis
(Gervais and Deville, 1853)

This small dolphin has an extensive distribution along the east coast of South and Central America and throughout much of the Amazon River basin. The Tucuxis that inhabit freshwater environments are significantly smaller than their marine counterparts, and the two groups are considered separate subspecies. The name Tucuxi (pronounced toó-koo-shee) originated in the Brazilian Amazon and has been adopted as the common name for the species in the scientific literature. Outside Brazil, in Spanish-speaking countries, these dolphins are locally called either *bufeos* (Peru) or *toninas* (Venezuela). Some scientists have proposed that the species be named the Estuarine Dolphin, because of its preference for estuarine environments. The specific name is derived from the Latin word for "riverine"; the origin of the genus name is unknown.

DESCRIPTION The Tucuxi is often described as looking like a diminutive bottlenose dolphin, and indeed the resemblance is enough to mislead a casual observer. The body is small and compact. The beak is moderate in length and the mouthline is straight. The dorsal fin, set at midback, is triangular, low in profile, and can be slightly hooked at the peak (often strongly reminiscent of the Harbor Porpoise's dorsal fin). It is not strongly falcate or recurved. There are 26 to 36 pairs of teeth in both the upper and lower jaws.

As they surface in the wild, Tucuxis appear to have subtle countershading; the body is evenly light gray with lighter undersides that are sometimes suffused with pink. On close inspection, the line of separation between the dorsal and ventral shades of gray is clear, running from the corner of the mouth, below the eye, and curving down toward the flipper.

RANGE AND HABITAT The Tucuxi is endemic to the rivers, lakes, and warm coastal marine waters of eastern South and Central America. The marine subspecies *(Sotalia fluviatilis guianensis)* appears to have a continuous distribution from at least as far north as Nicaragua (about 14°35'N, 83°15'W) south to Florianópolis, Brazil (27°35'S, 48°34'W). Its southern limit is approximately at the convergence of the warm, south-flowing Brazil Current and the cool, north-flowing Falkland Current, lending support to the theory that sea temperature helps to restrict this species'

- SMALL BODY
- LOW-PROFILE, TRIANGULAR DORSAL FIN
- WELL-DEFINED, MODERATE-LENGTH BEAK
- EVENLY LIGHT GRAY DORSAL COLORATION
- COASTAL DISTRIBUTION IN TROPICAL AND SUBTROPICAL WESTERN ATLANTIC, PLUS THROUGHOUT AMAZONIA

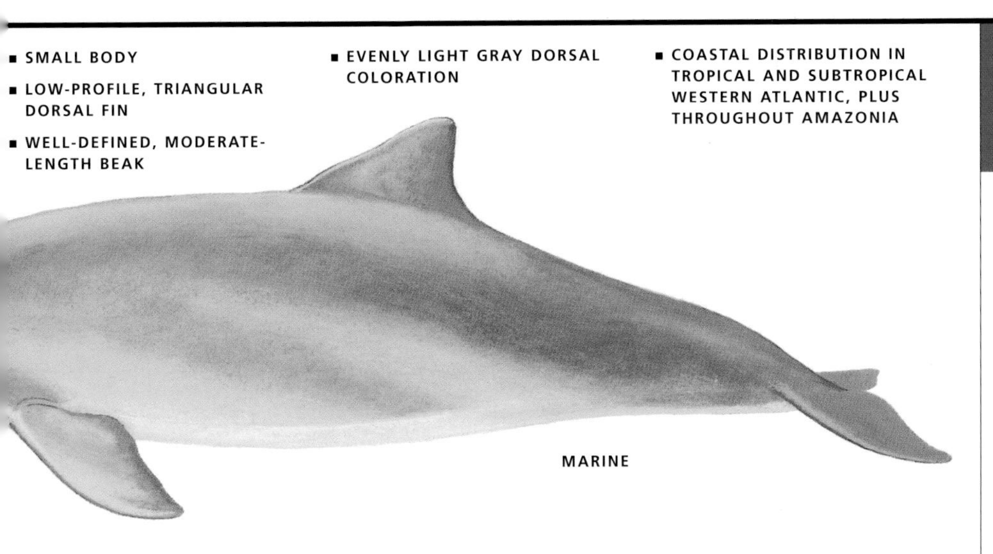

distribution. In some areas, marine Tucuxis show a preference for water deeper than about 20 feet (6 m), but are not known to range offshore of the continental shelf. They tend to concentrate in estuaries and large gulfs, including Lake Maracaibo in Venezuela and Guanabara Bay in Brazil. The freshwater subspecies *(S. f. fluviatilis)* in Amazonia inhabits a vast network of lakes, streams, and large rivers as far upstream as Peru, Ecuador, and Colombia. In the lower Orinoco, its presence has been confirmed only up to the Arature confluence (8°36'N, 60°53'W); it is not known to occur in the upper Orinoco. While no Tucuxis are known to migrate long distances, the freshwater animals make at least short-distance movements in response to seasonal changes in water level.

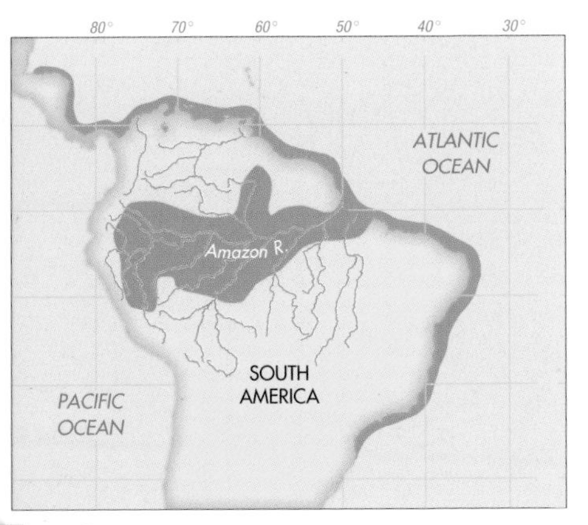

TUCUXI
FAMILY DELPHINIDAE

MEASUREMENTS AT BIRTH

LENGTH *Freshwater:* 28–33″ (70–83 cm)

WEIGHT Unavailable

MAXIMUM MEASUREMENTS

LENGTH **MALE** *Marine:* 6′3″ (1.9 m)
Freshwater: 4′11″ (1.5 m)
FEMALE *Marine:* 6′10″ (2.1 m)
Freshwater: 4′11″ (1.5 m)

WEIGHT **MALE** *Freshwater:* 104 lb (47 kg)
FEMALE *Freshwater:* 117 lb (53kg)

LIFE SPAN Unavailable

SIMILAR SPECIES The Franciscana overlaps with the Tucuxi in nearshore marine waters of southern Brazil; it is similar in size but has a longer, narrower beak, and its dorsal fin tends to be more broadly rounded at the peak. The Tucuxi's triangular dorsal fin and small size should distinguish it from the Common Bottlenose Dolphin, as well as other delphinids that occur in the same tropical and subtropical waters of the western Atlantic. Juvenile Amazon River Dolphins in the Amazon and lower Orinoco River basins may be difficult to distinguish from Tucuxis, particularly at a distance and when the viewing angle is poor. Adult river dolphins, however, are relatively easy to identify by their large bodies, long beaks, and less distinct dorsal fins.

Tucuxis are high-energy dolphins that frequently leap clear of the water. This individual, in a protected area on the southern coast of Brazil, exhibits the countershaded color pattern: dark dorsal coloration and light ventral coloration.

BEHAVIOR Tucuxis usually occur in close-knit pods ranging in size from two or three to about 10 or 12 individuals. In marine waters, schools of up to about 30 are sometimes seen. While at times members of a group surface synchronously, they can also alternate at the surface, making counting difficult. Nothing is known about the nature or duration of affiliations between individuals. Tucuxis are fairly demonstrative, often arcing clear of the water and sometimes somersaulting, spyhopping, and splashing the surface with their flukes or flippers. They can sometimes be seen "surfing" in the wakes of passing ships, although they do not bow ride.

REPRODUCTION Little is known about reproduction in either the marine or freshwater subspecies. Researchers studying freshwater Tucuxis in the central Amazon of Brazil have estimated the gestation period to be slightly more than 10 months. The Tucuxi's reproductive cycle appears to correspond with the annual flood cycle in the Amazon, ensuring that females give birth when ample food is available to support lactation. Thus, at least in Brazil, conception is thought to occur mainly in January and February, so that most births take place during the low-water season in October and November. At this time, fish are more concentrated in deep

pools and main river channels, presumably making them easier for the dolphins to find and catch.

FOOD AND FORAGING Tucuxis eat a large variety of fish. The most detailed studies of their diet have been conducted in the Brazilian Amazon, where at least 28 fish species from 11 families have been found in the stomachs of Tucuxis. In marine waters, they take schooling pelagic and bottom-dwelling fish, as well as coastal cephalopods. In nearshore coastal waters and mangrove-lined lagoons, they often follow prey into shallow water. Tucuxis sometimes surface with fish in their beaks, and it is possible to find groups of feeding dolphins by looking for flocks of seabirds circling and diving overhead, lured by the fish the dolphins bring to the water's surface.

STATUS AND CONSERVATION The Tucuxi is common over much of its range, but no reliable population estimates exist. Direct capture is apparently infrequent; however, there is concern about incidental entanglements in fishing nets, which are known to occur frequently in many parts of the species' range. In the absence of regular observation and reporting, it is impossible to provide even a rough estimate of how many animals are killed annually. In parts of Amazonia, the genital organs and eyes of these dolphins are sometimes sold as amulets, and in coastal areas their meat is eaten or used as shark bait. Thus, fishermen have an incentive to retain dolphins taken accidentally and on some occasions also to hunt or net them deliberately. The Tucuxi's riverine and nearshore distribution makes it vulnerable to a large variety of environmental threats, including river damming and other modifications, pollution, and underwater noise. Some areas inhabited by the marine subspecies, such as Lake Maracaibo and Guanabara Bay, experience heavy ship traffic, oil production, and fishing activity.

ABOVE: *Tucuxis often occur in close-knit pods of a few to a few dozen individuals. The dorsal fin is generally low in profile and triangular or slightly falcate.*
LEFT: *The marine form of the Tucuxi off the coast of Colombia.*

Common Bottlenose Dolphin

Tursiops truncatus
(Montagu, 1821)

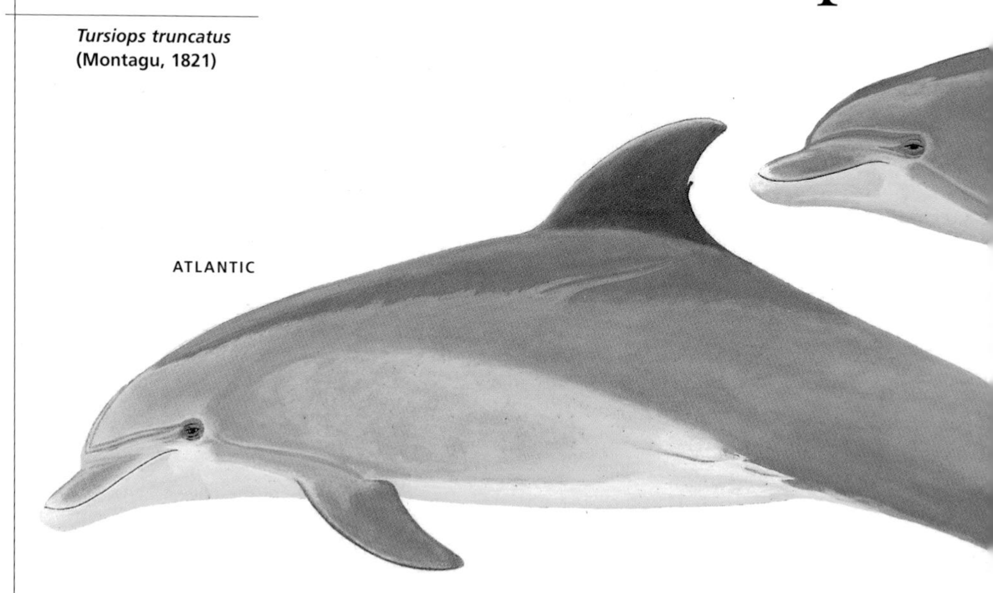

This is the archetypal dolphin, well known to the ancient Greeks and Romans because of its common nearshore presence throughout the Mediterranean Sea. Today it has achieved worldwide exposure as the star of the television series (and later film) *Flipper* and as the main attraction in many oceanariums. The Common Bottlenose Dolphin has only very recently been determined to be a species separate from the Indo-Pacific Bottlenose. Scientists were long aware of differences between the two types, but it wasn't until adequate analyses had been completed in the late 1990s that a consensus was reached about assigning the two forms separate species status.

DESCRIPTION Relative to most other dolphins, the Common Bottlenose has a wide head and body, a short stubby beak, long flippers, and a moderately tall, falcate dorsal fin. It has a marked crease between the melon and the beak. There is, however, considerable variability within the species. In most areas where they have been adequately studied, inshore and offshore forms have been found to differ consistently in their morphology. Offshore animals tend to be larger-bodied and darker in color, with smaller flippers, than their inshore counterparts. In both forms, males are somewhat larger than females in length, girth, and mass. The Common Bottlenose Dolphin has large teeth, with 20 to 26 pairs in the upper jaw and 18 to 24 pairs in the lower.

The color pattern consists mostly of gray tones, with strong countershading (dark dorsally, light ventrally) but no sharp demarcation between the elements. There is a dark dorsal cape that often begins at the apex of the melon and extends back past the dorsal fin. A light blaze may sometimes intrude onto the cape. A paler gray overlay covers the sides and flanks, and the belly may be off-white or pinkish. There is a muted eye-to-flipper stripe. The flippers, flukes, and dorsal fin are generally dark or medium gray. Tooth rakes (rake-like scratches or scars caused by other dolphins' teeth) are often present on various parts of the body.

RANGE AND HABITAT This dolphin is a cosmopolitan species that occurs in oceans and peripheral seas at tropical and temperate latitudes. It occupies a wide variety of habitats and is regarded as perhaps the most adaptable cetacean. There are coastal populations along the conti-

- ROBUST BODY WITH SHORT TO MEDIUM-LENGTH BEAK
- LARGE FALCATE DORSAL FIN
- BASICALLY GRAY BODY WITH MUTED COLOR PATTERN
- OFTEN SEEN VERY NEAR SHORE AND IN ESTUARIES
- COSMOPOLITAN DISTRIBUTION

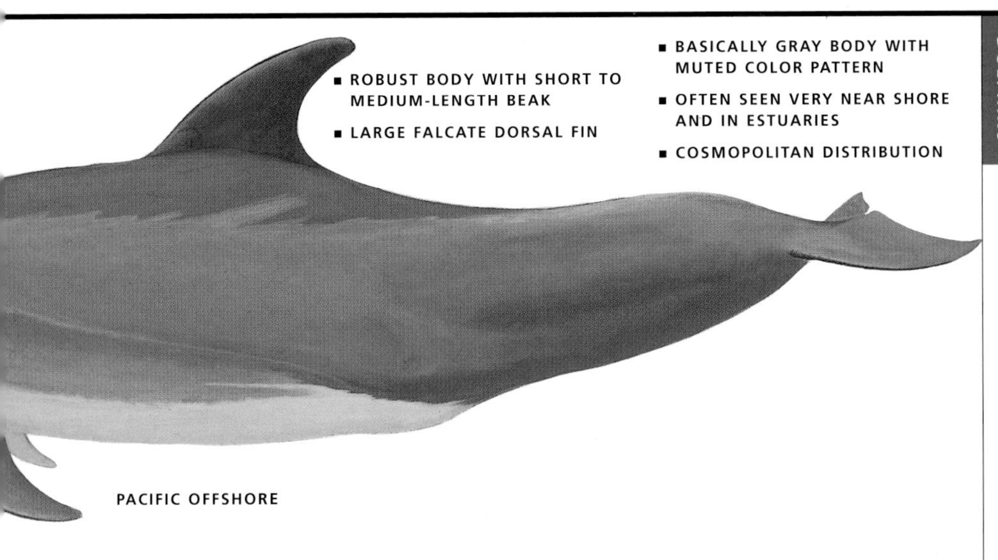

PACIFIC OFFSHORE

nents and around most oceanic islands and atolls. These animals often move into, or in some instances reside in, bays, estuaries, and the lower reaches of rivers. In addition, there are pelagic populations centered far offshore—for example, in the Gulf Stream of the North Atlantic and in the eastern tropical Pacific. In North American waters at least, Common Bottlenose Dolphins are found mainly where surface temperatures are in the range of 50 to 90°F (10–32°C). Off the Atlantic coast, the coastal population undertakes a seasonal north–south migration between New Jersey and North Carolina (or farther south). The range of the California coastal population has expanded several hundred miles northward since the El Niño warm-water incursion of 1982–1983.

SIMILAR SPECIES Common Bottlenose Dolphins can be confused with Rough-toothed, Risso's,

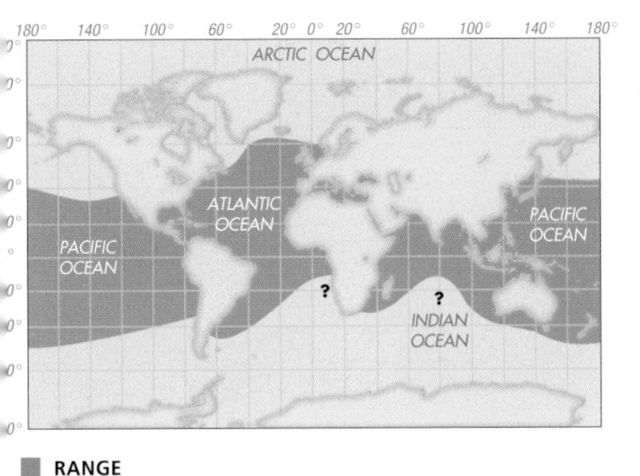

RANGE
? POSSIBLE RANGE

COMMON BOTTLENOSE DOLPHIN
FAMILY DELPHINIDAE

MEASUREMENTS AT BIRTH

LENGTH 33–55″ (84–140 cm)
WEIGHT 31–44 lb (14–20 kg)

MAXIMUM MEASUREMENTS

LENGTH **MALE** 8′–12′6″ (2.45–3.8 m)
FEMALE 7′10″–12′ (2.4–3.7 m)
WEIGHT **MALE** *Western North Atlantic:* 1,100 lb (500 kg)
FEMALE 570 lb (260 kg)

LIFE SPAN

MALE 40–45 years
FEMALE At least 50 years

Pantropical Spotted, and Atlantic Spotted Dolphins throughout those species' ranges, which means primarily offshore, relatively deep portions of the Common Bottlenose's range, plus, in the case of the Atlantic Spotted Dolphin, in tropical and warm temperate coastal waters. The Rough-toothed Dolphin is the only dolphin with flippers as large as those of bottlenose dolphins, but the absence of a definite crease between the melon and beak clearly distinguishes the Rough-toothed. Spotted dolphins, especially when not heavily spotted, are easy to confuse with bottlenose dolphins. They generally have a longer, slimmer beak and bolder pigmentation features. At a distance, the tall dorsal fin of Risso's Dolphin can lead to confusion, but closer in, the blunt forehead and white markings readily distinguish that species. In coastal waters from southern Brazil to Nicaragua, Tucuxis can be mistaken for small bottlenose dolphins. However, the Tucuxi's dorsal fin is not nearly as tall and recurved; it more nearly resembles that of a Harbor Porpoise than a bottlenose dolphin. Hump-backed dolphins overlap in shallow nearshore waters of the eastern Atlantic, western Pacific, and Indian Oceans, and often co-occur with bottlenose dolphins. Adult hump-backed dolphins tend to be lighter in color, with a longer, slimmer beak and a smaller dorsal fin that in some regions has a hump at its base.

BEHAVIOR Common Bottlenose Dolphins occur in groups that vary greatly in size, depending partly on habitat. Animals in bays form smaller groups (2 to 15 individuals) than those offshore (often many tens or hundreds). Composition and stability of these groups also varies. Bands of related females may stay together for many years, during which time they are visited briefly and occasionally by adult males. Pair bonds between males have been documented to last for 20 years or longer. Bottlenose dolphins often associate with other cetacean species, particularly pilot whales. These dolphins are classic bow riders and regularly approach powered vessels. They also surf in waves of all kinds, including those made by large whales or storms and coastal breakers. Individual bottlenose dolphins in a number of areas have developed lasting affinities with boaters, divers, and bathers. Contrary to their image as "friendly" animals, bottlenose dolphins can act aggressively toward other cetaceans. They routinely displace other delphinids from preferred bow-riding positions. In Scotland's Moray Firth, bottlenose dolphins have been seen chasing, butting, and propelling Harbor Porpoises clear of the water. Sharks are significant natural predators of bottlenose dolphins, and it is not unusual to see wounds or scars attributed to shark bites on the bodies of living dolphins.

Common Bottlenose Dolphins are active and agile, often leaping clear of the surface. It can seem as though they have adapted to perform in oceanariums, where their intelligence, curiosity, and energy make them extremely popular.

As foraging generalists, Common Bottlenose Dolphins employ an impressive variety of strategies for capturing prey. Here, a dolphin homes in on a school of fish off Roatán Island, Honduras, in the southwestern Caribbean Sea.

REPRODUCTION The reproductive system and cycle of the Common Bottlenose Dolphin are better known than those of any other cetacean, as they have been studied closely both in captivity and in the wild. Gestation lasts about a year. Calves can be born at any season, but few are born in the colder winter months in temperate regions. Lactation lasts at least a year. Calves are often not fully weaned until 18 to 20 months of age, and they may continue to associate with their mothers for several more years. The average calving interval is at least three years. Wild females as old as 45 years have given birth and raised a calf to weaning.

FOOD AND FORAGING Common Bottlenose Dolphins are usually described as catholic in their food habits. They prey on a large variety of organisms, depending on the habitat. Coastal animals tend to feed on fish and invertebrates that live on or near the bottom, while offshore animals eat pelagic or mesopelagic fish and squid. The presence of deep-sea fish in the stomachs of some offshore animals suggests that they dive to depths of more than 1,600 feet (500 m). Croakers, sea trout, mackerel, and mullet are typical prey in shelf waters. These dolphins are often attracted to fishing operations and learn to feed behind shrimp trawlers, taking discards and injured or disoriented escapees. Bottlenose dolphins forage individually and cooperatively. Among their more spectacular techniques for catching prey are "fish whacking" (striking a fish with the flukes and knocking it clear of the water) and driving schools of fish onto mudflats and partially beaching themselves to collect the fish.

STATUS AND CONSERVATION Overall, the Common Bottlenose Dolphin remains abundant and widely distributed. The general outlook for the species is reasonably good, but some regional and local populations are certainly at risk of disappearing because of habitat degradation, fishery conflicts, pollution, or overkilling. Large numbers were killed in fisheries in the eastern United States as recently as the 1920s and in the Black Sea until the 1980s or 1990s. The Black Sea population remains depleted, with little immediate prospect of recovery in its severely degraded environment. Major die-offs of these dolphins along the U.S. Atlantic and Gulf of Mexico coasts have been linked to viral outbreaks and acute exposure to toxins. There is speculation that heavy burdens of pollutants in the dolphins' bodies have weakened their immune systems. In the Mediterranean Sea, pollution, fish population collapses, and gillnet entanglement are viewed as serious threats to local populations. Japanese fishermen harpoon at least several hundred bottlenose dolphins each year, and these dolphins are also deliberately killed for food or bait in many other areas, including the West Indies, Venezuela, Peru, Chile, the Philippines, and Sri Lanka.

Indo-Pacific Bottlenose Dolphin

Tursiops aduncus
(Ehrenberg, 1833)

Scientists have long acknowledged the possible existence of this second species of bottlenose dolphin, but its co-occurrence with and general similarity to the Common Bottlenose made the two species difficult to distinguish using standard approaches. The science of molecular genetics seems to have broken the logjam, and a consensus has finally emerged that recognizes the two as separate species. The last word on the subject has yet to be written, though. Some genetic evidence indicates that the Indo-Pacific species may be more closely allied to the genus *Stenella* than to the other *Tursiops.* Its status is likely to remain in flux for many years to come. Moreover, it will take some time before the differences and similarities between the two bottlenose species are well described; many descriptions in the literature are based on combined samples and analyses.

DESCRIPTION The external appearance of the Indo-Pacific Bottlenose Dolphin is best described in relation to that of the Common Bottlenose. Indo-Pacific animals are somewhat smaller and less robust in form, with a longer, more slender beak, a less convex melon, and, at least in some areas, a proportionally larger dorsal fin and flippers. Indo-Pacific Bottlenoses tend to have more teeth, although the range of counts for the two species overlap; for example, off southern Africa, Indo-Pacific Bottlenoses have 23 to 29 pairs in both the upper and lower jaws, and Common Bottlenoses have 21 to 24.

The Indo-Pacific Bottlenose Dolphin has a basically gray color pattern that is darker dorsally. Indo-Pacific animals are usually generally lighter in body color than Common Bottlenose Dolphins. Dark ventral spotting or flecking is the most conspicuous distinguishing external feature of the Indo-Pacific species, but the spotting is age-related and variable between populations (for example, some groups off southeastern Australia are unspotted). Spotting begins on the belly region as the animal approaches sexual maturity and can eventually spread to much of the ventral half of the body, including the throat, flippers, and sides of the face.

RANGE AND HABITAT The Indo-Pacific Bottlenose Dolphin is found in coastal waters in tropical and subtropical latitudes. It occurs in the western North Pacific from southern Japan

- ROBUST BODY WITH MEDIUM LENGTH BEAK
- LARGE, FALCATE DORSAL FIN
- BASICALLY GRAY BODY, DARKER DORSALLY
- DARK SPOTTING ON BELLY AND SIDES OF ADULTS IN MOST REGIONS
- FOUND MAINLY NEAR SHORE IN WARM WATERS OF WESTERN PACIFIC AND INDIAN OCEANS

southward (including the Taiwan Strait), in the subtropical western South Pacific, in much of the Indonesian archipelago, along the north coast of Australia between New South Wales and Western Australia, and along the entire rim of the Indian Ocean including the Persian Gulf and Red Sea. Off the east coast of southern Africa, from about Cape Agulhas northward and eastward, these dolphins are common in the shallows (less than 100 feet/30 m deep) within a half mile (1 km) of shore. They are seen regularly just seaward of the surf zone and in the clear water just outside turbid estuarine plumes. The total range of the Indo-Pacific Bottlenose Dolphin is much more restricted than that of the Common Bottlenose Dolphin, which occurs both near shore and offshore and ranges beyond the tropics and subtropics into temperate regions.

SIMILAR SPECIES Hump-backed dolphins live alongside or in close proximity to Indo-Pacific Bottlenose Dolphins in many areas. They are of similar size and configuration, but hump-backed dolphins have shorter, less recurved dorsal fins

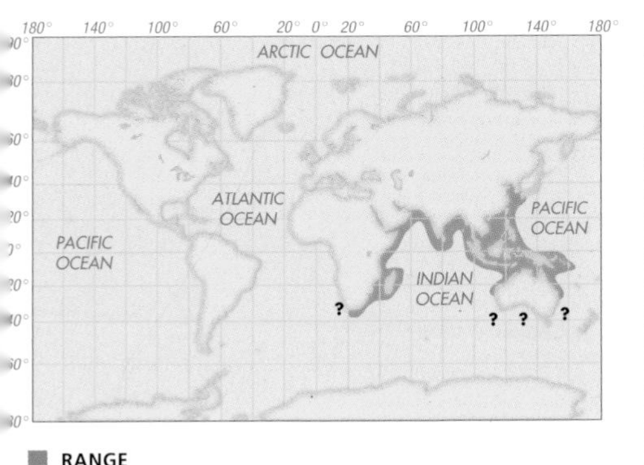

RANGE
? POSSIBLE RANGE

INDO-PACIFIC BOTTLENOSE DOLPHIN
FAMILY DELPHINIDAE

MEASUREMENTS AT BIRTH
LENGTH 33–44″ (84–112 cm)
WEIGHT 20–46 lb (9–21 kg)

MAXIMUM MEASUREMENTS
LENGTH **MALE** 8′6″ (2.6 m)
FEMALE 8′2″ (2.5 m)
WEIGHT **MALE** 510 lb (230 kg)
FEMALE 400 lb (180 kg)

LIFE SPAN
More than 40 years

The wild dolphins of Shark Bay, Western Australia, began during the 1960s to approach small fishing boats and "beg" for a portion of the day's catch. Since then, they have become so tame that they accept fish from vacationers who wade into the shallows near an old fishing camp. In addition to being tourist attractions, the Shark Bay dolphins have been the subjects of intensive scientific study.

and generally occur in smaller (6 to 8 or fewer individuals), less conspicuous groups. Hump-backed dolphins in eastern and southern Asia are white or pink as adults, often with dark spotting on the back. The Pantropical Spotted Dolphin occurs in the same latitudes but generally is found farther offshore than the Indo-Pacific Bottlenose and has white spotting on the back and sides.

BEHAVIOR Indo-Pacific Bottlenose Dolphins occur in groups ranging into the high tens and low hundreds, but the average number is about 5 to 15 individuals. In the Taiwan Strait around the Penghu Islands, it is not unusual to see both Indo-Pacific and Common Bottlenose Dolphins in mixed schools that frequently include other dolphin species as well. Although bottlenose dolphins in captivity have reproduced with other species (for example, with Rough-toothed and Risso's Dolphins and with False Killer and Short-finned Pilot Whales), there is little evidence of hybridization between the Indo-Pacific and Common species in the wild, or for that matter in captivity. Indo-Pacific Bottlenose Dolphins are seen in mixed groups with hump-backed dolphins off South Africa and Australia. The bottlenose animals tend to be dominant, sometimes aggressively chasing lone hump-backed dolphins and maintaining preferred positions in relation to trawl nets. At least in some parts of their range, Indo-Pacific Bottlenose Dolphins live in the midst of large shark populations. Nasty wounds and scars on the bodies of living dolphins, along with remains found in shark stomachs, provide abundant proof that large sharks actively prey on these dolphins. Youngsters appear to be the most vulnerable.

A group of Indo-Pacific Bottlenose Dolphins swim in echelon, casting shadows on the sandy ocean bottom off Australia. In much of their range, these dolphins inhabit shallow coastal waters, often within a mile of shore.

REPRODUCTION The Indo-Pacific Bottlenose Dolphin's main calving and mating period is late spring and summer, although reproductive activity occurs to some extent throughout much of the year. Gestation lasts about a year. Calves are weaned after 18 to 24 months of lactation, but they may continue to associate with their mothers for up to three years. The average calving interval is probably about three years.

FOOD AND FORAGING Generally catholic and opportunistic in its food habits, the Indo-Pacific Bottlenose may depend on only a few primary species in a given region. Its diet is dominated by bottom- or reef-dwelling cephalopods and fish but also includes pelagic inshore schooling species. Most prey are less than 8 inches (20 cm) long, but some can be considerably larger, including fish up to 24 inches (60 cm) long. Adult males consume the larger prey items and also may forage farther offshore than females and juveniles. Off KwaZulu-Natal, South Africa, some groups of dolphins regularly steal fish from commercial and sport fishing lines. In Moreton Bay, Australia, they commonly follow trawlers and feed on the bycatch.

STATUS AND CONSERVATION The Indo-Pacific Bottlenose Dolphin is not considered endangered, but its exclusively nearshore distribution makes it vulnerable to environmental degradation, direct exploitation, and fishery conflicts. In the recent past, large numbers of both bottlenose dolphin species have been driven ashore and killed by fishermen in Taiwan's Penghu Islands. Although this deliberate killing is now prohibited in Taiwan, gillnet mortality continues to be a problem there and throughout most of the species' range. Off KwaZulu-Natal, South Africa, and Queensland, Australia, large-mesh nets are set to protect bathing beaches from sharks, and these nets have been responsible for the deaths of many bottlenose dolphins. In addition, the dolphins living off KwaZulu-Natal, and probably in other areas, are exposed to continuing inputs of DDT as a result of malaria-control efforts on land. This has been cited as a potential threat to their reproductive fitness. With the rapid proliferation of oceanariums in southern Asia, the demand for live bottlenose dolphins is bound to increase, and unregulated collections could have a severe impact on local wild populations.

Like all cetacean mothers, female Indo-Pacific Bottlenose Dolphins are extremely solicitous of their newborn calves. In this photograph, the mother appears to be keeping an eye on the photographer while encouraging her young calf.

Pantropical Spotted Dolphin

Stenella attenuata
(Gray, 1846)

The Pantropical Spotted Dolphin is one of the most abundant cetaceans on the planet, even though its numbers have been seriously reduced in some areas by massive incidental killing. It was one of two species (the other being the Spinner Dolphin) killed by the hundreds of thousands annually during the early period of tuna purse seining in the eastern tropical Pacific Ocean. Perhaps the only benefit of all the wasteful killing is that scientists came to know these two pelagic dolphin species better than many common nearshore species. With thousands of fresh carcasses available for study, it was possible to delineate their geographical differences in color pattern and morphology, investigate their foraging habits, and estimate their life-history parameters. During lengthy cruises undertaken in the search for ways to reduce dolphin mortality in the tuna purse seine nets, behaviorists were able to observe and track these animals, and oceanographers to identify habitat factors that influence their distribution. There appear to be many geographical populations of the Pantropical Spotted Dolphin that differ in body size, coloration, and skull characteristics. Scientists have thus far described three subspecies.

DESCRIPTION The Pantropical Spotted Dolphin has a moderately slender body, a relatively small dorsal fin, and a moderately long, slender beak. Adult males have a thickened caudal peduncle and a noticeable bulge behind the anus. These features, however, vary geographically. The dorsal fin is strongly falcate, and the flippers are tapered. There are 35 to 48 pairs of teeth in the upper jaw and 34 to 47 pairs in the lower jaw.

These dolphins have a dark back and lighter belly, with a complex and extremely variable system of spotting and striping. The basic pattern includes a dark gray dorsal cape that dips low onto the sides below and forward of the dorsal fin. The lighter gray ventral color sweeps upward onto the caudal peduncle in a wide swath well behind the dorsal fin. The beak is white-tipped and conspicuous in some populations, making it a useful field mark. Adults have a black "mask," visible from the side or in front, as well as a dark jaw-to-flipper stripe. The dorsal fin, flippers, and flukes are dark gray, and the fin and flippers may become spotted on adults. The degree of spotting in general varies greatly between individuals and regional populations. The race of large-bodied Pantropical Spotted Dolphins in coastal waters of the

- WHITE-TIPPED BEAK
- PROMINENT DARK DORSAL CAPE THAT DIPS LOW ONTO SIDES
- LIGHT GRAY SWATH SWEEPING UPWARD ON CAUDAL PEDUNCLE
- VARIABLE DEGREE OF SPOTTING
- USUALLY OCCURS IN LARGE SCHOOLS
- TROPICAL TO WARM TEMPERATE DISTRIBUTION

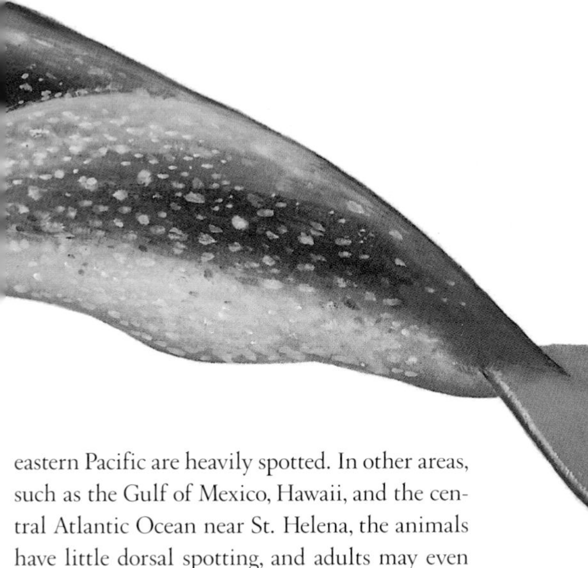

eastern Pacific are heavily spotted. In other areas, such as the Gulf of Mexico, Hawaii, and the central Atlantic Ocean near St. Helena, the animals have little dorsal spotting, and adults may even appear unspotted when seen from a distance. Calves are born unspotted; young dolphins begin to acquire dark spots on the ventral surfaces and, later, light spots on the dorsal surfaces.

RANGE AND HABITAT Pantropical Spotted Dolphins occur in all tropical to warm temperate oceanic waters between roughly 40°N and 40°S. Three subspecies are recognized. One inhabits nearshore waters around the Hawaiian Islands, another occurs in offshore waters of the eastern tropical Pacific, and a third, *Stenella attenuata graffmani,* occurs in coastal waters between Baja California and the northwestern coast of South America. Tagging data suggest that home ranges of Pantropical Spotted Dolphins in the eastern tropical

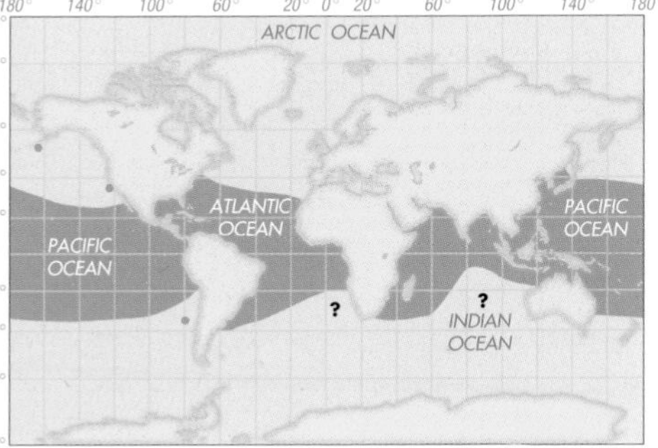

RANGE
EXTRALIMITAL RECORDS
? POSSIBLE RANGE

PANTROPICAL SPOTTED DOLPHIN
FAMILY DELPHINIDAE

MEASUREMENTS AT BIRTH
LENGTH 33″ (85 cm)
WEIGHT Unavailable

MAXIMUM MEASUREMENTS
LENGTH **MALE** 8′5″ (2.57 m)
FEMALE 7′10″ (2.4 m)
WEIGHT **MALE** 260 lb (120 kg)
FEMALE Unavailable

LIFE SPAN
MALE 40 years
FEMALE 46 years

Pacific are about 200 to 300 nautical miles in diameter, and that their overall distribution shifts toward shore in fall and winter and offshore in spring.

SIMILAR SPECIES The Atlantic Spotted Dolphin shares much of the Pantropical Spotted's range in the Atlantic; it has a heavier, more robust body, tends to be more densely spotted, and has a light shoulder blaze. Bottlenose dolphins are present throughout most of the worldwide range of Pantropical Spotted Dolphins. Both the Indo-Pacific and Common Bottlenose Dolphins tend to have more robust bodies and lack the well-defined dorsal cape of Pantropical Spotteds. Indo-Pacific Bottlenoses often have dark spots and flecks on the belly and sides, although their dorsal spotting is not as dense; they have a largely coastal distribution. The Rough-toothed Dolphin overlaps extensively but lacks the spotting of Pantropical Spotted Dolphins.

BEHAVIOR Pantropical Spotted Dolphins are among the most gregarious cetaceans. In offshore areas, they are often encountered in schools of hundreds or thousands. These large aggregations include numerous smaller social units, each containing fewer than 20 individuals that remain physically close to one another and surface and dive synchronously. The groups can consist entirely of adult females with young, only juveniles, or only adult males. Group membership can be fluid, although the bands of adult males tend to be stable and to move with stereotyped precision. Groups of Pantropical Spotted Dolphins often associate with schools of Spinner Dolphins.

Hawaiian fishermen call Pantropical Spotted Dolphins "leapers" because they often make high, arching leaps above the surface. They engage in various other aerial acrobatics, and they also bow ride, although in areas where they have been chased and encircled by tuna nets they generally flee from fast-moving vessels. Two trained Pantropical Spotteds were able to reach top speeds of slightly more than 21 knots in just 2 seconds. Natural predators of these dolphins include large sharks, Killer and False Killer Whales, and possibly even Pygmy Killer Whales.

REPRODUCTION Calves are born year-round, with several seasonal peaks. The gestation period is a little more than 11 months. Lactation lasts for at least a year, often longer than two years, although calves begin taking solid food at three to six months of age. The average calving interval is 2½ to nearly 4 years, depending on the population.

FOOD AND FORAGING Pantropical Spotted Dolphins prey on small pelagic fish from at least 18 families, cephalopods from at least eight families, and crustaceans. Most of their prey items occur either near the surface or in the water column. In the eastern tropical Pacific, these dolphins tend to associate with schools of tuna, Spinner Dolphins, and flocks of oceanic birds. These associations may improve their foraging efficiency, afford protection from predators, or

Pantropical Spotted Dolphins race explosively beside a research vessel in the Gulf of Mexico, revealing their white-tipped beaks, small flippers, and streamlined bodies.

simply provide them with a sense of orientation in the pelagic void. While there is considerable overlap in diet between yellowfin tuna and Pantropical Spotted Dolphins, there is much less between Spotteds and Spinners.

STATUS AND CONSERVATION There are estimated to be more than 2 million Pantropical Spotted Dolphins in the eastern tropical Pacific, at least another 400,000 in Japanese waters of the western Pacific, about 13,000 off the eastern United States, and several tens of thousands in the northern Gulf of Mexico. The species' world population is probably well in excess of 3 million. The Pantropical Spotted Dolphin is not in danger of extinction; however, the tuna fishery in the eastern tropical Pacific apparently reduced the northeastern offshore population during the 1960s and 1970s to only about 25 percent of its earlier abundance. The status of this population has been a central issue of those arguing for continued or tighter regulations on the tuna fishery. Pantropical Spotted Dolphins have been killed in large numbers by drive fisheries in Japan and the Solomon Islands, and they are frequent targets of dolphin hunters in Sri Lanka, the Lesser Antilles, Indonesia, and the Philippines, killed for both bait and human consumption.

OPPOSITE: *A Pantropical Spotted Dolphin flexes as it leaps high above the surface off Hawaii. The dark dorsal cape that dips low onto the sides below the dorsal fin and the light gray swath that sweeps high onto the caudal peduncle help identify the species.*
ABOVE: *This dorsal view of Pantropical Spotted Dolphins in the Pacific Ocean emphasizes the prominent dorsal cape and the conspicuous light gray area on the caudal peduncle. It also shows why dolphins are difficult to count, as some animals in this group are deep in the water column while others are near the surface.*

Atlantic Spotted Dolphin

Stenella frontalis
(G. Cuvier, 1829)

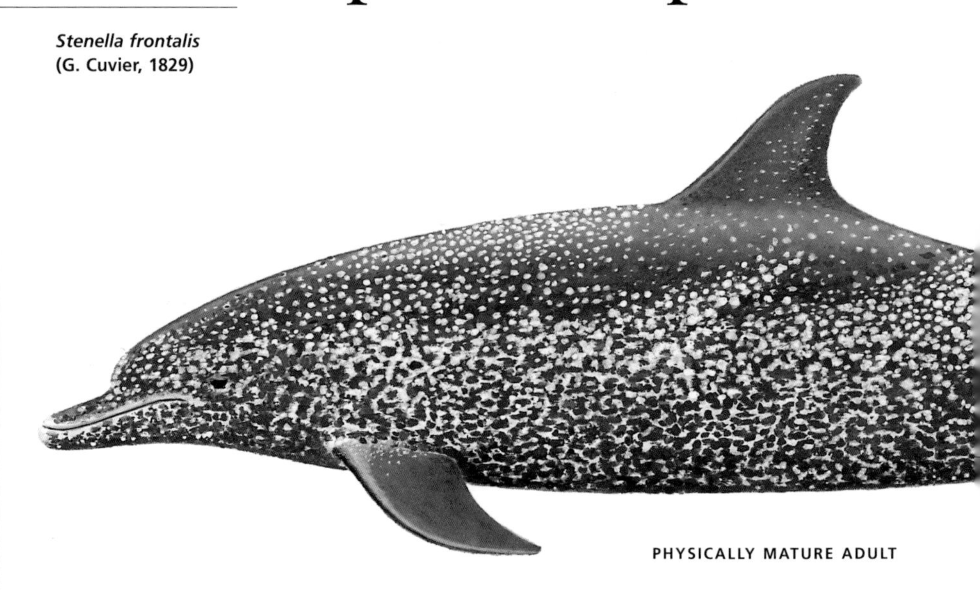

PHYSICALLY MATURE ADULT

The Atlantic Spotted Dolphin is one of two members of the genus *Stenella* that are endemic to tropical and warm temperate zones of the Atlantic Ocean; the other is the Clymene Dolphin. A large, heavily spotted form of Atlantic Spotted Dolphin that occurs in coastal waters of the southeastern United States may yet prove to be a valid subspecies. Considering the abundance of Atlantic Spotted Dolphins and their proximity to the southern and southeastern United States, researchers know surprisingly little about their biology. A community of Atlantic Spotted Dolphins in the Bahamas has been habituated to human divers and boats since the early 1970s. Tour companies and researchers have taken advantage of this exceptional opportunity, and a long-term study of the animals has provided many new insights about the species. Also, after an adult male Atlantic Spotted Dolphin stranded alive in Texas in 1995, it was rehabilitated, and its dorsal fin was fitted with a satellite-linked time-depth recorder and a VHF radio transmitter. The dolphin was released into the wild and has provided preliminary information on the species' diving and traveling patterns.

DESCRIPTION The Atlantic Spotted Dolphin is robustly built in comparison with the Pantropical Spotted Dolphin. The beak is moderate in length and thickness. The dorsal fin is fairly erect and strongly falcate, situated at midback. There are 32 to 42 pairs of teeth in the upper jaw and 30 to 40 pairs in the lower jaw.

Atlantic Spotted Dolphins have a three-part color pattern: a dark gray dorsal cape, a lighter gray zone on the sides that crosses over the back slightly forward of the middle of the caudal peduncle, and a white belly. A light gray shoulder blaze intrudes onto the cape below the front of the dorsal fin. The beak is often white-tipped. Adults have dark spots on the belly and light spots on the sides and back. The extent of spotting on adults is highly variable. Calves are born unspotted, and dark spotting begins to appear on the belly at approximately weaning age. Spotting becomes more intense with age.

RANGE AND HABITAT Atlantic Spotted Dolphins occur throughout much of the tropical and warm temperate Atlantic Ocean. In the western Atlantic they range from Cape Cod, Massachusetts, south to Rio Grande do Sul, Brazil, and in the eastern

- THREE-PART COLORATION: DARK GRAY BACK, LIGHTER SIDES, WHITE BELLY
- SPOTTING RANGES FROM ABSENT IN CALVES TO STRIKING IN SOME ADULTS
- COMBINATION OF SHOULDER BLAZE AND SPOTTING IS DIAGNOSTIC
- WHITE-TIPPED BEAK
- OCCURS ONLY IN TROPICAL AND WARM TEMPERATE ATLANTIC

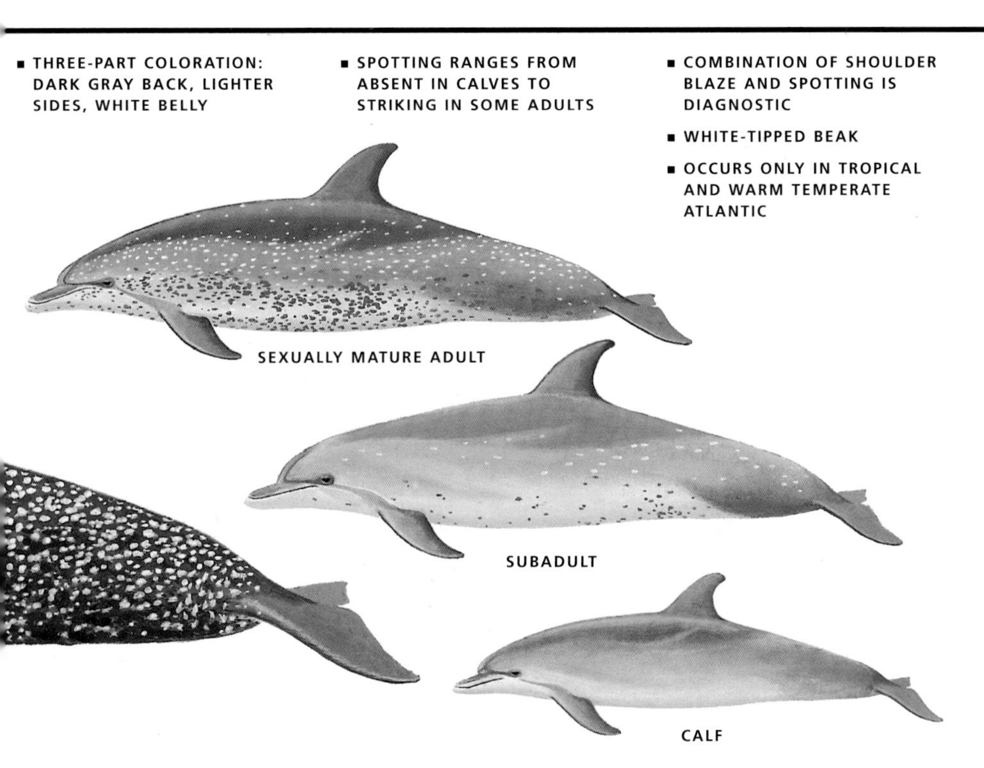

Atlantic from the Azores south to the Canary Islands, St. Helena, and Gabon, Africa. They are common in the Gulf of Mexico and the Caribbean Sea, but they are generally absent in the Mediterranean Sea. Although best known from coastal waters, they have also been seen far from land, in waters off Brazil and in mid-North Atlantic waters influenced by the Gulf Stream current. They are common inside the 250-meter depth contour, often occurring in water as shallow as 65 feet (20 m). They tend to be replaced by Common Bottlenose Dolphins in inshore and very nearshore waters. The sandflats on Little Bahama Bank, where divers swim with Atlantic Spotted Dolphins, are only 20 to 40 feet (6–12 m) deep. Resightings of the same individuals in the same place year after year indicate that at least some of the tame Atlantic Spotted Dolphins in the Bahamas are long-term residents.

SIMILAR SPECIES The Common Bottlenose Dolphin, which occurs throughout the range of the Atlantic Spotted Dolphin, is very similar in body size and form, especially to young Atlantic Spotteds whose dorsal cape, shoulder blaze, and spotting are subdued or absent. Bottlenose

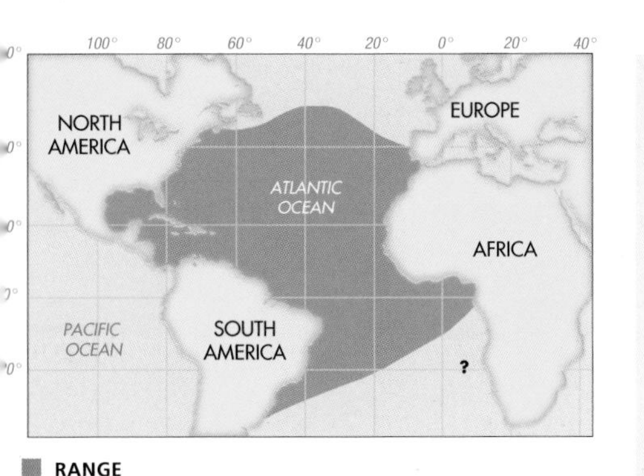

RANGE
? POSSIBLE RANGE

ATLANTIC SPOTTED DOLPHIN
FAMILY DELPHINIDAE

MEASUREMENTS AT BIRTH
LENGTH About 35–43″ (90–110 cm)
WEIGHT Unavailable

MAXIMUM MEASUREMENTS
LENGTH **MALE** 7′5″ (2.26 m)
FEMALE 7′6″ (2.29 m)
WEIGHT **MALE** 310 lb (140 kg)
FEMALE 290 lb (130 kg)

LIFE SPAN Unavailable

dolphins are unspotted and do not have a white-tipped beak, and their dorsal cape and shoulder blaze, if present at all, are generally much less distinct. The range of the Pantropical Spotted Dolphin overlaps the entire range of the Atlantic Spotted Dolphin but tends to be more oceanic. In the northern Gulf of Mexico, for example, most sightings of Pantropical Spotteds are seaward of the 200-meter depth contour, while Atlantic Spotted Dolphins are almost always seen along or inside that contour. Pantropical Spotted Dolphins are less robust and have a more sharply demarcated color pattern, with a clearer separation between the dark dorsal cape and the lighter sides. Also, the spotting of Pantropicals is less dense and prominent, and the dorsal fin appears more slender and delicate in profile.

BEHAVIOR Like other members of its genus, the Atlantic Spotted Dolphin is a fairly gregarious species. However, group size is rarely larger than about 50 individuals, and in coastal areas groups of only 5 to 15 are typical. These dolphins are willing bow riders, and they can be acrobatic at times. While a few Atlantic Spotted Dolphins have been trained to perform in Florida oceanariums, they generally do not acclimate well to captive conditions and are therefore not popular in the display industry. In the Bahamas, severe shark bite wounds on some individuals, especially calves and juveniles, indicate that shark predation is an ever-present danger for the species.

REPRODUCTION Little is known about reproduction in Atlantic Spotted Dolphins. Females nurse their calves for more than three years and sometimes for as long as five. Since the average calving interval is three or four years, it is not unusual for a female to be pregnant and lactating at the same time.

LEFT: *Atlantic Spotted Dolphins tend to be more densely spotted and have a heavier, more robust body than Pantropical Spotteds.* **ABOVE:** *Inquisitive and playful, Atlantic Spotted Dolphins play with a sea fan on the Bahama Banks. In this region, the dolphins appear to have grown accustomed to divers, facilitating both research and tourism.*

FOOD AND FORAGING Atlantic Spotted Dolphins prey on small fish, cephalopods, and benthic invertebrates. They sometimes follow trawlers to feed on debilitated or discarded organisms. The radio-tracked dolphin mentioned in the introduction remained in shallow waters near shore and rarely dove for longer than 2 minutes. On one occasion in the Gulf of Mexico, researchers observed a school of these dolphins coordinating their movements to trap and prey upon a fish school near the surface.

STATUS AND CONSERVATION While this species is not thought to be in danger, there is little available data on abundance or mortality rates. Abundance estimates are available only for small portions of the species' range, and some of these may be inaccurate due to the difficulty of distinguishing between Atlantic and Pantropical Spotted Dolphins. Somewhat more than 50,000 Atlantic Spotted Dolphins occur off the eastern United States, and at least a few thousand are present in the northern Gulf of Mexico. In the eastern and southern Caribbean, fishermen harpoon dolphins opportunistically, sometimes killing Atlantic Spotted Dolphins, but no information on numbers is available. These dolphins certainly become entangled and die in gillnets and purse seines, although this mortality is also poorly documented.

OPPOSITE: *Cetacean calves are nursed underwater, finding their mothers' paired teats on the posterior belly region. This Atlantic Spotted Dolphin calf is unspotted, while its mother exhibits the spotted adult pattern.*

Spinner Dolphin

Stenella longirostris
(Gray, 1828)

The Spinner Dolphin is named for its unique habit of leaping high above the surface and spinning on its longitudinal axis. The Spinner became well known during the late 1960s and the 1970s for its role in the eastern tropical Pacific tuna fishery. Fishermen use the dolphins as indicators of the presence of yellowfin tuna, then chase and encircle them, expecting that the tight bond between the fish and the mammals will ensure a good catch of tuna within the bag of the purse seine. Unfortunately, many of the dolphins that become entangled in the fishing gear die before the fishermen can release them. Direct mortality of dolphins in the tuna fishery has declined appreciably, but the effects of stress from capture are an ongoing concern. Much of what is known about the biology of Spinner Dolphins comes from studies related to the tuna fishery, supplemented by work with a few captive animals in Hawaii. In addition, a 25-year study of wild Spinner Dolphins in Kealakekua Bay, Hawaii, provided numerous insights about the species' behavior and social structure. The specific name is from the Latin for "long beak" and refers to the Spinner's long, narrow beak. Researchers have described four subspecies of Spinner Dolphin.

DESCRIPTION The Spinner Dolphin is said to have more regional variability in form and color pattern than any other cetacean. The body is usually slender but can be fairly robust, depending on the population. The basic body configuration always includes a relatively flat melon and a long, narrow, well-defined beak. The dorsal fin, positioned at midback, is erect and can be either slightly falcate or triangular. In some adult males in the eastern Pacific, the fin cants forward and appears as if it were "on backward." These males also have a prominent bulge on the ventral surface of the caudal peduncle. The flippers taper to a considerable degree and are almost pointed at the tips. There are 42 to 64 pairs of teeth in the upper jaw and 41 to 62 pairs in the lower jaw.

The basic color pattern includes a dark dorsal cape (not always present), lighter gray sides and flanks, and a white or light gray belly. The dorsal cape's lower margin runs almost parallel to the body axis and does not dip low onto the sides. The upper jaw is gray, and the lower jaw white; the beak has a dark tip, and there is a dark border along the mouthline. There is an eye-to-flipper stripe that extends forward from the eye to merge with the dark mouthline. This general color

- USUALLY SLENDER BUILD
- LONG, NARROW BEAK AND FLAT MELON
- DORSAL FIN ERECT, SOMETIMES CANTED FORWARD
- ENGAGES IN AERIAL SPINNING
- VENTRAL MARGIN OF DORSAL CAPE ALMOST PARALLEL TO BODY AXIS
- TROPICAL DISTRIBUTION

pattern can be completely obscured, as on Spinners in the far eastern Pacific, which appear almost solidly gray. Spinners farther offshore in the Pacific tend to be intermediate in appearance between the gray eastern animals and the animals with the standard color pattern (and in fact are thought to be hybrids).

RANGE AND HABITAT The Spinner Dolphin is generally regarded as a pantropical species, but it occurs throughout the tropics and subtropics in a number of discrete geographical populations. It is rarely encountered in latitudes higher than about 30°N and 30°S and is definitely most abundant between the Tropics of Cancer and Capricorn. Scientists have thus far described and named four subspecies. The most widespread, *Stenella longirostris longirostris,* is an offshore subspecies centered around oceanic islands of the tropical Atlantic, Indian, and western and central Pacific Oceans; in the Pacific, it generally does not range east of about 145°W. This subspecies is really a grab bag that subsumes many regional populations, including those

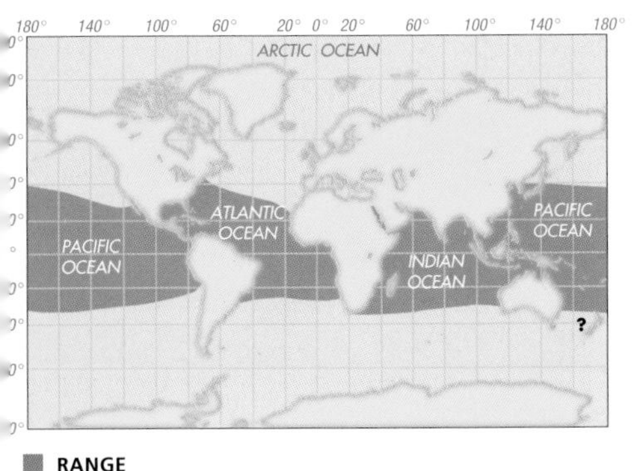

RANGE
? POSSIBLE RANGE

SPINNER DOLPHIN
FAMILY DELPHINIDAE

MEASUREMENTS AT BIRTH
LENGTH 30″ (76–77 cm)
WEIGHT Unavailable

MAXIMUM MEASUREMENTS
LENGTH **MALE** 7′9″ (2.35 m)
FEMALE 6′11″ (2.11 m)
WEIGHT **MALE** 172 lb (78 kg)
FEMALE 143 lb (65 kg)

LIFE SPAN
At least 20 years

A large school of Spinner Dolphins in the Pacific Ocean creates a frothy turmoil while speeding along the surface.

around Hawaii ("Hawaiian" Spinners). In the Pacific, east of 145°W and between roughly 24°N off Baja California and 10°S off Peru, the Eastern Spinner *(S. l. orientalis),* replaces *S. l. longirostris.* Intermediate forms or hybrids between these two subspecies are found throughout most of the offshore eastern tropical Pacific; they are sometimes known as "whitebelly" or "southwestern" Spinners. *(S. l. centroamericana),* the Central American Spinner, inhabits coastal shelf waters of the far eastern Pacific between southern Mexico and Costa Rica. Finally, *S. l. roseiventris,* the Dwarf Spinner, occurs in shallow coastal waters of Southeast Asia from Malaysia to northern Australia.

SIMILAR SPECIES The Clymene Dolphin is present throughout the Spinner's range in the tropical Atlantic, and the two species can be difficult to distinguish. The Clymene's dark gray eye-to-flipper stripe narrows anteriorly; its dorsal cape dips markedly at midbody; and it has a shorter, sharply patterned beak. The Pantropical Spotted Dolphin shares the Spinner's entire range. Apart from its spotted appearance (which can be either muted or prominent), the Pantropical Spotted has a shorter, thicker beak and a more falcate dorsal fin; also, its dorsal cape dips lower onto the sides below and anterior to the dorsal fin, and is interrupted behind the fin by a broad gray stripe that sweeps over the caudal peduncle.

BEHAVIOR There is no single description to adequately cover the social organization or activity pattern of this widely distributed and enormously flexible species. While Spinner Dolphins are certainly gregarious, group size depends on the circumstances. In the open ocean, they can occur in schools of several thousands that are composed of smaller groups. In coastal waters, however, they may live in groups of a few dozen to a few hundred individuals. Spinners in the Hawaiian Islands have what is called a "fission-fusion society." They "fuse" into large schools of hundreds as they move offshore at night to feed, and then split into much smaller groups of perhaps a few dozen while resting and socializing in inshore bays and reef-fringed lagoons during the day. In contrast to these large, island-centered populations that appear to mix freely, smaller Spinner populations at remote atolls tend to be more rigidly structured and closely knit, with no evidence of diurnal fission.

While other species of dolphins occasionally spin during a breach, this behavior is the Spinner's trademark. A single leap can involve as many as four revolutions. Most often, a series of spinning jumps begins with the highest—up to 10 feet (3 m) above the surface—and each successive jump is less vigorous. Spinners can make as many as 14 jumps in a series. Spinner Dolphins are subject to predation by large sharks, Killer Whales, and possibly False Killer Whales, Pygmy Killer Whales, and Short-finned Pilot Whales.

REPRODUCTION Like Pantropical Spotted Dolphins, Spinner Dolphins can give birth at any time of the year; however, all populations that have been studied exhibit one or more seasonal peaks. The timing of these peaks is variable across the geographical populations. The gestation period is about 10½ months, and lactation lasts one to two years. The average calving interval is about three years.

FOOD AND FORAGING In general, Spinner Dolphins prey on small mesopelagic fish, squid, and shrimp that are within about 650 to 1,000 feet (200–300 m) of the surface. In Hawaiian waters, where the foraging behavior of Spinner Dolphins has been studied in the greatest detail, they hunt mainly at night when the deep scattering layer moves toward the surface. The diminutive Spinners in Southeast Asia apparently differ in their foraging habits from the other populations that have been studied, feeding on benthic and coral reef organisms.

STATUS AND CONSERVATION The Spinner Dolphin is abundant and widespread. However, the population of Eastern Spinners is probably still less than half the size it was before the advent of tuna purse seining in the late 1950s. Recent estimates (from the late 1980s) of the two populations in the eastern tropical Pacific that were hardest hit by the tuna fishery are about 580,000 Eastern Spinners and close to 1 million "whitebelly" Spinners. "Hawaiian" Spinner Dolphins have not been affected by the tuna fishery and remain abundant throughout their range. The largely resident population along the Kona coast of Hawaii may number about 2,000. There are at least a few thousand in the northern Gulf of Mexico. No good abundance estimates exist from elsewhere in the species' range. Spinner Dolphins die in gillnets throughout the tropics, and in some areas their meat is used for bait or human consumption. Especially large numbers are taken in nets and by harpooning in Sri Lanka and the Philippines. Dwarf Spinners in the Gulf of Thailand frequently die accidentally in shrimp trawl nets, while those off northern Australia die in shark gillnets. Although the species as a whole may be secure, local populations are vulnerable to overexploitation, habitat degradation, and harassment.

This Spinner Dolphin in the northwestern Hawaiian Islands lives up to its name, spinning dramatically above the ocean surface.

Clymene Dolphin

Stenella clymene
(Gray, 1850)

The Clymene Dolphin was confirmed as a valid species only in 1981. Before then it had been viewed simply as a short-beaked variant of the Spinner Dolphin, which it closely resembles in coloration. Although it is now recognized as a distinct species, the Clymene Dolphin remains one of the least-known cetaceans.

DESCRIPTION The Clymene Dolphin is shorter than most ocean dolphins, with a robust body, medium-length beak, and slightly falcate or triangular dorsal fin. There are 36 to 49 pairs of teeth in the upper jaw and 38 to 48 pairs in the lower.

The body coloration can be divided into three zones: a dark gray dorsal cape, light gray sides, and a white belly. There is a noticeable dip in the ventral margin of the dark cape below the dorsal fin. The eye-to-flipper stripe is noticeably wider at the flipper end. The lips and tip of the beak are black. Viewed dorsally, the beak is marked by a black band connecting the tip with the melon. This band is bordered on both sides by a pale gray to white blaze. About halfway along the dark band, a dark gray, mustache-like marking is often present. Bites from cookie-cutter sharks are often present on the sides and abdomen.

RANGE AND HABITAT The Clymene Dolphin occurs only in deep tropical and subtropical Atlantic waters, including the Gulf of Mexico and the Caribbean Sea, but is not known to live in the Mediterranean. Strandings have been documented as far north as New Jersey and as far south as southern Brazil.

SIMILAR SPECIES The Spinner Dolphin occurs throughout the Clymene Dolphin's range, and the two species sometimes occur in mixed schools. They are very difficult to distinguish. The Spinner's dorsal cape usually parallels the gray side field, lacking the dip below the dorsal fin, and its eye-to-flipper stripe does not widen toward the flipper. Its beak is longer and narrower, and it lacks the "mustache" mark. Common dolphins, especially the Short-beaked, overlap much of its range (but not in the Gulf of Mexico). The "hourglass," or crisscross, effect on the sides of common dolphins is distinctive, particularly when the yellowish-tan patch on the foresides is clearly evident.

BEHAVIOR, REPRODUCTION, AND FORAGING Clymene Dolphins most commonly occur in

- DARK CAPE WITH ROUNDED DIP BELOW DORSAL FIN
- "MUSTACHE" MARK ON UPPER SURFACE OF BEAK
- EYE-TO-FLIPPER STRIPE THAT WIDENS TOWARD FLIPPER
- ENDEMIC TO TROPICAL AND SUBTROPICAL ATLANTIC

groups of a few individuals to about 50. Larger groups of up to several hundred are occasionally seen. They sometimes spin in much the same way as Spinner Dolphins do, although they may not rotate completely in the air and often land on their sides or backs. Clymene Dolphins sometimes associate with Spinner and common dolphins. Clymene Dolphins feed on small mesopelagic fish and squid. At least some of their prey inhabit the deep scattering layer. Some of these organisms migrate vertically, coming closer to the surface at night. Essentially nothing is known about reproduction in this species.

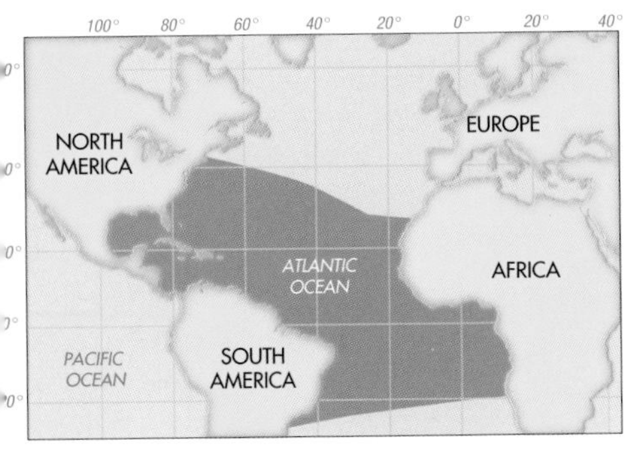

RANGE

STATUS AND CONSERVATION

The Clymene Dolphin may be naturally rare. The only area with an abundance estimate is the northern Gulf of Mexico, with about 5,500 individuals. These dolphins are harpooned at least occasionally at St. Vincent in the Lesser Antilles, and those caught incidentally in gillnets in Venezuela are either eaten or used for bait in the shark fishery. They are probably among the species taken in tuna purse seines off tropical western Africa.

CLYMENE DOLPHIN
FAMILY DELPHINIDAE

MEASUREMENTS AT BIRTH

LENGTH Unavailable
WEIGHT Unavailable

MAXIMUM MEASUREMENTS

LENGTH **MALE** 6′6″ (1.97 m)
FEMALE 6′3″ (1.9 m)
WEIGHT **MALE** 176 lb (80 kg)
FEMALE 165 lb (75 kg)

LIFE SPAN Unavailable

Striped Dolphin

Stenella coeruleoalba
(Meyen, 1833)

It is often assumed that the classical Greek and Roman reverence for dolphins came from contact with local bottlenose or common dolphins. Yet, as early Mediterranean seafarers ventured farther and farther from land, they would have begun to encounter large schools of Striped Dolphins. This species is still the most common cetacean in the Mediterranean Sea (even though more than 1,000 individuals died of a disease outbreak in 1990–1992). This active, gregarious, and beautifully marked dolphin is widespread and reasonably abundant in tropical and temperate regions of the world's oceans. The specific name, derived from the Latin *caeruleus* for "sky blue" and *albus* for "white," refers to its striking coloration.

DESCRIPTION The Striped Dolphin is a fairly robust dolphin with a moderately long, well-defined beak, short tapered flippers, and a prominent, falcate dorsal fin. There are 39 to 55 pairs of small teeth in both the upper and lower jaws.

The color pattern is a striking combination of bluish gray and white, but the boldness of the markings varies regionally and individually. The beak, appendages, and back are gray; the throat and belly are white. There are two narrow black stripes on each side of the body. They begin as a single stripe from the beak to the eye that diverges, one branch going from the eye to the flipper and the other continuing along the side from the eye to the anal region. A short, delicate subsidiary stripe sometimes branches downward from the side stripe. A distinctive light blaze is usually present below and anterior to the dorsal fin, interrupting the dark cape pigmentation.

RANGE AND HABITAT This is a cosmopolitan species that occurs in tropical and warm temperate waters. Its northern limits are Newfoundland, southern Greenland, Iceland, the Faeroe Islands, and Denmark in the Atlantic. In the Pacific the northern limits are the Sea of Japan, Hokkaido, Washington state, and along roughly 40° N across the western and central Pacific. The Striped Dolphin is found south to Buenos Aires, Cape Province, Western Australia, New Zealand, and Peru. There are numerous populations of Striped Dolphins that are more or less isolated from one another. The large Mediterranean population, for example, apparently does not mix with the

- WHITE TO LIGHT GRAY SHOULDER BLAZE, SWEEPING BACK AND UP TOWARD DORSAL FIN
- BOLD, NARROW BLACK STRIPES FROM EYE TO ANUS AND EYE TO FLIPPER
- FAIRLY GREGARIOUS AND ATTRACTED TO POWERED VESSELS
- DISTRIBUTION MAINLY SEAWARD OF CONTINENTAL SHELF IN TROPICAL AND WARM TEMPERATE SEAS

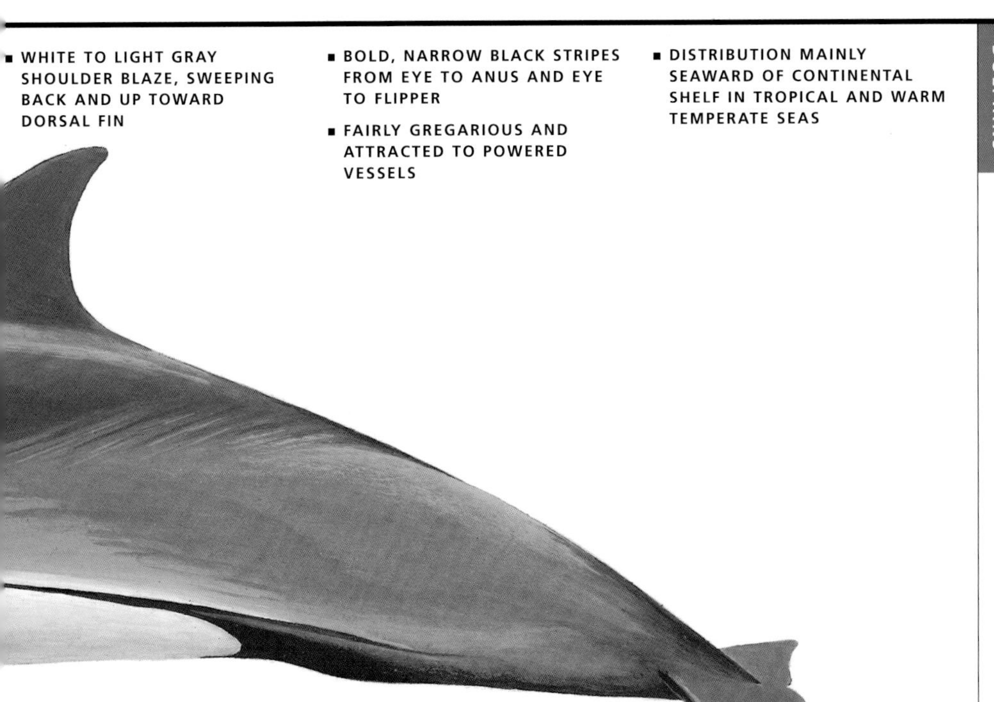

eastern Atlantic population. Striped Dolphins prefer highly productive oceanic waters. They are not seen often, or in large numbers, landward of the continental shelf edge but become common in deeper slope waters. Although these dolphins are not known to make strong or well-defined migrations, in the western Pacific Ocean they are associated with the shifting northern boundary of the warm Kuroshio Current, which reaches northward to 46°N in summer and retreats to 33°N in winter. Off South Africa, they associate with the warm Agulhas Current.

SIMILAR SPECIES

Common dolphins are the species most likely to be confused with Striped Dolphins. The combined distribution of the two common dolphins greatly overlaps that of the Striped Dolphin, and all three species occur in fairly dense schools. Close attention must be paid to the color pattern on the sides and flanks. At least in some regions, common dolphins have a highly distinctive "hourglass," or crisscross, pattern on the sides, and their dark dorsal "saddle" dips down in a

180° 140° 100° 60° 20° 0° 20° 60° 100° 140° 180°
ARCTIC OCEAN
ATLANTIC OCEAN
PACIFIC OCEAN
INDIAN OCEAN
PACIFIC OCEAN

■ RANGE

STRIPED DOLPHIN
FAMILY DELPHINIDAE

MEASUREMENTS AT BIRTH

LENGTH	32″–37″ (90–95 cm)
WEIGHT	15–24 lb (7–11 kg)

MAXIMUM MEASUREMENTS

LENGTH	MALE	8′8″ (2.65 m)
	FEMALE	7′10″ (2.4 m)
WEIGHT	MALE	350 lb (160 kg)
	FEMALE	330 lb (150 kg)

LIFE SPAN

57–58 years

V-shape. Common dolphins do not have a light-colored blaze intruding on the dark dorsal color in the shoulder region. With a good side view, it may also be possible to see that the flipper stripe of common dolphins originates at the middle of the lower jaw rather than the eye and that common dolphins lack the long eye-to-anus stripe. Fraser's Dolphin, another gregarious tropical species that occurs in many of the same areas as the Striped Dolphin, has a similar, but wider, eye-to-anus stripe. However, it has only a very short beak, a smaller and more triangular dorsal fin, and no shoulder blaze. Anywhere you find Striped Dolphins, you may encounter at least one of the spotted or bottlenose species. Bottlenose and spotted dolphins in some areas have a shoulder blaze reminiscent of the Striped Dolphin's, but these species lack the bold stripes and sharp demarcation between dorsal darkness and ventral lightness exhibited by Striped Dolphins.

The striped pattern of the Striped Dolphin is clearly shown in this photograph from the Gulf of Mexico. The bold narrow stripes from eye to flipper and eye to anus help to distinguish the species.

BEHAVIOR Striped Dolphins travel in dense schools that average about 100 animals but can contain as many as 500. (The cohesiveness of these schools makes it possible for hunters in Japan to drive large groups into shallow water where they can be killed.) The average school size off South Africa is about 75. In the western Mediterranean, it was about 25 before the large die-off in 1990 to 1992; it declined to about seven, then returned to the earlier level, apparently as animals regrouped into schools of the preferred size. School composition varies. Some schools have only adults, some only juveniles, and some both adults and juveniles. Schools are fairly conspicuous because the dolphins tend to churn the surface with their leaping. Striped Dolphins are willing bow riders. In the eastern tropical Pacific, they are known by fishermen and researchers as "streakers," a reference to their habit of dashing away from vessels at high speed. A few mass strandings, one involving approximately 100 individuals, have been reported.

REPRODUCTION Most of what is known about Striped Dolphin reproduction has come from studies associated with the intensive drive fishery in Japan or examinations of animals found stranded in the northwestern Mediterranean Sea. Striped Dolphin calves are born primarily in late summer and fall after a gestation period of a year or slightly longer. Most calving and calf nurturing takes place within large schools of at least 30 individuals. Such schools are composed of mature animals, calves, and a few juveniles. The calving interval is thought to be about four years, although in the heavily exploited Japanese population it has declined to less than three years. Mediterranean females are sexually mature by about 12 years of age. In Japan between 1956 and

1958, females attained sexual maturity at an average age of 9.4 years; in 1968 to 1970, at 7.5 years.

FOOD AND FORAGING The extensive distribution of Striped Dolphins indicates that they have a fairly diverse diet. They take a wide variety of shoaling fish and cephalopod species. Generally, the fish preferred by Striped Dolphins are less than 5 inches (13 cm) long, the cephalopods less than 8 inches (20 cm) long (in dorsal mantle length). They take prey anywhere in the water column as long as it occurs in large, dense schools. Many of their fish prey have luminous organs and live in the deep scattering layer.

STATUS AND CONSERVATION Striped Dolphins remain abundant on a global scale. There are close to 2 million in the eastern tropical Pacific, and recent estimates in U.S. waters alone suggest that there are about 5,000 in the northern Gulf of Mexico, close to 62,000 in the western North Atlantic, and 20,000 off California, Oregon, and Washington. An estimate for the western Mediterranean Sea, following the 1990 to 1992 die-off of more than 1,000 individuals, was in excess of 100,000 dolphins. In spite of these numbers, there is reason for concern about the status of the Striped Dolphin. In the Mediterranean, large numbers are killed incidentally in a pelagic drift-net fishery for tuna and swordfish; overfishing and habitat degradation have caused a widespread decline of fish and cephalopod resources; and the stress from food shortages and the high contaminant burdens in their own tissues may make Mediterranean Striped Dolphins especially susceptible to disease. In Japan, where recent abundance estimates for three concentration areas totaled about 570,000, Striped Dolphins were subjected to a very intensive drive hunt during the second half of the 20th century. One or more coastal populations there have been driven to low levels, perhaps even extirpated.

OPPOSITE: *Striped Dolphins are the most abundant cetaceans in the Mediterranean Sea despite a massive die-off in the early 1990s. They are not hunted in the Mediterranean, but large numbers die in gillnets, and tissue levels of some pollutants are alarmingly high.*

TOP: *A school of Striped Dolphins "runs" at the surface off one of the volcanic islands of the Azores. The lateral stripes are clearly seen, as is the distinctive light blaze that interrupts the dark dorsal cape ahead of the dorsal fin.*

Fraser's Dolphin

Lagenodelphis hosei
Fraser, 1956

This dolphin's existence was established fairly recently on the basis of a single skull that had been recovered from a beach in Sarawak, on the island of Borneo, in the late 19th century. Francis Fraser, a scientist at the British Museum, meticulously compared the skull to those of other dolphins and found it to be intermediate between the genera *Lagenorhynchus* and *Delphinus* —thus the name *Lagenodelphis.* After Fraser wrote his description of the species, which referred only to the skull, nothing further was learned about the animal until 1971. In that year, a number of dolphins that proved to be of this species were killed in tuna purse seines in the eastern tropical Pacific, and others were found stranded on the southeastern coasts of Africa and Australia. It soon became clear that the old name, Sarawak Dolphin, was inappropriate; the species is now thought to have a pantropical distribution.

DESCRIPTION Fraser's Dolphin has a robust body, a short beak, and markedly small flippers and flukes. Set at midbody, the dorsal fin is small and triangular, often slightly falcate; it is apparently more erect in males than females, and its height is highly variable. There are 36 to 44 pairs of teeth in the upper jaw and 34 to 44 pairs in the lower jaw.

The coloration is not particularly complex but often vivid. There is a dark, grayish-blue cape, the sides are lighter gray, and the belly and throat are whitish (cream to pink-tinged). The most conspicuous feature is a broad dark stripe from the face to the anus (not present on all animals). On the face, the stripe can be reminiscent of a bandit's mask. A dark, prominent flipper stripe originates at the lower jaw and merges with the side stripe. It is important to note that the darkness of the stripes and the amount of contrast are highly variable on these dolphins. Generally, in a school of sufficient size, at least some of the individuals will be clearly marked and recognizable as Fraser's Dolphins.

RANGE AND HABITAT Fraser's Dolphin occurs primarily in water deeper than 3,300 feet (1,000 m) in the tropics worldwide. It is associated with areas of upwelling in the eastern tropical Pacific. There have been occasional strandings, including mass strandings, of Fraser's Dolphins in temperate areas such as southern Australia, France, and Uruguay; researchers believe that these represent

- SHORT, BARELY NOTICEABLE BEAK
- SMALL, TRIANGULAR, SOMEWHAT FALCATE DORSAL FIN
- SMALL FLIPPERS
- BROAD DARK STRIPE FROM FACE TO ANUS
- TROPICAL, DEEPWATER DISTRIBUTION

extralimital movements related to oceanographic anomalies, such as El Niño. Fraser's Dolphins have also been seen off southeastern Africa at latitudes of 29°S to 34°S in summer; these occurrences are thought to be associated with southward extensions of the warm, subtropical Agulhas Current.

SIMILAR SPECIES Fraser's Dolphins can be readily confused with other tropical oceanic species that have stripe patterns with strongly contrasting colors, notably Striped Dolphins and common dolphins, which overlap with Fraser's throughout its range. These other species have much longer beaks and taller, more falcate dorsal fins. The Striped Dolphin's light shoulder blaze conspicuously disrupts the otherwise similar dorsal and side coloration, and its dark eye-to-anus stripe is much narrower than that of Fraser's and sometimes includes a short subsidiary stripe branching downward behind the eye. In common dolphins, the dark dorsal cape normally dips to a V-shape on the sides below the dorsal fin, and there is a tan or yellowish-tan patch on the side in front of the dorsal fin.

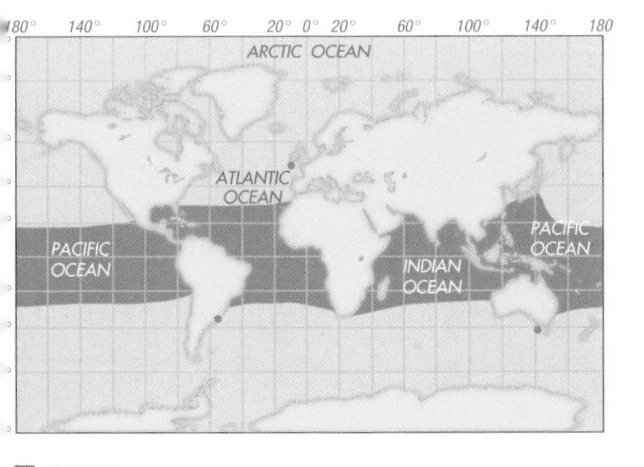

■ RANGE
● EXTRALIMITAL RECORDS

FRASER'S DOLPHIN
FAMILY DELPHINIDAE

MEASUREMENTS AT BIRTH
LENGTH 3′4″ (1 m)
WEIGHT Less than 42 lb (19 kg)

MAXIMUM MEASUREMENTS
LENGTH **MALE** 8′10″ (2.7 m)
FEMALE At least 8′8″ (2.64 m)
WEIGHT **MALE** 460 lb (210 kg)
FEMALE At least 350 lb (160 kg)

LIFE SPAN
At least 16 years (probably much longer)

BEHAVIOR Schools of Fraser's Dolphins typically range in size from about 100 to 1,000 individuals, although groups of just a few to a few tens are also seen. Judging by the composition of mass strandings, researchers have concluded that adult males and females as well as juveniles occur in the same schools. Frequently, Fraser's Dolphins travel in mixed schools with Melon-headed Whales, somewhat less often with False Killer Whales. They occasionally approach vessels to ride the bow wave, but usually only briefly. In the eastern tropical Pacific where they have been subjected to chasing by tuna fishermen, and off the Philippines where they have been hunted, they consistently flee from ships and boats. On the tuna grounds, the dolphins tend to group closely as a vessel approaches, then explosively and rapidly begin swimming away, creating much froth and foam. During such episodes, they change course abruptly, and this makes them difficult to follow.

REPRODUCTION Essentially nothing is known about the reproduction of Fraser's Dolphin.

FOOD AND FORAGING In the eastern tropical Pacific (and probably elsewhere), Fraser's Dolphins feed mainly on mesopelagic fish, shrimp, and squid. They do most of their foraging well below the surface, primarily at depths of 800 to 1,600 feet (250–500 m) and thus catch much, if not most, of their prey in conditions of near or total darkness, regardless of time of day. They appear to feed selectively on three particular types of fish: silvery, deep-bodied species with large eyes; dark, thick-bodied species with medium-size eyes; and solitary, slender, fast-moving predators. The first two groups are not known to be attracted to the surface at night but rather remain at approximately the same depth throughout the 24-hour cycle. Although rarely observed in association with tuna schools or flocks of birds in the

Fraser's Dolphins, seen here off Taiwan, have a short beak that is barely noticeable when seen at a distance. This lack of a prominent beak distinguishes them from most other oceanic dolphins found in the tropics.

eastern tropical Pacific, Fraser's Dolphins have been seen feeding at the surface with large numbers of terns in the Indian Ocean.

STATUS AND CONSERVATION Fraser's Dolphins are killed in many kinds of fishing nets, including purse seines, drift nets, and anti-shark barrier nets. They are also harpooned by artisanal whalers in at least the Lesser Antilles, Indonesia, Sri Lanka, and the Philippines. Japanese (and in previous years Taiwanese) fishermen occasionally take them in mixed-species drive hunts. Nowhere, however, are these dolphins known to be caught regularly or in large numbers, presumably because they are pelagic and naturally rare near most coastlines. It has been estimated that several hundred thousand Fraser's Dolphins inhabit the eastern tropical Pacific. The species is not considered to be immediately threatened in any part of its range.

TOP: *Few sights can compare with that of an excited school of Fraser's Dolphins in full flight from a tuna vessel.* **LEFT:** *Fraser's Dolphins have small dorsal fins and flippers. The dark lateral stripe, although variable in intensity, continues onto the side of the face, giving an appearance somewhat reminiscent of a bandit's mask.*

Short-beaked Common Dolphin

Delphinus delphis
Linnaeus, 1758

Sleek, handsomely marked, and gregarious, the Short-beaked Common Dolphin ranges throughout tropical and warm temperate regions of the Atlantic and Pacific Oceans. Fishermen and some scientists in the western North Atlantic refer to it as the Saddleback Dolphin because of its dark, saddle-shaped marking, while tuna fishermen in the eastern tropical Pacific call it the White-bellied Porpoise, even though it is not a porpoise. Some uncertainty remains about how many species should be recognized in the genus *Delphinus.* While it is a long-standing convention to regard the short-beaked offshore form as a single cosmopolitan species, the longer-beaked regional variants, confined mainly to warm coastal waters, continue to vex systematists. Much of the literature published prior to the last few years treats all common dolphins as a single species.

DESCRIPTION The Short-beaked Common Dolphin has a slender and evenly proportioned body, with a moderately long beak, a tall, somewhat falcate dorsal fin, and moderately large, tapered flippers. Males are about 5 percent longer than females. Tooth counts range from 41 to 54 pairs in both jaws, with the upper jaw usually having one or two more pairs than the lower jaw.

The color pattern is relatively complex. The entire dorsal surface from the front of the melon to well behind the dorsal fin is very dark gray to black. The belly is white. The dark dorsal coloration dips low onto the sides below the dorsal fin in a V-shaped "saddle" configuration, resulting in an "hourglass," or crisscross, pattern on the sides. The anterior segment, called a "thoracic patch" by scientists, is relatively light gray to medium golden-yellow, and it contrasts sharply with the dark dorsal color. The posterior segment of the hourglass is a dirty gray that sweeps over the caudal peduncle. A dark eye patch is continuous with a dark stripe that extends forward and joins the blackness of the lips. The upper surface of the beak is often white or gray, but the tip is dark. Although basically dark, the dorsal fin of adults is often light gray to white in the middle and darker around the borders. The flippers are similar in color to the dorsal fin. A bold dark stripe connects the flipper with the lower jaw; it narrows noticeably anterior to the eye. The ventral whiteness extends forward above the flipper stripe to at least under the eye.

- V-SHAPED DARK SADDLE, WITH DOWNWARD-ORIENTED POINT ON SIDE DIRECTLY BELOW DORSAL FIN
- LIGHT GRAY OR YELLOWISH PATCH ON SIDES FORWARD OF DORSAL FIN
- LIGHT GRAY FLANKS AND CAUDAL PEDUNCLE
- TALL DARK DORSAL FIN, OFTEN WITH GRAY OR WHITISH AREA IN CENTER
- OFTEN IN FAIRLY LARGE GROUPS, ACTIVE AT THE SURFACE
- WIDESPREAD IN TROPICAL AND TEMPERATE WATERS OF ATLANTIC AND PACIFIC OCEANS

RANGE AND HABITAT The total distribution of the Short-beaked Common Dolphin encompasses continental shelf and pelagic waters of the Atlantic and Pacific Oceans. It is definitely present in the western Atlantic from Newfoundland to Florida; the eastern Atlantic from the North Sea to Gabon (including the Mediterranean and Black Seas); the southwestern Pacific around New Caledonia, Tasmania, the southern half of Australia, and New Zealand; the western North Pacific around Japan and eastward to 160°W between 28°N and 43°N; and the tropical and warm temperate eastern Pacific from southern California to central Chile westward to 135°W. It probably occurs in other areas, but potential confusion with the Long-beaked Common Dolphin creates uncertainty. The Short-beaked Common Dolphin is especially common along shelf edges and in areas with sharp bottom relief such as seamounts and escarpments. It shows a strong affinity for areas with warm, saline surface waters. Off the eastern United States, it is particularly abundant in continental slope waters from the Northeast Peak of Georges Bank southward to about 35°N. In at least the

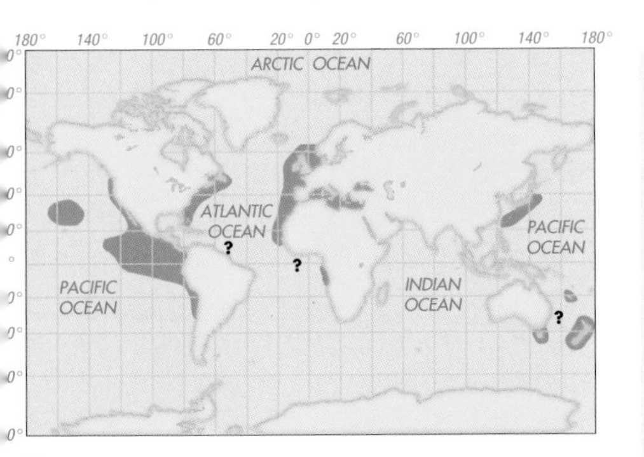

■ RANGE

? POSSIBLE RANGE

SHORT-BEAKED COMMON DOLPHIN

FAMILY DELPHINIDAE

MEASUREMENTS AT BIRTH

LENGTH 31–39″ (80–100 cm)

WEIGHT Unavailable

MAXIMUM MEASUREMENTS

LENGTH **MALE** 8′10″ (2.7 m)

FEMALE 8′6″ (2.6 m)

WEIGHT Probably about 330 lb (150 kg)

LIFE SPAN Unavailable

These Short-beaked Common Dolphins in Monterey Bay, California, display the species' unmistakable tan "thoracic patch" and facial striping.

eastern North Pacific, these dolphins apparently shift their distribution northward in warm-water years. They can occur near shore and have been known to stray far up rivers occasionally. Individuals can range over considerable distances in a short time. For example, a radio-tagged animal traveled at least 270 nautical miles in 10 days.

SIMILAR SPECIES Off southern California and probably elsewhere, Short-beaked and Long-beaked Common Dolphins are often seen near each other, so there is ample scope for confusion. In addition to its noticeably longer beak, the Long-beaked has a longer, narrower head, and its melon is less rounded and thus flatter in profile. The overall impression is of a more slender, longer-bodied animal. The color differences are subtle, but the "thoracic patch" of the Long-beaked tends to be darker and to contrast less with the dark back color. The Long-beaked has a dark gray stripe from flipper to anus that separates the ventral whiteness from the "hourglass" segment. Striped Dolphins have a broadly similar global distribution and are similar in size and overall body shape to the Short-beaked Common Dolphin. Good views of the color pattern are necessary to distinguish them at sea. The bold eye-to-anus and eye-to-flipper stripes, together with the white shoulder blaze of the Striped Dolphin, are distinctive. In the North Atlantic, the Atlantic White-sided Dolphin has a complex pigmentation pattern of white, shades of gray, and an area of yellowish tan that can lead to confusion with common dolphins. However, the Atlantic White-sided Dolphin has a more robust body, much shorter beak, and an unmistakable bright white patch below the dorsal fin; and its tan patch is on the rear portion of its side rather than the front. In the tropical and subtropical Atlantic, Clymene Dolphins can be easily confused with Short-beaked Common Dolphins. Their dark cape is rounded rather than pointed at its apex, and they lack the hourglass pattern on the sides.

BEHAVIOR Short-beaked Common Dolphins typically gather in schools of hundreds or thousands, although the schools generally consist of smaller groups of 30 or fewer. These dolphins are eager bow riders and are active at the surface.

REPRODUCTION The reproductive activity of Short-beaked Common Dolphins in tropical waters may be relatively nonseasonal, but in higher latitudes, calving peaks in late spring or early summer. Gestation lasts 10 or 11 months, lactation for at least 10 months. The calving interval is at least two years. Females tend to remain in lower latitudes (or in the case of the Black Sea, farther offshore) during calving and lactation periods.

FOOD AND FORAGING Squid and small schooling fish (for example, anchovies, lanternfish, hake, deep-sea smelt) are the principal prey of Short-beaked Common Dolphins. Their foraging off southern California (and probably elsewhere) is attuned to the nighttime vertical migration of the deep scattering layer. A large school of dolphins often begins to disperse into small groups in the late afternoon, anticipating the ascent of prey during the evening hours. After a night of purposeful, deep diving, these groups coalesce in the early morning hours and stop their foraging. The daylight hours are spent resting and socializing before the nighttime feeding cycle resumes.

STATUS AND CONSERVATION Short-beaked Common Dolphins are killed accidentally in industrial trawls, gillnets, and many other types of fisheries throughout their range. Although they were fairly abundant in the northern basin of the western Mediterranean Sea as recently as the early 1970s, they are rare there today. Similarly, they seem to have disappeared from the coast of northeastern Florida in the last few decades. In the Black Sea, where in some years more than 100,000 dolphins and porpoises (of three species, including common dolphins) were killed before hunting ceased in the 1980s, populations are thought to be seriously depleted. The exact causes of the declines are not known, but particularly in the Mediterranean and Black Seas, habitat degradation from human activities is likely involved. There are at least tens of thousands of Short-beaked Common Dolphins in shelf and slope waters off the eastern United States, close to 100,000 off the British Isles and France, and hundreds of thousands off the west coast of North America. Populations in the central and western North Pacific and the northeastern Atlantic were hard hit by high-seas driftnet fisheries for flying squid and tuna, but the United Nations drift-net moratorium has helped reduce this threat. Although small numbers are still killed each year in tuna purse seines, recent abundance estimates in the eastern tropical Pacific total about 3 million, divided among three populations. Large incidental kills in industrial trawl fisheries off western Europe and eastern South America are a relatively new and ongoing concern.

Short-beaked Common Dolphins are energetic animals that form groups of hundreds or even thousands of individuals.

Long-beaked Common Dolphin

Delphinus capensis
Gray, 1828

This species exemplifies the never-ending process of taxonomic re-evaluation and revision. The small, widely distributed oceanic dolphins of the genus *Delphinus* exhibit substantial variation in body size and form, skull features, and color pattern. Scientists studying these dolphins off California in the 1960s concluded that there were two types: a long-beaked form and a short-beaked form. In the mid-1990s, a thorough comparison of color patterns, external morphology, and skull characteristics, coupled with a genetic evaluation, showed that the Long-beaked and Short-beaked Common Dolphins represent well-differentiated species. A problem remained with an extremely long-beaked form found in the Red Sea and Indian Ocean (often denoted as *D. tropicalis*). Some scientists believe that it represents a third species of *Delphinus*, but it is presently regarded as a regional variant of the Long-beaked Common Dolphin.

DESCRIPTION The Long-beaked Common Dolphin has a slender, evenly proportioned body, with a long beak and a tall, somewhat falcate dorsal fin. The melon is flatter and the angle where it meets the beak is more gradual than in the Short-beaked Common Dolphin. Males are about 5 percent longer than females in both *Delphinus* species. Tooth counts range from 47 to 67 pairs in both the upper and lower jaws.

The color pattern is similar to that of the Short-beaked Common Dolphin but is generally more muted. On the Long-beaked, the ventral whiteness usually does not extend above the flipper stripe and forward to at least the eye as it does on the Short-beaked; the eye patch is not nearly as dark; and light blazes on the dorsal fin and flippers are less common. A gray stripe connects the area immediately above the flipper with the anal region. This flipper-to-anus stripe, much wider at both ends than in the middle, serves to demarcate the ventral whiteness from the side pigmentation.

RANGE AND HABITAT The Long-beaked Common Dolphin occurs in nearshore tropical and warm temperate waters of some oceans. It is found in the eastern Pacific from central California (Point Conception) south to northern Chile, including the Gulf of California; along the western Pacific coasts of Korea and southern Japan; in the western Atlantic from Venezuela south to the vicinity of the Rio de la Plata estuary in

- MUTED CRISSCROSS PATTERN ON SIDES
- TAN OR YELLOWISH-TAN PATCH ON SIDES FORWARD OF DORSAL FIN
- PALE GRAY FLANKS AND CAUDAL PEDUNCLE
- FLIPPER STRIPE RELATIVELY WIDE, WELL DEMARCATED FROM WHITE THROAT BUT NOT FROM GRAY SIDE
- TALL DARK DORSAL FIN
- OFTEN IN FAIRLY LARGE GROUPS, ACTIVE AT SURFACE
- FOUND IN WARM COASTAL WATERS OF ATLANTIC, PACIFIC, AND INDIAN OCEANS

TROPICALIS FORM

Argentina; and in the eastern Atlantic off the west coast of Africa. It occurs in waters influenced by the Agulhas Current off southeastern Africa (the specific name *capensis* refers to the Cape of Good Hope, where the type specimen was collected in the early 19th century); and probably also in Madagascan waters. The *tropicalis* form of the Long-beaked Common Dolphin occurs in coastal waters from central China southward and westward along the southern coast of Asia and throughout the western islands of the Indo-Malay archipelago, and along the rim of the Indian Ocean.

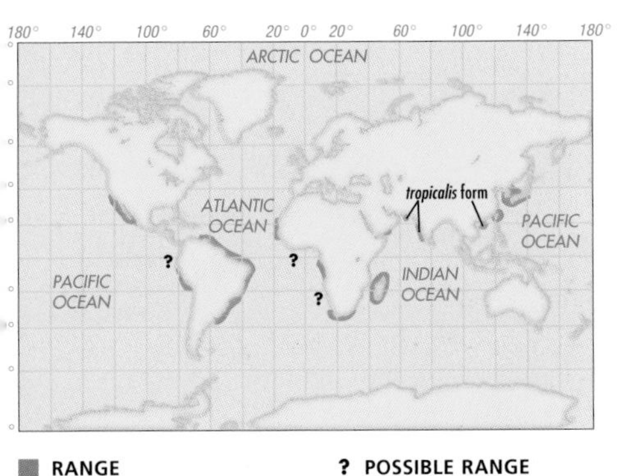

RANGE
RANGE OF *TROPICALIS* FORM
? POSSIBLE RANGE

SIMILAR SPECIES The Short-beaked Common Dolphin occurs with the Long-beaked off southern California and possibly other areas, and the two are easily confused. The clearest difference may be in the white abdominal field, which extends over the flipper and flipper stripe in the Short-beaked but not in the Long-beaked. Striped Dolphins are similar to common dolphins in size and overall body shape but tend to occur farther offshore than Long-beaked Common Dolphins.

BEHAVIOR School sizes can range from fewer than 10 to as many as several thousand individuals. In

LONG-BEAKED COMMON DOLPHIN
FAMILY DELPHINIDAE

MEASUREMENTS AT BIRTH
LENGTH 31–39″ (80–100 cm)
WEIGHT Unavailable

MAXIMUM MEASUREMENTS
LENGTH **MALE** 8′4″ (2.54 m)
FEMALE 7′4″ (2.24 m)
WEIGHT Probably about 330 lb (150 kg)

LIFE SPAN
Possibly close to 40 years

Long-beaked Common Dolphins in the Gulf of California, Mexico. This group appears to be in a relatively tranquil mode, despite the breaching animal in the foreground.

one ship survey off southern California, the average group size was close to 200. Long-beaked Common Dolphins are eager bow riders and are active at the surface.

REPRODUCTION The Long-beaked Common Dolphin's reproductive activity in tropical waters is probably less seasonal than in more temperate regions, where there may be a greater tendency to calve in the warmer months. There is no evidence of any major reproductive differences with Short-beaked Common Dolphins.

FOOD AND FORAGING The principal prey of Long-beaked Common Dolphins are small schooling fish (for example, pilchards, hake, sardines, and anchovies) and squid.

STATUS AND CONSERVATION Although their global distribution is generally less extensive than that of Short-beaked Common Dolphins, and their aggregate abundance is likely much lower, Long-beaked Common Dolphins are not known to face major immediate threats. In several areas, most notably the west coast of Africa and the east and west coasts of South America, the documentation of abundance and removals is too sketchy to evaluate the status of populations. However, there is growing concern about the large number killed in Peru and used for human food or shark bait. There are an estimated 15,000 to 20,000 common dolphins, most or all of them Long-beaked, off the east coast of southern Africa, and at least a few thousand off southern California.

Although the two species of common dolphin have similar overall color patterns, there are a few consistent differences in the details. This close-up view of Long-beaked Common Dolphins off the west coast of Baja California shows some of these subtleties.

White-beaked Dolphin

Lagenorhynchus albirostris
(Gray, 1846)

- SHORT THICK BEAK, USUALLY WHITE
- PROMINENT, FALCATE DORSAL FIN
- DIFFUSE COLORATION WITH BROAD, GRAYISH-WHITE BLAZES ON SIDES
- GRAYISH-WHITE "SADDLE" ON BACK BEHIND DORSAL FIN
- LIMITED TO COOL WATERS OF NORTH ATLANTIC

This robust dolphin occurs in a broad band across the northern rim of the North Atlantic and into ice-free subarctic and arctic waters during the summer. Its common occurrence near western European shores meant that it was among the first of the *Lagenorhynchus* dolphins to be recognized and described. The term *Lagenorhynchus* is derived from the Latin *lagenos* for "bottle" or "flask" and *rhynchos* for "beak" or "snout," and thus means "bottle-beak." Indeed, some of the early descriptions of species in the genus called them "bottlenoses," a term now reserved for the *Tursiops* dolphins. Some early taxonomists referred to the *Lagenorhynchus* group as "plowshare-headed dolphins," noting that the beak resembled a plowshare, or the blade of a plow.

DESCRIPTION The White-beaked Dolphin has a robust body and a short thick beak. The dorsal fin is tall and strongly falcate, and the caudal peduncle is thickly keeled dorsally and ventrally. There are 23 to 28 pairs of teeth in the upper jaw and 22 to 28 pairs in the lower, but the first few in each row are often concealed within the gum. The color pattern is diffuse and highly variable. The entire beak is often white, continuous with the white of the throat and belly, and flecked with gray. There is a dark gray dorsal cape that is variable in extent but often continues forward onto the head. The dorsal fin is also dark gray. Large dark zones are interrupted by conspicuous grayish-white blazes high on the sides and on the flanks. These light blazes are sometimes broken and sometimes continuous with one another. The light gray flank color extends onto the dorsal surface behind the dorsal fin, making the front part of the caudal peduncle look white when viewed from above and through the water.

RANGE AND HABITAT White-beaked Dolphins are endemic to high-latitude shelf waters of the North Atlantic, including northern portions of the Gulf of St. Lawrence and the North Sea (they are rare in the Baltic Sea today). They are more common in European than American waters. Populations in eastern and western Atlantic waters are morphologically distinct and therefore probably do not often mix. The southern limits of normal distribution are Cape Cod in the west and France in the east; the northern limits are the

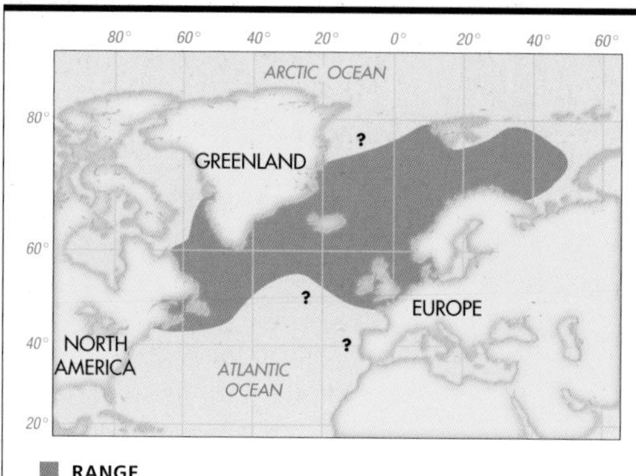

WHITE-BEAKED DOLPHIN

FAMILY DELPHINIDAE

MEASUREMENTS AT BIRTH

LENGTH 3′7″–3′8″ (1.1–1.2 m)

WEIGHT Unavailable

MAXIMUM MEASUREMENTS

LENGTH **MALE** 10′ (3.1 m)
FEMALE 9′3″ (2.83 m)

WEIGHT **MALE** 770 lb (350 kg)
FEMALE 680 lb (310 kg)

LIFE SPAN Unavailable

central Davis Strait in the west, and Svalbard, Norway (to about 80°N), and the Barents Sea in the east. Although they move in summer into subarctic and arctic waters that are ice-covered, or at least ice-infested, in winter, they are not ice-adapted in the same sense that Belugas and Narwhals are. White-beaked Dolphins are driven out of enclosed areas and away from shore by winter ice formation. Tens and sometimes hundreds of these animals are victims of ice entrapment in Newfoundland in some years. They are crushed, suffocated, or forced to strand.

SIMILAR SPECIES Throughout its range, the White-beaked Dolphin can be confused with the Atlantic White-sided Dolphin, although in general, the latter is more pelagic and occurs in somewhat lower latitudes. Atlantic White-sided Dolphins are most readily distinguished at sea by the absence of light gray or white on the back behind the dorsal fin (the entire upper surface of the caudal peduncle is dark). In addition, the white ventral field sweeps forward cleanly well above the flippers to just below the eye and onto the chin and lower jaw in Atlantic White-sided Dolphins. The zones of coloration on Atlantic White-sided Dolphins are more sharply delineated, with a bright white elongated patch on the side below the dorsal fin and a yellowish-tan elongated patch high on the flank.

BEHAVIOR White-beaked Dolphins are typically seen in groups of 5 to 50 and occasionally in schools of several hundred. Studies of the compositions of ice-entrapped groups suggest that there is some degree of segregation into age groups, and juveniles may form separate schools from adults with calves. White-beaked Dolphins are attracted to powered vessels, are active bow riders, and can be acrobatic above the surface. They frequently associate with feeding Fin and Humpback Whales, and sometimes with other dolphin species.

REPRODUCTION Little is known about reproduction in the White-beaked Dolphin. Calving apparently takes place from spring to autumn (May–September).

FOOD AND FORAGING Some eastern Canadian fishermen call White-beaked Dolphins "squid-hounds." However, these dolphins eat a fairly wide variety of fish, ranging from small schooling species like herring and capelin to larger shallow-water, bottom-dwelling species such as cod, haddock, and whiting, and invertebrates such as octopus and benthic crustaceans. They have been seen foraging cooperatively, circling and trapping fish schools against the surface. Apparently they also forage on bottom-dwelling organisms.

Framed by the snow-capped coast of Iceland, a White-beaked Dolphin springs clear of the surface, offering an ideal profile.

STATUS AND CONSERVATION A patchwork of abundance estimates suggests that there are at least many tens of thousands and possibly a few hundred thousand individuals throughout the White-beaked Dolphin's range. A pronounced decrease in abundance has occurred since the early 1970s off the northeastern United States, while there seem to have been increases in some areas off northwestern Europe. These dolphins are at least occasionally killed by hunters and fishermen in Newfoundland, Labrador, Greenland, and the Faeroe Islands. These catches are not recorded systematically, so it is uncertain whether local groups of dolphins have been depleted. Some mortality occurs in fishing gear (trawls and gillnets) throughout the species' range, but nowhere is it thought to be high enough to deplete populations.

OPPOSITE: *White-beaked Dolphins, such as these off Husavik, Iceland, have a large, strongly falcate dorsal fin. The fin and center of the back are black.*

LEFT: *Underwater view of White-beaked Dolphins in the Gulf of St. Lawrence. The dark areas in the center of the back and on the sides contrast with the light blazes on the caudal peduncle and sides.*

Atlantic White-sided Dolphin

Lagenorhynchus acutus
(Gray, 1828)

The Atlantic White-sided Dolphin is one of the most colorfully marked of all the cetaceans. While many other dolphins have stripes, blazes, and patches on their sides and flanks, none has such a combination of bold, well-defined white and yellowish-tan patches on the sides. Gregarious and vigorously playful, like many other oceanic dolphins, these animals come to the bows of vessels and often leap high above the water's surface. The Latin name *acutus* means "sharp" or "pointed" and refers to this dolphin's dorsal fin. Researchers in the western North Atlantic often call these animals "white-sides," and sometimes just "lags," short for *Lagenorhynchus.*

DESCRIPTION The Atlantic White-sided Dolphin has a robust body, with a short thick beak. It has a moderately tall and falcate dorsal fin, and a keeled caudal peduncle. There are 29 to 40 pairs of small teeth in the upper jaw and 31 to 38 pairs in the lower.

The color pattern is complex and sharply demarcated. The entire dorsal surface, including the upper jaw, melon, and dorsal fin, is dark gray or black. On the side, the gray is interrupted by a narrow, bright white patch beginning below the dorsal fin and extending posteriorly, and a narrow yellowish-tan patch that abuts the white patch and continues back almost to the flukes. A dark patch surrounds the eye, and a light gray stripe extends from just in front of the eye to the flipper. The belly and throat are white; the white extends up onto the side above the flipper.

RANGE AND HABITAT The Atlantic White-sided Dolphin is endemic to the temperate North Atlantic. Its northern limits are the Davis Strait in the west and the Norwegian Sea in the east; the southern limits are North Carolina (35°N) in the west and the Celtic Sea, and possibly the Azores (37–39°N), in the east. This dolphin occurs in shelf waters as well as deep slope and canyon waters. It concentrates in areas with high seafloor relief. Off eastern North America, this species exhibits a seasonal trend of increasing density southward along the shelf edge in winter and spring in association with cool, relatively fresh waters from the Gulf of Maine. It occurs in the higher latitudes of its range mainly in warmer months. It may also tend to move inshore in summer and offshore in winter.

- SHORT, INCONSPICUOUS BEAK
- BOLD WHITE PATCH ON SIDE
- YELLOWISH-TAN STREAK ON FLANK JUST ABOVE WHITE PATCH
- SHARPLY DEMARCATED PIGMENTATION ZONES
- WHITE OF BELLY EXTENDS ONTO SIDE ABOVE FLIPPER
- LIMITED TO TEMPERATE WATERS OF NORTH ATLANTIC

SIMILAR SPECIES The distribution of the Atlantic White-sided Dolphin significantly overlaps with those of the White-beaked Dolphin and the Short-beaked Common Dolphin. The Atlantic White-sided is less common than the White-beaked Dolphin in the North Sea and higher latitudes of the northeastern Atlantic but much more common in shelf waters of eastern North America from about 39°N (New York–New Jersey) to the Gulf of Maine, and also in the Gulf of St. Lawrence. The white beak (although it is not always white), the grayish-white saddle behind the dorsal fin, and the more diffuse grayish-white markings high on the sides ahead of the dorsal fin distinguish the White-beaked Dolphin. The Short-beaked Common Dolphin has a much more prominent, slender beak and a slighter build overall. The border of its dark cape forms a V (rather than a straight line as in the Atlantic White-sided Dolphin) on the side below the dorsal fin, and its tan or yellowish-tan patch is wider and farther forward, anterior to the dorsal fin.

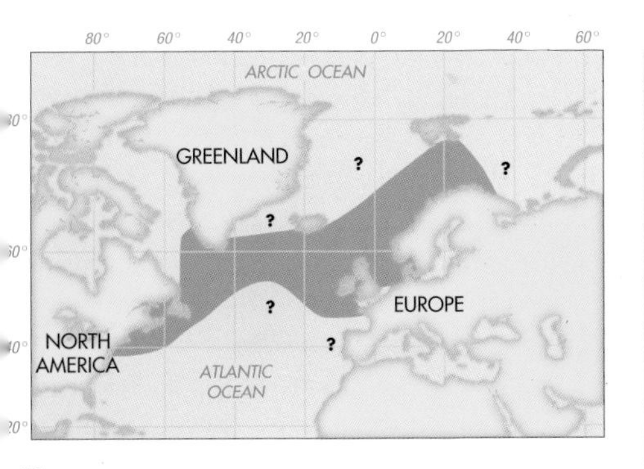

ATLANTIC WHITE-SIDED DOLPHIN

FAMILY DELPHINIDAE

MEASUREMENTS AT BIRTH

LENGTH 3′7″–4′ (1.08–1.22 m)

WEIGHT 44 lb (20 kg)

MAXIMUM MEASUREMENTS

LENGTH **MALE** 9′3″ (2.82 m)
FEMALE More than 8′ (2.43 m)

WEIGHT **MALE** More than 510 lb (230 kg)
FEMALE More than 400 lb (180 kg)

LIFE SPAN

At least 17 years

BEHAVIOR Groups of Atlantic White-sided Dolphins vary in size from a few individuals to schools of several hundred. Large aggregations apparently form when many cohesive groups of 5 to 15 animals merge to socialize, migrate, or perhaps take advantage of a concentrated food source. Mass strandings of groups of tens to more than 100 animals, and drive catches of hundreds at once, suggest that the animals in these large aggregations are socially affiliated to a considerable degree. Data from strandings indicate that juveniles, at least part of the time, live in separate schools from adults. Atlantic White-sided Dolphins are often seen in association with Fin and Humpback Whales in the Gulf of Maine, and they associate with various other odontocetes as well. Mixed schools of White-beaked and Atlantic White-sided Dolphins have been observed in the North Sea and elsewhere off western Europe.

REPRODUCTION June and July appear to be the peak calving months for this species. The gestation period is 11 months, and lactation probably lasts about 18 months. Females in their prime may give birth every other year.

FOOD AND FORAGING Herring, hake, and squid are staple food items, but these dolphins consume a fairly large variety of prey, including small mackerel and various bottom fish. They are sometimes seen cooperatively surrounding fish schools near the surface. There was a dramatic increase in the numbers of Atlantic White-sided Dolphins in the Gulf of Maine in the late 1970s and the 1980s that correlated with a precipitous decline in herring and a corresponding irruption of sand lances. Researchers have inferred that sand lances are a major food source, although studies of stomach contents have not confirmed this.

STATUS AND CONSERVATION The drive hunt in the Faeroe Islands, best known for killing more than 1,000 Long-finned Pilot Whales in most years, also takes large numbers of Atlantic White-sided Dolphins. Catches in recent years have ranged from about 150 to nearly 550. There is no information on abundance for the population (or populations) being exploited in this region. Considerable numbers are killed in midwater trawls, including trawls that operate along the shelf edge off southwestern Ireland. These animals may belong to the same population as that being hunted in the Faeroe Islands. Estimates in U.S. shelf and shelf-edge waters suggest a population of about 30,000 animals. In addition, close to 12,000 have been estimated to summer in the Gulf of St. Lawrence. Set against these numbers is the incidental kill in gillnets and trawls, estimated at more than 200 per year in U.S. waters alone during the late 1990s. There are no estimates of the annual kill in Canadian waters. In spite of the mortality in parts of its range, the species, overall, is not considered immediately threatened.

OPPOSITE BOTTOM: *A breaching Atlantic White-sided Dolphin shows the two distinguishing narrow patches on the side. The anterior patch is stark white, the posterior patch yellowish tan.*

OPPOSITE TOP: *The range of the Atlantic White-sided Dolphin overlaps that of the White-beaked Dolphin in cold temperate latitudes of the North Atlantic Ocean. Even in dappled lighting underwater, the Atlantic White-sided can be distinguished by the more continuously dark dorsal surface.*

ABOVE: *Atlantic White-sided Dolphins, seen here in the Gully, a protected area off Nova Scotia, are abundant and widely distributed in deep waters along the east coast of North America from the Gulf of Maine to the Gulf of St. Lawrence.*

Pacific White-sided Dolphin

Lagenorhynchus obliquidens
Gill, 1865

These abundant, gregarious dolphins are familiar to mariners who spend time in temperate waters of the North Pacific. They are the only *Lagenorhynchus* species in those waters, and most field researchers refer to them simply as "lags." Their handsome markings led to their being called "striped dolphins" by some early researchers, but in recent decades that name has been reserved for *Stenella coeruleoalba*. Conservationists raised the alarm in the 1980s upon learning that thousands of Pacific White-sided Dolphins were being killed each year in high-seas drift-net fisheries for squid, tuna, and billfish. The United Nations ban on this type of fishing, which took effect in 1993, reduced the immediacy of the bycatch problem. However, hundreds are still taken each year in salmon fisheries within the national waters of some countries.

DESCRIPTION The Pacific White-sided Dolphin has a robust body and a barely noticeable beak. The dorsal fin is unusually large, considering the overall body size, and the flippers are large, although not exceptionally so. The dorsal fin is set at a low angle relative to the back; its shape is highly variable, from sharply falcate to lobate (broadly rounded at the peak). There are 23 to 36 pairs of teeth in both the upper and lower jaws.

The color pattern is complex and sharply demarcated. The back and sides are dark gray, interrupted dorsally only by a long, light gray or whitish stripe. Seen from above, the two stripes, one on each side of the body, resemble a pair of suspenders. On each side, the stripe begins on the side of the face ahead of the eye, sweeps up onto the back, and then branches down toward the light gray on either side of the caudal peduncle. A broad, conspicuous zone of light gray (or off-white) dominates the side anterior to the dorsal fin. Black coloration on the lips is continuous with a narrow black line passing under the eye to the flipper. The belly is white and sharply demarcated from the gray sides by a black line between the flipper and anus. The dorsal fin is dark along the front edge, but the rear two-thirds are light gray. The flippers also often have lighter areas.

RANGE AND HABITAT The total range of the Pacific White-sided Dolphin consists of cold temperate waters in a broad swath across the northern rim of the Pacific Ocean, extending

- INCONSPICUOUS BEAK
- LARGE, STRONGLY FALCATE, BICOLORED DORSAL FIN
- LIGHT GRAY "SUSPENDERS" ALONG BACK
- PROMINENT, LIGHT GRAY PATCH ON FORESIDE
- BLACK LIPS AND BOLD BLACK LINE DEMARCATING WHITE BELLY
- CONFINED TO MEDIUM LATITUDES OF THE NORTH PACIFIC

along continental coasts to the South China Sea in the west and the Gulf of California in the east. It includes the Sea of Japan, the southern Sea of Okhotsk, and the southern Bering Sea. Seaward of continental shelves and slopes, this dolphin is found mainly between the latitudes 38°N and 47°N. A winter sighting of 40 animals in the eastern tropical Pacific was tentatively assigned to this species, and there are a few questionable reports from around the Revillagigedo Islands, off Mexico at about 19°N, but there is no definitive evidence that these dolphins normally occur farther south than Taiwan in the west or the tip of Baja California (about 23°N) in the east. Pacific White-sided Dolphins commonly occur in some deep inshore channels of Alaska, British Columbia, and Washington. They become most abundant in shelf waters off southern California during winter (November–April) and off Oregon and Washington during late spring (May). Since few animals are seen in California offshore waters during summer, it appears that there is a north–south, rather than inshore–offshore, sea-

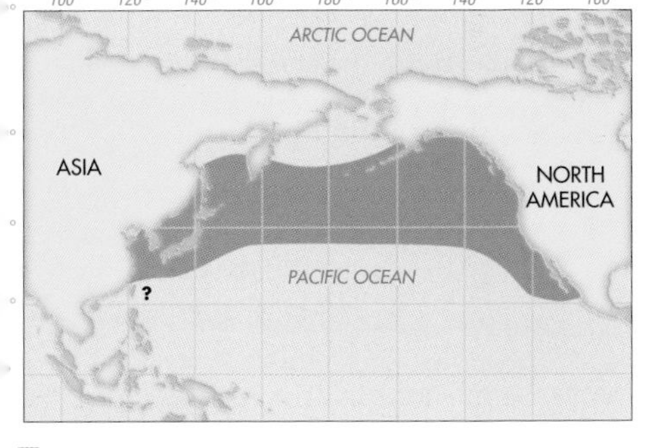

RANGE
? POSSIBLE RANGE

PACIFIC WHITE-SIDED DOLPHIN
FAMILY DELPHINIDAE

MEASUREMENTS AT BIRTH

LENGTH	3′ (92 cm)
WEIGHT	Unavailable

MAXIMUM MEASUREMENTS

LENGTH	MALE	8′2″ (2.5 m)
	FEMALE	7′9″ (2.36 m)
WEIGHT	MALE	440 lb (200 kg)
	FEMALE	330 lb (150 kg)

LIFE SPAN

More than 40 years

sonal migration in the eastern North Pacific. However, the picture is complicated by the existence of several geographically separate and morphologically distinct populations. Also, numbers clearly increase in the Inside Passage of British Columbia during winter months, suggesting an onshore movement in that season.

SIMILAR SPECIES Pacific White-sided Dolphins are strikingly similar in external appearance to Dusky Dolphins of the Southern Hemisphere, but this creates difficulty only in cases where unidentified dolphins are encountered in the tropical Pacific. Otherwise, within most of its range, there is little likelihood of confusing the Pacific White-sided Dolphin with any other species except the equally gregarious Short- and Long-beaked Common Dolphins and occasionally with Dall's Porpoise. Common dolphins usually do not occur north of central California, and their much more prominent beaks and V-shaped saddle markings should help distinguish them. Dall's Porpoise shares most of the Pacific White-sided Dolphin's range, and the greatest difficulty arises with fast-moving animals that create lots of spray as they race along the surface. The much more modest, triangular dorsal fin and stark black-and-white coloration of Dall's Porpoise should distinguish it. It is also much less inclined than the Pacific White-sided Dolphin to leap clear of the water or engage in aerial acrobatics.

BEHAVIOR Schools of thousands of Pacific White-sided Dolphins are occasionally seen, but estimates of average group size range from 10 to 50 to about 100. Observers have noticed close-knit groups of five or fewer heavily scarred individuals with extremely falcate dorsal fins—evidently adult males—swimming together within large schools, possibly indicating that males form lasting alliances. Pacific White-sided Dolphins engage in bow riding and aerial behavior. They are often seen in mixed-species aggregations with other cetaceans, pinnipeds, and seabirds, and have a particular association with Northern Right Whale Dolphins and Risso's Dolphins in some offshore areas. Killer Whales have been observed successfully preying on Pacific White-sided Dolphins.

REPRODUCTION Surprisingly little is known about reproduction for such a common species. Gestation probably lasts about a year. Most calving appears to take place in late spring and summer (April–August), and the lactation period is at least six months. These dolphins do not give birth more often than every other year, and possibly much less frequently than that.

A Pacific White-sided Dolphin swims off San Diego, California, with a kelp frond draped over its flipper. The species' two-toned dorsal fin coloration and the light stripes on the side and back are well shown.

A 29-year-old pregnant female has been reported, indicating that these dolphins have a fairly long reproductive life.

FOOD AND FORAGING The Pacific White-sided Dolphin is a versatile and opportunistic feeder. Small schooling fish (herring, anchovies, capelin, sardines) are important prey in relatively shallow inshore waters of British Columbia. Squid prevail off California, and small mesopelagic fish and cephalopods associated with the deep scattering layer are preferred in offshore and very deep coastal waters. These dolphins are not considered deep divers, so feeding on organisms of the deep scattering layer presumably takes place primarily at night when prey are nearest the surface. When foraging during daytime, they can be seen working cooperatively to corral a tightly balled school of fish, often with seabirds in attendance. The dolphins individually penetrate the fish school to catch their prey, but they also pick off fish while swimming along the periphery.

STATUS AND CONSERVATION Estimates of abundance for the entire North Pacific total close to 1 million individuals. However, the movements of Pacific White-sided Dolphins toward vessels (to bow ride) have a tendency to inflate shipboard estimates, so a dose of skepticism must be applied to such high numbers. An aerial survey estimate for California shelf and slope waters was greater than 100,000. Increased numbers of these dolphins in inshore waters of British Columbia and southeastern Alaska during the 1990s coincided with major shifts in fish distribution. It is likely, therefore, that the local increase represents a change in dolphin distribution rather than a population increase. The total kill of Pacific White-sided Dolphins in the Japanese, Korean, and Taiwanese drift-net fisheries from 1978 to 1990 was estimated at 49,000 to 89,000. They are killed occasionally in harpoon and drive hunts in Japan, and at least 466 were killed in the late 1970s in a government-supported culling operation to protect fisheries in western Japan. At present, the level of overall mortality is not thought to be high enough to endanger the species.

ABOVE: *Pacific White-sided Dolphins commonly leap clear of the water while swimming. These dolphins have unusually large, falcate dorsal fins.*
BELOW: *Pacific White-sided Dolphins often associate with other cetaceans. The paths of the two dolphins in this photograph are about to intersect those of two Humpback Whales.*

Dusky Dolphin

Lagenorhynchus obscurus
(Gray, 1828)

This Southern Hemisphere dolphin has an irregular distribution centered in shelf waters along discrete sections of continental and island coastlines. There are a number of separate populations, and animals from southwestern Africa, South America, and New Zealand are geographically and genetically isolated from one another. It is unclear whether the groups around oceanic islands in the western South Pacific, South Atlantic, and Indian Ocean are also discrete or regularly mix with animals in other areas. There has been considerable confusion about the Dusky Dolphin's identity and taxonomic affiliations over the years. Apart from the fact that it has certain skull features reminiscent of the genus *Stenella,* its pigmentation and general appearance are very similar to those of the Pacific White-sided Dolphin, and it can be difficult to distinguish at sea from Peale's Dolphin.

DESCRIPTION The Dusky Dolphin's body is robust in the middle and tapers considerably behind the dorsal fin. It has a thick, stubby, scarcely noticeable beak. The dorsal fin is tall and falcate. There are about 32 pairs of teeth in both the upper and lower jaws.

The pigmentation pattern is very similar to that of the Pacific White-sided Dolphin. The back is dark, and the dorsal fin is bicolored, black in the front, light gray in back. The throat and belly are white, with a sharp border separating the light gray side pigmentation from the white throat. A long patch of light gray on the foreside extends forward over the eye onto the side of the head. The beak is dark. A light gray flank patch extends forward in two branches, creating prominent blazes on the side below the dorsal fin.

RANGE AND HABITAT The Dusky Dolphin is generally found in coastal and shelf waters of portions of South America (north to 11°S on the west coast, 30°S on the east coast), southwestern Africa (between Walvis Bay, Namibia, and Hout Bay, South Africa), and New Zealand. It also occurs near the following oceanic islands: Campbell, Auckland, and Chatham in the western South Pacific, Falklands and Tristan da Cunha in the South Atlantic, and Prince Edward, New Amsterdam, and St. Paul in the Indian Ocean. It occurs at least sporadically off Tasmania and southern Australia. Waters around the Kerguelen Islands in the southern Indian Ocean are not part

- PROMINENT, FALCATE, BICOLORED DORSAL FIN
- INCONSPICUOUS BEAK
- POINTED GRAY BLAZES ON SIDES BELOW DORSAL FIN
- GREGARIOUS AND ACROBATIC
- DISTRIBUTED IN COOL TEMPERATE WATERS OF SOUTHERN HEMISPHERE

of the confirmed range, contrary to many statements in the literature on this species. The Dusky Dolphin occurs in the Tierra del Fuego and Cape Horn region mainly in summer. It is sighted occasionally south of the Antarctic Convergence (to 60°30'S). There is a low-density hiatus in the distribution along the coast of Chile between 36°30'S and 46°S. In New Zealand, the Dusky is common all around South Island except the southern half of the west coast and is found around the southern third of North Island to Hawke Bay in the east and New Plymouth (39°S) in the west. There are density increases in the northern parts of the range in winter and the southern parts in summer, but migratory behavior is not well understood.

SIMILAR SPECIES The Dusky Dolphin is easily confused with Peale's Dolphin in much of the South American part of its range, especially

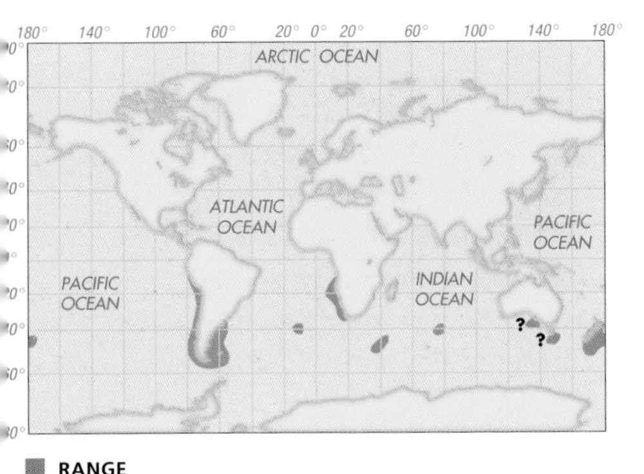

DUSKY DOLPHIN

FAMILY DELPHINIDAE

MEASUREMENTS AT BIRTH

LENGTH	*Peru:* 3′ (91 cm)
WEIGHT	*Peru:* 21 lb (9.5 kg)

MAXIMUM MEASUREMENTS

LENGTH	**MALE**	*South Africa:* 5′11″ (1.81 m) *Peru:* 6′11″ (2.11 m)
	FEMALE	*South Africa:* 6′4″ (1.93 m) *Peru:* 6′9″ (2.05 m)
WEIGHT	**MALE**	*New Zealand:* 187 lb (85 kg)
	FEMALE	*New Zealand:* 172 lb (78 kg)

LIFE SPAN

At least 35 years

High-flying jumps are typical of Dusky Dolphins, shown here off the rugged coastline of New Zealand.

south of 47°S in summer. Peale's Dolphin has a wider gray foreside patch that terminates abruptly at the eye rather than extending forward onto the side of the head as in the Dusky Dolphin. Often Peale's Dolphin is a more sluggish swimmer and is found in smaller groups closer to shore. The Dusky Dolphin can also be confused with the Hourglass Dolphin, especially offshore in the southern portions of the Dusky's range. The Hourglass Dolphin has a prominent dark zone behind the flipper, more nearly symmetrical and sharply demarcated foreside and flank patches, and a broader-based dorsal fin.

BEHAVIOR Dusky Dolphins are commonly seen in groups of 15 to 100 and occasionally in schools of up to at least 500. These dolphins are known for their high-energy aerial acrobatics, including somersaults and body slams. They are attracted to boats to bow ride and are popular objects of dolphin-watching tourism in New Zealand, even participating in swim-with-wild-dolphins tours. Large groups of Dusky Dolphins off New Zealand often have smaller groups of Short-beaked Common Dolphins in their midst during summer. Animals with intermediate features have been observed, and it is likely that the two species hybridize occasionally. Elsewhere, Dusky Dolphins have been seen in mixed schools with Southern Right Whale Dolphins and pilot whales. Killer Whales frighten Dusky Dolphins toward shore, sometimes into water less than 3 feet (1 m) deep. Killer Whales have been seen off Kaikoura attacking and catching Dusky Dolphins, sometimes flipping them high into the air.

REPRODUCTION In Peru, Dusky Dolphins mate in late winter or early spring (September–October). About 11 to 13 months of gestation and a year of lactation follow. Small calves are seen in New Zealand and Argentina mainly in late spring and summer (November–January). Little else is known about the species' reproduction.

FOOD AND FORAGING Anchovies are important prey for Dusky Dolphins in South America, while squid and mesopelagic fish (lanternfish and hoki) are important in the deep waters off New Zealand. Off South Africa, seven cephalopod species and 16 fish species have been identified in stomach contents, and some of these species live in the deep scattering layer. The Dusky Dolphin's foraging behavior appears to differ by region or season. At times, the dolphins cooperate to surround and concentrate schooling fish near the surface. At other times or in other places (certainly off Kaikoura, New Zealand), they tend to rest during the day and feed at night on vertically migrating organisms.

A Dusky Dolphin playing with kelp off New Zealand.

STATUS AND CONSERVATION Although Dusky Dolphins are considered abundant, few population estimates are available. Using data collected in aerial surveys off central Patagonia in the mid-1990s, researchers estimated that the local population was at least 7,000. The annual incidental catch in midwater fishing trawls off Patagonia in the mid-1980s, estimated at roughly 400 to 600 dolphins, primarily females, was high enough to raise concerns. Mortality apparently has decreased since the 1980s as midwater trawling for shrimp has become more tightly regulated. Several hundred Dusky Dolphins continue to die each year in various types of fishing gear off Argentina. Some animals are also taken in beach seines and purse seines, and by harpooning off South Africa, but the number is not thought to be large. In Peru, by contrast, large numbers (up to 10,000 per year in the mid-1980s) have been taken in drift gillnets and by harpooning. Conservationists suspect that the species has suffered a serious decline in abundance, and substantial mortality continues despite a prohibition on deliberate dolphin hunting instituted in Peru in 1993. The annual incidental kill of Dusky Dolphins in fishing gear off Kaikoura is estimated to be within the range of 50 to 150 animals, based on observations during the mid-1980s.

This close-up view of the face of a Dusky Dolphin near Kaikoura, New Zealand, shows the dark beak, the large light gray area that dominates the side and face, and the sharply demarcated white zone on the throat.

Peale's Dolphin

Lagenorhynchus australis
(Peale, 1848)

Peale's Dolphins have a close association with kelp forests along the southern coasts of South America, including the convoluted channels of Tierra del Fuego. They have long been regarded as endemic to this region, and they appear to have the most restricted range of any species in their genus. However, in 1988, observers aboard a tour vessel watched and photographed a group of bow-riding dolphins near the steep shelf edge of Palmerston Atoll, westernmost of the Cook Islands in the tropical South Pacific. Of all known species, these animals were judged to most closely resemble Peale's Dolphin. But they were in water deeper than 12,000 feet (3,600 m) and some 1,500 miles (2,500 km) west and north of this species' known normal range. If in fact they were Peale's Dolphins, they probably would have had to ride the Humboldt Current into the tropics and then move westward in the South Equatorial Current. Other possibilities are that these mystery dolphins either were Dusky Dolphins or belonged to a warm-water *Lagenorhynchus* species yet to be described.

DESCRIPTION Peale's Dolphin has a robust body and a very abbreviated beak that is hardly noticeable at sea. Set at midback, the dorsal fin is tall and falcate. There are 29 to 34 pairs of teeth in both the upper and lower jaws.

The color pattern is complex and individually variable. The face, including the lips and chin, and the dorsal body surface are dark. The dorsal fin is dark in front and, although somewhat lighter on the rear third, does not appear sharply bicolored. A dark stripe originating in the anterior part of the back sweeps diagonally across the side at midbody, separating two light gray areas—a larger one on the side in the front of the body and a smaller one on the flank. The ventral surface is white, and the whiteness extends above the axilla (the "armpit" region at the posterior base of the flipper), creating a distinctive axillary mark. A dark band runs from the corner of the mouth to the anus, separating the white belly from the gray sides.

RANGE AND HABITAT Peale's Dolphin is a coastal species, occurring from about 38°S in the southwestern Atlantic, around Cape Horn, north to 33°S in the southeastern Pacific. It is not common north of about 44°S and is seen as far south as Drake Passage at 59°S. Peale's is among the most common cetaceans along the southernmost

- PROMINENT, FALCATE DORSAL FIN
- DARK FACE
- LIGHT AREAS ON FLANK AND FORESIDE, SEPARATED BY WIDE DIAGONAL DARK STRIPE
- WHITE PATCH AT BASE OF FLIPPER
- OFTEN ASSOCIATED WITH KELP
- COASTAL AND INSHORE DISTRIBUTION OFF SOUTHERN SOUTH AMERICA

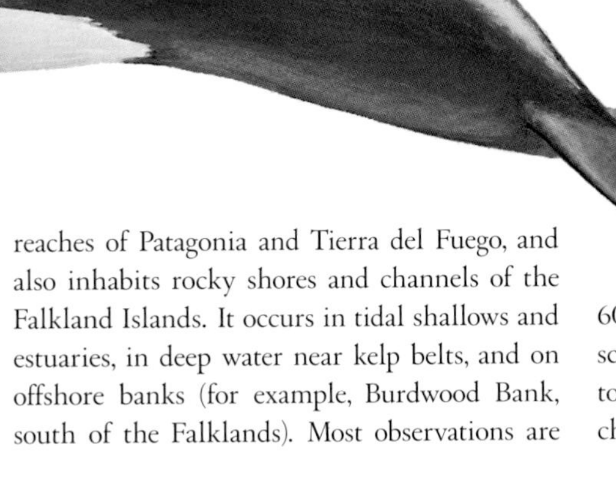

reaches of Patagonia and Tierra del Fuego, and also inhabits rocky shores and channels of the Falkland Islands. It occurs in tidal shallows and estuaries, in deep water near kelp belts, and on offshore banks (for example, Burdwood Bank, south of the Falklands). Most observations are made in areas with sea surface temperatures of less than 60°F (15°C). Peale's Dolphins have been described as "entrance" animals because they tend to be encountered in swift-flowing confluences, channel entrances, and narrows.

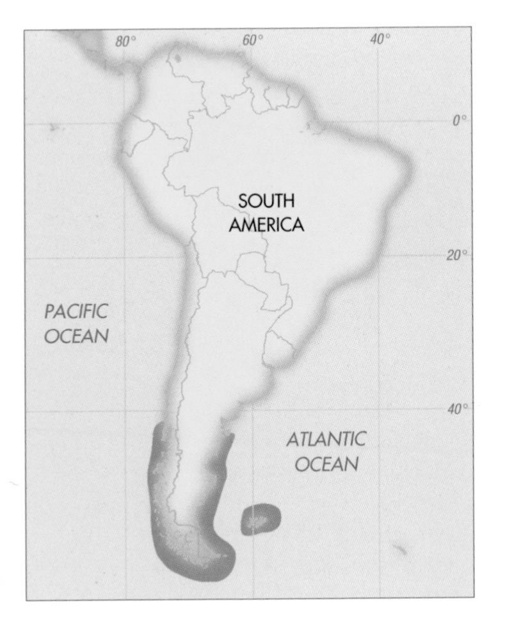

RANGE

PEALE'S DOLPHIN
FAMILY DELPHINIDAE

MEASUREMENTS AT BIRTH
LENGTH About 3′ (1 m)
WEIGHT Unavailable

MAXIMUM MEASUREMENTS
LENGTH **MALE** 7′1″ (2.16 m)
FEMALE 6′11″ (2.1 m)
WEIGHT Unavailable

LIFE SPAN Unavailable

A Peale's Dolphin surfaces slowly, with almost no splash.

SIMILAR SPECIES Peale's Dolphins are easily confused with Dusky Dolphins throughout their range. In general, Dusky Dolphins occur in extreme southern South America primarily in summer, and then usually in large open bays and offshore. They also tend to be more gregarious than Peale's Dolphins, often forming schools of many tens or hundreds of individuals. The most noticeable difference in appearance at sea is in the patch on the foresides: On the Dusky Dolphin, the patch is much lighter in color (almost white); it is slightly narrower and continues onto the head and thus seems longer. The double-streaked flank patch on the Dusky (as opposed to the single patch on Peale's) is usually seen only secondarily. Dusky Dolphins have a less-robust appearance, a lighter face, and no dark line demarcating the flank patch from the white abdomen. Although Dusky Dolphins are generally quicker and more active, Peale's Dolphins do, at times, become equally demonstrative. The Hourglass Dolphin may overlap with Peale's off southern Chile, Argentina, and the Falkland Islands, although it generally occurs in more offshore waters. It is distinguished by its bold, black-and-white color pattern and its comparatively quick, energetic movements.

BEHAVIOR Peale's Dolphins typically occur in small groups of a few (2–7) to about 20 individuals, but they sometimes form larger aggregations of up to 100 during summer and fall. They are often observed patrolling parallel to shore in shallow water. Although ponderous and slow-moving much of the time, they can also be active

These two Peale's Dolphins appear to be interacting vigorously, as one of them swipes its tail across the other's bow. The dolphin's tall, erect dorsal fin and the light striping pattern on the sides are characteristic of this species.

and demonstrative at the surface, breaching, spy-hopping, and head slapping. They playfully interact with vessels, bow riding, surfing on waves, and quickly zigzagging around and below the hull. They often swim and surface synchronously with Commerson's Dolphins.

REPRODUCTION Almost nothing is known about reproduction in Peale's Dolphin. The calving season may span the period from spring to autumn (October to at least February).

FOOD AND FORAGING Peale's Dolphins appear versatile in their food habits, in terms of the sizes and types of prey and their methods of catching them. They eat cephalopods (squid and octopus), various fish (especially bottom and near-bottom species), and crustaceans (shrimp and crabs). Young dolphins may even take quantities of tunicates (salps) during the course of being weaned. Peale's Dolphins are strongly associated with kelp forests, foraging along edges and while weaving among the fronds, and may search for octopus among kelp holdfasts and in rock crevices. These dolphins sometimes feed cooperatively in straight lines or in large, circular formations. They also may come together in a "starburst" or "flower" formation, flailing the surface and creating a whirlpool as they feed.

STATUS AND CONSERVATION Peale's Dolphins are accidentally killed in nets set from shore and in midwater trawls, but the bycatch is not currently thought to be large enough to represent a conservation problem. In the 1970s and 1980s, Peale's Dolphins were commonly harpooned and used as crab bait in both Chile and Argentina. With much of the local fishing effort being redirected from crab to sea urchins in recent years, the scale of this hunting has declined considerably. There is, however, little effort to control the activities of fishermen or to document cetacean abundance in the coastal waters of southern South America. This makes it exceedingly difficult to evaluate the conservation status of Peale's Dolphin.

A Peale's Dolphin leaps from placid waters off the Falkland Islands, in the South Atlantic. Peale's Dolphin can be distinguished from the Dusky, with whose range it overlaps, by the large light area on the side, anterior to the dorsal fin, that stops abruptly at the eye rather than extending forward onto the face.

Hourglass Dolphin

Lagenorhynchus cruciger
(Quoy and Gaimard, 1824)

Whalers and researchers plying antarctic and subantarctic waters have reported frequent sightings of Hourglass Dolphins but have rarely had the opportunity for sustained observations. There has never been a directed hunt, or even an appreciable bycatch, of this species, so very few scientific specimens have been available for examination. In fact, the Hourglass Dolphin was described formally, and accepted as a valid species, only on the basis of rough drawings made at sea. The French zoologists who christened it had no skull or other artifact in hand to prove the creature's distinctiveness from other cetaceans. Nevertheless, modern zoologists have had no cause to quibble with their diagnosis. The Latin name *cruciger* literally means "cross-bearing." It apparently refers, however inaptly, to the coloration pattern, which is only vaguely reminiscent of either a cross or an hourglass. Early American whalers and sealers, perhaps more appropriately, called this dolphin "sea skunk," in recognition of the combination of black and white stripes on the sides.

DESCRIPTION The Hourglass Dolphin has a compact body with a slight beak but very prominent appendages. The large dorsal fin is set at midback and is variably shaped, from moderately erect and rounded or hooked to low-angled and strongly falcate. There are about 30 pairs of teeth in both the upper and lower jaws.

The highly distinctive color pattern gives the animal a bold black-and-white appearance. The pattern is dominated by two large white patches on the sides, connected by a narrow, almost imperceptible white line. The forward patch begins on the face, passes over the eye, and narrows to a point just below the front insertion of the dorsal fin. The rear patch, on the flank, begins as a narrow line on the side below the dorsal fin and widens posteriorly before narrowing as it approaches the flukes. The rest of the body (including the beak and appendages) is black, except for the belly and throat, which are light gray. The ventral color merges with the forward white patch in front of the eyes.

RANGE AND HABITAT The Hourglass Dolphin is the only truly polar species of dolphin in the Southern Hemisphere (not counting Killer and Long-finned Pilot Whales as "dolphins"). Its range is centered in offshore waters between 45°S

- SHORT STUBBY BEAK
- LARGE, CENTRALLY PLACED DORSAL FIN, SOMETIMES STRONGLY HOOKED
- STRIKING BLACK-AND-WHITE COLORATION, SHARPLY DEMARCATED
- EAGER BOW RIDER AND FAST SWIMMER
- HIGH-LATITUDE, OFFSHORE DISTRIBUTION IN SOUTHERN HEMISPHERE

and 60°S. The northernmost confirmed records in the South Atlantic and South Pacific are, respectively, at about 36°S and 33°S; the southernmost are near 66°S and 68°S. The Hourglass Dolphin occurs both north and south of the Antarctic Convergence and in cool waters associated with the West Wind Drift. It is observed mainly in areas where surface temperatures range between 31 and 45°F (0.3°–7°C), but has also been seen where the surface temperature was as high as 55°F (13°C). Hourglass Dolphins reach their southernmost distribution in the South Pacific sector of the Antarctic (150°E–150°W). Although their distribution is probably circumpolar, sighting densities are relatively low in two portions of the range: in the South Pacific between 80°W and 150°W, and in the South Atlantic between 0° and 40°W.

SIMILAR SPECIES There has been much confusion in the older literature because of the difficulty of distinguishing among the three Southern Hemisphere *Lagenorhynchus* species, the Dusky, Peale's, and Hourglass Dolphins. At present, the three species are well described, and their differences are clear. Generally, the other two

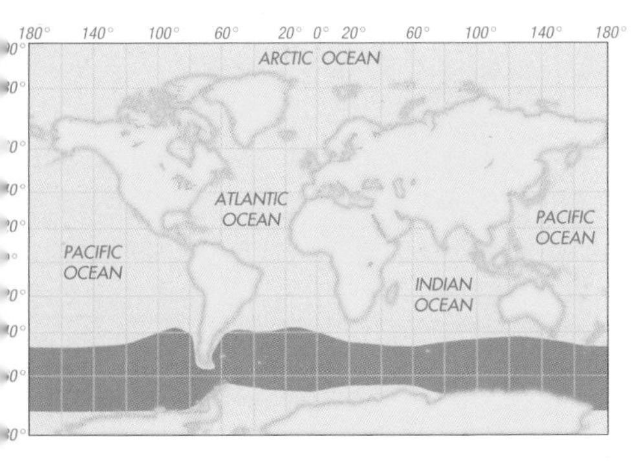

HOURGLASS DOLPHIN
FAMILY DELPHINIDAE

MEASUREMENTS AT BIRTH
LENGTH Unavailable
WEIGHT Unavailable

MAXIMUM MEASUREMENTS
LENGTH **MALE** At least 6′2″ (1.87 m)
FEMALE At least 6′ (1.83 m)
WEIGHT **MALE** Unavailable
FEMALE Unavailable

LIFE SPAN Unavailable

An Hourglass Dolphin races in Drake Passage, between Cape Horn and the South Shetland Islands.

species occur closer to shore and in lower latitudes than the Hourglass Dolphin, but there is potential overlap with Peale's in southern Chile and Argentina and around the Falkland Islands, and with the Dusky in these areas as well as off southern Australia, Tasmania, New Zealand, and the oceanic islands in or near the Antarctic Convergence. Neither Dusky nor Peale's Dolphins, with their dark faces and more muted color patterns, exhibit the bold flashes of white (the foreside and flank patches) characteristic of Hourglass Dolphins, or the black lips starkly contrasting with the lightly pigmented throat and white face. Another species that may cause confusion in coastal waters of southern South America and around the Falkland and Kerguelen Islands is the smaller Commerson's Dolphin. This black-and-white dolphin is mainly coastal, and the white area on the front half of the body is much more extensive, continuing onto the back ahead of the dorsal fin and ventrally onto the belly.

BEHAVIOR Hourglass Dolphins occur in groups that average about four to seven animals but are occasionally seen in schools of 60 or even 100. They are sometimes seen with other cetaceans, especially Fin Whales and Long-finned Pilot Whales, and whalers sometimes used the presence of Hourglass Dolphins as an indication that

Groups of Hourglass Dolphins are encountered regularly in the Antarctic. When viewed clearly, their black-and-white color pattern is unmistakable.

Fin Whales were in the vicinity. Concentrations of feeding seabirds often draw attention to schools of Hourglass Dolphins. These animals are eager bow riders and often play in the wake of the stern. They are known for their quick, energetic movements, surfing on large waves, and skimming along the surface with a noticeable splash.

REPRODUCTION Nothing is known about reproduction in the little-studied Hourglass Dolphin.

FOOD AND FORAGING Other than the fact that it eats small fish and squid, nothing is known about this dolphin's feeding habits.

STATUS AND CONSERVATION Data from research cruises indicate that there are about 140,000 Hourglass Dolphins in the Antarctic. These surveys took place in summer (January) and covered only waters south of the Antarctic Convergence, so the estimate probably does not apply to the entire species range. A few animals have been taken in gillnets in the South Pacific, but the scale of mortality caused by human activities is thought to have been small.

Right Whale Dolphins

Northern Right Whale Dolphin, *Lissodelphis borealis* Peale, 1848
Southern Right Whale Dolphin, *Lissodelphis peronii* (Lacépède, 1804)

A streamlined, almost eel-like body and the complete lack of a dorsal fin or ridge make the right whale dolphins unique among the small oceanic cetaceans. The only obvious difference between the Northern and Southern Right Whale Dolphins, apart from their separate distributions, is their black-and-white color pattern. Some taxonomists have considered them subspecies, pending a proper analysis of skeletal and other features. Right whale dolphins presumably got their common name from whalers, who identified them at a distance by the absence of a dorsal fin, a trait shared by the right whales. The genus name comes from the Greek word *lissos,* for "smooth," and refers to the species' smooth back, uninterrupted by either a fin or ridge. *Borealis* is from the Latin word for "northern," and *peronii* recognizes the French naturalist, François Péron, who in 1802 observed and described Southern Right Whale Dolphins south of Tasmania. Herman Melville dubbed the southern species the "mealy-mouthed porpoise," finding its white face reminiscent of an animal that has "just escaped from a felonious visit to a meal bag."

DESCRIPTION The right whale dolphins have unusually streamlined, slender bodies. Their bodies are compressed dorso-ventrally (unlike many cetaceans, whose bodies are laterally compressed) and taper dramatically behind the midsection. Southern Right Whale Dolphins appear to have consistently larger girths than the northern species. The beak of both species is short, with a long, straight mouthline, and is set off from the melon by a shallow crease. The flippers and flukes are small in both species, although the flippers of the Southern may be somewhat larger than the Northern's. There are 37 to 54 pairs of teeth in both the upper and lower jaws.

Northern Right Whale Dolphins are black with white ventral patches. There is a small white patch behind the tip of the lower jaw and a wide white patch on the chest that narrows behind the flippers and continues along the ventral midline to the fluke notch. The flukes have large white patches on the ventral surface and are gray dorsally. Southern Right Whale Dolphins are black dorsally and white ventrally. The ventral whiteness sweeps up from the throat and across the melon to encompass most of the face, and it also encompasses the flippers and extends far up the sides at midbody. There is a thin black band

- SLENDER BODIES WITH NO DORSAL FIN
- STARK BLACK-AND-WHITE COLOR PATTERNS (MORE WHITE ON SOUTHERN SPECIES)
- MAKE LOW-ANGLE LEAPS
- USUALLY OCCUR IN SIZEABLE GROUPS
- LIMITED TO COLD TEMPERATE WATERS OF NORTH PACIFIC (NORTHERN SPECIES) AND SOUTHERN HEMISPHERE (SOUTHERN SPECIES)

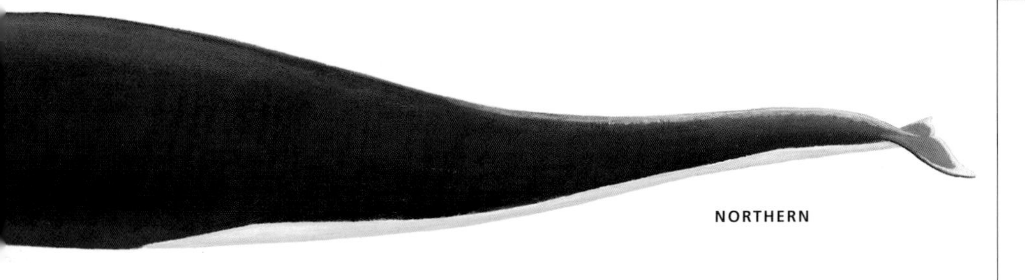

NORTHERN

SOUTHERN

on the trailing edge of the flippers. The flukes are gray above and white below, with a thin black line along the rear margin.

RANGE AND HABITAT The Northern Right Whale Dolphin inhabits cool temperate and subarctic waters across the entire North Pacific. It ranges in the west from approximately the Kuril Islands south to the Sanriku coast of Honshū, Japan, and in the east from the Gulf of Alaska and Washington state south to Baja California. In midocean, it occurs mainly between 34°N and 47°N. It generally does not enter the Sea of Japan, the Sea of Okhotsk, or the Gulf of California. The Southern Right Whale Dolphin has a circumpolar distribution between about 40°S and 55°S. It is present in the Great Australian Bight, the Tasman Sea, and around the Chatham Islands. It

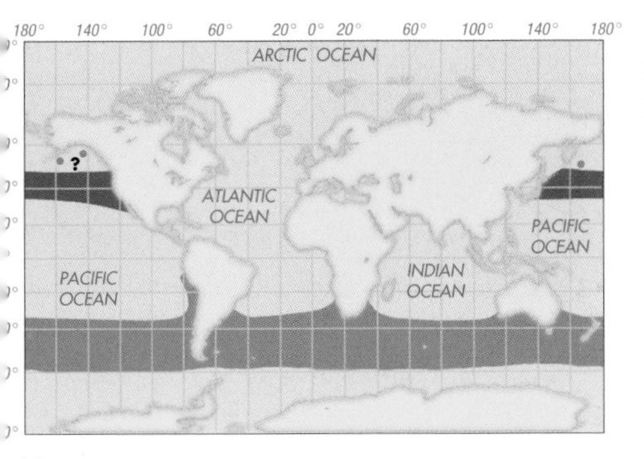

■ RANGE OF NORTHERN RIGHT WHALE DOLPHIN
■ RANGE OF SOUTHERN RIGHT WHALE DOLPHIN
? POSSIBLE RANGE
● EXTRALIMITAL RECORDS

RIGHT WHALE DOLPHINS
FAMILY DELPHINIDAE

MEASUREMENTS AT BIRTH

LENGTH	Probably about 35–40″ (90–100 cm)
WEIGHT	Unavailable

MAXIMUM MEASUREMENTS

LENGTH	**MALE**	*Northern:* 10′ (3.07 m) *Southern:* 9′9″ (2.97 m)
	FEMALE	*Southern:* 7′6″ (2.28 m)
WEIGHT	**MALE**	*Northern:* 240 lb (113 kg) *Southern:* 260 lb (116 kg)
	FEMALE	Unavailable

LIFE SPAN

42 years *(Northern)*

occasionally moves as far north as 25°S off São Paulo, Brazil, 23°S off southwestern Africa, and 12°30'S off Peru. These northern incursions are usually associated with cold, north-flowing currents, including the Benguela Current in the eastern Atlantic and the Humboldt Current in the eastern Pacific.

Right whale dolphins are pelagic, normally in waters colder than 70° F (20° C). They generally come near shore only in areas where the continental shelf is narrow or where productivity on the shelf is especially high, such as the upwelling zone off southwestern Africa. Sighting surveys suggest that, at least in the eastern North Pacific, they make seasonal inshore–offshore and north–south shifts that are presumably related to prey availability. However, nothing definite is known about movements of individuals.

SIMILAR SPECIES Their slender, streamlined bodies and lack of a dorsal fin make the right whale dolphins relatively easy to distinguish

TOP: *A school of Southern Right Whale Dolphins, off Kaikoura, New Zealand, engage in their characteristic low-angle leaping.* **LEFT:** *Northern Right Whale Dolphins, such as this individual in Monterey Bay, California, appear mostly black. They lack the white face and large white patches that extend high onto the sides of their southern counterparts, although they do have a small white patch behind the tip of the lower jaw.* **ABOVE RIGHT:** *This Southern Right Whale Dolphin off Kaikoura exhibits the bold white face and flippers that help distinguish its species.*

from all other cetaceans. At a distance, however, porpoising sea lions or fur seals may cause some confusion in identification.

BEHAVIOR Right whale dolphins are generally gregarious, and schools of several hundred to more than a thousand are not unusual. Individuals and small groups sometimes travel with larger groups of other species. They associate frequently with Pacific White-sided Dolphins in the North Pacific and with pilot whales or Dusky Dolphins in the Southern Hemisphere. Very fast swimmers, these dolphins may erupt into states of high excitement, with much breaching, tail slapping, and bursts of energetic swimming. At other times, their deliberate, low-profile surfacing rolls make them difficult to detect in any but the calmest seas. While they can be willing bow riders, they also sometimes avoid or flee from ships. Entire schools have been known to remain submerged for up to 6½ minutes. No mass strandings of Northern Right Whale Dolphins have been recorded. On one occasion, 77 Southern Right Whale Dolphins stranded on Chatham Island in the South Pacific.

REPRODUCTION The gestation period for Northern Right Whale Dolphins is a little more than a year, and the peak calving season in the central North Pacific is in July and August. The average calving interval is at least two years, but a small percentage of female Northern Right Whale Dolphins have been found to be simultaneously lactating and pregnant. Little is known about reproduction in Southern Right Whale Dolphins.

FOOD AND FORAGING Right whale dolphins live in close association with the deep scattering layer, a community of organisms that migrate vertically in the water column. These dolphins' staple prey are small fish, especially lanternfish and squid.

STATUS AND CONSERVATION The populations of both right whale dolphins are generally considered to be secure, and neither species is hunted on a significant scale in any part of its range. There are an estimated 14,000 Northern Right Whale Dolphins off the west coast of North America, but no abundance estimates exist for other areas. During the late 1980s, tens of thousands of Northern Right Whale Dolphins were killed annually in pelagic drift gillnets set primarily for squid. Their populations had been greatly depleted by 1993, when a United Nations moratorium on high-seas drift gillnetting came into force. The recovery and long-term security of Northern Right Whale Dolphins undoubtedly depends on continued restrictions on drift gillnetting. Hundreds of right whale dolphins die each year in gillnets set inside the 200-nautical-mile Exclusive Economic Zones of various countries, particularly those that target billfish, squid, and tuna. A rapidly developing swordfish fishery off northern Chile may represent an emergent threat to the southern species.

Risso's Dolphin

Grampus griseus
(G. Cuvier, 1812)

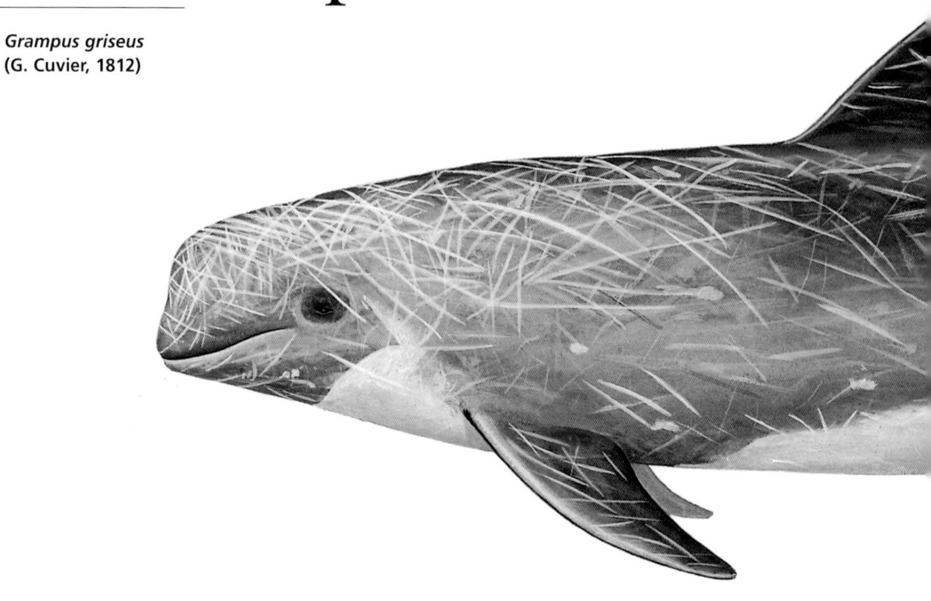

Many researchers call Risso's Dolphin "grampus," a term also once applied to the Killer Whale. The word may have evolved from the Latin *grandis* ("large" or "great") combined with *piscis* ("fish"), or possibly from the French *grand poisson* ("large fish") or *gras poisson* ("fat fish"). The specific name *griseus,* meaning "grizzled, mottled with gray," refers to the body color. Three dolphins that stranded together on the west coast of Ireland in 1933 defied attempts by researchers to identify them. They had unusual characteristics that placed them outside the range of variation of any known species. After a thorough evaluation, it was suggested that they were likely *Grampus* x *Tursiops* hybrids. Indeed, since that time, there have been numerous births in Japanese oceanariums involving a Risso's father and bottlenose dolphin mother. Had modern tools of genetic analysis been available in 1933, the mystery of the wild dolphins' parentage probably could have been quickly resolved.

DESCRIPTION Risso's Dolphin has a distinctive, beakless head shape and a body that is noticeably more robust in the front half than in the back. The melon is broad, squarish in profile, and creased in front by a characteristic longitudinal furrow. The upper jaw projects slightly beyond the lower. The dorsal fin is tall, erect, and moderately falcate. The flippers are long and sickle-shaped. There are two to seven pairs of teeth in the front portion of the lower jaw, and occasionally small vestigial teeth in the upper jaw.

The coloration is highly variable but always striking. The dorsal surface ranges within a single school from pale buff to dark brown to gray. Young calves are gray to brown dorsally and cream ventrally; they become silvery gray, then darken to almost black, then lighten as they age. The lip margins and chin are often white. In some adults, a narrow white lip line is bordered ventrally by a dark band, the contrast giving a clown-like appearance to the face. The eye area is dark-shaded. All adults have light gray or white markings on the undersides. These typically consist of an anchor-shaped chest patch, connected by a light streak to a large, diffuse zone of white between the umbilicus and the anus. Extensive scarring, concentrated on the back and sides, makes many adults appear almost completely white except for the dark dorsal fin and flippers. Most of the linear scars are assumed to be made

- BROAD HEAD WITH NO BEAK
- CLEFT MELON
- TALL, ERECT DORSAL FIN
- EXTENSIVE LINEAR SCARRING ON ADULTS
- LIGHT, ALMOST WHITE APPEARANCE OF OLDER INDIVIDUALS
- OCCURS WORLDWIDE IN TROPICAL AND WARM TEMPERATE WATERS

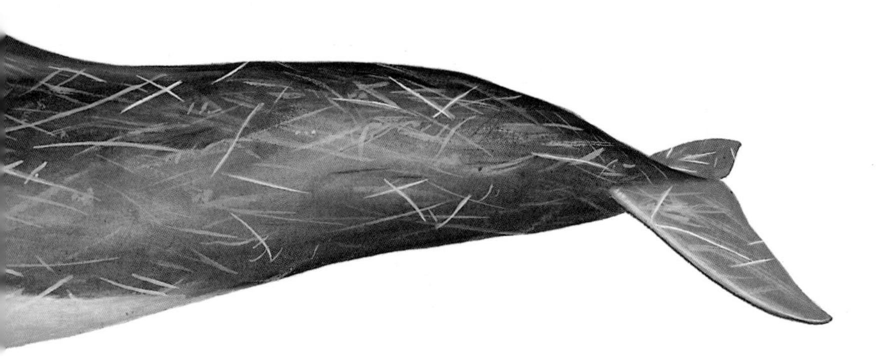

by the teeth of other Risso's Dolphins or by their squid prey.

RANGE AND HABITAT Risso's Dolphin has an extensive distribution in tropical and warm temperate waters of all oceans and large seas, including the Mediterranean and Red Seas, but not the Black Sea. It is found primarily in waters with surface temperatures of 50 to 82°F (10–28°C). In some areas, or possibly seasonally, Risso's Dolphins occupy a very narrow niche, best described as the steep upper continental slope, where water depths usually exceed 1,000 feet (300 m). They also move onto the shelf at times, presumably in response to squid availability. Although seasonal shifts in density occur, clear migratory patterns have not been defined. There is suggestive evidence that changed ecological conditions in two areas at least (northern Gulf of Mexico and off southern California) resulted in Risso's Dolphins moving into waters previously occupied by Short-finned Pilot Whales.

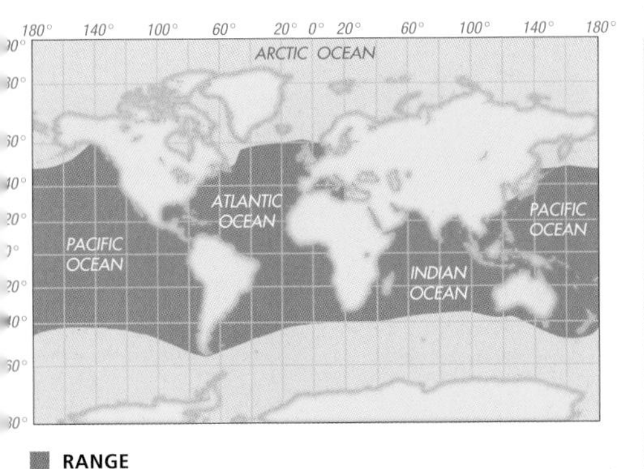

RISSO'S DOLPHIN
FAMILY DELPHINIDAE

MEASUREMENTS AT BIRTH
LENGTH 3′7″–4′11″ (1.1–1.5 m)
WEIGHT Unavailable

MAXIMUM MEASUREMENTS
LENGTH **MALE** 12′6″ (3.83 m)
FEMALE 12′ (3.66 m)
WEIGHT **MALE** Unavailable
FEMALE Unavailable

LIFE SPAN
More than 30 years

SIMILAR SPECIES At a reasonably close distance, Risso's Dolphin should be easy to distinguish from any other species. Other delphinids with tall, falcate dorsal fins—in particular, bottlenose dolphins, False Killer Whales, and Killer Whales—share the entire range of Risso's Dolphins and can be mistaken at a distance (only females and juveniles in the case of Killer Whales). Bottlenose dolphins differ in having a definite beak and usually a less-scarred and less-whitish coloration than Risso's. The False Killer Whale is much darker overall, and its dorsal fin is generally shorter and less erect. Killer Whales have strongly contrasting black and white color zones and are generally larger. The overall light gray or whitish appearance of adult Risso's Dolphins can lead to confusion with Belugas that have strayed south of their normal high-latitude range, but the latter lack a dorsal fin and are more evenly gray or white.

BEHAVIOR Risso's Dolphins are usually seen in

The tall, falcate dorsal fin of the Risso's Dolphin generally remains dark even as the rest of the body lightens and becomes plastered with scars. Many of the scars are made by the teeth of companions, but others are thought to be made by lampreys, cephalopods, and cookie-cutter sharks.

OPPOSITE TOP: *These adult Risso's Dolphins near the Azores, in the North Atlantic, appear almost ghost-like underwater. The body's whitish cast comes from a lightening with age, in combination with extensive scarring.*
ABOVE: *In profile, Risso's Dolphin has a blunt head, tapered flippers, and a prominent dorsal fin. Although superficially similar to the pilot whales, it has a more erect dorsal fin with a narrower base.*

groups of 12 to 40 individuals, averaging 25. Loose aggregations of 100 to 200, or even several thousand, are seen occasionally. On one occasion in 1990, almost 600 animals from a herd estimated at 3,000 stranded on an island off southern Japan. These dolphins are often encountered in mixed schools with various other odontocetes. They can be playful and acrobatic during interludes of rest near the surface, with breaching and tail slapping fairly common. Dives may last half an hour. They usually do not bow ride, but a famous individual called "Pelorus Jack" routinely escorted passenger vessels crossing Cook Strait, New Zealand, from 1888 to 1912.

REPRODUCTION Scientists know essentially nothing about reproduction in this species.

FOOD AND FORAGING Risso's Dolphins are squid specialists. They occasionally consume other cephalopods (octopus and cuttlefish) as well, but there is little evidence that they regularly eat fish or crustaceans. They eat anchovies to some extent off southern Africa. Much of their feeding takes place at night, possibly because some prey species migrate toward the surface then.

STATUS AND CONSERVATION Risso's Dolphin appears abundant, widely distributed, and not immediately threatened globally. It is taken as a bycatch in many kinds of fishing gear, including drift gillnets, longlines, and trawls, and it is also killed in drive hunts in Japan and the Solomon Islands and by harpooning in Indonesia, the Lesser Antilles, and Japan. There are estimated to be close to 30,000 Risso's Dolphins off the east coast of the United States, more than 2,000 in the northern Gulf of Mexico, 30,000 off the U.S. west coast, 175,000 in the eastern tropical Pacific, and 85,000 in the western North Pacific and East China Sea. Local declines may have occurred in areas such as Sri Lanka and Japan, where catches are high relative to abundance.

Melon-headed Whale

Peponocephala electra
(Gray, 1846)

The Melon-headed Whale is a widespread tropical species that is poorly known. Much of what scientists know about it comes from studies of mass strandings, which can involve hundreds of individuals. Until the mid-1960s, the species had been assigned to the genus *Lagenorhynchus* as *L. electra,* but in 1966 its genus name was changed to *Peponocephala*. The authors who coined the new genus name thought that *pepo* was Latin for "melon." However, Dale Rice, an American cetologist, pointed out recently that *pepo* actually refers to the pumpkin *(Cucurbita pepo)*. Thus, although the christeners intended to call this animal the Melon-headed Whale, they in fact ended up giving it a name that, according to Rice, is best translated as "pumpkin-headed whale."

DESCRIPTION The Melon-headed Whale has a fairly robust body that tapers noticeably at both ends. The head is small, with little or no beak and a rounded melon that slopes downward at approximately a 45° angle. The head can appear triangular or conical, depending on the view, and the melon can be somewhat bulbous in older individuals. The dorsal fin is tall (up to 1 ft/30 cm), falcate, pointed, and positioned at midback. The flippers are moderately long and taper to a point. Once known as the "many-toothed blackfish," the Melon-headed Whale has 20 to 26 pairs of teeth in both the upper and lower jaws, compared with fewer than 15 pairs per jaw in other "blackfish."

Overall, the body appears almost completely dark gray or black, except for variable light gray or white ventral markings and often light gray, pink, or white lips. On close inspection in good lighting, the color pattern on the head, back, and sides is much more complicated. An indistinct dark dorsal cape dips low onto the sides at midbody, and there are dark, mask-like areas on the sides of the face.

RANGE AND HABITAT The Melon-headed Whale is oceanic and pantropical, occurring mainly between 20°N and 20°S. Its occasional occurrence in temperate regions probably represents straying in association with warm currents. It is relatively common in the Philippines, especially near Cebu Island where the shelf is narrow and deep water occurs close to shore. Other areas where this whale is most often seen include the

- SMALL BODY SIZE, WITH LITTLE OR NO BEAK
- PROMINENT, ERECT DORSAL FIN
- TAPERED, POINTED FLIPPERS
- DARK BODY COLOR
- USUALLY OCCURS IN LARGE SCHOOLS, OFTEN TOGETHER WITH FRASER'S DOLPHINS
- MAINLY PELAGIC, TROPICAL DISTRIBUTION

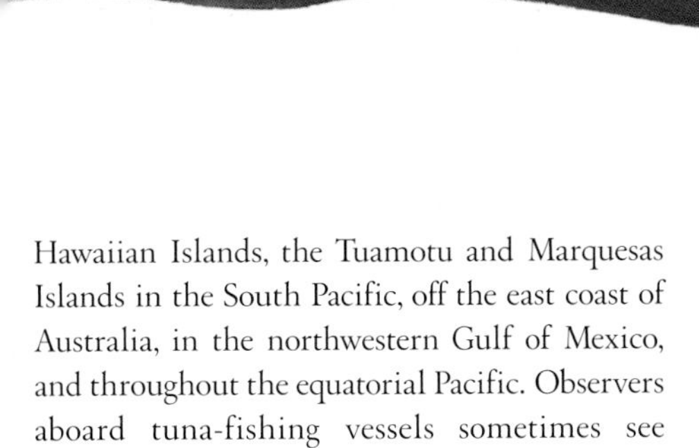

Hawaiian Islands, the Tuamotu and Marquesas Islands in the South Pacific, off the east coast of Australia, in the northwestern Gulf of Mexico, and throughout the equatorial Pacific. Observers aboard tuna-fishing vessels sometimes see Melon-headed Whales in the Gulf of Panama.

SIMILAR SPECIES The Pygmy Killer Whale and False Killer Whale occur throughout the Melon-headed Whale's range and so closely resemble it in color and the appearance of the dorsal fin that there is little hope of distinguishing between these species at a distance. The Pygmy Killer Whale generally occurs in smaller schools of 50 or fewer, and at close quarters is seen to have a rounded, relatively short head and less-pointed flippers. The adult False Killer Whale is much larger than the Melon-headed Whale and can also be distinguished by the less-prominent dorsal fin, the conspicuous hump on the leading edge of the flippers, a more-prominent melon that protrudes

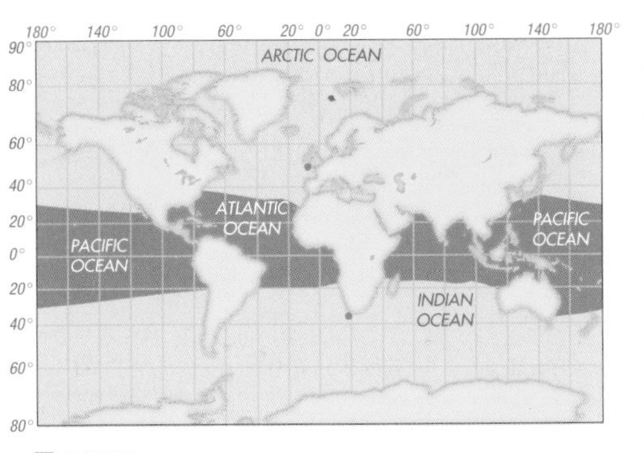

RANGE
EXTRALIMITAL RECORDS

MELON-HEADED WHALE

FAMILY DELPHINIDAE

MEASUREMENTS AT BIRTH

LENGTH About 3′4″ (1 m)
WEIGHT Less than 33 lb (15 kg)

MAXIMUM MEASUREMENTS

LENGTH **MALE** 8′8″ (2.65 m)
FEMALE 9′ (2.75 m)
WEIGHT **MALE** 460 lb (210 kg)
FEMALE At least 460 lb (210 kg)

LIFE SPAN

MALE More than 22 years
FEMALE More than 30 years

somewhat beyond the tip of the jaw, and an even more muted dorsal cape pattern. Neither of these two species is known to associate with Fraser's Dolphin, as Melon-headed Whales typically do.

BEHAVIOR Melon-headed Whales are extremely gregarious, usually occurring in schools of several hundred and sometimes more than 1,000 animals. The schools appear to consist of many smaller, compact groups whose movements and activities are coordinated. The fact that many hundreds strand together indicates a strong social bond within the school as a whole. Traveling schools make considerable splash, with the tightly bunched animals generally breaking the surface in a flat trajectory and pushing crescents of water ahead of them. They occasionally ride the bows of passing vessels. There is a strong and consistent association between Melon-headed Whales and Fraser's Dolphins. On most occasions, the Melon-heads are on the periphery of or trailing a school of Fraser's Dolphins. They are also sometimes seen in the vicinity of Rough-toothed Dolphins.

REPRODUCTION Very little is known about reproduction in this species. Mass strandings have been the only source of useful data. Gestation probably lasts about a year, and there is no reason to believe that calving is strongly seasonal.

OPPOSITE: *A young Melon-headed Whale, separated from its mother and others of its kind near Hawaii.*
ABOVE: *A group of Melon-headed Whales spyhopping, perhaps to inspect their surroundings.*
RIGHT: *A Melon-headed Whale breaching. The dorsal view highlights the conical shape of its head.*

FOOD AND FORAGING As a high-seas, deep-water species, the Melon-headed Whale preys entirely on pelagic fish, squid, and occasionally crustaceans. Many of its prey organisms inhabit water up to 5,000 feet (1,500 m) deep, and it is thought to at least occasionally feed fairly deep in the water column.

STATUS AND CONSERVATION There are about 45,000 Melon-headed Whales in the eastern tropical Pacific, and about 2,000 in the northwestern Gulf of Mexico. Their abundance elsewhere is unknown. No particular conservation problem has been identified for the species. The traditional drive hunt at the Solomon Islands may have targeted Melon-headed Whales. Teeth were the main object of the hunt and were used for making "porpoise tooth" necklaces that served as currency and dowry items. This hunt was very active in the 1960s and continued in some parts of the Solomons until as recently as the early 1990s. Small numbers are taken in gillnets and by harpooning throughout the tropics. Large schools are sometimes seen on the dolphin-hunting grounds off Taiji, Japan (on the southeastern coast of Honshū); however, hunters usually do not drive Melon-headed Whales, as their meat is not considered palatable.

Pygmy Killer Whale

Feresa attenuata
Gray, 1874

Two skulls obtained by the British Museum during the 19th century were the only evidence of the Pygmy Killer Whale's existence until the early 1950s. In 1954, Japanese scientist Munesato Yamada published a description of a "rare porpoise" killed by whale and dolphin hunters at Taiji on the southeastern coast of Honshū. He noted that the animal's skull matched the museum specimens attributed to the genus *Feresa* but that it also had similarities to the Killer Whale, including the number and form of its teeth, a tall dorsal fin, and white ventral markings. He thus proposed that it be called the "Lesser" or "Pygmy" Killer Whale. A few of these whales brought into captivity for brief periods in Hawaii and South Africa proved to be pugnacious, attacking and even killing their tank-mates.

DESCRIPTION The Pygmy Killer Whale is moderately robust in the front half of its body, while the back half is slender. There is no beak, and the rounded melon extends somewhat forward of the front of the mouth. The dorsal fin, positioned at or slightly behind midback, is tall, upright, and falcate. The flippers are moderately long and tapered. There are 8 to 12 pairs of teeth in the upper jaw and 10 to 13 pairs in the lower.

As with other "blackfish," the body appears mostly dark, with white areas only on the lips and belly. In good lighting conditions, it is possible to see a well-defined, dark dorsal cape that dips somewhat below the dorsal fin. Lighter gray zones on the sides extend from the eyes to the flukes. There are irregular white ventral markings centered along the midline in the rear half of the body, which sometimes extend high enough onto the flanks to be noticeable in a side view.

RANGE AND HABITAT The Pygmy Killer Whale has a pantropical range. In the Atlantic, it occurs at least as far north as Florida, the Gulf of Mexico, and Senegal, and as far south as northern Argentina and southern Africa. In the Indian Ocean, it is known to occur in the Arabian Sea and waters around Sri Lanka. In the Pacific, it occurs from Honshū (Japan), Hawaii, and the Gulf of Tehuantepec in Mexico, south to Queensland (Australia) and Peru. It is an oceanic species and does not normally enter shallow, nearshore waters.

- SMALL, MOSTLY DARK BODY
- ROUNDED HEAD WITH WHITE LIPS
- LARGE, ERECT, FALCATE DORSAL FIN
- MUTED DARK DORSAL CAPE AND SLIGHTLY LIGHTER SIDES
- MAINLY OFFSHORE, PANTROPICAL DISTRIBUTION

SIMILAR SPECIES Melon-headed and young False Killer Whales both occur throughout the Pygmy Killer's range and are difficult to distinguish from Pygmy Killers. When seen bow riding or from above, the Melon-headed Whale has a triangular, relatively narrow head and more pointed flippers. The False Killer Whale has a smaller, relatively narrow head and a long, slender body compared to the Pygmy Killer; its dorsal fin is smaller relative to its size; and there is a distinctive hump on the leading edge of its flippers.

BEHAVIOR Groups of Pygmy Killer Whales usually contain fewer than about 50 individuals, but they occasionally contain more than 100. These whales are considerably less active and animated than most other oceanic dolphins. In fact, when compared to species such as the Melon-headed Whale, they can seem slow and lethargic. However, they occasionally bow ride and sometimes jump high above the surface. The pugnacious nature of Pygmy Killer Whales, as evidenced by observations of a few captive individuals and of animals enclosed within tuna purse seines with

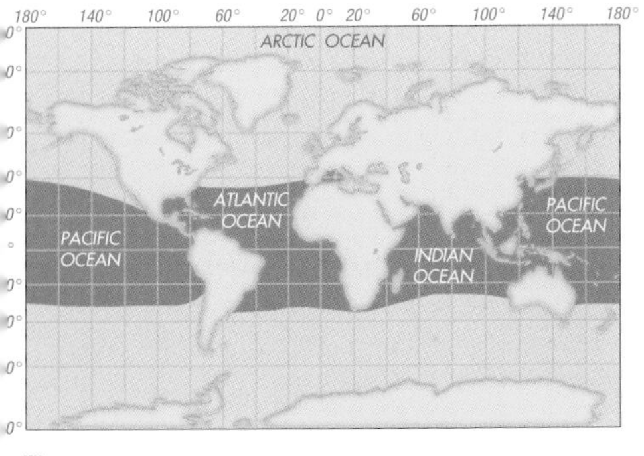

RANGE

PYGMY KILLER WHALE
FAMILY DELPHINIDAE

MEASUREMENTS AT BIRTH
LENGTH 31″ (80 cm)
WEIGHT Unavailable

MAXIMUM MEASUREMENTS
LENGTH **MALE** 8′6″ (2.6 m)
FEMALE At least 8′2″ (2.5 m)
WEIGHT Unavailable

LIFE SPAN Unavailable

These Pygmy Killer Whales, swimming near a Humpback Whale, display their species' robust body shape, with a rounded head, long tapered flippers, and prominent dorsal fin.

dolphins, may explain why they are rarely seen with other small cetaceans. They do, however, sometimes bow ride alongside Rough-toothed Dolphins in Hawaii.

REPRODUCTION Almost nothing is known about reproduction in this species. Most calves are presumably born in summer.

FOOD AND FORAGING Very little is known about the diet or feeding habits of Pygmy Killer Whales. Squid beaks and parts of fish have been found in the few stomachs examined from stranded or killed animals.

STATUS AND CONSERVATION The only area where there is an estimate of abundance for the Pygmy Killer Whale is the eastern tropical Pacific, where there are thought to be about 40,000. Small numbers are taken in various hunts and gillnet fisheries around the world. Fishermen in the Cape Verde Islands, Lesser Antilles, Indonesia, Sri Lanka, the Philippines, and possibly elsewhere harpoon them opportunistically. Since the species may be naturally rare, even small numbers of removals could be a cause for concern.

Pygmy Killer Whales appear to be naturally rare. A group swims off the Kona coast of Hawaii, one area where they can be seen with some regularity.

False Killer Whale

Pseudorca crassidens
(Owen, 1846)

- TALL, ERECT, FALCATE DORSAL FIN
- SMALL, ROUNDED OR CONICAL HEAD WITH NO BEAK
- DARK COLORATION

- HUMP ON LEADING EDGE OF FLIPPERS
- PRIMARILY TROPICAL TO WARM TEMPERATE DISTRIBUTION

The False Killer Whale's existence was first established on the basis of subfossil remains found in The Fens of Lincolnshire, England, in 1846. Scientists thought it to be an extinct species until 1861, when a school of False Killer Whales stranded in Germany. As the common and genus names imply (*pseudo* is Latin for "false"), this whale resembles the Killer Whale in some respects. The specific name, *crassidens,* is from the Latin words *crassus* and *dens,* meaning "thick" or "dense" and "tooth," respectively. False Killer Whales are much less inclined than Killer Whales to use their teeth in attacks on mammalian prey, although a few have been observed doing so. Unlike Killer Whales, False Killers have not been the subjects of focused, long-term field studies. Much of what scientists know about them comes from strandings. The strandings frequently involve hundreds of animals and may result in the eradication of entire schools. False Killer Whales are notorious in the fishing industry for stealing such commercially valuable fish as tuna and yellowtail from baited longlines.

DESCRIPTION The False Killer Whale is slender-bodied, with a small, rounded or bluntly conical head and a long straight mouthline. The melon overhangs the tip of the lower jaw, particularly in adult males. The slender dorsal fin is erect and can be more than 15 inches (40 cm) high. It is falcate and positioned approximately at midback. The flippers are wider at the base than the tip and have a distinctive hump, or bulge, midway on the leading edge. There are 7 to 12 pairs of large, conical teeth in both the upper and lower jaws.

The body is almost entirely dark, except for light areas on the throat and chest and along the ventral midline. A large, anchor-shaped blaze across the rear of the throat and chest is variably gray to nearly white. This blaze narrows to a thin stripe that may extend to the genital slit. There may also be light gray areas on the sides of the head. Some animals have white, star-shaped scars on the body, presumably caused by bites from cookie-cutter sharks.

RANGE AND HABITAT False Killer Whales have an extensive oceanic range. They occur in all tropical and warm temperate waters, including the Mediterranean and Red Seas, the Gulf of Mexico, and the Sea of Japan. In the Pacific Ocean, their range extends from Japan and

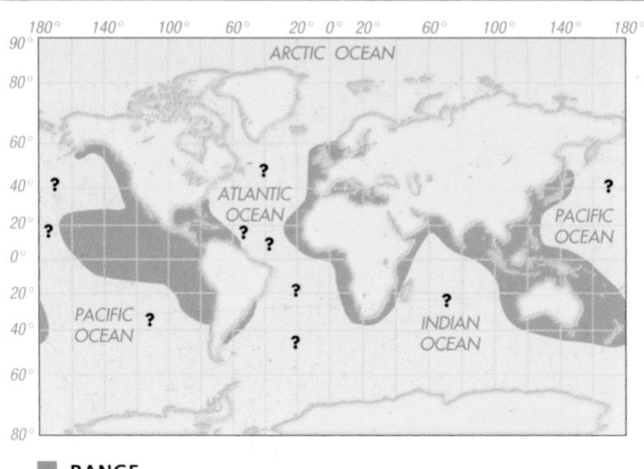

FALSE KILLER WHALE

FAMILY DELPHINIDAE

MEASUREMENTS AT BIRTH

LENGTH 5′3″–6′3″ (1.6–1.9 m)

WEIGHT Unavailable

MAXIMUM MEASUREMENTS

LENGTH **MALE** 19′6″ (6 m)
FEMALE 16′6″ (5 m)

WEIGHT **MALE** At least 3,000 lb (1,360 kg)
FEMALE Unavailable

LIFE SPAN

MALE 58 years
FEMALE 63 years

British Columbia south to New Zealand (including the Chatham Islands), South Australia, Tasmania, and Chile. In the Atlantic, they range from Maryland and Scotland south to Chubut, Argentina, and southern Africa. They are usually encountered in waters deeper than 3,300 feet (1,000 m).

SIMILAR SPECIES Pygmy Killer, Melon-headed, and Short-finned Pilot Whales and Risso's Dolphins occur throughout warm-water regions inhabited by the False Killer Whale (there is less overlap with the higher-latitude Long-finned Pilot Whale). All of these species lack the False Killer's distinctive bulge on the leading edge of the flippers. The Pygmy Killer and Melon-headed Whales are much smaller—only about half the size of the False Killer—and often have white lips. The dorsal fins of pilot whales are positioned far forward on the body, rise from the back at a low angle, and have a long base. Risso's Dolphin has a tall dark dorsal fin and a squarish head profile, and adults are heavily scarred and lighter in color than False Killers.

BEHAVIOR The False Killer Whale is a gregarious species. It typically occurs in groups of 10 to 20, and these groups usually belong to larger schools consisting of hundreds of individuals. The strong social affiliations of the species are evident from the large numbers that frequently strand together. False Killers are active at the surface and frequently bow ride. It is not unusual to see bottlenose dolphins traveling with them. False Killer Whales chase and attack smaller dolphins during tuna purse-seining operations in the eastern tropical Pacific, although this is not necessarily typical behavior. They have also been reported attacking large whales. On one occasion near the Galápagos Islands, a school of female and immature Sperm Whales reacted defensively when a mixed group of False Killer Whales and Common Bottlenose Dolphins rapidly approached them. From the chunks of flesh floating

OPPOSITE: *This close-up view of a group of False Killer Whales provides a good look at the rounded and somewhat conical head and the long straight mouthline. The dorsal fin is erect and falcate.*

LEFT: *False Killer Whales feeding on fish in their underwater realm off Hawaii. The hump on the front edge of the tapered flippers is a distinctive feature.*

in the water during the encounter, researchers surmised that the False Killers had bitten at least the Sperm Whales' flippers and flukes.

REPRODUCTION The False Killer Whale has a low reproductive rate. The calving interval is relatively long, estimated from one study to be nearly seven years. In addition, females older than about 45 years of age are barren. The species maintains its numbers by offsetting its low reproduction with high survival; this strategy, however, makes it particularly vulnerable to overexploitation. Females give birth after a gestation period of 14 to 16 months and nurse their calves for 1½ to 2 years.

FOOD AND FORAGING False Killer Whales are versatile predators, known—from both direct observations and from examinations of stomach contents—to consume a fairly large variety of fish and cephalopods. They specialize in particular types of prey, depending on the region. Off southern Africa, for example, they forage over the continental slope and take mainly squid, while in Hawaiian waters they often eat mahi mahi and yellowfin tuna. At least when eating tuna, they tear the flesh and ignore the internal organs, gills, and tail. They have been known to remove tuna as large as 70 pounds (30 kg) from fishing lines. False Killer Whales often travel in a broad band up to several miles wide, presumably to increase their chances of finding prey.

STATUS AND CONSERVATION False Killer Whales are hunted at least opportunistically in Indonesia and the West Indies, and they are also killed incidentally in various fisheries. In most years, small numbers (a few tens) are killed in Japan, either by driving or harpooning. Some of the animals caught in the drives are sold to oceanariums. In the 1960s, Japanese fishermen at Iki Island complained that False Killer Whales were interfering with their yellowtail fishery. As a result, more than 900 False Killer Whales were killed at Iki Island between 1965 and 1980. There are an estimated 16,500 False Killer Whales in the western Pacific north of 25°N, 40,000 in the eastern tropical Pacific, at least 500 off Hawaii, and a few hundred in the northern Gulf of Mexico.

Killer Whale

Orcinus orca
(Linnaeus, 1758)

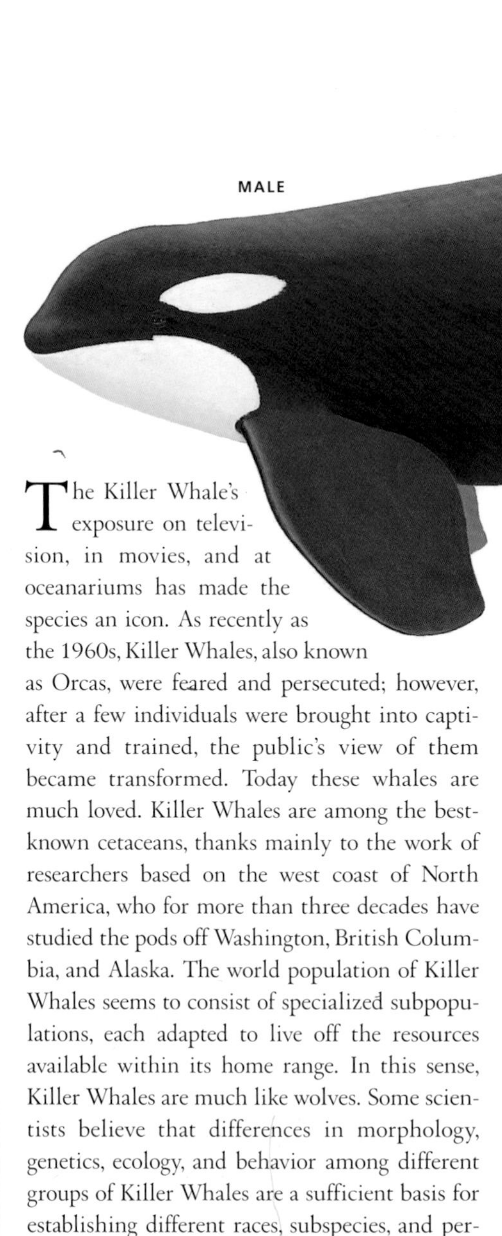

The Killer Whale's exposure on television, in movies, and at oceanariums has made the species an icon. As recently as the 1960s, Killer Whales, also known as Orcas, were feared and persecuted; however, after a few individuals were brought into captivity and trained, the public's view of them became transformed. Today these whales are much loved. Killer Whales are among the best-known cetaceans, thanks mainly to the work of researchers based on the west coast of North America, who for more than three decades have studied the pods off Washington, British Columbia, and Alaska. The world population of Killer Whales seems to consist of specialized subpopulations, each adapted to live off the resources available within its home range. In this sense, Killer Whales are much like wolves. Some scientists believe that differences in morphology, genetics, ecology, and behavior among different groups of Killer Whales are a sufficient basis for establishing different races, subspecies, and perhaps even species.

DESCRIPTION The Killer Whale's body is extremely robust; it is the largest delphinid. The head is conical and lacks a well-defined beak. The dorsal fin, situated at midback, is large, prominent, and highly variable in shape: falcate in females and juveniles, erect and almost spike-like in adult males. On males, the dorsal fin can reach a height of 3 to 6 feet (1–1.8 m). The flippers are large, broad, and rounded, very different from the typically sickle-shaped flippers of most delphinids. There are 10 to 14 pairs of large pointed teeth in both the upper and lower jaws.

The color pattern consists of highly contrasting areas of black and white. The white ventral zone, continuous from lower jaw to anus, narrows between the all-black flippers and branches behind the umbilicus. The ventral surface of the flukes and adjacent portion of the caudal peduncle are also white. The back and sides are black, except for white patches on the flanks that rise from the uro-genital region and prominent oval white patches slightly above and behind the eyes. There is a highly variable, gray to white saddle marking on the back behind the dorsal fin.

- TALL, ERECT DORSAL FIN, MORE PROMINENT IN ADULT MALE
- LARGE ROUNDED FLIPPERS
- DISTINCTIVE BLACK-AND-WHITE COLOR PATTERN
- LARGE SIZE RELATIVE TO OTHER DOLPHINS
- COSMOPOLITAN DISTRIBUTION

FEMALE

RANGE AND HABITAT Considered the most widespread cetacean, the Killer Whale is truly cosmopolitan and is not limited by such habitat features as water temperature, or depth. It occurs in highest densities at high latitudes, especially in areas with an abundance of prey. Its movements generally appear to track those of favored prey species or to take advantage of pulses in prey abundance or vulnerability, such as during times and in areas of fish spawning and seal pupping. In the Antarctic during summer, most Killer Whales position themselves near the ice edge and in channels within the pack ice, where they prey on baleen whales, penguins, and seals. It is uncertain how far, or where, they migrate. Some may remain in antarctic waters year-round. In the Arctic, Killer Whales rarely move close along or into the pack ice. Researchers studying Killer Whales in Washington and British Columbia have identified "resident" and "transient" pods,

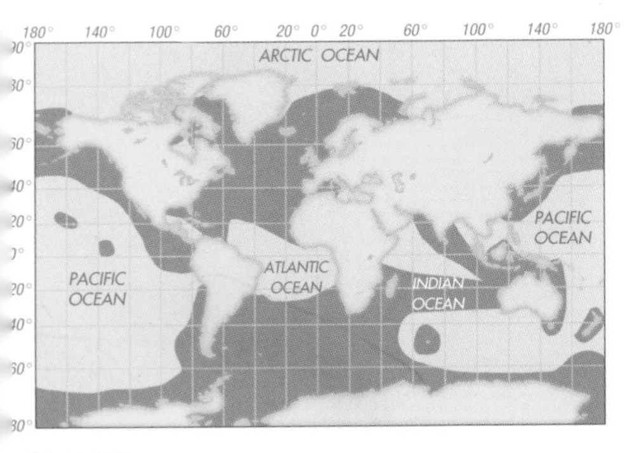

RANGE

KILLER WHALE
FAMILY DELPHINIDAE

MEASUREMENTS AT BIRTH

LENGTH 7′3″–8′6″ (2.2–2.6 m)

WEIGHT About 350 lb (160 kg)

MAXIMUM MEASUREMENTS

LENGTH **MALE** 30′ (9 m)
FEMALE 26′ (7.9 m)

WEIGHT **MALE** At least 12,000 lb (5,600 kg)
FEMALE At least 8,400 lb (3,800 kg)

LIFE SPAN

MALE 50–60 years
FEMALE 80–90 years

Killer Whales evoke strong responses from people in part because they are at once large, intimidating, and playful. Here a young breaching animal displays the species' broad flippers and white ventral markings, while a larger animal in the foreground shows the impressive dorsal fin and the distinctive light "saddle" marking on the back immediately behind the fin.

although both types of pod are present year-round. Some individuals occupy very large ranges. For example, photo-identification studies have shown that some Killer Whales move between Alaska and California. (The range map for this species shows areas where Killer Whales are known to occur but probably under-represents the total range of the species.)

SIMILAR SPECIES The Killer Whale is among the easiest of the cetaceans to identify. However, at a distance, the relatively prominent dorsal fins of the False Killer Whale and Risso's Dolphin can cause confusion. Both species overlap with Killer Whales in tropical and temperate waters.

BEHAVIOR The basic social unit of resident Killer Whales in Washington and British Columbia is the matrilineal group, consisting of two to four generations of two to nine related individuals. Matrilineal groups are stable over long periods, and all members may contribute to calf rearing. A number of groups that spend much of their time together constitute a pod. The largest resident pod in the area of Washington and British Columbia contains close to 60 individuals. Resident pods greet one another by facing off in two tight lines, then mingling in a relaxed manner, as if to reaffirm their social bonds. While adult females tend to be associated with one or more pods, adult males are sometimes solitary.

Killer Whales often breach, spyhop, and slap the surface with their flukes or flippers. They exhibit varied responses to vessels, ranging from indifference to curiosity. Mass strandings occur occasionally, and pods sometimes become trapped in tidal ponds or inlets. Wind-blown or fast-forming ice can be a hazard for Killer Whales in the Arctic and Antarctic, forcing them to remain in small pools of open water for prolonged periods.

REPRODUCTION In the resident population off Washington and British Columbia, calving occurs year-round, with a peak between autumn and spring. The average calving interval is five years. Females usually stop reproducing after about 40 years of age. Studies of whales in captivity suggest that gestation lasts 15 to 18 months. Although Killer Whales begin eating solid food

RIGHT: *These spyhopping Killer Whales belong to one of the populations that visit or reside in inshore waters of Washington state and British Columbia.*
BELOW: *This group of Killer Whales includes three adult males, each of them readily identifiable by the tall, triangular dorsal fin. The animals in the center of the group are either females or juvenile males.*

at a very young age, they continue to nurse for at least a year and may not be fully weaned until close to two years of age.

FOOD AND FORAGING Killer Whales eat a diet ranging from small schooling fish and squid to large baleen and sperm whales. Their prey items include sea turtles, otters, sirenians, sharks, rays, and even deer or moose, which they catch swimming across channels. Pods tend to specialize. For example, some depend largely on salmon, tuna, or herring, while others patrol pinniped haulouts or follow migratory whale populations, much as wolves follow caribou herds. Killer Whales obviously need to use cooperative hunting to harass and subdue large prey items, but they also cooperate to consolidate and maintain tight balls of baitfish, taking turns slicing through the schools to feed. Killer Whales also steal fish from longlines, scavenge on discarded fishery bycatch, and selectively eat the tongues of baleen whales. Prey may be strongly influenced by their fear of Killer Whales; pinnipeds flee from the water onto land or ice and whales and dolphins move into nearshore shallows or hide in cracks in pack ice.

STATUS AND CONSERVATION While as a species the Killer Whale is not endangered, whaling or live-capture operations have depleted some regional populations. Resident and transient populations off Washington and British Columbia number only in the low hundreds, and are threatened by pollution, heavy ship traffic, and possibly reduced prey abundance. There is concern that intensive whale-watching operations may influence the behavior of Killer Whales, and that the loud "seal-scarers" used to protect salmon pens from predation by pinnipeds may be driving Killer Whales away from their preferred inshore resting and foraging waters. About 8,500 Killer Whales are thought to occur in the eastern tropical Pacific, at least 850 in Alaskan waters, possibly close to 2,000 off Japan, and about 80,000 in the Antarctic during summer. Estimates from most other areas are in the hundreds or low thousands. Whalers in Japan, Indonesia, Greenland, and the West Indies continue to hunt Killer Whales; while the whales are killed in only small numbers, the effects of hunting on local populations could be substantial.

Long-finned Pilot Whale

Globicephala melas
(Traill, 1809)

The pilot whales are among the most familiar whales because of their extensive global distribution and general abundance, their proclivity for coming ashore and stranding in large numbers, and the practice in some areas of driving them to shore for slaughtering. The name "pilot whale" refers to the belief (unproven and probably not true) that a single member of a pod pilots, or leads, the group, and that the others continue to follow, even when it means certain death. The scientific name derives from the Latin word *globus,* meaning "globe," and the Greek word *kephale,* for "head," a clear reference to the bulbous head shape. Newfoundland fishermen appropriately applied the name "pothead" to this species, and people in the Canadian Maritimes still use it. Commercial whalers referred to pilot whales as "blackfish" (the name *melas* is derived from the Greek for "black"), a useful but ambiguous term that applies equally to Killer, False Killer, Pygmy Killer, and Melon-headed Whales. Three subspecies of Long-finned Pilot Whales are recognized by some researchers: one in the North Atlantic, one in the Southern Hemisphere, and one (extinct) in the North Pacific.

DESCRIPTION The Long-finned Pilot Whale has a long but robust body. The melon ranges from bulbous to squarish and can protrude beyond, or overhang, the front of the mouth. There is a very short, barely noticeable beak, and the mouthline slants up toward the eye. The dorsal fin has a characteristic profile: set ahead of midbody, it is long at the base relative to its height, rises at a shallow angle, and is falcate in shape. The flippers are long (about one-quarter of the body length) and tapered. There are usually 9 to 12 pairs of teeth in both the upper and the lower jaws.

The color pattern is fairly simple. The body is basically dark gray or black to dark brown. There is a tapering white or light gray streak behind the eye, extending up and back from behind each eye toward the front of the dorsal fin. There is also a large white or gray saddle on the back behind the dorsal fin. Both features are consistently and conspicuously present on Southern Hemisphere whales, while only the saddle is apparent on most Northern Hemisphere animals. On the belly, a broad white or light gray throat patch (often described as anchor-shaped) is joined by a narrow stripe to a wider light area in the uro-genital region.

- BULBOUS MELON WITH NO NOTICEABLE BEAK
- BROAD-BASED, LOW-PROFILE, FALCATE DORSAL FIN POSITIONED FAR FORWARD ON BODY
- BLACK OR DARK BROWN BODY COLOR
- WHITE OR LIGHT GRAY "SADDLE" ON BACK BEHIND DORSAL FIN
- WHITE OR LIGHT GRAY BLAZE ABOVE AND BEHIND EYE, ESPECIALLY IN SOUTHERN HEMISPHERE
- COOL TEMPERATE TO SUBPOLAR DISTRIBUTION IN NORTH ATLANTIC AND SOUTHERN HEMISPHERE

RANGE AND HABITAT The Southern subspecies of Long-finned Pilot Whale is circumpolar, with its northern limits approximately at São Paulo (Brazil), Cape Province (South Africa), Crozet and Heard Islands, southern Australia, New Zealand, and around 19°S off Chile. It occurs regularly at least as far south as the Antarctic Convergence (47°S–62°S) and occasionally to 67°S to 68°S in the South Pacific. The North Atlantic subspecies is boreal and subarctic, ranging from North Carolina, the Azores, Madeira, and Mauritania northward to Davis Strait and the Greenland and Barents Seas. It occurs in the western Mediterranean, the North Sea, and the Gulf of St. Lawrence. The North Pacific subspecies, known from archaeological evidence from the 8th to the 12th centuries, is now extinct, possibly "replaced" by the Short-finned Pilot Whale. In the western North Atlantic at least, Long-finned Pilot Whales are pelagic, occurring in especially high densities in winter and spring over the continental slope, then moving inshore and onto the

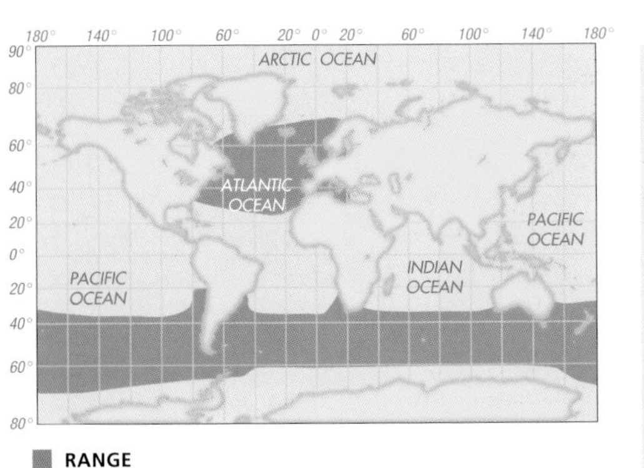

LONG-FINNED PILOT WHALE

FAMILY DELPHINIDAE

MEASUREMENTS AT BIRTH

LENGTH 5′3″–6′7″ (1.6–2 m)

WEIGHT 165 lb (75 kg)

MAXIMUM MEASUREMENTS

LENGTH **MALE** 21′ (6.3 m)
FEMALE 15′6″ (4.7 m)

WEIGHT **MALE** 5,000 lb (2,300 kg)
FEMALE 2,900 lb (1,300 kg)

LIFE SPAN

MALE 45 years

FEMALE 60 years

shelf in summer and autumn. These movements follow squid and mackerel populations.

SIMILAR SPECIES The Long-finned Pilot Whale is nearly impossible to distinguish from the Short-Finned Pilot Whale in areas of range overlap (at roughly 35°N–40°N and 25°S in the western Atlantic; 20°N–45°N in the eastern Atlantic; 14°S–25°S in the eastern Pacific; off southern Africa at 34°S–35°S; and off Tasmania). The False Killer Whale may overlap with the Long-finned Pilot Whale in the lower latitudes of the latter's range, but its melon is more tapered and not bulbous; its dorsal fin is more slender, more upright, and farther back on the body; it lacks the light dorsal saddle and streak behind the eye; and its behavior is more dolphin-like (it rides bow waves and often leaps above the surface).

BEHAVIOR School size is difficult to assess because large schools containing hundreds of Long-finned Pilot Whales are often dispersed, with small, close-knit groups of perhaps 10 to 20 scattered across a large expanse of ocean. These small pods are thought to be formed around adult females and their offspring. Strandings involving hundreds of individuals are not unusual, and these demonstrate that the large schools have a high degree of social cohesion. Smaller scattered groups may coalesce when driven by whalers or before approaching shore to strand. The behavior of Long-finned Pilot Whales near the surface varies from quiet rafting or milling to purposeful diving to bouts of playfulness. These whales often spyhop but rarely breach and do not ride bow waves or approach vessels. Mature males fight and sometimes seriously injure or even kill one another by ramming. Long-finned Pilot Whales are well known for associating with other cetacean species, both dolphins and large whales.

REPRODUCTION Long-finned Pilot Whales in the North Atlantic mate and calve mainly between late spring and autumn (April–September). Gestation lasts about a year, and lactation lasts for up to two years. As females are rarely pregnant and lactating at the same time, the total reproductive cycle usually lasts at least three years and often five. North Atlantic females can become pregnant even when 55 years old, although pregnancy after 40 years of age is rare. Little is known about reproduction in the Southern Hemisphere; births have been recorded in spring, summer, and autumn (October–early April). Sexual activity is common within pods of pilot whales, but effective breeding (leading to conception) usually involves males from outside the female-centered pods.

FOOD AND FORAGING In the North Atlantic, Long-finned Pilot Whales feed mainly on squid and mackerel. They also ingest shrimp (particularly young whales) and various other fish species occasionally. These whales probably take most of their prey at depths of 600 to 1,650 feet

The dorsal fins of these Long-finned Pilot Whales in the Southern Hemisphere display the broad-based, low profile typical of the species. The large animal in the foreground has a particularly massive dorsal fin.

(200–500 m), although they can almost certainly forage deeper if necessary. Some of their prey migrate vertically and thus might be hunted mainly at night when the deep-living organisms come nearer the surface. In the Southern Hemisphere, fish appear to be at least as important as squid in the diet.

STATUS AND CONSERVATION There are at least 10,000 Long-finned Pilot Whales in the western North Atlantic and at least a few hundred thousand in the central and eastern North Atlantic. Fishermen of northern regions have a long tradition of driving pilot whales ashore and killing them; these hunts occurred historically at Cape Cod, Newfoundland, the Shetland and Orkney Islands, and Ireland. American sperm whalers during the 19th and early 20th centuries also often harpooned these whales. In recent years, Long-finned Pilot Whales have been hunted regularly only in the Faeroe Islands. There, the annual catch (by driving) increased from an average of about 1,500 in the early 1970s to nearly 2,500 in the 1980s, and declined to approximately 1,000 to 1,500 in the 1990s. In addition to this deliberate mortality, many tens or hundreds of Long-finned Pilot Whales are known to die each year from entanglement in longlines, trawls, and gillnets. With the exception of a poorly documented drive fishery in the Falkland Islands, Long-finned Pilot Whales do not seem to have been exploited on a significant scale in southern South America or indeed elsewhere in their Southern Hemisphere range. An estimated 200,000 pilot whales are present in summer south of the Antarctic Convergence.

TOP LEFT: *This Long-finned Pilot Whale in the Ligurian Sea near Italy has the long tapered flippers as well as the light ventral markings characteristic of its species.* **TOP RIGHT:** *Long-finned Pilot Whales are abundant in the North Atlantic Ocean, particularly near the edges of continental shelves. This light-colored calf off Nova Scotia, Canada, attempts to keep pace with its mother.* **RIGHT:** *Long-finned Pilot Whales have a light, anchor-shaped throat patch that is visible on this spyhopping animal in Italy's Ligurian Sea.*

Short-finned Pilot Whale

Globicephala macrorhynchus
Gray, 1846

This pilot whale's tropical and warm temperate distribution complements that of its cool-water relative, the Long-finned Pilot Whale. Apart from different distributions, the characteristics distinguishing the two species are subtle. Although the flippers of Long-finned Pilot Whales tend to be proportionally longer, and they generally have two or three more pairs of teeth, there is some overlap even in these characteristics, and analysis of skull features is necessary for a definitive diagnosis. In the 1980s, Japanese coastal whalers increased their catch of Short-finned Pilot Whales following a rise in the price of whale meat. Scientists who examined the caught whales discovered that there were two distinct forms of the species, one living off northern Japan and one off southern Japan. The two groups differ morphologically and genetically, and they associate with different oceanic current systems. It is possible that similar population differences exist in other parts of the species' range but have yet to be documented.

DESCRIPTION The Short-finned Pilot Whale is almost identical to the Long-finned Pilot Whale in form. It has a long but robust body and a bulbous to squarish melon that can protrude beyond the front of the mouth. There is a very short, barely noticeable beak. The mouthline slants up toward the eye. Set ahead of midbody, the falcate dorsal fin has a long base relative to its height and rises at a shallow angle. The flippers are shorter, on average, than those of the Long-finned, usually no more than one-sixth of the body length. There are normally seven to nine pairs of teeth in both the upper and lower jaws.

The color pattern is fairly simple. The body is basically black or dark brown, with a light gray throat patch that is joined by a narrow stripe to a wider light area in the uro-genital region. There is usually a light gray saddle on the back behind the dorsal fin, but it is variable in brightness and shape. There is often a light streak behind the eye, beginning at the eye and extending up and back toward the front of the dorsal fin, but it may be muted and barely noticeable. There may also be a light chevron-shaped mark behind the blowhole.

RANGE AND HABITAT The Short-finned Pilot Whale is widespread and abundant throughout the world's tropical and warm temperate marine waters. Its southern limits are about 25°S along

- BULBOUS OR SQUARISH MELON WITH NO DISCERNIBLE BEAK IN ADULTS
- BROAD-BASED, LOW-PROFILE, FALCATE DORSAL FIN POSITIONED FAR FORWARD ON BODY
- BLACK OR DARK BROWN BODY COLOR
- LARGE, OFTEN CONSPICUOUS LIGHT GRAY SADDLE BEHIND DORSAL FIN
- PANTROPICAL AND WARM TEMPERATE DISTRIBUTION

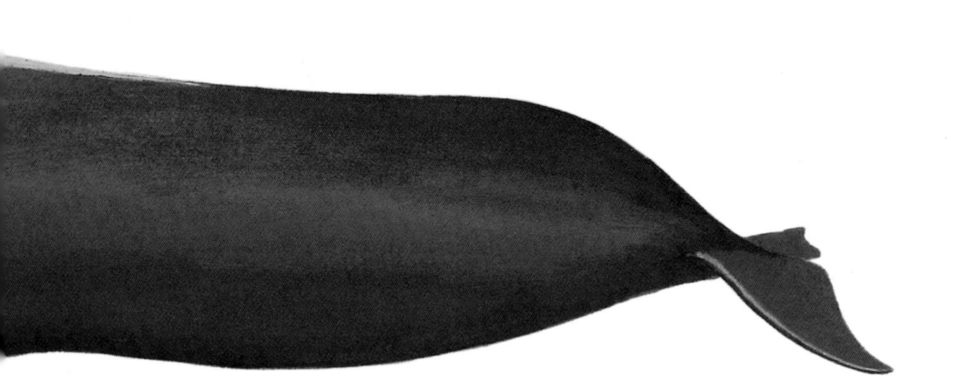

both the east and west coasts of South America, Cape Province (South Africa), western Australia, Tasmania, and North Island (New Zealand). The northern limits are New Jersey and the central Bay of Biscay in the Atlantic, and Hokkaido (Japan) and Vancouver Island (Canada) in the Pacific. It occurs all around the northern rim of the Indian Ocean. It is common around the Canary Islands, present near Madeira and the Azores, and common in the Caribbean, but absent from the Mediterranean. In Japan, southern-form whales occur mainly in Kuroshio Current waters with surface temperatures of more than 75°F (24°C) in summer and more than 68°F (20°C) in winter, while northern-form whales occur in cooler mixed waters of the cold Oyashio and the

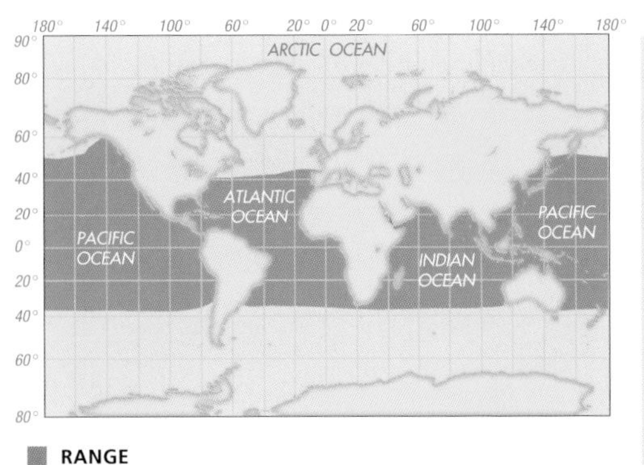

RANGE

SHORT-FINNED PILOT WHALE

FAMILY DELPHINIDAE

MEASUREMENTS AT BIRTH

LENGTH *Northern Japan:* 6′1″ (1.85 m)
Southern Japan: 4′7″ (1.4 m)

WEIGHT *Northern Japan:* 183–185 lb (83–84 kg)
Southern Japan: 82–84 lb (37–38 kg)

MAXIMUM MEASUREMENTS

LENGTH **MALE** *Northern Japan:* 24′ (7.2 m)
Southern Japan: 17′6″ (5.3 m)
FEMALE *Northern Japan:* 17′ (5.1 m)
Southern Japan: 13′6″ (4.1 m)

WEIGHT Unavailable

LIFE SPAN

MALE 46 years

FEMALE 63 years

warmer Kuroshio and Tsugaru Currents, with temperatures of less than 75°F (24°C) in summer and 46 to 70°F (8–21°C) in winter. Changes in distribution have been noted in the northern Gulf of Mexico and off southern California, where pilot whales were common for some years and then all but disappeared, apparently replaced by Risso's Dolphins. This variability in distribution is probably associated with warm current systems, which influence prey populations.

SIMILAR SPECIES There is some overlap in distribution of the Short-finned and Long-finned Pilot Whale species, at roughly 35°N to 40°N and 25°S in the western Atlantic; 20°N to 45°N in the eastern Atlantic; 14°S to 25°S in the eastern Pacific; off southern Africa at 34°S to 35°S; and off Tasmania. They cannot be reliably distinguished at sea. The False Killer Whale overlaps with the Short-finned Pilot Whale throughout its entire range and is of similar size, color, and shape. The False Killer Whale's head is narrower and more tapered, and its dorsal fin more slender, more upright, and farther back on the body; it lacks the light gray saddle and streak behind the eye, and has a broad hump on the front margin of its flipper; and its behavior is more dolphin-like (for example, it rides bow waves and often leaps above the surface).

BEHAVIOR Schools of Short-finned Pilot Whales generally contain 15 to 50 animals of mixed sex and age. There are usually several mature males in a school, in approximately the ratio of one mature male to eight mature females. Males generally leave their natal school, while females may remain in theirs for life. Males die at a younger age than females, but the cause of earlier death is unknown; they are not heavily scarred and show little evidence of battle wounds. When traveling

TOP LEFT: *The genus name of the pilot whales,* Globicephala, *refers to the globe-shaped, or bulbous, head. This close-up view of a Short-finned Pilot Whale confirms the aptness of the name.*

RIGHT: *An exceptional underwater view of a mother Short-finned Pilot Whale, apparently nursing her calf, off the Kona coast of Hawaii. The dorsal fin is situated well ahead of the midpoint of the back.*

OPPOSITE RIGHT: *This Short-finned Pilot Whale falls onto its side after what appears to have been a half-breach. This species has an almost squarish head in profile as well as sharply tapered flippers.*
RIGHT: *An adult Short-finned Pilot Whale accompanied by a calf off Tenerife in the Canary Islands. A substantial whale-watching industry in the Canaries depends on the consistent presence of pilot whales.*

and presumably searching for food, groups of pilot whales are often oriented in ranks more than a mile long and only a few animals deep. Such formations are sometimes described as "chorus lines." A trained pilot whale in the open ocean dived routinely to depths greater than 1,000 feet (300 m) and occasionally to 1,650 feet (500 m). Its longest submergence time was close to 15 minutes. Short-finned Pilot Whales frequently associate with other cetaceans, especially bottlenose dolphins.

REPRODUCTION Gestation lasts about 15 months, and lactation for at least two years. The calving interval averages from five to almost eight years, with older females giving birth less frequently than younger ones. Females continue to ovulate until almost 40 years of age and may still be lactating in their late 40s. Some may nurse their last calf for as long as 15 years. Off Japan, the northern- and southern-form pilot whales have different breeding seasons. The northern mates mainly from August to January, with a peak in October and November; most calves are born in midwinter (December–January). The southern population mates in May and calves in midsummer (July–August).

FOOD AND FORAGING The diet of Short-finned Pilot Whales, like that of their longer-finned relatives, is dominated by squid. In some areas, the occurrence of pilot whales coincides with the appearance of tuna, and it is assumed that they are both preying on squid. Nearshore observations of pilot whales, and indeed some mass strandings, have sometimes been associated with inshore spawning migrations of squid. Short-finned Pilot Whales also take octopus and fish, at least occasionally.

STATUS AND CONSERVATION There are about 150,000 Short-finned Pilot Whales in the eastern tropical Pacific and perhaps 1,000 in shelf waters off California. Populations off Japan have been estimated at about 4,000 to 5,000 northern-form and 25,000 southern-form whales. Pilot whales off Japan have been hunted for centuries. The direct kill in Japan, by both driving and harpooning, totaled close to 2,300 from 1985 to 1989, and ranged from about 200 to nearly 500 per year during the 1990s. Short-finned Pilot Whales have also been hunted for decades in the Lesser Antilles, especially in St. Vincent and St. Lucia, where the combined catch was in the hundreds annually until at least the mid-1970s. It is impossible to say how large the catches have been in recent years because no reliable statistics are kept. Short-finned Pilot Whales are also hunted in Indonesia and Sri Lanka, and are taken incidentally in fishing gear in many other areas. Overall, the species remains relatively abundant although some regional populations, especially those off Japan and in the eastern Caribbean, are probably depleted because of intensive exploitation.

Irrawaddy Dolphin

Orcaella brevirostris
(Owen *in* Gray, 1866)

This small Asian dolphin is probably in serious danger of disappearing over much of its range, but until recently it has received little attention from researchers and conservationists. Its distribution near shore and in rivers and estuaries means that it regularly comes in contact with people. In Laos and Cambodia, some villagers view dolphins as reincarnated humans and expect them to rescue drowning people and protect bathers from crocodiles. In Myanmar (formerly Burma), where there is a longstanding reverence for Irrawaddy Dolphins, some fishermen "call" dolphins by tapping on their canoes with wooden pins, then rely on the animals to herd schools of fish near enough to be caught in throw nets. This cooperative fishing phenomenon stands in stark contrast to the situation in many other areas, where dolphins that become entangled in gillnets are either killed outright or sold to oceanariums. The Irrawaddy Dolphin superficially resembles the Beluga, with similar light coloring, rounded head, and lack of a beak. In fact, during the 1970s and 1980s several researchers proposed classifying the species as a monodontid. However, there is now a strong consensus, based on morphology and genetics, that the Irrawaddy belongs with the Delphinidae family. Its common name refers to the large Irrawaddy River (also spelled Ayeyarwady) that bisects Myanmar.

DESCRIPTION The Irrawaddy Dolphin is small and slender, with a small, triangular dorsal fin positioned at the center of the back. The fin is variably shaped: the peak bluntly rounded or pointed and the rear margin straight or slightly concave. The flippers are disproportionately large and broad, with rounded tips. The head shape is smoothly rounded, with a fairly bluff melon and no beak. The mouthline is straight and angles up toward the eye. An unusual feature of the Irrawaddy is that only its first two cervical vertebrae are fused, so that, like the Beluga, its neck is flexible and often set off by a noticeable crease. There are 16 to 20 pairs of teeth in the upper jaw and 15 to 19 pairs in the lower jaw.

The body often appears uniformly light or dark gray with only subtle patterning, including a lighter ventral surface. However, Irrawaddy Dolphins off northeastern Australia have a muted, three-tone pattern: a broad, gray or bluish-gray dorsal cape, lighter or brownish-gray sides, and a white ventral field.

- HEAD ROUNDED IN PROFILE
- SMALL TRIANGULAR DORSAL FIN
- EVEN GRAY COLORATION
- ENDEMIC TO SOUTHEASTERN ASIA, INDONESIA, NORTHERN AUSTRALIA

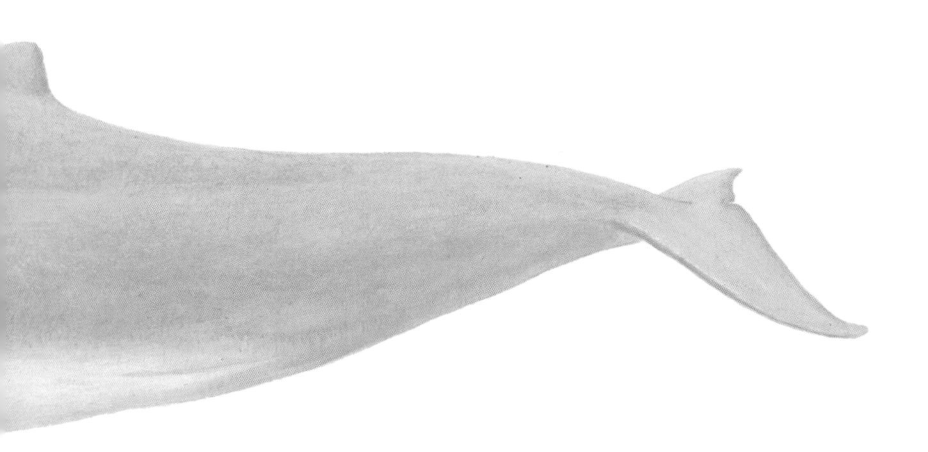

RANGE AND HABITAT The Irrawaddy Dolphin occupies certain rivers and shallow coastal marine waters in southeastern Asia, Indonesia, and northern Australia. Along the mainland of southern Asia, it occurs from the central west coast of the Bay of Bengal east to the southern South China Sea; in the Irrawaddy River as far upstream as Bhamo, some 800 miles (1,300 km) from the Andaman Sea; in the Sundarbans (the area around the mouths of the Ganges River); and in at least two partially isolated brackish lagoons: Chilka Lake, in India, and Songkhla Lake, in Thailand. It is present along much of the Malay Peninsula; the Indonesian islands of Sumatra, Java, and Borneo; New Guinea; and Australia's Northern Territory and Queensland. The animals in Australia and Papua New Guinea differ morphologically from those in Southeast Asia, and the differences may prove great enough to warrant recognition of two separate species. A freshwater population inhabits the Mahakam River and its associated lakes in Kalimantan, Borneo. Another small group of at least a few dozen animals is present in the shallow upper

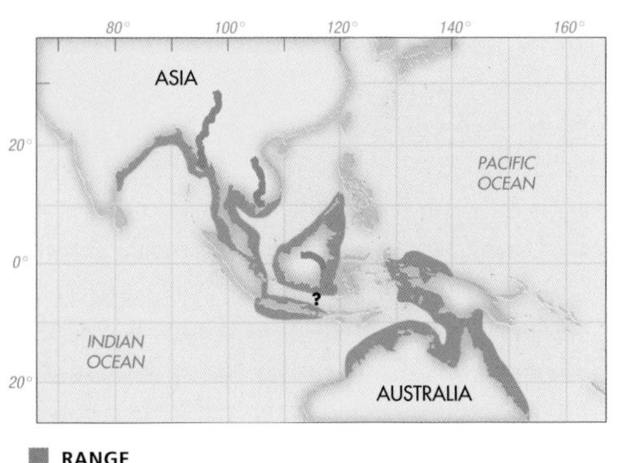

IRRAWADDY DOLPHIN
FAMILY DELPHINIDAE

MEASUREMENTS AT BIRTH

LENGTH About 35–40″ (90–100 cm)
WEIGHT 22–26 lb (10–12 kg)

MAXIMUM MEASUREMENTS

LENGTH **MALE** 8′ (2.75 m)
FEMALE 7′7″ (2.32 m)
WEIGHT **MALE** More than 290 lb (130 kg)
FEMALE Unavailable

LIFE SPAN
About 30 years

reaches of Malampaya Sound, Palawan, Philippines. The Irrawaddy was formerly common in the mouth of the Mekong River and upstream to southern Laos, just below the Khone waterfalls, but its abundance in these areas is thought to have declined considerably.

The marine distribution appears to be discontinuous, with concentrations of animals in brackish, turbid deltas and mangrove creeks. In their marine environment, Irrawaddy Dolphins are not generally found in water deeper than 60 feet (18 m) or farther than about 3 miles (5 km) from shore. In Chilka Lake, they have been seen almost stranding on sandbanks in water less than 3 feet (1 m) deep. Within river systems, Irrawaddy Dolphins tend to occupy deep pools, to 230 or even 330 feet (70–100 m). Their movements within the river systems are influenced by fish migrations, which in turn are likely associated with seasonal changes in water level.

SIMILAR SPECIES The Finless Porpoise occurs throughout the Irrawaddy's marine distribution, with the exception of the eastern islands of Indonesia, New Guinea, and Australia, and is perhaps the species most likely to cause confusion. It is considerably smaller and lacks a dorsal fin. Bottlenose dolphins (one or both species) and the Indo-Pacific Hump-backed Dolphin also occur throughout the Irrawaddy's marine distribution. These other species are generally more active and have prominent beaks and differently shaped dorsal fins that are larger and more falcate on the bottlenose dolphins and with a longer, thicker base on the hump-backed dolphin. The pinkish coloration of some hump-backed dolphins also distinguishes them.

BEHAVIOR Most groups of Irrawaddy Dolphins are no larger than about 7 to 10 individuals, and pairs and trios are commonly encountered. These animals tend to swim slowly and deliberately, except when actively chasing prey. Some observers have described them as "sluggish" or "unobtrusive." When surfacing, they often expose little of their body; however, their exhalations can be sonorous and their blow is at least faintly visible in proper lighting. Irrawaddy Dolphins have an unusual habit of "spitting" water in an arc above the surface; the function of this behavior is not known, but it has been reported to occur only in fresh water. At the surface, Irrawaddy Dolphins regularly roll onto their side, waving a flipper above the water, and they occasionally breach and spyhop. A dive typically lasts two minutes but can be as long as six minutes. There is no evidence that Irrawaddy Dolphins regularly associate with other cetacean species.

OPPOSITE: *Irrawaddy Dolphins have a small dorsal fin that can vary from triangular to slightly falcate. They occur alone or in small groups usually containing fewer than 10 individuals.*
LEFT: *The head of the Irrawaddy Dolphin is smoothly rounded. In this exceptional close-up view, the dark eye and corner of the mouth can be seen clearly.*

REPRODUCTION Little reliable information exists on reproduction in Irrawaddy Dolphins. Based on observations of two captive births, the gestation period is estimated to be 14 months. A captive-born Irrawaddy began eating solid food at six months old and was weaned by two years.

FOOD AND FORAGING Irrawaddy Dolphins eat a wide variety of fish and also some crustaceans (shrimp) and cephalopods (squid, cuttlefish, and octopus). In marine waters off Australia, they take small schooling fish (such as herring, anchovies, halfbeaks, and ponyfish) and solitary bottom fish (cardinalfish, bream, grunt, halibut, and lizardfish). In rivers, they are known to feed on various species of catfish, sometimes discarding the heads and upper bodies.

STATUS AND CONSERVATION Although its total range is extensive, the Irrawaddy Dolphin has a fragmented distribution, and its abundance appears to be low in all areas where it has been studied. Its occupation of nearshore and freshwater areas makes it exceptionally vulnerable to the harmful effects of many human activities. In particular, mortality caused by entanglement in gillnets and by explosives used for fishing may have eliminated or drastically depleted local populations in some coastal areas, such as in Thailand and Vietnam. The freshwater populations in the Mahakam and Mekong River basins and the tiny population in Malampaya Sound are in immediate danger of extinction. In the Mekong, the planned construction of hydroelectric dams threatens to block the movements of these dolphins and their migratory prey. Continued removals for oceanariums add to the problems of fishery bycatch and habitat degradation. In Australian waters, some efforts have been made to monitor the dolphins' populations and protect them from deliberate killing; however, they remain threatened by barrier nets set to protect swimmers from sharks and by gillnet fisheries for barramundi and salmon.

Porpoises

The term "porpoise," derived from the Latin *porcus* for "pig" and *piscus* for "fish," has often been used colloquially to refer to various smaller-size cetaceans, including dolphins and small whales. However, the true porpoises (family Phocoenidae) are in fact well differentiated from other cetaceans. The phocoenids are small cetaceans with relatively small flippers and no prominent beak. Their teeth are laterally compressed, with spatulate crowns that together form a cutting edge. Most porpoises inhabit shallow, nearshore waters.

Although some authorities have divided the Phocoenidae into two subfamilies, this issue is unresolved at present. There are three well-established genera in the family. The most diverse, *Phocoena,* contains four species, while the other two, *Neophocaena* and *Phocoenoides,* each contain one. Porpoises seem especially prone to entanglement in fishing nets, and their numbers have declined considerably in some areas.

GENUS *NEOPHOCAENA* The Finless Porpoise is thought to be the oldest and least-specialized living phocoenid. As its name indicates, the Finless Porpoise has no dorsal fin but rather a groove or low ridge along the center of the back. The species is evenly colored, from light gray to almost black, depending on the area. It is found in nearshore marine waters of the Indian Ocean and western Pacific, with one freshwater population in the Yangtze River of China.

Dall's Porpoise

GENUS *PHOCOENA* Of the four species of *Phocoena*, the Harbor Porpoise is by far the best known. Although it often appears dull gray when observed at sea, the overall color pattern is complex and individually distinctive. The ventral white and dorsal gray merge on the sides of the front half of the body to create a light gray zone, and some individuals have a dark eye patch and flipper stripe. The Harbor Porpoise's small, triangular dorsal fin makes it relatively easy to identify as it rolls quickly at the surface. Harbor Porpoises occur in temperate to subarctic waters along the rims of the North Atlantic and North Pacific, and in the Black Sea, but generally not the Mediterranean.

The Vaquita has probably the most limited range of any living cetacean, confined to the upper quarter of the Gulf of California. It is most closely related to the Southern Hemisphere species Burmeister's Porpoise (found 3,000 miles/5,000 km away), despite the fact that the Harbor Porpoise is actually the nearest living phocoenid (about 1,500 miles/2,500 km away). Vaquitas have a much higher dorsal fin than other porpoises, and both the dorsal fin and flippers are pointed at the tips. The facial pigmentation pattern is bold, with large black eye and lip patches and prominent flipper stripes. With its limited range and frequent entanglements in fishing gear, the Vaquita is at risk of extinction.

The two remaining members of the genus occur only in temperate to subantarctic waters of the Southern Hemisphere. Burmeister's is coastal,

Harbor Porpoises

like its northern relatives, and its distribution appears to be continuous along the southeastern and southwestern shores of South America. The color pattern is muted, and when seen at sea, this porpoise appears uniformly dark. Its dorsal fin is distinctive—set far back on the body and rising at a shallow angle, with a blunt peak. The Spectacled Porpoise is black and white, with a sharp line of demarcation separating the dark dorsal and white ventral surfaces. The white ring around the black eye patch explains the species' name. It occurs in cold waters around southern South America and in offshore portions of the Southern Ocean.

Porpoises are exceptionally vulnerable to entanglement in gill nets. As a result, the Vaquita (left), which occurs only in the northern Gulf of California, Mexico, has been brought close to extinction. The status of its closest living relative, Burmeister's Porpoise (right) of coastal southern South America, is less certain, but it has a much more extensive distribution and is probably much more abundant.

GENUS *PHOCOENOIDES* The genus *Phocoenoides* includes a single species, Dall's Porpoise, an oddly shaped cetacean with a tiny head and deep body. The dorsal fin is triangular, wide-based, and sometimes recurved at the tip. These porpoises display a striking combination of black-and-white coloration: The dorsal fin and flukes are "trimmed" or "frosted" with white, and all animals have prominent white patches on the sides. There are two well-defined color morphs, one with a white patch from the caudal peduncle to below the front edge of the dorsal fin, the other with a patch extending the length of the body from the caudal peduncle to just forward of the flipper. Dall's Porpoises live only in the North Pacific Ocean. Unlike most porpoises, they inhabit deep water and therefore tend to occur offshore.

Harbor Porpoise *page 460*

Dall's Porpoise *page 470*

Finless Porpoise

Neophocaena phocaenoides
(G. Cuvier, 1829)

This small porpoise is endemic to eastern and southern Asia, occurring from the Persian Gulf in the west to Japan in the east. The species was originally described from a skull that had mistakenly been reported as coming from the southern tip of Africa. The Finless Porpoise is known by a variety of local names, most notably *sunameri* in Japan and *jiangzhu* (meaning "river pig") along China's Yangtze River, where a unique freshwater population shares the Yangtze River Dolphin's habitat. Three subspecies have generally been recognized, but a recent analysis suggests that there may be two species—one tropical and one temperate—and that both of them have several geographic forms. Throughout their range, Finless Porpoises inhabit shallow, nearshore waters, ensuring close contact with people and making them exceptionally vulnerable to conflict with fisheries and the effects of pollution.

DESCRIPTION The slender gray body of this small porpoise gives the appearance of compact simplicity. The head is rounded, with no beak. The flippers are of moderate length and gently taper toward the tips. As the name indicates, this porpoise lacks a dorsal fin; however, all members of the species have a ridge along the middle of the back, variably covered with lines of small denticles or tubercles. This ridge varies geographically in both width and height. Animals from Japan, the East China Sea, and the Yangtze River have a narrow ridge (less than ½ in/1.3 cm wide), while those from the South China Sea and westward to the Persian Gulf have a much wider ridge (1¼–4¾ in/3–12 cm). The ridge can be as much as 2¼ inches (5.5 cm) high but is very low (usually less than ¾ in/1.5 cm) in porpoises from the Yangtze River and the South China Sea. The teeth range in number from 12 to 22 pairs in the upper jaw and 14 to 23 pairs in the lower.

Color patterns appear to differ regionally. Finless Porpoises in Japan and northern China are dark gray at birth and lighten considerably as they age, sometimes becoming almost white as adults. Those in the South China Sea and elsewhere are light gray to cream at birth and darken to charcoal gray as adults. The belly is generally somewhat lighter than the back, and there can be pale areas around the mouth and throat.

RANGE AND HABITAT Finless Porpoises occur from the Persian Gulf in the west to Japan in the

- SMALL BODY SIZE
- NO DORSAL FIN
- UNIFORM GRAY COLORATION
- SMALL GROUP SIZE AND LOW SURFACING PROFILE
- USUALLY SEEN NEAR SHORE IN SHALLOW WATER
- ENDEMIC TO EASTERN AND SOUTHERN ASIA

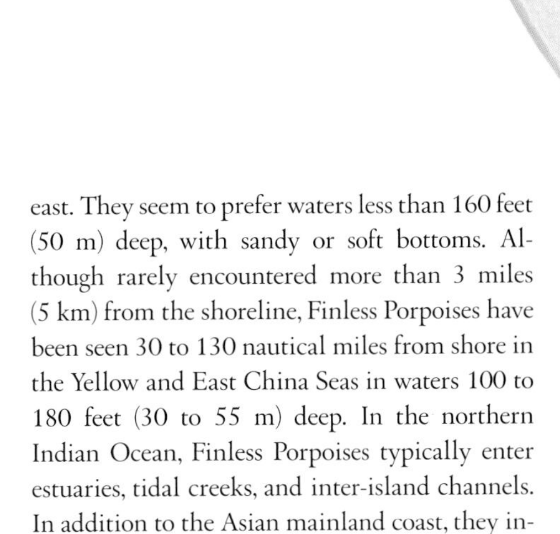

east. They seem to prefer waters less than 160 feet (50 m) deep, with sandy or soft bottoms. Although rarely encountered more than 3 miles (5 km) from the shoreline, Finless Porpoises have been seen 30 to 130 nautical miles from shore in the Yellow and East China Seas in waters 100 to 180 feet (30 to 55 m) deep. In the northern Indian Ocean, Finless Porpoises typically enter estuaries, tidal creeks, and inter-island channels. In addition to the Asian mainland coast, they inhabit coastal waters of several Indonesian islands (for example, Java and Borneo), Hong Kong, and Taiwan. They occur year-round in China in the middle and lower reaches of the Yangtze River, and in the large freshwater lakes Poyang and Dongting. They probably occur year-round in all tropical portions of their range. The northern limit is at roughly 38°N on the Pacific coast of Japan. Some populations undertake at least short-range seasonal movements, such as migrating through Naruto Strait between the Inland Sea of Japan and the Pacific Ocean.

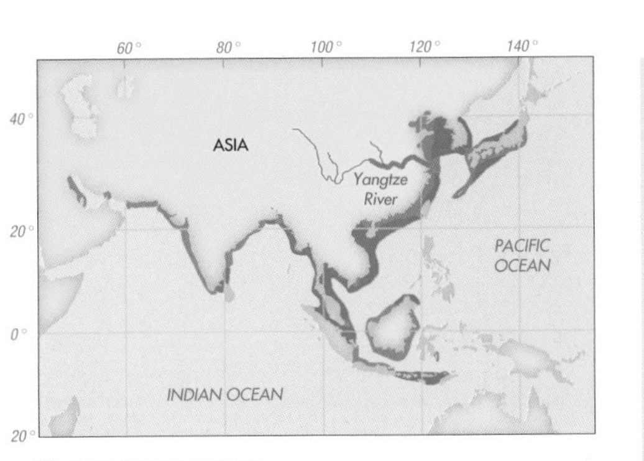

FINLESS PORPOISE
FAMILY PHOCOENIDAE

MEASUREMENTS AT BIRTH

LENGTH *Yangtze River:* 27–29″ (68–74 cm)
Japan: 30–31″ (75–80 cm)

WEIGHT About 13 lb (6 kg)

MAXIMUM MEASUREMENTS

LENGTH **MALE** 6′7″ (2 m)
FEMALE *Indian Ocean:* 5′1″ (1.55 m)
Yellow Sea/Bohai area: 6′7″ (2 m)

WEIGHT More than 159 lb (72 kg)

LIFE SPAN

At least 33 years

SIMILAR SPECIES Finless Porpoises occur in many of the same areas as Irrawaddy Dolphins and Dugongs, both of which have gray bodies and rounded heads. Irrawaddy Dolphins overlap only in Southeast Asia and the Bay of Bengal. The presence of a dorsal fin and the somewhat larger size of Irrawaddy Dolphins should make them readily distinguishable from Finless Porpoises. The Yangtze River Dolphin overlaps in the Yangtze River; it is larger and lighter and has a long narrow beak. Dugongs occur as far west as the southeastern coast of Africa and as far north as the southern South China Sea. Although adult Dugongs grow considerably larger than porpoises, young Dugongs could easily be confused with Finless Porpoises. However, Dugongs have two nostrils at the very front of the head, and as bottom grazers their mouth is oriented downward.

BEHAVIOR These animals are not highly gregarious, typically occurring in groups of only two to five individuals. Aggregations of several tens are sometimes seen, but these probably represent the convergence of several smaller groups on a rich food source rather than social affiliation. Finless Porpoises do not bow ride and are usually difficult to approach. Occasionally they follow fast-moving vessels and ride in the stern wake. They generally do not leap from the water, but rather simply roll inconspicuously at the surface as they breathe. Average dive times are only about 20 seconds, and they rarely stay submerged for longer than a few minutes. Sounds produced by Finless Porpoises are limited to high-frequency clicks, presumably for echolocation. Remains of Finless Porpoises have been found in the stomachs of large sharks and Killer Whales.

REPRODUCTION Nothing is known about their mating behavior, but Finless Porpoises are seasonal breeders. Populations in tropical areas, in the South China Sea, and off western Kyushu,

OPPOSITE LEFT: *An exceptional view of Finless Porpoises surfacing in Hong Kong waters, with sun glittering off their round heads. These small porpoises can be cryptic and difficult to detect or follow at sea.* **OPPOSITE RIGHT:** *One animal in this small group of Finless Porpoises off Hong Kong provides a good view of the flukes, which are fairly wide relative to the narrow caudal peduncle and have a concave rear margin.* **LEFT:** *The light coloration of this Finless Porpoise in a Japanese oceanarium contrasts sharply with the darker adult coloration of Finless Porpoises in areas such as the South China Sea.*

Japan, are reproductively active in autumn and winter. Those in more temperate regions, including the Inland Sea and the Pacific coast of Japan, produce most of their calves in spring and early summer. Gestation lasts about 11 months. Young porpoises start taking solid foods at about six months and are probably weaned at about seven months. Most females probably give birth every two years.

FOOD AND FORAGING The Finless Porpoise has been described as an opportunistic feeder. Its varied diet includes fish, shrimp, squid, cuttlefish, and octopus. Porpoises in captivity eat the equivalent of about 5 to 6 percent of their body weight per day.

STATUS AND CONSERVATION Finless Porpoises have been severely reduced, perhaps even eliminated, from some fairly large segments of their historic range. The principal cause is incidental capture in fishing gear, especially gillnets. Various other factors suspected of contributing to declines include pollution, habitat degradation, and reduced prey availability due to overfishing. Dangerously high levels of some contaminants have been found in the bodies of Finless Porpoises from Japan and Hong Kong. The best evidence for a declining trend comes from Japan, where two sets of surveys were conducted in the middle and eastern Inland Sea—one in the late 1970s and the other in 1999 and 2000. The results indicate that in some areas the abundance of Finless Porpoises has plummeted. A rapid and continuing decline in the Yangtze River has also been reported by Chinese researchers. Unfortunately, even rudimentary data on abundance and trends are lacking for most of the species' historic range. Two areas in Japan were estimated to have a total of about 3,000 Finless Porpoises in the mid-1990s, and there are at least 250 in Hong Kong and surrounding waters.

Harbor Porpoise

Phocoena phocoena
(Linnaeus, 1758)

Interest in these small, inconspicuous inhabitants of North Atlantic and North Pacific coastlines was limited to a few scientific specialists until as recently as the late 1970s. By the 1990s, after a period of intense focus on the incidental mortality of dolphins and porpoises in gillnets, Harbor Porpoises had become high-profile subjects of concern, prompting international agreements, extensive (and costly) research efforts, and controversial changes in fishing regulations. In Europe, where they are called Common Porpoises, these animals were once the targets of large directed hunts. In Lille Belt, a narrow strait connecting the Baltic and North Seas, 17th- and 18th-century fishermen caught as many as 3,000 porpoises in a single winter season. They intercepted the migrating animals with a row of boats and, beating the water with sticks, drove them far into fjords. There the porpoises were trapped in nets, hauled onto the beach, and killed with knives. Today the species persists in much of its historic range, but the Baltic population is only a tiny remnant of what it once was. Fishermen in the Canadian Maritimes and New England call Harbor Porpoises "puffing pigs" or just "puffers."

DESCRIPTION The Harbor Porpoise has a robust body and a short, poorly demarcated beak. It reaches maximum girth just ahead of the dorsal fin, where girth equals about two-thirds of body length in adults. The medium-size dorsal fin is triangular or slightly falcate and is set at midbody. Several rows of small tubercles line the fin's leading edge. There are usually 21 to 29 pairs of teeth in the upper jaw and 20 to 29 pairs in the lower.

The color pattern is subtle but complex. A simple dark gray cape is overlaid on a much lighter gray dorsal field, with variable dark gray flecking in the light gray area. The throat and belly are white, and there may be gray streaking on the throat. There are often a dark chin patch, a dark eye ring, and a dark stripe (sometimes several) from the corner of the mouth to the flipper. At sea, Harbor Porpoises are usually seen only briefly and partially as they roll at the surface to breathe, appearing evenly charcoal gray with a wash of light gray or white on the sides above the flippers.

RANGE AND HABITAT As their common name implies, Harbor Porpoises are coastal, often found in fjords, bays, estuaries, and harbors. They are limited to northern temperate and subarctic

- GRAY BACK, LIGHTER SIDES AND BELLY
- SMALL BODY SIZE
- TRIANGULAR DORSAL FIN AT CENTER OF BACK
- DARK MOUTH-TO-FLIPPER STRIPE
- RAPID ROLL WITH LITTLE OR NO SPLASH WHEN SURFACING
- LIMITED TO COOL TEMPERATE AND SUBARCTIC NORTH PACIFIC AND NORTH ATLANTIC, MAINLY COASTAL

waters. In the western Atlantic they are found from Cape Hatteras north to West Greenland, including the Gulf of St. Lawrence but not Hudson Bay, and in the eastern Atlantic around Iceland and the Faeroe Islands and from western Africa north to the Barents Sea, including the North, Baltic, and Black Seas but not the Mediterranean. In the Pacific they occur from Monterey Bay and central Japan (around 34°N) north to the Chukchi Sea, including the Aleutian chain, Sea of Okhotsk, and northern Sea of Japan. Their distribution is generally discontinuous, resulting in numerous geographical populations. North Pacific and North Atlantic populations are entirely separate, as is the population in the Black Sea and Sea of Azov; a few animals occur in the northern Aegean Sea as well.

SIMILAR SPECIES Harbor Porpoises are considerably smaller than most other cetaceans in the Northern Hemisphere, which makes them relatively easy to identify at sea. Larger bodies and taller, usually more falcate dorsal fins distinguish coastal dolphins from Harbor Porpoises. Dall's

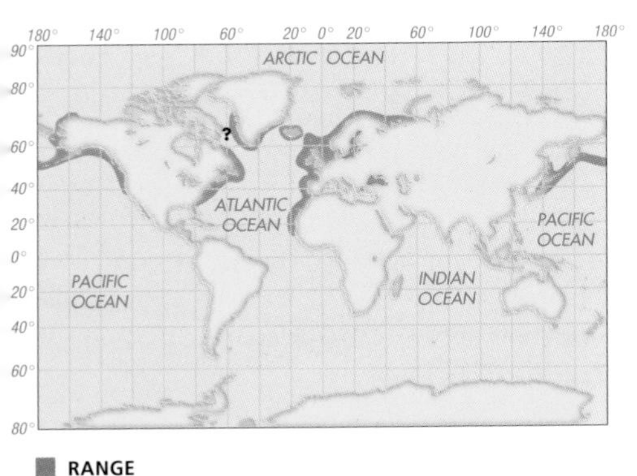

RANGE
? POSSIBLE RANGE

HARBOR PORPOISE
FAMILY PHOCOENIDAE

MEASUREMENTS AT BIRTH
LENGTH 28–30″ (70–75 cm)
WEIGHT 11–13 lb (5–6 kg)

MAXIMUM MEASUREMENTS
LENGTH **MALE** 5′2″ (1.57 m)
FEMALE 5′6″ (1.68 m)
WEIGHT **MALE** 134 lb (61 kg)
FEMALE 168 lb (76 kg)

LIFE SPAN
24 years

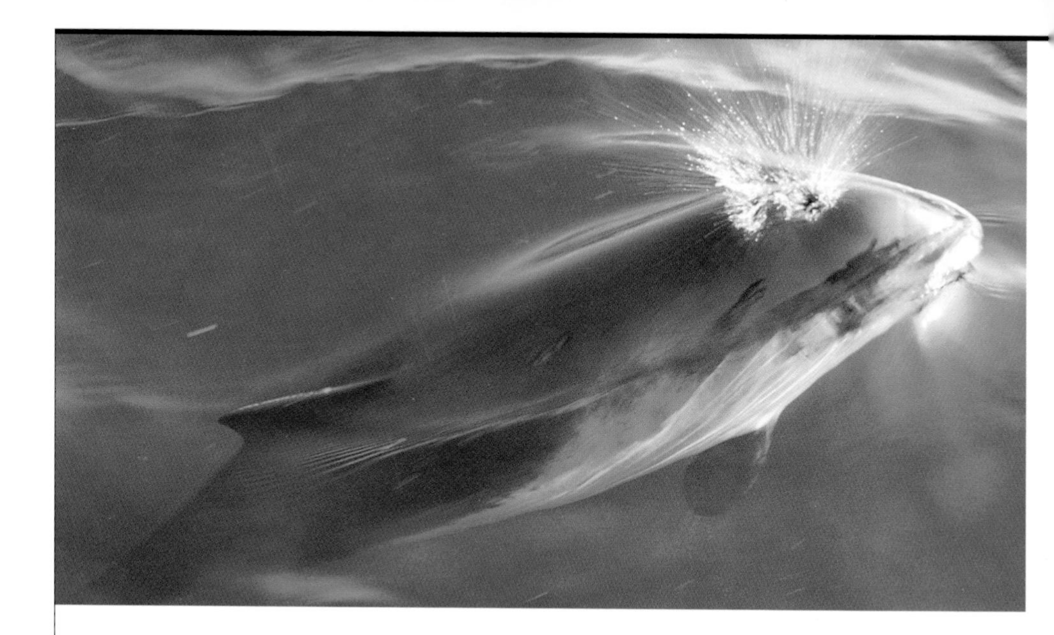

Porpoise is perhaps the species most likely to cause confusion, but only in the North Pacific where there is substantial range overlap. Dall's Porpoises generally cause a "rooster tail" of spray as they break the surface to breathe, but Harbor Porpoises sometimes create a similar splash upon surfacing. Also, both species "slow roll" at the surface at times. The dorsal fin of Dall's Porpoise, although similar in size, shape, and placement to that of the Harbor Porpoise, is black and white and thus easily distinguished.

BEHAVIOR Harbor Porpoises are generally perceived as solitary and nonsocial, and are usually seen alone or in small groups of two to five individuals. Group sizes tend to increase toward the end of summer, but little is known about social affiliations and interactions. Individuals are highly mobile, using home ranges of thousands of square miles and often traveling many miles per day. These porpoises tend not to exhibit curiosity toward vessels and are hard to approach and follow. They may arc clear of the water ("porpoise") occasionally but otherwise are not inclined to aerial activity. At sea, they are usually seen only briefly and partially as they roll at the surface to breathe. They are subject to predation by large sharks and Killer Whales.

ABOVE: *Harbor Porpoises are rarely seen at such close range in the wild. However, in the lower Bay of Fundy, eastern Canada, the animals are extremely common in summer and fall.*
RIGHT: *This Harbor Porpoise is in captivity in Europe. In recent years, a number of porpoises removed from fishing gear have been brought into facilities specially designed for experiments to find ways to reduce their accidental capture in nets.*

LEFT: *Three Harbor Porpoises surfacing together in Frederick Sound, southeastern Alaska.*
BELOW: *This exceptional underwater photograph of a Harbor Porpoise in Sognefjorden, Norway, shows the species' somewhat conical head shape, three-part color pattern, and triangular dorsal fin.*

REPRODUCTION Although short-lived by cetacean standards, Harbor Porpoises can be highly productive. Females, after reaching sexual maturity at three to four years of age, may become pregnant annually for at least several years in a row. This means, however, that they are simultaneously lactating and pregnant, and thus under nutritional stress, for much of their adult lives. In all areas where the Harbor Porpoise has been studied, its reproduction is strongly seasonal. Ovulation and conception occur within a brief time span during the late spring or early summer. Gestation lasts for about 10 to 11 months. Most calves in the Gulf of Maine and Bay of Fundy are born in May, and those in western European waters from June to July. Calves may be nursed for 8 to 12 months.

FOOD AND FORAGING Harbor Porpoises usually feed individually. Large aggregations, seen occasionally, are likely adventitious rather than indications of cooperative foraging. Much of the prey of Harbor Porpoises is found near the seafloor, but they also forage in the water column. Schooling fish less than 16 inches (40 cm) long, such as herring, capelin, sprat, and silver hake, form the bulk of the diet. Maine fishermen call these porpoises "herring hogs." Harbor Porpoises also eat cephalopods (squid and octopus). Calves often ingest small crustaceans during the early phases of weaning.

STATUS AND CONSERVATION In total, there are hundreds of thousands of Harbor Porpoises, but many geographical populations are substantially reduced from historical levels. The Black and Baltic Sea populations are among the most threatened. Incidental mortality in fisheries, especially bottom-set gillnets, is a threat throughout the species' range. It has been demonstrated that the porpoise echolocation system is capable of detecting the net fibers. Therefore, they must either have the system "switched off" when they blunder into the nets or fail to recognize them as dangerous. An active means of alerting porpoises to the presence of a net and keeping them away from it is needed. Over the past decade, considerable effort has been devoted to developing and testing acoustic deterrents ("pingers") for use with gillnets. Initial trials show promise, but some scientists are skeptical of the long-term efficacy and worry about the side effects of introducing more noise into the underwater environment.

Vaquita

Phocoena sinus
Norris and McFarland, 1958

- TALL, TRIANGULAR, "DOLPHIN-LIKE" DORSAL FIN
- LITTLE OR NO BEAK
- EVENLY DARK GRAY ON BACK AND LIGHTER ON SIDES, WITH NO SHARP DEMARCATION BETWEEN COLORS
- BRIEF, CRYPTIC SURFACING BEHAVIOR
- VERY LIMITED DISTRIBUTION, CONFINED TO UPPER QUARTER OF GULF OF CALIFORNIA

This little porpoise is considered the most critically endangered of all the marine cetaceans, with the possible exception of the Northern Hemisphere right whales. Found only in the upper Gulf of California, Mexico, it also has the most limited distribution of any marine cetacean. Indeed, the Latin name *sinus*, meaning "pocket," "recess," or "bay," refers to its highly localized range. The Vaquita ("little cow" in Spanish) was unrecognized as a species until 1958, when it was described on the basis of three beach-cast skulls. Scientists did not obtain fresh specimens showing the pigmentation pattern and general external morphology or begin conducting successful sighting surveys until 1985. Even though the Vaquita is geographically nearer the Harbor Porpoise (found off central California), it is more closely related to Burmeister's Porpoise (occurring from Peru and southward). Presumably, it evolved from an ancestral population that moved northward into the Gulf of California during the Pleistocene epoch, perhaps 1 million years ago.

DESCRIPTION The Vaquita has a typical porpoise body shape, with its girth up to 68 percent of its body length. It has little or no beak, but there is a very slight protrusion of the upper jaw at the base of the melon. The dorsal fin is unusually large (relative to that of other porpoises) and very upright, with a straight vertical or slightly falcate rear margin. There are 16 to 22 pairs of teeth in the upper jaw and 17 to 20 in the lower.

The color pattern is subtle but relatively complex. The dark gray cape extends about halfway down onto the sides, where it meets the lighter gray lateral field. The throat and belly are white, streaked with gray. The demarcation between the color fields can appear diffuse in field conditions but is clear on close inspection. There is a prominent black eye patch and a variably well-defined dark stripe from the front of the flipper to the middle of the lower jaw. Black covers the lips and a patch on the chin.

RANGE AND HABITAT The Vaquita is endemic to the northwestern corner of the Gulf of California (often called the Sea of Cortez). Most confirmed observations have been in water less than 130 feet (40 m) deep and within about 25 miles (40 km) of shore.

SIMILAR SPECIES Only two other odontocetes occur regularly in the upper Gulf of California: the Common Bottlenose Dolphin and the Long-beaked Common Dolphin. Both are considerably larger, more gregarious, and more active at the surface than the Vaquita. They also have well-defined beaks and usually show more of the body when surfacing.

BEHAVIOR, REPRODUCTION, AND FORAGING The Vaquita occurs singly, in pairs, or in small groups of up to about seven. It does not approach vessels and is generally undemonstrative at the surface, emerging only briefly and cryptically. Because of this behavior, Vaquitas can be effectively surveyed only when conditions are ideal, with calm or light winds and good lighting. This species is subject to predation by large sharks and possibly Killer Whales, at least occasionally. Little is known about reproduction in this porpoise, but it is thought to be broadly similar to the Harbor Porpoise in most respects. This would mean that gestation lasts about 11 months, and sexual maturity occurs at three to six years of age. Most births occur in late winter and early spring (February–April), with a peak in late March and early April. The Vaquita preys mainly on small fish and squid that live on or near the bottom in relatively shallow water. Studies of stomach contents have found a variety of species represented, suggesting that Vaquitas are fairly unselective about what they eat.

STATUS AND CONSERVATION There is no history of direct exploitation of the Vaquita, nor is there any reliable indication of how abundant the species might have been in the past. However, its fate has been inadvertently linked to that of another upper Gulf of California endemic species, a large sea bass called the totoaba. A commercial fishery for totoabas began in the 1920s and peaked in 1942. The large-mesh gillnets used to catch totoabas proved deadly to Vaquitas, and the bycatch, although undocumented, must have been extremely high. The commercial totoaba fishery was closed in the 1970s, but illegal and "experimental" fishing have continued, as has gillnetting for sharks and rays. Vaquitas also die in mackerel and shrimp trawls. The current best estimate of Vaquita abundance is 500 to 600. Although Mexico has established a biosphere reserve in the upper Gulf explicitly to protect Vaquitas, totoabas, and other threatened wildlife, the problem fisheries are still not adequately managed. Consequently, the Vaquita remains critically endangered.

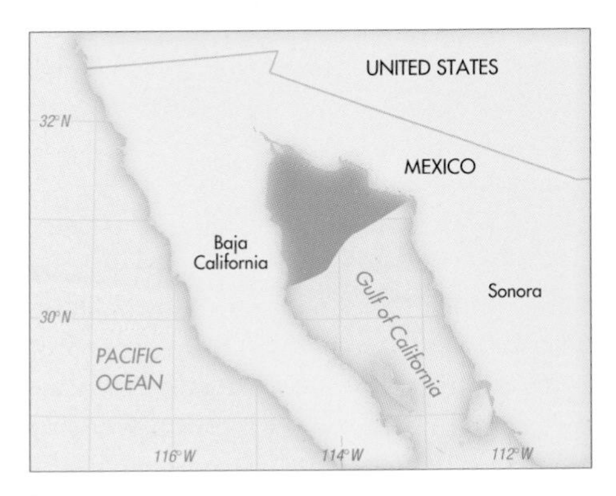

RANGE

VAQUITA
FAMILY PHOCOENIDAE

MEASUREMENTS AT BIRTH

LENGTH 28–31″ (70–78 cm)
WEIGHT More than 17 lb (7.5 kg)

MAXIMUM MEASUREMENTS

LENGTH **MALE** 4′9″ (1.45 m)
FEMALE 4′11″ (1.5 m)
WEIGHT About 99–110 lb (45–50 kg)

LIFE SPAN

At least 21 years

Burmeister's Porpoise

Phocoena spinipinnis
Burmeister, 1865

- SMALL BODY SIZE
- LACK OF DISTINCT BEAK
- LOW-PROFILE DORSAL FIN WITH CONVEX REAR MARGIN, SET FAR BACK ON BODY
- ALL-DARK COLORATION
- DISTRIBUTION LIMITED TO COASTAL WATERS OF SOUTHERN SOUTH AMERICA

This endemic South American porpoise is surprisingly poorly known, considering its extensive distribution and considerable abundance in coastal waters. Most of what is known about it comes from studies of stranded carcasses or dead animals killed incidentally in fishing gear. Its specific name, derived from *spina,* for "thorn," and *pinna,* for "fin," refers to rows of small blunt tubercles, or projections, on the front margin of its dorsal fin. Most of this porpoise's range along the southern cone of South America falls within Spanish-speaking jurisdictions. It is variously known by the Spanish names *marsopa espinosa* ("thorny porpoise"), *chancho marino* ("sea pig"), and the more generic *tonina* or *tonino* (applied to various small cetaceans). In southern Brazil it is known as *boto de dorsal espinhosa,* Portuguese for "dolphin with a thorny dorsal fin."

DESCRIPTION Burmeister's Porpoise displays the typically robust porpoise body shape (its girth is as much as two-thirds of its body length). It has a conical head and no distinct beak. The flippers are large and broad at the base. It differs from all other phocoenids in the position of its dorsal fin, set well behind the middle of the back. The small fin's shape is distinctive, with a rounded peak and convex trailing edge. Its front edge rises at an unusually low angle, so that the fin appears to point tailward rather than vertically as in most cetaceans. Two to four rows of tiny tubercles are present on the front edge of the fin. There are 10 to 22 pairs of teeth in the upper jaw and 16 to 25 in the lower.

The coloration is generally dark gray dorsally, but with a subtle, muted pattern that is not observable in the field. The head can be a lighter gray with a dark longitudinal streak from behind the blowhole to the tip of the upper jaw, where it converges with a dark lip patch. There is a dark eye patch and often a whitish eye ring. The belly and throat are light gray to white. Dark gray asymmetrical stripes run from the chin to the flippers.

RANGE AND HABITAT Burmeister's Porpoise ranges, possibly continuously, along the southern half of South America from southern Brazil (about 28°S–29°S) to northern Peru (about 5°S), including Beagle Channel, Tierra del Fuego, and the vicinity of Cape Horn. It is primarily coastal, sometimes seen inside the kelp line and in estuaries and the lower reaches of rivers. It also occurs up to tens of miles offshore in depths to at least 200 feet (60 meters). Scientists have interpreted trends in incidental catches and sightings of Burmeister's Porpoises to mean that the animals move inshore in summer and offshore in winter. Their distribution is thought to be related to the cool, northward-flowing Humboldt Current in the Pacific and the Falkland Current in the Atlantic. These porpoises live in a fairly broad range of surface temperatures, from the cool subantarctic waters off southern South America to the warmer inshore waters farther north.

SIMILAR SPECIES The similar-size Franciscana overlaps the coastal distribution of Burmeister's in Brazil and northern Argentina, but its long beak and more upright, triangular dorsal fin distinguish it. The Spectacled Porpoise occurs in the eastern part of the range of Burmeister's, from southern Uruguay to Tierra del Fuego. It has a more upright, triangular dorsal fin placed nearer the center of the back, and a sharply demarcated, black-and-white color pattern. In the western part of its range, especially along the Chilean coast, Burmeister's Porpoise overlaps extensively with the similar-size and dark-colored Chilean Dolphin, best distinguished by its dorsal fin, which is more upright and rounded at the peak.

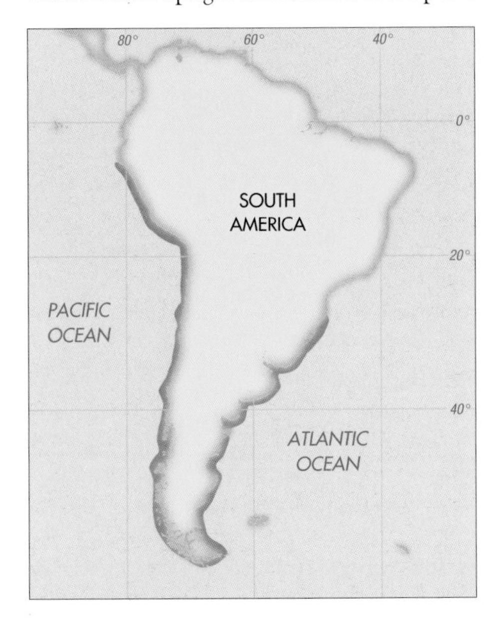

BEHAVIOR, REPRODUCTION, AND FORAGING Burmeister's Porpoises are usually seen alone or in pairs, less often in groups of up to eight individuals. They are sometimes seen in larger aggregations of 10 or more, but these are probably a result of concentrated food sources. This porpoise is extremely cryptic, showing little of its body when surfacing, and is difficult to observe other than in calm conditions. Its dives last one to three minutes. Small groups have been seen surfing on coastal breakers, then retreating with the waves and following them in again. Like the Harbor Porpoise, Burmeister's appears to have a high reproductive rate, with about two-thirds of adult females pregnant at any one time, and about a third of these simultaneously nursing the previous year's calf. Most conceptions and births occur in summer (February–March). The gestation period may be 11 to 12 months. Burmeister's Porpoises principally eat small schooling fish such as sardines, anchovies, and hake, but also take squid, mysid shrimp, and krill.

STATUS AND CONSERVATION Even though it is probably equally vulnerable to incidental mortality in fishing gear (especially gillnets), Burmeister's Porpoise has received much less conservation attention than the Harbor Porpoise. Burmeister's Porpoises are used for human food and for bait, and in at least some places they are hunted deliberately with guns and harpoons. No reliable abundance estimates are available.

BURMEISTER'S PORPOISE
FAMILY PHOCOENIDAE

MEASUREMENTS AT BIRTH

LENGTH	34″ (86 cm)
WEIGHT	Unavailable

MAXIMUM MEASUREMENTS

LENGTH	**MALE**	6′ (1.82 m)
	FEMALE	6′3″ (1.91 m)
WEIGHT	**MALE**	172 lb (78 kg)
	FEMALE	250 lb (105 kg)

LIFE SPAN
More than 12 years

■ **RANGE**

Spectacled Porpoise

Phocoena dioptrica
Lahille, 1912

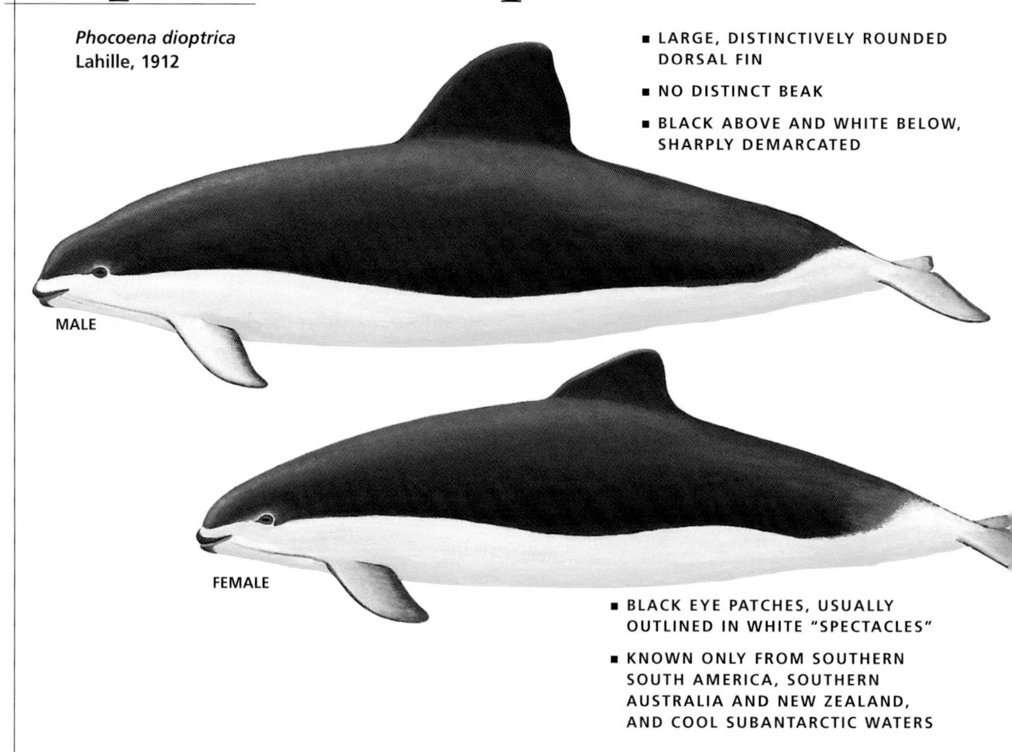

The Spectacled Porpoise is one of the least-known cetaceans. The vast majority of records consist of weathered carcasses and bones collected from the windswept shores of Tierra del Fuego at the southern tip of South America. Fernando Lahille, who christened the species in 1912, was so struck by the black eye patches rimmed with white that he chose the name *dioptrica,* from the Latin word for "spectacled." Known in South America primarily as *marsopa de anteojos,* the literal Spanish translation of the English name, this porpoise may also be called *tonina,* a generic term for small cetaceans. For a number of years after the mid-1980s, the species was assigned to a separate genus, *Australophocaena,* but is now back where it was from the start, in the genus *Phocoena.*

DESCRIPTION The small robust body has the typical porpoise form, with no distinct beak and a maximum girth of more than half the body length. The dorsal fin is roughly triangular, but with a broadly rounded peak, a long base, and straight or convex front and rear margins. The fin is significantly higher and longer in males than in females. The flippers are small relative to those of other *Phocoena* species. There are 16 to 26 pairs of teeth in the upper jaw and 16 to 22 pairs in the lower. Even in adults, some of the rear upper-jaw teeth are unerupted.

The coloration is highly distinctive: The jet-black back and upper lateral surfaces are sharply demarcated from the white lower lateral and ventral surfaces. The ventral whiteness may extend onto the dorsal surface just ahead of the flukes in some specimens. The dorsal fin is always black; the upper surfaces of the flippers and flukes can be either white or black. There is a black eye patch, usually outlined in white. The lips are black.

RANGE AND HABITAT The Spectacled Porpoise occurs in both coastal and offshore waters of the

Southern Hemisphere and is possibly circumpolar. Strandings and confirmed at-sea observations have occurred primarily along the southeastern coast of South America (eastern Tierra del Fuego and Patagonia), in offshore waters to the south of Cape Horn and New Zealand/Australia, and around offshore islands of the circumpolar West Wind Drift, specifically Heard and Kerguelen in the Indian Ocean, Auckland and Macquarie in the South Pacific, and South Georgia and the Falklands in the South Atlantic. It appears to be associated with cold temperate waters; sightings in the subantarctic have been in surface temperatures of 40 to 50°F (5–10°C). The northernmost record was a stranding in southern Brazil at 32°S, an area influenced by the cold Falkland Current.

SIMILAR SPECIES The Spectacled Porpoise overlaps with Burmeister's Porpoise along the eastern and southern coasts of South America, with Commerson's Dolphin along the same coasts and around the Falkland and Kerguelen Islands, and with the Chilean Dolphin along the continent's southern and western coasts. All four species are similar in size and lack a well-defined beak, although the Chilean Dolphin has a more elongated head. Unlike the Spectacled Porpoise, Burmeister's Porpoise appears completely dark in coloration, and its dorsal fin is set farther back and has a more pointed peak and lower profile. Commerson's Dolphin, with a rounded dorsal fin reminiscent of that of the Spectacled Porpoise, has a large area of white on its back and sides in front of the dorsal fin and a slighter build, and is gregarious, quick, and acrobatic. The Chilean Dolphin's dorsal fin is also large and rounded at the peak but is curved back rather than straight or convex as in the Spectacled Porpoise.

BEHAVIOR, REPRODUCTION, AND FORAGING The few groups of Spectacled Porpoises that have been observed have contained only two or three individuals. They are presumably not very gregarious. While nothing is known with certainty about its reproduction, the Spectacled Porpoise is thought to calve mainly in late spring or summer (November–February). The diet and feeding behavior of the Spectacled Porpoise are also essentially unknown. Anchovies and small crustaceans were found in the stomach of one stranded animal.

STATUS AND CONSERVATION Spectacled Porpoises die in coastal gillnets and in bottom and midwater trawls. They are also shot for food or bait at least occasionally. Essentially nothing is known about their abundance or conservation status.

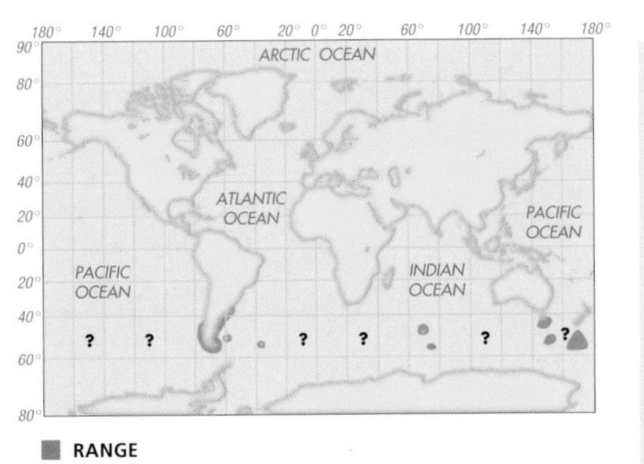

SPECTACLED PORPOISE
FAMILY PHOCOENIDAE

MEASUREMENTS AT BIRTH

LENGTH	35–40″ (90–100 cm)
WEIGHT	Unavailable

MAXIMUM MEASUREMENTS

LENGTH	**MALE**	7′5″ (2.25 m)
	FEMALE	6′9″ (2.05 m)
WEIGHT	**MALE**	More than 250 lb (115 kg)
	FEMALE	More than 187 lb (85 kg)

LIFE SPAN Unavailable

Dall's Porpoise

Phocoenoides dalli
(True, 1885)

Its unique body form readily distinguishes Dall's Porpoise from the rest of the true porpoises and indeed from any other cetacean. It is in some ways ecologically more like an oceanic dolphin than a coastal porpoise. Oddly, however, a fetus taken from a dead Dall's Porpoise in British Columbia was determined (through DNA sequencing) to have been fathered by a Harbor Porpoise. This discovery offers a possible explanation for the atypically pigmented porpoises occasionally seen swimming with and behaving like Dall's Porpoises around southern Vancouver Island and elsewhere. Dall's Porpoises came to world attention in the mid-1970s as conservationists learned of the high mortality rates associated with drift-net fishing for salmon in the temperate North Pacific. More recently, Dall's Porpoise populations have suffered from the moratorium on commercial whaling as Japanese consumers have turned to this species to help "compensate" for the reduced availability of minke and other whale meat.

DESCRIPTION Dall's Porpoise is very thick-bodied and robust, with a tiny head. It has little or no beak. The caudal peduncle is strongly keeled above and below, posterior to the anal region. The flippers are small and positioned far forward. The small dorsal fin is wide-based, triangular, and positioned at midbody. Adult males have a deeper caudal peduncle than females and a pronounced hump immediately behind the anus. Their dorsal fin is canted forward, and the trailing edge of their flukes is markedly convex. There are 21 to 28 pairs of very small teeth in both the upper and lower jaws.

DALL'S PORPOISE
FAMILY PHOCOENIDAE

MEASUREMENTS AT BIRTH

LENGTH	3′4″ (1 m)
WEIGHT	24 lb (11 kg)

MAXIMUM MEASUREMENTS

LENGTH	**MALE** 7′3″–7′10″ (2.2–2.39 m)
	FEMALE 6′11″ (2.1 m)
WEIGHT	**MALE** 370–440 lb (170–200 kg)
	FEMALE 400 lb (180 kg)

LIFE SPAN

Less than 15 years (extreme case 22 years)

- TINY HEAD, ROBUST BODY, DEEP CAUDAL PEDUNCLE
- TRIANGULAR DORSAL FIN, BLACK WITH WHITE "FROSTING"
- SHARPLY DEMARCATED BLACK-AND-WHITE COLORATION
- RAPID SWIMMING AND "ROOSTER TAIL" OF SPRAY WHEN SURFACING
- LIMITED TO COOL TEMPERATE NORTH PACIFIC

DALLI-TYPE MORPH, FEMALE

This porpoise is mostly black, but with bright white patches on the flanks and belly. There are two consistent and well-defined color morphs. In the *dalli*-type morph, the white flank patch extends forward to approximately the level of the front of the dorsal fin; in the *truei*-type morph the patch extends farther forward, reaching the flipper. The flank patch is significantly smaller on *dalli*-type porpoises in the Sea of Japan and the Sea of Okhotsk than on *dalli*-type porpoises elsewhere. Both the dorsal fin and the flukes become "frosted" with light gray, then white, as the animal ages; the white "frosting" on the dorsal fin usually has dark flecking.

RANGE AND HABITAT Dall's Porpoise is endemic to the cool temperate North Pacific. It is common in shelf, slope, and offshore waters from approximately the U.S.–Mexico border (32°N) and central Japan (35°N) north to the Bering and Okhotsk seas (but absent from the shallow northeastern Bering Sea). It occurs in oceanic waters of the central North Pacific north of 41°N. The *truei*-type is common only in the western Pacific between 35°N and 54°N; it is absent or very rare in the eastern Pacific. The *dalli*-type occurs throughout the species' range, including the Seas of Japan and Okhotsk. Dall's Porpoise has a preference for deep (more than 600 ft/180 m), cool

TRUEI-TYPE MORPH

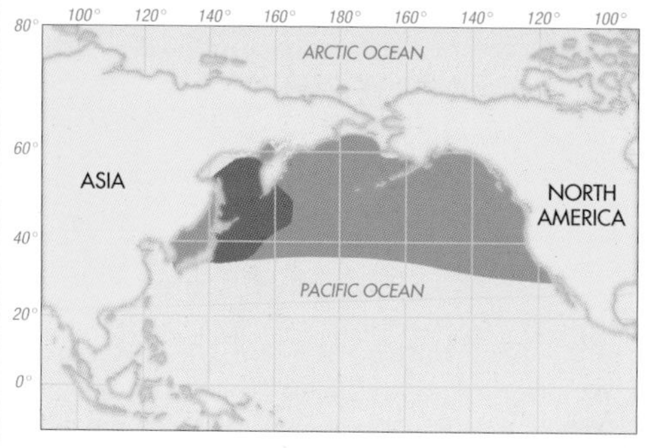

(less than 63°F/17°C) waters. In some areas it shifts seasonally from north and offshore in summer to south and inshore in winter.

SIMILAR SPECIES Pacific White-sided Dolphins overlap extensively with Dall's Porpoises in distribution and often cause a similar "rooster tail" of spray as they surface. The tall, strongly falcate dorsal fin and complex, light gray striping pattern on the sides and back make them relatively easy to distinguish at close range. Harbor Porpoises overlap with Dall's Porpoise in coastal portions of its range and also occasionally splash as they surface. They generally inhabit shallower water, swim more slowly, do not bow ride, are more likely to be solitary or in pairs, and lack the sharply contrasting black-and-white color pattern.

BEHAVIOR Dall's Porpoises occur in small (most often 2 to 12) and fluid groups; loosely associated groups can form feeding aggregations involving tens to hundreds, rarely thousands, of individuals. This porpoise's movements are very quick; it is thought to be the fastest of the small cetaceans. When surfacing, it typically creates "rooster tails" of spray that can almost obscure the animal itself. In inshore waters, its movements may be more deliberate, and the keeled caudal peduncle may be evident above the surface as the animal "slow rolls." Dall's Porpoises are avid bow riders. They often weave in and out of bow waves, making jerky movements. Mothers with young calves avoid vessels. This species is subject to predation by Killer Whales, but its ability to swim at high speeds may minimize the threat of shark attack.

REPRODUCTION The sexual dimorphism in body size and shape, and the smallness of the testes, suggest that male Dall's Porpoises compete for exclusive access to females. To achieve dominance they would need to be adept at fighting or, alternatively, capable of intimidating one another with a visual or other type of display. Males in

Flukes

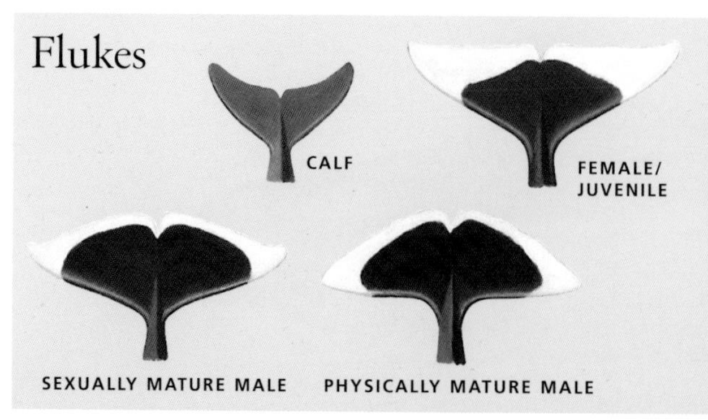

The flukes of Dall's Porpoise calves have a concave trailing edge and are entirely dark-colored. As the animals grow, their flukes change, with the trailing edges becoming straight in mature females and markedly convex in mature males. As the shape of the flukes changes, so does the color, with white "frosting" appearing on the trailing edges.

OPPOSITE: *A classic view of a "rooster tail," the term used to describe the spray of water created as a Dall's Porpoise slices along the surface. This animal is near the northern limit of the species' range, in the vicinity of the Pribilof Islands, eastern Bering Sea.*

ABOVE LEFT: *The life of a Dall's Porpoise is filled with hazards, not the least of which is the possibility of being attacked by Killer Whales. A Killer Whale in the Chatham Strait of Alaska tosses its victim high into the air.* **ABOVE RIGHT:** *This* dalli*-type Dall's Porpoise in Alaskan waters appears to be resting. The white patch on the side of the* dalli *type ends approximately just below the front edge of the dorsal fin, in contrast to the patch on* truei*-type Dall's Porpoises, which extends forward to the flipper.*

such situations have a reduced need to produce large quantities of sperm. Gestation in Dall's Porpoise lasts 10 to 11 months, and births take place between early spring and early fall, mainly June to August. The lactation period lasts at least two months, but for how much longer is uncertain. Females in their prime probably give birth annually.

FOOD AND FORAGING Dall's Porpoise mainly eats small schooling fish (for example, herring, pilchards, hake) and squid. A high proportion of its diet consists of deepwater, vertically migrating species. It is uncertain whether Dall's Porpoises obtain most of their prey at night, when the prey organisms come closest to the surface, or also dive deep to forage during the daytime. Their blubber is thin for a cold-water species, so they must maintain a relatively high metabolic rate and thus a high and regular caloric intake.

STATUS AND CONSERVATION Dall's Porpoise is abundant and widely distributed. There may be close to 50,000 individuals off California, Oregon, and Washington; at least 80,000 in Alaskan waters; more than 100,000 off the Pacific coast of Japan; and several hundred thousand in the Sea of Okhotsk. Their tendency to approach vessels, however, makes it difficult to avoid overestimating their abundance. Many animals are killed each year in gillnets, trawls, and other types of fishing gear. Nevertheless, current levels of incidental mortality are probably not high enough to reduce the large, productive populations in most areas. The greatest known threat is the porpoise hunt in Japan, where the annual harpoon kill of Dall's Porpoises increased in the 1980s from thousands to several tens of thousands. More than 40,000 were reportedly landed in 1988. Since then, a national quota of about 18,000 per year has prevented the kill from approaching earlier levels. However, there is still great concern about the future of Dall's Porpoises in the western Pacific and adjacent seas. In addition to the directed take, unknown but possibly large numbers of porpoises die in Japanese drift nets set in Russian waters of the Sea of Okhotsk.

Sirenians

The order Sirenia belongs to an unusual group of four mammalian orders known as the "subungulates," which includes the elephants (order Proboscidea), hyraxes (order Hyracoidea), and aardvarks (order Tubulidentata). Subungulates share several anatomical features, including dental characteristics and the lack of a clavicle, nails, or hooves. There are four living sirenians classified in two families: the Trichechidae, with three species of manatees, and the Dugongidae, the Dugong. A second dugongid, Steller's Sea Cow, was hunted to extinction in the 18th century.

Sirenians are well adapted to aquatic life. They have torpedo-shaped bodies with no dorsal fin, no hindlimbs, sparse hair, and paddle-like forelimbs that they use to maneuver by pivoting, sculling, or paddling. Manatees have a large, rounded tail shaped somewhat like that of a beaver, while the Dugong's V-shaped tail is similar to the dolphin's. Buoyancy is aided by their thick, heavy bones and the positioning of their lungs and diaphragm. Sirenians have two nostrils, situated on the tip of their muzzle, that are covered by flaps of skin when they submerge.

Manatees and the Dugong are herbivorous and feed on seagrasses and other aquatic vegetation. Their thick, movable lip pads are equipped with stiff bristles that aid in grasping and moving food toward the mouth. Sirenians have evolved an unusual method of tooth

Dugong, Australia

replacement to compensate for the wear caused by their abrasive diet. Manatees have only molars, which are continually replaced as the teeth move horizontally from the back of the jaw to the front. When the teeth reach the front of the jaw, the roots are absorbed and the teeth eventually fall out. Dugongs have three pairs of molars and premolars in each jaw. As the animals age, they lose the first set of premolars, along with the first pair of molars. The remaining molars continue to grow throughout the Dugong's life. Males have two large incisors (tusks) that erupt through the gums after puberty.

Male sirenians have no external genitalia, and the genital aperture is located near the middle of the belly. In females the aperture is near the anus, slightly forward of the caudal peduncle. Sirenians usually have a single offspring after a gestation of over a year.

All sirenian populations have declined over the past few centuries, and they have been extirpated in many areas. The primary causes of population decline are related to human activities. Dugongs and manatees have been hunted throughout their ranges. Despite measures to protect them, hunting persists in some areas, and sirenian meat can still be found in some markets. Sirenians, particularly manatees, have also been subject to increasing mortality caused by collisions with watercraft, being crushed in flood gates and locks, and entanglement in fishing

Amazonian Manatee, Brazil

gear. Dugongs have succumbed to fishing trawls, shark and turtle nets, and dynamite fishing, and are vulnerable to oil spills, cyclones, storm surges, parasites, and predation by sharks and Killer Whales.

FAMILY TRICHECHIDAE Manatees are primarily tropical in distribution and are found along the Atlantic Basin. The Florida subspecies of

Steller's Sea Cow

Hydrodamalis gigas
(Zimmermann, 1780)

In addition to the living Dugong, the only other recently extant member of the Dugongidae was Steller's Sea Cow, which was found in the Bering Sea. Steller's Sea Cow had a large body (up to 26 ft/8 m long), a small head, a short neck, stubby foreflippers, and a split tail. Its skin was very rough, and it totally lacked finger bones and teeth. It fed primarily on kelp and other algae. It is believed that hunting by indigenous people eliminated the species from much of its range. When it was discovered in the early 18th century, it lived only around Bering and Copper Islands in the Commander Islands. Only 27 years after its discovery, sealers and Sea Otter hunters had hunted it to extinction.

the West Indian Manatee is the only manatee that ranges into subtropical and more temperate areas, but it must find refuge in warm water from springs or industrial effluents during periods of colder weather. West Indian and West African Manatees occupy similar habitats. They sometimes ascend far up rivers but also frequent estuaries and nearshore coastal areas where there is plentiful food and calm water. The Amazonian Manatee is found only in the freshwater Amazon River and its tributaries.

FAMILY DUGONGIDAE The Dugong is the only surviving member of the once diverse and widespread family Dugongidae. Disjunct populations of Dugongs are found from eastern Africa and Madagascar through the coastal regions of the Indo-Pacific to as far north as Okinawa and east to the Philippines and Palau. Numbers appear low throughout most of their range except in Australian waters. Dugongs are strictly marine. They occur in nearshore coastal areas but sometimes can be found in deeper offshore waters.

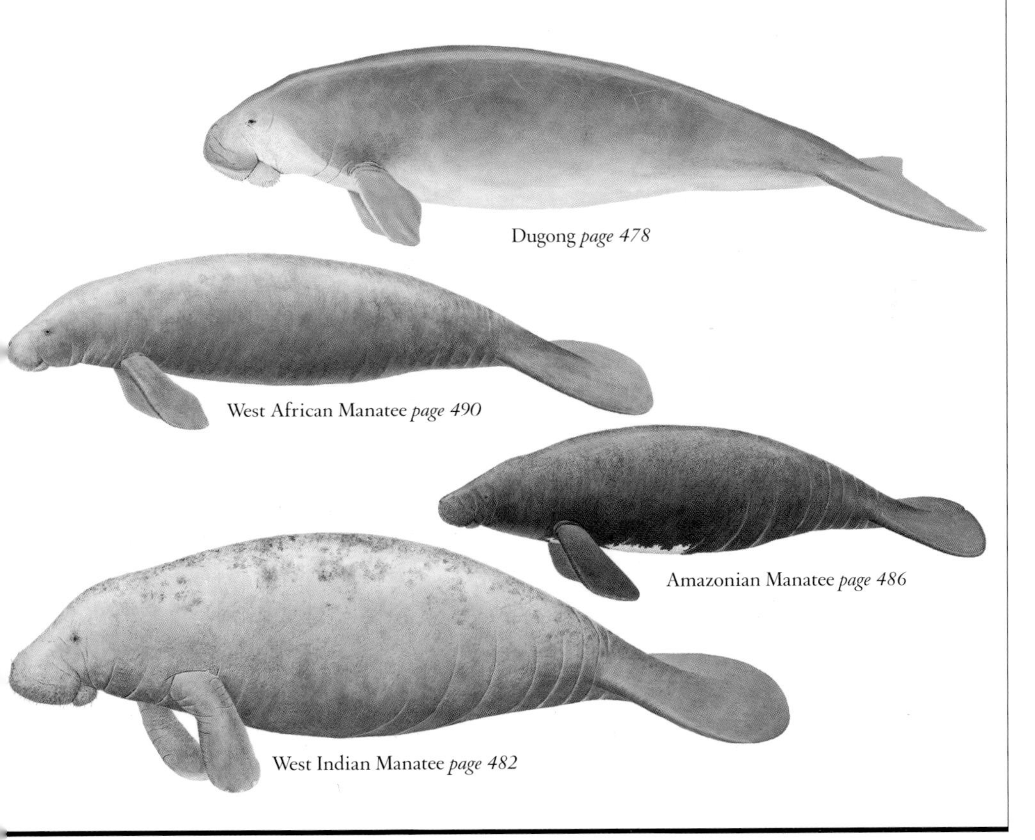

Dugong *page 478*

West African Manatee *page 490*

Amazonian Manatee *page 486*

West Indian Manatee *page 482*

Dugong

Dugong dugon
(Müller, 1776)

Dugongs are similar to manatees in their herbivorous diets, but they differ in their preference for marine habitats. They are the only sirenians found in the Indian and Pacific Oceans. Dugongs have a fluked tail like dolphins and whales, and tusks like their distant relatives the elephants. Although they have a wide distribution and are abundant in some parts of their range, they are threatened by continued hunting, habitat degradation, and incidental deaths in fishing gear. Dugongs are known locally as "sea cows," "sea pigs," or "sea camels."

DESCRIPTION The Dugong is large and fusiform, with smooth, brownish to dark gray skin that is sparsely covered with hair. The caudal peduncle is laterally compressed and ends in a fluked tail. The paddle-like flippers are moveable and lack nails. Dugongs lack external ears and have small eyes. The nostrils are located on the tip of the muzzle and have flaps that close when the animal submerges. The broad, flattened rostral disk at the end of the large muzzle is distinctly turned downward, an adaptation for bottom feeding. Massive lip pads also aid the Dugong in feeding. Two large incisors (tusks) erupt through the gums in males, and rarely in females, after puberty; the tusks are not visible when the mouth is closed. There are six premolars and six molars in both the upper and lower jaws, although all the teeth are never present at one time. As Dugongs age, they lose all the premolars and the first pair of molars; the last two pairs of molars grow during the Dugong's life. This differs from the manatees, which have continual molar replacement through forward migration of the teeth.

RANGE AND HABITAT The Dugong is a marine species, occurring along the coasts of the western Pacific and Indian Oceans, usually in areas where there are abundant seagrasses. Although considered a shallow-water species, Dugongs are occasionally observed in deeper waters, to about 75 feet (23 m). Their distribution is disjunct, reflecting habitat availability and human activities. Their historic range is reduced. Currently Dugongs occur off the east coast of Africa, from the Persian Gulf and the Red Sea south to Durban, South Africa, and including Madagascar.

- SMOOTH, BROWNISH TO DARK GRAY SKIN
- SMALL TUSKS ON UPPER JAW, PRIMARILY IN MALE
- NOSTRILS LOCATED AT TIP OF MUZZLE
- FLUKED TAIL, NO DORSAL FIN
- FOUND IN MARINE HABITATS OF THE INDO-PACIFIC REGION

Dugong distribution is discontinuous in Asian waters. They are found along the coasts of the Indian Ocean, including Sri Lanka, although they are considered to be rare in most areas. The northern extent of their range is around Okinawa, in the Ryukyu Islands, and at similar latitudes off the coasts of China and Taiwan. Other populations occur in the Philippines; Palau; Papua New Guinea and its associated islands, including West New Britain and the Manus Islands; New Caledonia; and the Vanuatu Islands. There may be more than 12,000 Dugongs in the Torres Strait, between New Guinea and Australia. They occur in the northern half of Australia from Shark Bay to Moreton Bay, although sightings have been noted as far south as Perth on the west coast and around Sydney on the east coast.

SIMILAR SPECIES No other species resemble Dugongs in their Indo-Pacific range. Dugongs may be confused with small cetaceans, particularly Finless Porpoises and Irrawaddy Dolphins; however, their lack of a dorsal fin and the location of their nostrils at the tip of the snout are distinguishing characteristics.

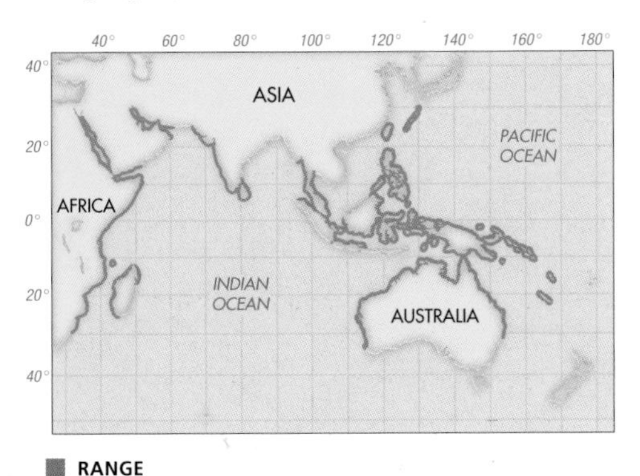

RANGE

DUGONG

FAMILY DUGONGIDAE

MEASUREMENTS AT BIRTH

LENGTH About 3′9″ (1.15 m)

WEIGHT 55–77 lb (25–35 kg)

MAXIMUM MEASUREMENTS

LENGTH About 11′ (3.3 m); maximum recorded 13′6″ (4.1 m)

WEIGHT 2,200 lb (1,000 kg)

LIFE SPAN

Maximum recorded age 73 years

BEHAVIOR Dugongs appear to have more elaborate social behaviors than the manatees. They commonly form small groups of up to six but may sometimes form large groups of several hundred animals, the function of which is unknown. Their movements are leisurely, with an average swimming speed of 6 miles per hour (10 km/h). Dugongs surface about every one to two minutes, but sometimes remain submerged for up to eight minutes. They surface with the tip of their muzzle out of the water and may roll like a dolphin when making deep dives or traveling. Daily movements of 16 miles (25 km) are common, and individuals may travel hundreds of miles during the course of a year. During the winter months, Dugongs may use deeper waters as a thermal refuge from cooler inshore waters. Dugong vocalizations consist of squeaks and squeals similar to those of the manatees. Sharks, Killer Whales, and crocodiles prey on Dugongs.

REPRODUCTION The Dugong has a polygamous mating system. Its mating behavior (except that reported from Shark Bay, Australia) appears similar to that of other sirenians. A female in estrus attracts a group of males that compete to mate with her. Researchers have observed several stages in the mating process, beginning with a "following phase," in which males form a tight group around a female as she swims in an effort to evade them. Next comes a period of intense activity, with males splashing, tail-thrashing, lunging, and twisting. This is followed by the "mounting phase," in which a male clasps the

ABOVE: *A Dugong in search of seagrass off Australia, accompanied by golden pilot jacks.*

BELOW: *Dugongs tend to be found in shallow waters where seagrasses are abundant.*

Dugongs are commonly seen in small groups but may sometimes gather in groups of several hundred.

female from underneath, while other males jostle for position and also grasp the female. In Shark Bay males have been observed to establish territories where they display to attract females. Calving occurs after a gestation estimated to last about 12 to 14 months; calving has been observed only in shallow water. Calves are weaned at about 18 months old, although they may remain with their mothers for several years. Breeding and calving can occur year-round in tropical areas, although there seem to be seasonal peaks. The calving interval ranges from 2½ to 7 years.

FOOD AND FORAGING Dugongs feed only on bottom vegetation, consuming the leaves, roots, and rhizomes of a variety of seagrasses. They do not use their tusks for feeding, rather manipulating food toward the mouth with their large, prehensile rostral disk. As they forage for bottom vegetation, they sometimes leave trails of cropped plants that can be seen when seagrass meadows are exposed during low tide. Dugongs are generally diurnal feeders, but in areas where there is human disturbance they tend to feed at night.

STATUS AND CONSERVATION As with manatees, the primary causes for Dugong population declines are related to human activities. From bones found in archaeological sites in the United Arab Emirates, scientists know that Dugongs have been hunted for at least 4,000 years. Historical harvesting of Dugongs has drastically reduced their numbers in some areas. For example, from the 17th to the 19th centuries, hundreds were killed in the waters off Madagascar for meat. Hunting for Dugongs continues throughout much of their range, and Dugong meat is sometimes openly sold in markets. Dugongs are protected in Australia but can still be legally taken by aboriginal hunters in certain areas. In addition to hunting, Dugongs are incidentally killed in fishing trawls, shark nets set along beaches to protect swimmers, and turtle nets, and by dynamite fishing. Dugongs are also vulnerable to oil spills; mortalities were observed after the 1983 *Nowruz* oil spill in the Persian Gulf and during the Gulf War. Natural threats to Dugongs include cyclones, storm surges, parasites, and predation by sharks and Killer Whales. Australia probably has the highest population of Dugongs, estimated at 85,000.

West Indian Manatee

Trichechus manatus
Linnaeus, 1758

West Indian and West African Manatees are unusual among marine mammals in that they move freely between fresh and salt water. They seem to prefer rivers with true grasses, or calm estuarine environments where there are plentiful seagrasses to eat and accessible sources of fresh water. These slow-moving animals can be difficult to observe since they usually expose just the tip of the muzzle when they surface to breathe. They have few, if any, natural predators; however, they continue to be illegally hunted in some areas and also suffer death or injury from exposure to cold, collisions with watercraft, and inhaling or ingesting brevetoxin (red tide) organisms. Scientists recognize two subspecies of the West Indian Manatee: The Florida Manatee occurs in the southeastern United States, and the Antillean Manatee is found throughout the Greater Antilles and along the Gulf of Mexico and the Atlantic coast of Central and South America.

DESCRIPTION Like all manatees, the West Indian Manatee has a fusiform shape with a broad back, no dorsal fin, and a small head. The body color is gray, although individuals may appear brown, reddish, whitish, or black due to algae and barnacles on the skin. The skin is rough, thick, and toughened, with sparse, evenly distributed hairs. Females tend to be larger than males. Manatees have only molars, which are continually replaced by horizontal migration along the jaw.

RANGE AND HABITAT Florida Manatees *(Trichechus manatus latirostris)* inhabit coasts, estuaries, and major rivers of Florida year-round. During warmer months they may range along the Atlantic coast as far north as Rhode Island and west along the Gulf coast past Louisiana. In winter, they concentrate in Florida, seeking refuge in warm-water springs or near industrial facilities that discharge heat into the water. Antillean Manatees *(T. m. manatus)* occur in the Greater Antilles (Cuba, Jamaica, Hispaniola, and Puerto Rico) and Trinidad, and also along the Atlantic coast and up rivers of Central and South America, south to Alagoas, Brazil. They have not been documented in the mouth of the Amazon River. West Indian Manatees appear to prefer calm waters with abundant vegetation and access

- BROWN TO GRAY BODY WITH SPARSE, EVENLY DISTRIBUTED HAIRS
- BROAD BACK, ROUNDED TAIL, NO DORSAL FIN
- USUALLY SOLITARY, SOMETIMES IN GROUPS OF UP TO ABOUT 20
- GATHERS AT WARM-WATER SITES DURING COLD WEATHER
- INHABITS SHALLOW NEARSHORE WATERS, RIVERS, OR LAGOONS
- OCCURS IN SOUTHEASTERN U.S., GREATER ANTILLES, AND ATLANTIC COAST OF CENTRAL AND SOUTH AMERICA

to fresh water. Their common nearshore habitats include bays, lagoons, and estuaries, and they ascend rivers for hundreds of miles. Antillean Manatees visit offshore freshwater springs within their range, presumably to drink. Areas where there are high-energy coastlines with little or no submergent vegetation available are largely devoid of manatees. West Indian Manatees can traverse deep oceanic passages. The Florida subspecies has been seen in offshore waters between the Dry Tortugas and Florida and in the Bahamas, while the Antillean subspecies has occurred at the Turneffe Islands, about 20 miles (30 km) off the coast of Belize.

SIMILAR SPECIES West Indian Manatees have not been documented in the lower Amazon River where Amazonian Manatees occur. However, their disjunct distribution both north and south of the Amazon River suggests that there may be some areas near the river mouth where both species could be found. Amazonian Manatees have

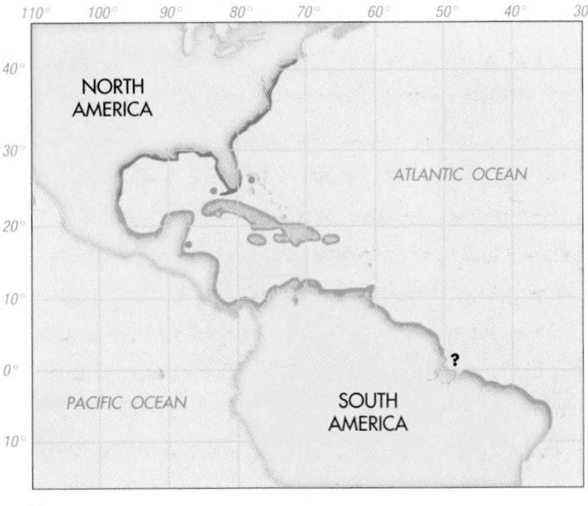

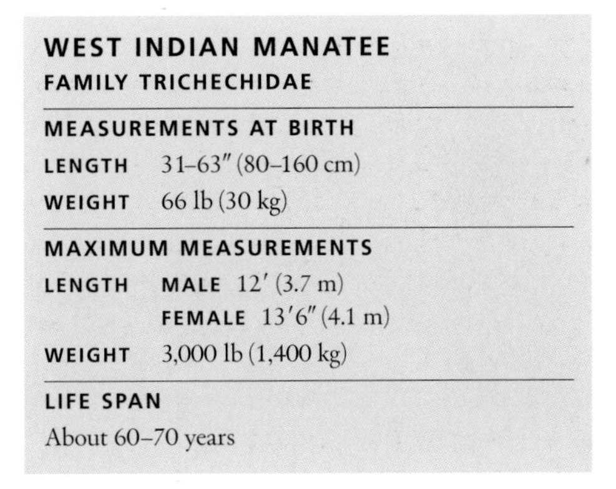

WEST INDIAN MANATEE
FAMILY TRICHECHIDAE

MEASUREMENTS AT BIRTH

LENGTH	31–63″ (80–160 cm)
WEIGHT	66 lb (30 kg)

MAXIMUM MEASUREMENTS

LENGTH	MALE 12′ (3.7 m)
	FEMALE 13′6″ (4.1 m)
WEIGHT	3,000 lb (1,400 kg)

LIFE SPAN
About 60–70 years

smooth rather than roughened skin, and white or pink ventral patches; they lack nails on the flippers.

BEHAVIOR West Indian Manatees may form ephemeral groups that are usually associated with mating or shared resources such as food, fresh water, or warm water. The basic social unit of this species, however, is the cow and calf. Mothers and calves communicate with high-pitched squeaks and squeals. If danger is present, mothers may position themselves between their calf and the danger. On rare occasions, mothers may try to push away a persistent intruder, such as a diver, but more typically a cow and calf will flee together while vocalizing back and forth. West Indian Manatees spend about six to eight hours a day feeding. They rest for two to four hours a day in warm weather and up to eight hours in colder weather. They demonstrate social facilitation, with members of a group coordinating behaviors such as breathing, resting, and traveling. Tagging studies in Belize suggest that females move very little while some males move extensively within a range of about 60 miles (100 km); some studies have shown similar patterns in Florida. In Florida, manatees can move 30 miles (50 km) in a day and traverse distances of over 300 miles (500 km) between their winter and summer grounds.

REPRODUCTION A female West Indian Manatee in estrus attracts a mating herd of males. These mating herds may stay together for several weeks, and the composition and number of males are dynamic. Researchers have observed as many as 22 males in one herd, vigorously jostling for a position close to the female. Individuals have been seen engaged in sex play, gripping each other with their foreflippers and rolling. Females give birth usually to one calf after a gestation period of 12 to 14 months; twins have been recorded in about 1.8 percent of births in some populations. Calves can swim soon after birth, but they remain close to their mothers until weaning, usually stationed

OPPOSITE RIGHT: *West Indian Manatees usually have a single calf, but two, either twins or a new calf with an older one, are not uncommon.* **ABOVE:** *The West Indian Manatee's thick bristly lip pads help to grasp plants and pull them into its mouth.* **LEFT:** *West Indian Manatees are highly tactile animals. They frequently clasp, touch, and roll with each other in what biologists call "cavorting behavior."*

just behind a foreflipper. Weaning occurs when calves are about 18 months old. Florida Manatees can breed year-round, although most calves are born in spring and summer (March–August), probably because the energetic costs to the female are lessened by the greater availability of vegetation. The average calving interval is about 2½ years.

FOOD AND FORAGING West Indian Manatees feed on a huge variety of aquatic and semiaquatic vegetation, and occasionally on bank growth within reach of the water. Their prehensile lips, aided by the presence of thick bristles, are adapted to wrapping around and grasping plants, and they use their highly flexible lip pads to manipulate the plants into the mouth. They may also hold or maneuver vegetation with their flippers. In Jamaica, manatees have reportedly taken fish from nets on rare occasions.

STATUS AND CONSERVATION West Indian Manatees are legally protected everywhere they occur. They have been protected in Florida since the 19th century, currently under the Florida Manatee Sanctuary Act of 1978, as well as the U.S. Marine Mammal Protection Act of 1972. Illegal hunting of this species nevertheless continues, for its meat as well as its oil and bones, which are used in various folk remedies. The ribs, similar in texture to ivory, are illegally carved and sold as jewelry in some areas. West Indian Manatees are also threatened by increased development in coastal areas, collisions with motorboats, entanglement in nets and fishing lines, ingestion of fishing gear, and crushing or drowning in locks and flood-control structures. In 1999, 82 (30 percent) of the 268 manatee carcasses recovered in Florida had been killed by watercraft, and watercraft-related deaths continue to increase at a faster rate than any others. In Belize, watercraft are an emerging cause of death for manatees. Natural threats to the species include exposure to brevetoxins (red tide), disease, and stress from cold.

While water turbidity and overhanging vegetation make it difficult to count manatees, aerial surveys conducted in 2001 estimated that there are at least 3,200 Florida Manatees on the Atlantic and Gulf coasts. There are at least 340 Antillean Manatees in Belize, where the highest density of this subspecies has been recorded. Manatees are particularly vulnerable because of their low rate of reproduction. Recent population models for Florida Manatees suggest that deaths may be close to, if not already exceeding, births in some regions. Florida Manatee populations appear to have increased over the past 30 years, but it is uncertain whether this trend will continue, due to increasing human-related threats.

Amazonian Manatee

Trichechus inunguis
(Natterer, 1883)

The Amazonian Manatee is the smallest sirenian. It is distinctive among the manatees with its more fusiform body, smooth skin, white or pinkish ventral markings, and lack of nails on the flippers. It occurs only in fresh water, and its distribution is limited to the rivers, lakes, and tributaries of the Amazon Basin. When water levels are high, Amazonian Manatees move into floodplain lakes and flooded forests, and during dry periods, where water levels fluctuate, they seek refuge in deep pools and lakes. Researchers know little about the social behavior of Amazonian Manatees because of their shy nature and the dark waters they inhabit. They are preyed upon by jaguars, caimans, and sharks, and are also hunted commercially. Commercial hunting may have resulted in a large reduction in the population of this species. The Amazonian Manatee is called *peixe-boi* ("ox-fish") in Brazil and *vaca marina* ("sea cow") or *manati* in Spanish-speaking countries.

DESCRIPTION The Amazonian Manatee is similar to other manatees in its fusiform shape, small head, broad back, and rounded tail. However, it has several distinguishing features. These include a smaller, more fusiform body, smooth skin, a longer narrower rostrum, smaller teeth, a white or pinkish ventral patch, and proportionally longer flippers that lack nails. The body is dark gray to black.

RANGE AND HABITAT The Amazonian Manatee occurs in the rivers and tributaries of the Amazon basin. In Brazil, it occupies the Amazon Solimões, Tocantins, Xingu, Tapajós, Nhamundá, Madeira, Negro Branco, and Tacatu Rivers; in Colombia, the Amazon Marañón, Putumayo, Caquetá and lower Apaporis Rivers; in Peru, the Napo, Tigre, Marañón, Samiria, Pacaya, Ucayali, and Huallaga Rivers; and in Ecuador, the Aguarico and Cuyabeno Rivers. Amazonian Manatees ascend rivers as far as rapids allow, but they prefer floodplain lakes and channels where vegetation is plentiful and water temperatures range from 77 to 86°F (25–30°C). In June when water levels are highest, the manatees are widely dispersed through lakes, rivers, and seasonally inundated forests and wetlands. As water levels drop in July and August, they move to permanent bodies of water or pools.

- DARK GRAY TO BLACK COLORATION, WITH WHITE OR PINK VENTRAL PATCH
- SMOOTH SKIN
- BROAD BACK, ROUNDED TAIL, NO DORSAL FIN
- LACKS NAILS ON FLIPPERS
- INHABITS RIVERS, TRIBUTARIES, AND LAKES OF AMAZON BASIN

SIMILAR SPECIES West Indian Manatees have not been documented around the mouth of the Amazon River, but their presence north and south of the Amazon suggests there are some areas where they may occur with Amazonian Manatees. West Indian Manatees have nails on the foreflippers and lack the light ventral patch and smooth skin of the Amazonian Manatee. Two cetaceans that occur in the Amazon River, the Amazon River Dolphin and the Tucuxi, can be distinguished by their fluked tail, dorsal ridge (river dolphin) or fin (Tucuxi), and blowhole on top of the head.

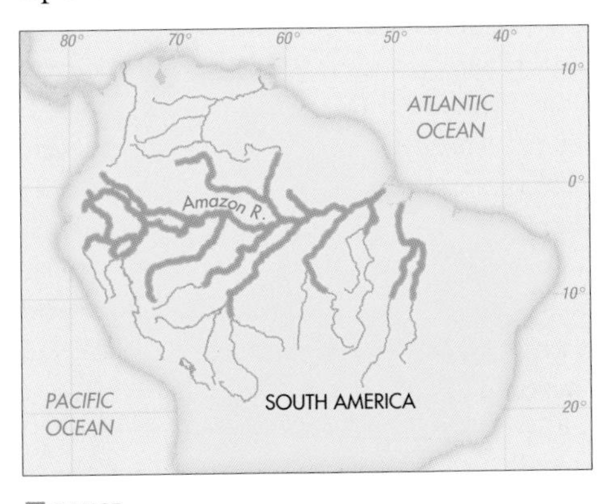

RANGE

BEHAVIOR While the strongest social bonds in this species are between mothers and calves, Amazonian Manatees may also form loose social groups while feeding, traveling, or mating. Feeding groups of less than 10 animals are seen; larger groups were once more common. Manatees communicate with high-pitched squeals and squeaks. In studies of captive Amazonian Manatees, individuals were equally active day and night, spending about eight hours feeding, four hours resting, and 12 hours moving around. In the wild, the activities of Amazonian Manatees are strongly tied to seasonal water levels, which

AMAZONIAN MANATEE
FAMILY TRICHECHIDAE

MEASUREMENTS AT BIRTH

LENGTH Probably about 31–37″ (75–85 cm)

WEIGHT 22–33 lb (10–15 kg)

MAXIMUM MEASUREMENTS

LENGTH 9′2″ (2.8 m)

WEIGHT 1,100 lb (480 kg)

LIFE SPAN

Probably about 60–70 years

fluctuate 35 to 50 feet (10–15 m) annually. During the dry season they conserve energy by dramatically decreasing most activities, including feeding. As water levels start to rise in December, calves are born, mating begins, and the manatees begin to restock their fat reserves. Amazonian Manatees are preyed upon by jaguars, caimans, and sharks.

REPRODUCTION Breeding and calving seasons for the Amazonian Manatee are influenced by seasonal changes in water levels. While births can occur any time between December and July, most take place between February and May, when water levels begin rising. Presumably females have better access to food at this time and are thus less stressed by the energetic demands of lactation. Captive calves grow about a half-inch (1.5 cm) and add about 2 pounds (1 kg) a week. Breeding occurs during the same period as calving, but in alternate years. Individuals have been seen engaged in sex play similar to that of West Indian Manatees, gripping each other with their fore-flippers and rolling. Gestation lasts about 12 months.

FOOD AND FORAGING Amazonian Manatees eat a large variety of aquatic and semiaquatic vascular plants. In the Amazon basin, seasonal changes in water levels affect food availability. During the rainy season from December to June, water levels are high and Amazonian Manatees forage in seasonally inundated areas where food is abundant. As water levels drop in July and August, they migrate to deep pools and rivers. When water levels are lowest, usually in November and December, manatees unable to find living vegetation may feed on dead vegetation or may fast for extended periods. A low metabolic rate and ample fat reserves enable them to survive without eating for up to 200 days. During periods of extended dry weather and low water levels, manatees may ingest clay, which may result in bowel obstructions and death. Captive Amazonian Manatees consume 20 to 35 pounds (9–15 kg) of vegetation a day. They feed for six to eight hours, usually in bouts of 30 to 90 minutes but sometimes for as long as 2½ hours.

STATUS AND CONSERVATION Amazonian Manatees are legally protected everywhere they

OPPOSITE: *The white or pinkish ventral markings of the Amazonian Manatee distinguish it from the other manatees.*
RIGHT: *As its name suggests, the Amazonian Manatee is found in the rivers and tributaries of the Amazon River basin in South America.*

occur. The largest populations occur in Brazil. They are considered to be uncommon in Peru, Colombia, and eastern Ecuador. While protected areas exist throughout the species' range, the remote terrain and lack of officers makes enforcement difficult. Hunting, usually with harpoons, continues to be a problem, and there are reports of both subsistence hunting by indigenous tribes and sales of manatee meat to provision the military. Between 1935 and 1954, thousands of manatee skins and tons of meat were exported from Brazil, depleting populations. In addition, catastrophic mortalities have occurred from natural events; the exceptionally dry weather during 1963 resulted in hundreds of manatees dying around Tefé, Coari, and Manacapuru. Habitat destruction due to deforestation, pollution from oil exploration, and mining also threatens Amazonian Manatees. Deforestation of the Amazon River basin is dramatically increasing. This has resulted in altered and reduced water flows that negatively impact both food availability and the dry-season retreats of Amazonian Manatees. Pollution, particularly siltation, results in reduced food resources and may also negatively affect the health and reproduction of the species. As is true for other sirenians, the Amazonian Manatee's low reproductive potential makes it particularly vulnerable to hunting and wide-scale habitat destruction and degradation.

The Amazonian Manatee's diet includes a large variety of aquatic and semiaquatic vascular plants.

West African Manatee

Trichechus senegalensis
Link, 1795

West African Manatees are well known among the inhabitants of coastal Africa as a source of food, oil, folk remedies, and folklore. Often referred to as "mami-wata," they are the basis for a variety of mermaid myths throughout western and central Africa. Christopher Columbus was the first European reputed to have seen manatees in the New World. Interestingly, he recognized them from West African Manatees he had encountered during his earlier voyages to Africa. West African Manatees have a secretive nature, probably a result of continued hunting pressures. While hunting has reduced populations in many areas, the species has maintained its historical range.

DESCRIPTION The West African Manatee has a typical manatee form, including a fusiform body with a broad back and no dorsal fin. It is gray, with thick, tough, slightly roughened skin and sparse hair that is evenly distributed over the body. Individuals inhabiting coastal waters may have barnacle growth. As in other manatees, the head is small relative to the body. The eyes are small and protrude slightly (unlike the West Indian Manatee, which has more deep-set eyes), and there are no external ears. The muzzle is blunter than the West Indian Manatee's, with a shorter, less downturned rostrum that may be an adaptation for feeding on floating plants, emergent vegetation, and bank growth. There are three to four fingernails on the outer edge of the flippers. Manatees have only molars, which are continually replaced.

RANGE AND HABITAT West African Manatees occur in most coastal waters and adjacent rivers from the Senegal River and its tributaries in southern Mauritania south to the Cuanza River in Angola. They ascend most major rivers until rapids, waterfalls, or shallow waters prevent their progress. They occur 1,250 miles (2,000 km) up the Niger River in the inland delta at Mali, where they are essentially landlocked upstream of rapids. Manatees are also landlocked in the Logone and Chari Rivers that flow into Lake Chad. Some pass the dry season in permanent lakes that become cut off from rivers with which they are connected during high water levels, such as lakes along the Benue River, a tributary of the Niger. They are found 45 miles (75 km) offshore among the shallow coastal flats and mangrove creeks of

- GRAY COLORATION WITH SPARSE, EVENLY DISTRIBUTED HAIRS
- BROAD BACK, ROUNDED TAIL
- LACKS DORSAL FIN
- USUALLY SOLITARY, BUT MAY OCCUR IN LOOSE AGGREGATIONS
- INHABITS SHALLOW NEARSHORE WATERS, RIVERS, OR LAGOONS OF WESTERN AFRICA

the Bijagós Archipelago of Guinea-Bissau. West African Manatees prefer quiet estuarine coastal lagoons and river mouths, but they are highly adaptive and can be found in practically any accessible area. Common habitats include large and small rivers, estuaries, freshwater and saltwater lagoons, quiet bays, and lakes.

SIMILAR SPECIES No other sirenians occur within the range of the West African Manatee. Atlantic Hump-backed and Common Bottlenose Dolphins occur with the West African Manatee in some estuaries and coastal areas. These cetaceans can be distinguished by the presence of a dorsal fin and blowhole.

BEHAVIOR West African Manatees are normally solitary. Individuals may form loose associations while resting, feeding, and mating, but the basic social unit of this species is the cow and calf. While not much is known about vocalizations, mothers and calves probably communicate through high-pitched squeaks and squeals. Manatees from Ivory Coast show strong daily activity

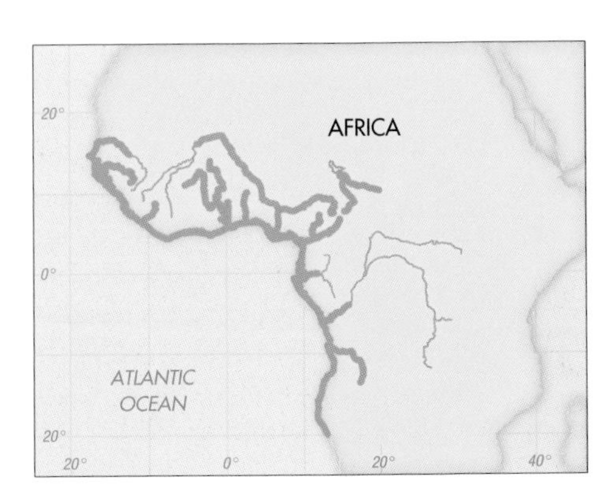

RANGE

WEST AFRICAN MANATEE
FAMILY TRICHECHIDAE

MEASUREMENTS AT BIRTH
LENGTH About 31–59″ (80–150 cm)
WEIGHT About 66 lb (30 kg)

MAXIMUM MEASUREMENTS
LENGTH **MALE** 11′ (3.3 m)
FEMALE 10′ (3 m)
WEIGHT About 3,000 lb (1,400 kg)

LIFE SPAN
May be more than 60 years

patterns, resting during the day and feeding and moving primarily at night. The increased nocturnal activity may be a response to hunting pressure. Manatees tend to rest in the middle of lagoons and rivers or underneath mangrove roots and overhanging vegetation, where they are more difficult to see when surfacing. They meander and also make seasonal movements between lagoons, along coastlines, and far up rivers, sometimes traveling hundreds of miles. Some have been known to move 45 miles (75 km) in two days.

REPRODUCTION There is virtually no information on the reproductive behavior of the West African Manatee, although researchers suspect that it is similar to that of the West Indian Manatee. Based on reports by local hunters as well as a single sighting by researchers, females in estrus are thought to attract mating herds of males. Aggregations in some areas during the beginning of the rainy season may also be reproductive in nature. Manatees inhabiting rivers and lakes in the region below the Sahel desert are reported to breed and give birth during periods of increased rains, presumably because the lactating females have better foraging opportunities and access to bank growth when water levels are high. Females give birth to a single calf, after a gestation of probably about a year. Calves are thought to remain with their mothers for more than a year.

FOOD AND FORAGING West African Manatees are mainly herbivorous. They eat a variety of floating, submergent, and emergent plants and bank growth (although where water is poorly lit or turbid, submerged vegetation is nonexistent or limited). In most areas manatees prefer emergent grasses such as those from the genera *Vossia, Echinochloa,* and *Paspalum.* Other important plant species include *Typha* species, *Phragmites* species, *Pennisetum purpureum, Rhizophora racemosa, Polygonum* species, *Alternanthera sessilis, Pistia stratiotes, Eichhornia crassipes, Cymodocea nodosa, Ruppia maritima,* and *Halodule wrightii.* West African Manatees have also been seen feeding on forest fruit that has fallen into the water and on the discarded peels of manioc (cassava), and they raid cultivated rice fields. Not strictly herbivorous, West African Manatees sometimes take fish from nets in rivers and estuaries of Sierra Leone and in the Senegal River, and there are reports of manatees occasionally eating bivalves. During the dry season when they are isolated in deep pools or lakes, they are reported to eat fine clay, algae, and detritus to survive.

STATUS AND CONSERVATION No good population estimates of the West African Manatee exist, although there are reports that the species is reduced. Based on informal surveys by fishermen and residents, the densest populations appear to occur in Senegal, Guinea-Bissau, Ivory Coast, Cameroon, and Gabon. The West African Manatee is legally protected everywhere it occurs; however, there has been little effort, with some local exceptions, to enforce existing laws. Manatees are prized for their meat and oil, and residents continue to hunt them with harpoons, traps, nets, and hooks. Evidence of this exists everywhere the manatees have been surveyed. West African Manatees are further threatened by habitat destruction, including the cutting of mangroves and the damming of rivers. They die in the turbines of hydroelectric dams and are sometimes fatally trapped in lakes, pools, and ponds during periods of lower water levels. A number of countries, in particular Senegal, Guinea-Bissau, Ivory Coast, Nigeria, Cameroon, and Gabon, have made efforts to establish protected areas for manatees. Manatees trapped in temporary ponds along the Senegal River are rescued and moved back into the main river.

Appendices

Regional Assemblages 494

Marine Mammal Morphology 499

Illustrated Glossary 500

Photo Credits 518

Index 522

Contributors 526

Acknowledgements 527

Regional Assemblages

The following lists group species of marine mammals by the region of the world in which they occur. There are seven regions: Arctic, North Pacific, North Atlantic, Antarctic, South Pacific, South Atlantic, and Indo-Pacific. The map below shows the boundaries of each region. Some species occur in more than one region (and on more than one list), and a few are found in all seven regions. These lists will help you narrow down the marine mammals that you may encounter in a particular part of the world and will facilitate species identification. Keep in mind that most species are found in only a portion of a given region. Refer to the range text and map for each species to see exactly where it occurs. A number of mammals in this guide do not inhabit marine environments; species that occur in freshwater and landlocked habitats are listed at the end of this section. Species are listed in the same order in which they appear in the species accounts.

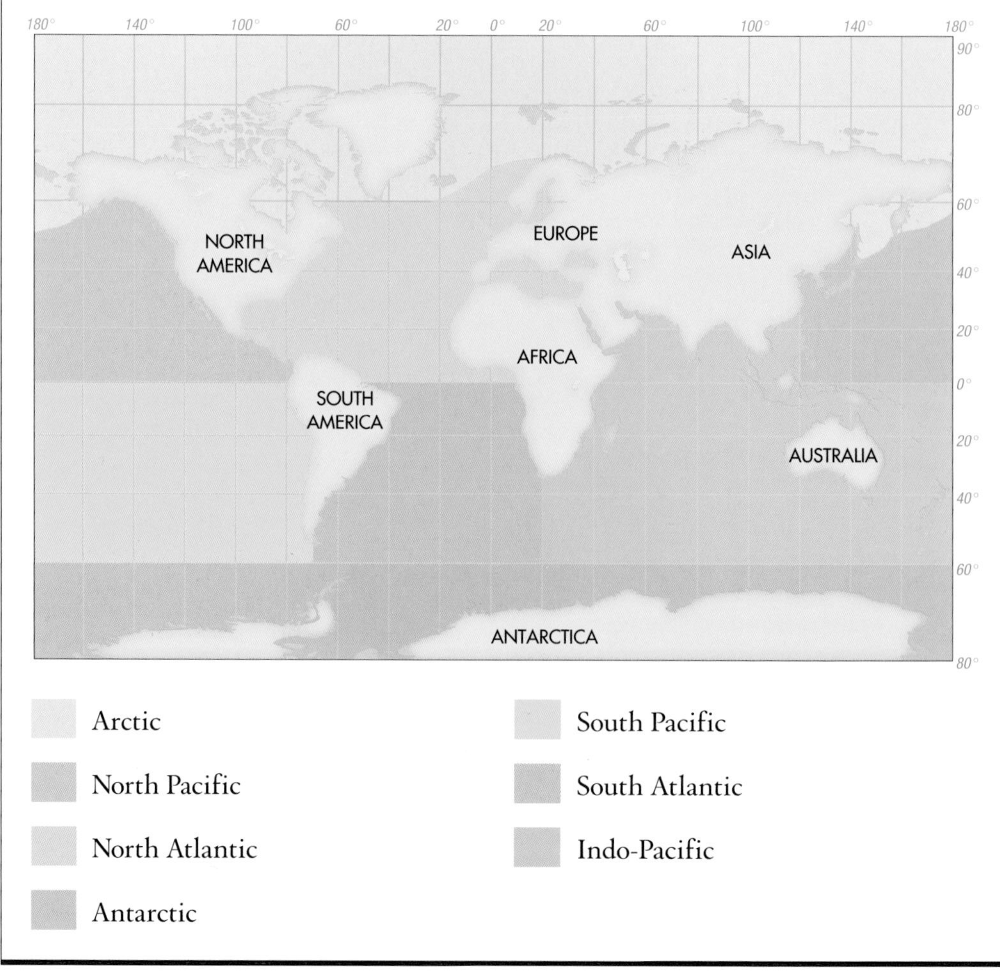

Arctic Region

POLAR BEAR AND OTTERS
Polar Bear *(p. 38)*

PINNIPEDS
Northern Fur Seal *(p. 86)*
Northern Sea Lion *(p. 94)*
Walrus *(p. 110)*
Bearded Seal *(p. 114)*
Harbor Seal *(p. 118)*
Largha Seal *(p. 122)*
Ringed Seal *(p. 125)*
Ribbon Seal *(p. 128)*
Gray Seal *(p. 138)*
Harp Seal *(p. 142)*
Hooded Seal *(p. 146)*

BALEEN WHALES
North Pacific Right Whale *(p. 190)*
Bowhead Whale *(p. 198)*
Gray Whale *(p. 204)*
Humpback Whale *(p. 208)*
Minke Whale *(p. 212)*
Sei Whale *(p. 226)*
Fin Whale *(p. 230)*
Blue Whale *(p. 234)*

SPERM WHALES
Sperm Whale *(p. 240)*

BEAKED WHALES
Baird's Beaked Whale *(p. 260)*
Northern Bottlenose Whale *(p. 268)*
Sowerby's Beaked Whale *(p. 280)*
Stejneger's Beaked Whale *(p. 296)*

BELUGA AND NARWHAL
Beluga *(p. 318)*
Narwhal *(p. 322)*

OCEAN DOLPHINS
White-beaked Dolphin *(p. 395)*
Atlantic White-sided Dolphin *(p. 398)*
Pacific White-sided Dolphin *(p. 402)*
Killer Whale *(p. 436)*
Long-finned Pilot Whale *(p. 440)*

PORPOISES
Harbor Porpoise *(p. 460)*
Dall's Porpoise *(p. 470)*

North Pacific Region

POLAR BEAR AND OTTERS
Sea Otter *(p. 42)*

PINNIPEDS
Guadalupe Fur Seal *(p. 72)*
Galápagos Fur Seal *(p. 82)*
Northern Fur Seal *(p. 86)*
Galápagos Sea Lion *(p. 90)*
California Sea Lion *(p. 90)*
Northern Sea Lion *(p. 94)*
Bearded Seal *(p. 114)*
Harbor Seal *(p. 118)*
Largha Seal *(p. 122)*
Ringed Seal *(p. 125)*
Ribbon Seal *(p. 128)*
Hawaiian Monk Seal *(p. 154)*
Northern Elephant Seal *(p. 162)*

BALEEN WHALES
North Pacific Right Whale *(p. 190)*
Gray Whale *(p. 204)*
Humpback Whale *(p. 208)*
Minke Whale *(p. 212)*
Bryde's Whale *(p. 222)*
Sei Whale *(p. 226)*
Fin Whale *(p. 230)*
Blue Whale *(p. 234)*

SPERM WHALES
Sperm Whale *(p. 240)*
Pygmy Sperm Whale *(p. 244)*
Dwarf Sperm Whale *(p. 244)*

BEAKED WHALES
Cuvier's Beaked Whale *(p. 254)*
Baird's Beaked Whale *(p. 260)*
Longman's Beaked Whale *(p. 266)*
Pygmy Beaked Whale *(p. 284)*
Hubbs' Beaked Whale *(p. 288)*
Ginkgo-toothed Beaked Whale *(p. 290)*
Blainville's Beaked Whale *(p. 294)*
Stejneger's Beaked Whale *(p. 296)*

BELUGA AND NARWHAL
Beluga *(p. 318)*
Narwhal *(p. 322)*

OCEAN DOLPHINS
Rough-toothed Dolphin *(p. 346)*
Indo-Pacific Hump-backed Dolphin *(p. 350)*
Common Bottlenose Dolphin *(p. 358)*
Indo-Pacific Bottlenose Dolphin *(p. 362)*
Pantropical Spotted Dolphin *(p. 366)*
Spinner Dolphin *(p. 374)*
Striped Dolphin *(p. 380)*
Fraser's Dolphin *(p. 384)*
Short-beaked Common Dolphin *(p. 388)*
Long-beaked Common Dolphin *(p. 392)*
Pacific White-sided Dolphin *(p. 402)*
Northern Right Whale Dolphin *(p. 418)*
Risso's Dolphin *(p. 422)*
Melon-headed Whale *(p. 426)*
Pygmy Killer Whale *(p. 430)*
False Killer Whale *(p. 433)*
Killer Whale *(p. 436)*
Short-finned Pilot Whale *(p. 444)*

PORPOISES
Finless Porpoise *(p. 456)*
Harbor Porpoise *(p. 460)*
Vaquita *(p. 464)*
Dall's Porpoise *(p. 470)*

SIRENIANS
Dugong *(p. 478)*

North Atlantic Region

POLAR BEAR AND OTTERS
Polar Bear *(p. 38)*

PINNIPEDS
Walrus *(p. 110)*
Bearded Seal *(p. 114)*
Harbor Seal *(p. 118)*
Ringed Seal *(p. 125)*
Gray Seal *(p. 138)*
Harp Seal *(p. 142)*
Hooded Seal *(p. 146)*
Mediterranean Monk Seal *(p. 150)*
Weddell Seal *(p. 166)*

BALEEN WHALES
North Atlantic Right Whale *(p. 190)*
Bowhead Whale *(p. 198)*
Humpback Whale *(p. 208)*
Minke Whale *(p. 212)*
Bryde's Whale *(p. 222)*
Sei Whale *(p. 226)*
Fin Whale *(p. 230)*
Blue Whale *(p. 234)*

SPERM WHALES
Sperm Whale *(p. 240)*
Pygmy Sperm Whale *(p. 244)*
Dwarf Sperm Whale *(p. 244)*

BEAKED WHALES
Cuvier's Beaked Whale *(p. 254)*
Longman's Beaked Whale *(p. 266)*
Northern Bottlenose Whale *(p. 268)*
True's Beaked Whale *(p. 276)*
Gervais' Beaked Whale *(p. 278)*
Sowerby's Beaked Whale *(p. 280)*
Blainville's Beaked Whale *(p. 294)*

BELUGA AND NARWHAL
Beluga *(p. 318)*
Narwhal *(p. 322)*

OCEAN DOLPHINS
Rough-toothed Dolphin *(p. 346)*
Atlantic Hump-backed Dolphin *(p. 350)*
Tucuxi *(p. 354)*
Common Bottlenose Dolphin *(p. 358)*
Pantropical Spotted Dolphin *(p. 366)*
Atlantic Spotted Dolphin *(p. 370)*
Spinner Dolphin *(p. 374)*
Clymene Dolphin *(p. 378)*
Striped Dolphin *(p. 380)*
Fraser's Dolphin *(p. 384)*
Short-beaked Common Dolphin *(p. 388)*
Long-beaked Common Dolphin *(p. 392)*
White-beaked Dolphin *(p. 395)*
Atlantic White-sided Dolphin *(p. 398)*
Risso's Dolphin *(p. 422)*
Melon-headed Whale *(p. 426)*
Pygmy Killer Whale *(p. 430)*
False Killer Whale *(p. 433)*
Killer Whale *(p. 436)*
Long-finned Pilot Whale *(p. 440)*
Short-finned Pilot Whale *(p. 444)*

PORPOISES
Harbor Porpoise *(p. 460)*

SIRENIANS
West Indian Manatee *(p. 482)*
West African Manatee *(p. 490)*

Antarctic Region

PINNIPEDS
Antarctic Fur Seal *(p. 62)*
Southern Elephant Seal *(p. 158)*
Weddell Seal *(p. 166)*
Crabeater Seal *(p. 170)*
Ross Seal *(p. 174)*
Leopard Seal *(p. 177)*

BALEEN WHALES
Humpback Whale *(p. 208)*
Dwarf Minke Whale *(p. 216)*
Antarctic Minke Whale *(p. 218)*
Sei Whale *(p. 226)*
Fin Whale *(p. 230)*
Blue Whale *(p. 234)*

SPERM WHALES
Sperm Whale *(p. 240)*

BEAKED WHALES
Arnoux's Beaked Whale *(p. 258)*
Southern Bottlenose Whale *(p. 272)*
Gray's Beaked Whale *(p. 282)*

OCEAN DOLPHINS
Commerson's Dolphin *(p. 332)*
Hourglass Dolphin *(p. 414)*
Killer Whale *(p. 436)*
Long-finned Pilot Whale *(p. 440)*

South Pacific Region

POLAR BEAR AND OTTERS
Marine Otter *(p. 46)*

PINNIPEDS
Juan Fernández Fur Seal *(p. 69)*
New Zealand Fur Seal *(p. 76)*
South American Fur Seal *(p. 79)*
Galápagos Fur Seal *(p. 82)*
Galápagos Sea Lion *(p. 90)*
South American Sea Lion *(p. 106)*

BALEEN WHALES
Southern Right Whale *(p. 194)*
Pygmy Right Whale *(p. 202)*
Humpback Whale *(p. 208)*
Dwarf Minke Whale *(p. 216)*
Antarctic Minke Whale *(p. 218)*
Bryde's Whale *(p. 222)*
Sei Whale *(p. 226)*
Fin Whale *(p. 230)*
Blue Whale *(p. 234)*

SPERM WHALES
Sperm Whale *(p. 240)*
Pygmy Sperm Whale *(p. 244)*
Dwarf Sperm Whale *(p. 244)*

BEAKED WHALES
Cuvier's Beaked Whale *(p. 254)*
Arnoux's Beaked Whale *(p. 258)*
Shepherd's Beaked Whale *(p. 264)*
Longman's Beaked Whale *(p. 266)*
Southern Bottlenose Whale *(p. 272)*
Hector's Beaked Whale *(p. 274)*
Gray's Beaked Whale *(p. 282)*
Pygmy Beaked Whale *(p. 284)*
Andrews' Beaked Whale *(p. 286)*
Ginkgo-toothed Beaked Whale *(p. 290)*
Strap-toothed Whale *(p. 292)*
Blainville's Beaked Whale *(p. 294)*

OCEAN DOLPHINS
Commerson's Dolphin *(p. 332)*
Chilean Dolphin *(p. 336)*
Rough-toothed Dolphin *(p. 346)*
Common Bottlenose Dolphin *(p. 358)*
Pantropical Spotted Dolphin *(p. 366)*
Spinner Dolphin *(p. 374)*
Striped Dolphin *(p. 380)*
Fraser's Dolphin *(p. 384)*
Short-beaked Common Dolphin *(p. 388)*
Long-beaked Common Dolphin *(p. 392)*
Dusky Dolphin *(p. 406)*
Peale's Dolphin *(p. 410)*
Hourglass Dolphin *(p. 414)*
Southern Right Whale Dolphin *(p. 418)*
Risso's Dolphin *(p. 422)*
Melon-headed Whale *(p. 426)*
Pygmy Killer Whale *(p. 430)*
False Killer Whale *(p. 433)*
Killer Whale *(p. 436)*
Long-finned Pilot Whale *(p. 440)*
Short-finned Pilot Whale *(p. 444)*

PORPOISES
Burmeister's Porpoise *(p. 466)*
Spectacled Porpoise *(p. 468)*

South Atlantic Region

POLAR BEAR AND OTTERS
Marine Otter *(p. 46)*

PINNIPEDS
South African Fur Seal *(p. 58)*
Antarctic Fur Seal *(p. 62)*
Subantarctic Fur Seal *(p. 66)*
South American Fur Seal *(p. 79)*
South American Sea Lion *(p. 106)*
Southern Elephant Seal *(p. 158)*
Leopard Seal *(p. 177)*

BALEEN WHALES
Southern Right Whale *(p. 194)*
Pygmy Right Whale *(p. 202)*
Humpback Whale *(p. 208)*
Dwarf Minke Whale *(p. 216)*
Antarctic Minke Whale *(p. 218)*
Bryde's Whale *(p. 222)*
Sei Whale *(p. 226)*
Fin Whale *(p. 230)*
Blue Whale *(p. 234)*

SPERM WHALES
Sperm Whale *(p. 240)*
Pygmy Sperm Whale *(p. 244)*
Dwarf Sperm Whale *(p. 244)*

BEAKED WHALES
Cuvier's Beaked Whale *(p. 254)*
Arnoux's Beaked Whale *(p. 258)*
Shepherd's Beaked Whale *(p. 264)*
Southern Bottlenose Whale *(p. 272)*
Hector's Beaked Whale *(p. 274)*
Gervais' Beaked Whale *(p. 278)*
Gray's Beaked Whale *(p. 282)*
Andrews' Beaked Whale *(p. 286)*
Strap-toothed Whale *(p. 292)*
Blainville's Beaked Whale *(p. 294)*

RIVER DOLPHINS
Franciscana *(p. 314)*

OCEAN DOLPHINS
Commerson's Dolphin *(p. 332)*
Heaviside's Dolphin *(p. 339)*
Rough-toothed Dolphin *(p. 346)*
Atlantic Hump-backed Dolphin *(p. 350)*
Tucuxi *(p. 354)*
Common Bottlenose Dolphin *(p. 358)*
Pantropical Spotted Dolphin *(p. 366)*
Atlantic Spotted Dolphin *(p. 370)*
Spinner Dolphin *(p. 374)*
Clymene Dolphin *(p. 378)*
Striped Dolphin *(p. 380)*
Fraser's Dolphin *(p. 384)*
Short-beaked Common Dolphin *(p. 388)*

Continued

South Atlantic Region, continued

Long-beaked Common Dolphin *(p. 392)*
Dusky Dolphin *(p. 406)*
Peale's Dolphin *(p. 410)*
Hourglass Dolphin *(p. 414)*
Southern Right Whale Dolphin *(p. 418)*
Risso's Dolphin *(p. 422)*
Melon-headed Whale *(p. 426)*
Pygmy Killer Whale *(p. 430)*
False Killer Whale *(p. 433)*
Killer Whale *(p. 436)*
Long-finned Pilot Whale *(p. 440)*
Short-finned Pilot Whale *(p. 444)*

PORPOISES

Burmeister's Porpoise *(p. 466)*
Spectacled Porpoise *(p. 468)*

SIRENIANS

West Indian Manatee *(p. 482)*
West African Manatee *(p. 490)*

Indo-Pacific Region

PINNIPEDS

South African Fur Seal *(p. 58)*
Australian Fur Seal *(p. 58)*
Antarctic Fur Seal *(p. 62)*
Subantarctic Fur Seal *(p. 66)*
New Zealand Fur Seal *(p. 76)*
Australian Sea Lion *(p. 98)*
New Zealand Sea Lion *(p. 102)*
Southern Elephant Seal *(p. 158)*
Weddell Seal *(p. 166)*
Crabeater Seal *(p. 170)*
Ross Seal *(p. 174)*
Leopard Seal *(p. 177)*

BALEEN WHALES

Southern Right Whale *(p. 194)*
Pygmy Right Whale *(p. 202)*
Humpback Whale *(p. 208)*
Dwarf Minke Whale *(p. 216)*
Antarctic Minke Whale *(p. 218)*
Bryde's Whale *(p. 222)*
Sei Whale *(p. 226)*
Fin Whale *(p. 230)*
Blue Whale *(p. 234)*

SPERM WHALES

Sperm Whale *(p. 240)*
Pygmy Sperm Whale *(p. 244)*
Dwarf Sperm Whale *(p. 244)*

BEAKED WHALES

Cuvier's Beaked Whale *(p. 254)*
Arnoux's Beaked Whale *(p. 258)*
Shepherd's Beaked Whale *(p. 264)*
Longman's Beaked Whale *(p. 266)*
Southern Bottlenose Whale *(p. 272)*
Hector's Beaked Whale *(p. 274)*
True's Beaked Whale *(p. 276)*
Gray's Beaked Whale *(p. 282)*
Pygmy Beaked Whale *(p. 284)*
Andrews' Beaked Whale *(p. 286)*
Ginkgo-toothed Beaked Whale *(p. 290)*
Strap-toothed Whale *(p. 292)*
Blainville's Beaked Whale *(p. 294)*

OCEAN DOLPHINS

Commerson's Dolphin *(p. 332)*
Hector's Dolphin *(p. 342)*
Rough-toothed Dolphin *(p. 346)*
Indo-Pacific Hump-backed Dolphin *(p. 350)*
Common Bottlenose Dolphin *(p. 358)*
Indo-Pacific Bottlenose Dolphin *(p. 362)*
Pantropical Spotted Dolphin *(p. 366)*
Spinner Dolphin *(p. 374)*
Striped Dolphin *(p. 380)*
Fraser's Dolphin *(p. 384)*
Short-beaked Common Dolphin *(p. 388)*
Long-beaked Common Dolphin *(p. 392)*
Dusky Dolphin *(p. 406)*
Hourglass Dolphin *(p. 414)*
Southern Right Whale Dolphin *(p. 418)*
Risso's Dolphin *(p. 422)*
Melon-headed Whale *(p. 426)*
Pygmy Killer Whale *(p. 430)*
False Killer Whale *(p. 433)*
Killer Whale *(p. 436)*
Long-finned Pilot Whale *(p. 440)*
Short-finned Pilot Whale *(p. 444)*
Irrawaddy Dolphin *(p. 448)*

PORPOISES

Finless Porpoise *(p. 456)*
Spectacled Porpoise *(p. 468)*

SIRENIANS

Dugong *(p. 478)*

Freshwater and Landlocked Habitats

Several species of "marine" mammals occur in freshwater or landlocked habitats. Some species occupy both marine and non-marine environments and are mentioned in the above lists as well as here.

SOUTH AMERICA

Amazon River Dolphin *(p. 306)*
Tucuxi *(p. 354)*
Amazonian Manatee *(p. 486)*

AFRICA

West African Manatee *(p. 490)*

ASIA

Caspian Seal *(p. 131)*
Baikal Seal *(p. 134)*
Ganges and Indus River Dolphins *(p. 302)*
Yangtze River Dolphin *(p. 310)*

Marine Mammal Morphology

PARTS OF A PINNIPED

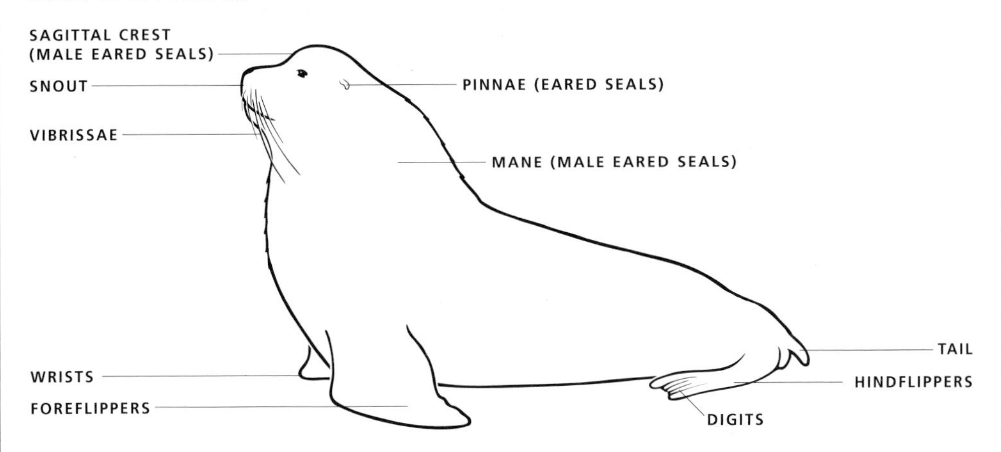

PARTS OF A BALEEN WHALE

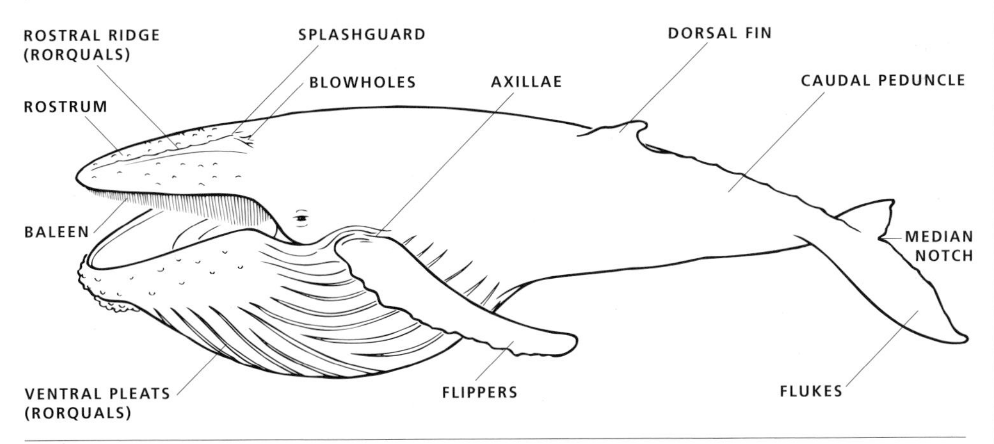

PARTS OF A TOOTHED CETACEAN (DOLPHINS, BEAKED WHALES, PORPOISES, ETC.)

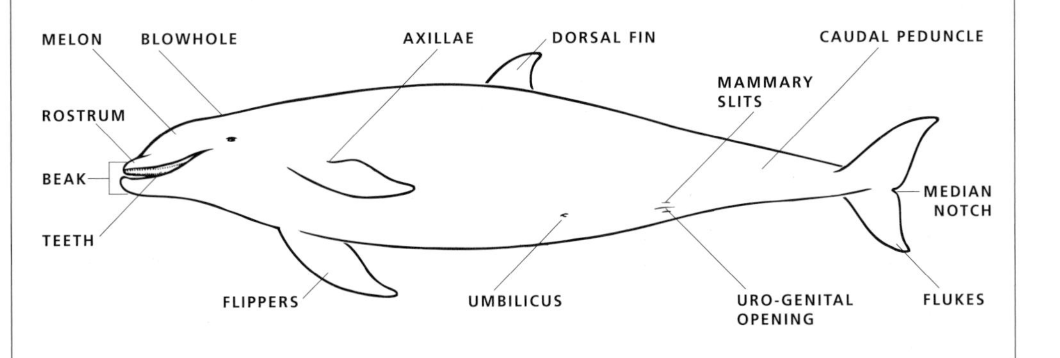

Illustrated Glossary

TERMS THAT ARE ILLUSTRATED WITH A PHOTOGRAPH HAVE A FOLLOWING THEIR DEFINITIONS.

ABORIGINAL HUNTING Hunting by people who are indigenous to a region. It is generally assumed that such hunting is low-impact ("sustainable") and traditional, and that products are used mainly for subsistence.

AGGREGATION A group of animals that converge on an area, usually for reasons of reproduction or feeding.

AMPHIPODS Crustaceans belonging to the order Amphipoda. Amphipods are a food source for some baleen whales.

ANO-GENITAL AREA The area on the ventral surface of the body generally near the anus and the genital opening.

ANTARCTIC CONVERGENCE A dynamic water-mass boundary, marked by sharp surface temperature gradients, that divides cold Antarctic waters from the more temperate water masses to the north; generally where the sea surface temperature ranges between about 35 and 40°F (1.5–5°C). Also called the Antarctic Polar Front.

ANTARCTIC REGIONS Generally in the vicinity of Antarctica; south of the Antarctic Convergence in latitudes greater than about 60°S. The marine Antarctic is sometimes defined as the area of sea adjoining the continent and covered by pack ice during a significant portion of the year. *See* map on back endpaper.

ANTERIOR Refers to areas on or toward the front part of the body.

AQUATIC VEGETATION Plant life that occurs entirely or primarily in water, including submerged, floating, or emergent forms.

ARCTIC BASIN Referring to the Arctic Ocean, or generally to waters north of the Arctic Circle (66°33'N latitude). *See* map on back endpaper.

ARTISANAL WHALING Shore-based, low-technology whaling; intermediate between subsistence and industrial whaling, with limited local distribution of products.

ATOLL An organic reef that surrounds a lagoon and is bordered by open sea, often with low sand islands.

AUSTRAL Southern; herein generally refers to the Southern Hemisphere.

AXILLA (*plural:* axillae) The "armpit" region at the posterior base of the flipper of a cetacean.

BALAENOPTERID A species in the baleen whale family Balaenopteridae; rorqual.

BALEEN A horny, keratinous substance that occurs as a series of comb-like plates suspended from the upper jaws of mysticete (baleen) whales (suborder Mysticeti); fibrous fringes along the inner surfaces of the plates filter and trap prey (zooplankton and fish) inside the mouth.

BARNACLES A common name for crustacean species in the subclass Cirripedia; in the adult stage they form a hard shell and remain attached to a submerged surface, which for some species can be a whale.

AGGREGATION *South American Sea Lions*

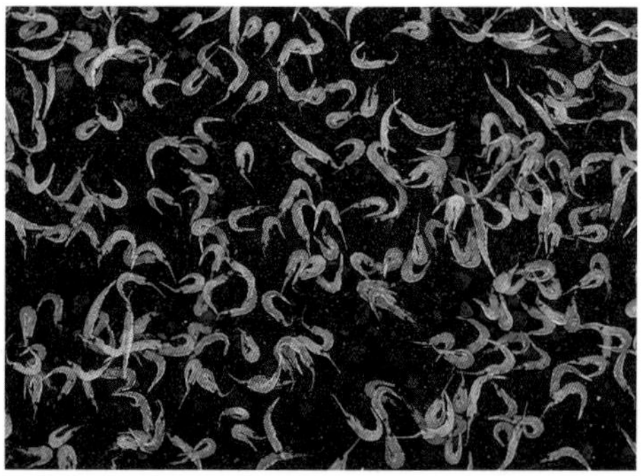

AMPHIPODS

BALEEN *Gray Whale*

BARNACLES *Gray Whale*

BARRAGE A term applied in the Indian Subcontinent to low gated dams built in the plains and used primarily to divert water for irrigation and flood control, as distinct from high dams built primarily in canyons and upper reaches to impound water for hydroelectric power generation and flow regulation.

BATHYPELAGIC Deep-living in offshore waters, especially deeper than about 2,000 feet (600 m).

BEAK The forward-projecting jaws of certain toothed cetaceans.

BENTHIC Living at or in, or associated with the bottom of a body of water.

BIOLUMINESCENT Describes animals that emit visible light.

BIOSPHERE Term referring collectively to all living things of the earth and the areas that they inhabit. The Biosphere Reserve Program of the United Nations Educational, Scientific, and Cultural Organization (UNESCO) encourages establishment of zoned, multiple-use areas for nature protection and sustainable use.

BLACKFISH A colloquial term adopted from American whalers and sometimes applied to pilot whales and other superficially similar species, including False Killer, Pygmy Killer, and Melon-headed Whales.

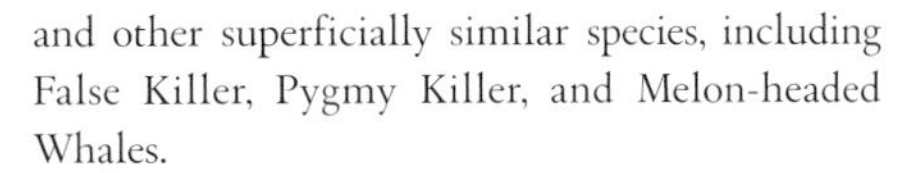

BLAZE A streak-like marking on the body of an animal; usually white or light-colored set against a dark background.

BLOW The visible cloud of warm moist air expelled from a whale's lungs as it surfaces. It can appear tall and columnar or low and bushy, depending on the species.

BLOWHOLE Nostril, or respiratory opening, of a cetacean; odontocetes have one, mysticetes two.

BLUBBER Layer of fatty tissue located immediately beneath the skin of most marine mammals. It insulates the body core and stores energy.

BLUFF Used to describe a forehead or melon of a cetacean that has a steep broad front.

BODY SLAM When a whale or dolphin falls onto the water surface following a breach and causes a loud report and conspicuous splash.

BOREAL Northern; herein generally refers to the Northern Hemisphere. In a broader biogeographical context, refers specifically to the temperate and subtemperate zones of the Northern Hemisphere.

BEAK *Indo-Pacific Bottlenose Dolphin*

BLOW *Humpback Whales*

BOW RIDING *Short-beaked Common Dolphins with Fin Whale*

BREACHING *Humpback Whale*

BOW RIDING A habit of some toothed cetaceans of placing themselves immediately ahead of a vessel or large whale to experience the assisted locomotion provided by the pressure wave.

BOW WAVE Crest of water that precedes a vessel as it moves along the surface.

BRACKISH Refers to water with salt concentrations intermediate between fresh and sea water (about 0.5 to 35 parts salt per 1,000 parts water).

BREACHING A whaler's term widely used to refer to cetaceans leaping or jumping clear of the water. A single leap is called a breach.

BREAKER A wave breaking on the shore or over a reef.

BREATHING HOLE A hole in the ice through which some species of seals surface to breathe. Seals use their teeth or the claws of their foreflippers to maintain the holes.

BUBBLE FEEDING A cooperative feeding technique of Humpback Whales, in which they use "clouds" or "nets" of bubbles to corral and trap small fish or invertebrates.

BULL A male seal, whale, or sirenian, especially an adult male.

BYCATCH The portion of a catch, or harvest, that consists of non-target organisms; animals taken incidentally during fishing operations.

CALF An infant cetacean, Walrus, or sirenian; an unweaned youngster.

CALLOSITIES Raised, thickened, and roughened patches of skin tissue on the head of a right whale. Callosities generally harbor large colonies of whale lice. The number, size, and configuration of callosities differ between individuals, facilitating photo-identification for research.

CALVING INTERVAL The period of time from one birth to the next, generally applicable only to cetaceans and sirenians.

CANINE The "dog" or "eye" tooth of mammals, usually conical and pointed, one on each side of the upper and lower jaws, between the incisors and premolars; used primarily for grasping prey.

CANYON A V-shaped submarine depression, or valley, with relatively steep sides and progressive deepening in a direction away from shore.

CAPE Dark region on the back of a toothed whale that begins anterior to the dorsal fin and often dips onto

BREATHING HOLE *Harp Seal*

BUBBLE FEEDING *Humpback Whales*

CALLOSITIES *North Atlantic Right Whale*

CAPE *Pantropical Spotted Dolphin*

the sides to varying degrees and in varying formations; sometimes confused with saddle.

CATHOLIC Refers to feeding habits that are generalized or unspecialized; unselective.

CAUDAL PEDUNCLE That portion of a cetacean's or sirenian's body behind the dorsal fin (only cetaceans) and anterior to the flukes. Also called the tail stock.

CEPHALOPOD A benthic or swimming mollusk that possesses a large head, large eyes, and a circle of arms or tentacles around the mouth; the shell can be external, internal, or absent, and an ink sac is usually present. Examples are squid and octopus.

CETACEAN A species in the mammalian order Cetacea, which includes whales, dolphins, and porpoises.

CHEEK TOOTH *See* Postcanine.

CHEVRON V-shaped, light-colored marking on the back or side of a cetacean.

CIRCUMGLOBAL Ranging all the way around the world.

CIRCUMPOLAR Ranging all the way around high northern or high southern latitudes.

COASTAL FLAT A level stretch of beach.

COLONY A highly integrated group of animals; herein refers specifically to land-breeding pinnipeds.

COMMERCIAL HUNTING Killing wild animals for profit.

CONFLUENCE Where two or three large rivers converge or where a tributary joins a main channel.

CONTINENTAL SHELF Submerged, relatively flat and gently sloping part of a continent, from shore to a depth of about 600 feet (180 m), or 100 fathoms; average shelf width worldwide is about 30 miles (50 km), with a range from zero to more than 800 miles (1,300 km).

CONTINENTAL SLOPE That part of the ocean floor beyond the edge of the continental shelf where the bottom drops down to the deep seafloor.

CONVENTION FOR THE CONSERVATION OF ANTARCTIC SEALS (CCAS) A part of the Antarctic Treaty System intended to regulate seal harvesting, conserve seal populations, and improve scientific knowledge; came into force in 1978.

CONVENTION ON INTERNATIONAL TRADE IN ENDANGERED SPECIES (CITES) OF WILD FAUNA AND FLORA Signed in Geneva, Switzerland, in 1973; came into force in 1975; its purpose is to regulate and monitor trade in threatened or endangered animals and plants.

COOKIE-CUTTER SHARK (*Isistius brasiliensis*) A small tropical shark with lips that function like suckers and allow it to attach temporarily to cetaceans. The resulting bite wounds appear as round or oval lacerations, with radiating groove patterns.

COPEPODS Minute, shrimp-like crustaceans. Copepods are a food source of some baleen whales.

COSMOPOLITAN Occurring essentially worldwide, in all major oceanic regions.

COUNTERSHADING A form of camouflage exhibited by many fish and cetaceans, with dark upper body surfaces and lighter undersides. When viewed from above the darker dorsal surface blends in with the water; from below the lighter ventral surface matches the light coming from the sky, making the animal hard to see.

COW A female whale, seal, or sirenian, especially an adult female accompanied by a calf.

CRUSTACEAN An invertebrate that breathes via gills or similar structures and has a segmented body, commonly covered by a shell; includes barnacles, crabs, shrimps, and lobsters.

CRYPTIC Tending to be concealed and thus difficult to detect and observe.

CURRENT Prevailing directional flow of water that determines oceanic circulation, maintained or influenced by, for example, winds, tides, or density

DORSAL FIN *Risso's Dolphins*

ECHELON *Sperm Whales*

gradients. The currents carry warm water away from the Equator and cold water away from the poles. Where these currents brush against continents or each other, or where they pass over underwater features, such as canyons and seamounts, they generate upwelling. *See* map on back endpaper.

DDT Dichlorodiphenyltrichloroethane; an organic pesticide banned in the United States in the early 1970s, in part due to its toxic effect on vertebrates. Still used in other parts of the world, especially in South America and Asia.

DEEP SCATTERING LAYER (DSL) A stratified population of organisms in ocean waters that causes sound to scatter when recorded on an echo sounder. The layers, from 160 to 650 feet (50–200 m) thick, occur within 650 feet (200 m) of the surface at night and at depths of about 1,000 feet (300 m) during the day. In other words, these animals migrate vertically in the water column in response to light conditions.

DELAYED IMPLANTATION A normal variation of the reproductive process exhibited in some mammals, including most species of pinnipeds. The embryo develops for only a short period after the egg is fertilized and then remains quiescent until it attaches to the uterine wall about two to three months later, after which growth resumes. Also called embryonic diapause.

DELPHINID A species in the toothed whale family Delphinidae; ocean dolphin.

DEMERSAL Refers to animals that live near the bottom of a body of water.

DENNING Entering a den or lair to rest, hibernate, or give birth and nurse young.

DENTITION The array of teeth possessed by an animal; *differentiated* dentition means that the teeth are not all of the same form; most toothed cetaceans have an *undifferentiated* dentition, meaning that their teeth are all the same.

DEPTH CONTOUR A contour line based on depth rather than elevation; an imaginary line connecting points of equal depth. Also called an isobath.

DEPTH GRADIENT The rate of increase or decrease in depth.

DIATOMS Microscopic phytoplankton characterized by a wall of overlapping halves ("valves") impregnated with silica. Diatoms sometimes coat the bodies of Blue Whales, appearing as a yellowish sheen on the undersides.

DIMORPHISM Occurrence of two distinct morphological types in a single population. *See* Sexual dimorphism.

DISJUNCT Describes a geographical distribution of a species that is characterized by major gaps; discontinuous.

DISPLAY Any behavior that conveys information, usually to members of the same species or to predators; often used during mating or territory defense.

DISTRIBUTION The range of a species or population; where animals occur.

DIURNAL Daily; pertains especially to actions that are completed within approximately 24 hours and that recur on a 24-hour cycle. Also used to mean active by day, as opposed to nocturnal.

DOLPHIN A term applied to small toothed cetaceans with conical teeth.

DORSAL Pertaining to the upper surface of the back or other body parts.

DORSAL CAPE *See* Cape.

DORSAL FIN The fin along the midline of the back of a cetacean. Most cetaceans have a dorsal fin.

DORSAL MANTLE LENGTH The mantle is the fleshy, membranous covering of a mollusk, responsible for secreting the shell (if present); the dorsal length of the mantle is used as an index of body size.

FALCATE *Pacific White-sided Dolphin*

DORSO-VENTRALLY COMPRESSED When diameter from top to bottom is less than from side to side; flattened in the horizontal plane.

DRIFT NET Fishing net suspended in the water vertically so that drifting or swimming animals will become trapped or entangled in the mesh; a gillnet that is not anchored.

DRIVE FISHERY A hunt for toothed cetaceans in which the animals are driven toward shore, usually by noise making or shooting and involving multiple boats; the animals are killed in shallow nearshore waters or exposed intertidal beach areas.

EAR FLAPS *See* Pinnae.

ECHELON Step-like formation of individuals in which each one is to the left or right of and parallel to the one in front of it.

ECHELON FEEDING When a group of whales move in coordinated fashion, side by side, while feeding (such as skim feeding by Bowhead Whales).

ECHOLOCATION Production of sounds that allow an animal to orient itself and locate objects by means of the returning sound waves or echoes.

ECOSYSTEM An ecological community and its physical environment considered as a whole.

EL NIÑO Short for El Niño Southern Oscillation (ENSO), a series of oceanographic events that occur in the eastern Pacific Ocean every few years, characterized by changes in prevailing current patterns, increases in sea temperature, and declines in upwelling. El Niños result in reduced food supplies for marine mammals.

EMERGENT Describes aquatic plants that reach above the water surface, as opposed to submerged or floating vegetation.

ENDANGERED SPECIES ACT A law passed by the United States Congress in 1973 to establish a comprehensive program to identify, protect, and facilitate the recovery of plant and animal species in danger of extinction. The U.S. Endangered Species List includes numerous marine mammals.

ENDEMIC Restricted to a particular region.

EPIBENTHIC Living just above the bottom of a body of water.

ESCARPMENT A steep submarine cliff that separates two relatively level areas of different depth.

ESTRUS (*adjective:* estrous) Period in which a female mammal ovulates and is receptive to mating. Also called heat.

ESTUARY A semi-enclosed tidal coastal body of at least partially salt water with free connection to the sea; or the lower end of a freshwater river.

EUPHAUSIID *See* Krill.

EXCLUSIVE ECONOMIC ZONE (EEZ) The area, extending from shore to 200 nautical miles offshore, in which nations have the exclusive sovereign right to manage resources. The EEZ concept was established in principle during the 1970s and became officially recognized in 1983 under the United Nations Convention on the Law of the Sea.

EXTANT Refers to a population or species currently in existence; not extinct.

EXTINCTION The death of all animals of a population or species.

EXTIRPATE To hunt or fish to extinction; exterminate; normally used to refer to a geographical population rather than a species.

EXTRALIMITAL Outside the normal range of a species or population.

FALCATE Shaped like a sickle; back-curved; refers to a dorsal fin with a concave rear margin.

FAMILY A taxon that is a subset of an order and that contains one or more genera. Latin names of mammal families end in "-idae." *See* Taxonomy.

FAST ICE Sea ice that is attached to land, sometimes extending out from shore for many miles. Also called landfast ice.

FATHOM A linear measure used mainly for marine water depths that is equal to 6 feet (1.83 m).

FILTER FEEDING The process by which baleen whales and Crabeater Seals trap prey inside the mouth by straining sea water through their baleen and teeth, respectively.

FISSIPED An obsolete technical term, still used informally to refer to animals in the order Carnivora other than the pinnipeds; formerly a member of the suborder Fissipeda or Fissipedia (no longer recognized).

FLIPPERS *Humpback Whale*

FLIPPERS *Northern Fur Seal*

FLANK Side of the body; used mainly to refer to the side of the posterior half of the body.

FLIPPERS Variably shaped, often paddle-like limbs of a cetacean, pinniped, or sirenian.

FLOODPLAIN Level area near a river channel; may be submerged by floodwaters.

FLUKES The two horizontally flattened, fin-like structures that comprise a cetacean's tail.

FOREFLIPPERS Fin-like anterior limbs or appendages of a pinniped, used by otariids for balance and propulsion in the water and for supporting their weight on land; used by phocids for steering in the water and for scratching and excavating on land and ice. *See* Flippers.

FORESIDE A side of the anterior half of the body.

FRAGMENTED Describes a distribution or population that is noncontinuous or disjunct, usually with the connotation that this condition was caused either directly or indirectly by human activities.

FUSIFORM Tapering at both ends; spindle-shaped.

GAPE The margin-to-margin distance between open jaws.

GENETIC VARIABILITY Amount of diversity within the genetic composition of a population; a population's potential for adaptation in response to changing environmental conditions.

GENUS (*plural:* genera) A taxon that is a subset of a family and that contains one or more species. *See* Taxonomy.

GESTATION The process of carrying young in the uterus from conception to birth.

GILLNET A net that is suspended vertically in the water column so that fish (and other organisms) swim into it and become entangled by their gills or other body structures as they try to back out.

GREGARIOUS Tending to occur in fairly large groups, rather than as solitary individuals or in small groups; sociable.

GROUNDSWELL An undulation in the ocean, with deep rolling waves, often caused by a distant storm.

GROWTH LAYER GROUP (GLG) Consists of an adjacent pair of light and dark bands in the dentine or cementum of a sectioned tooth; in most species, approximately one GLG is deposited per year; used to determine age.

GUARD HAIRS Relatively long, stiff, flattened hairs that provide an outer protective covering for the underfur of a mammal.

GULP FEEDING Method of feeding by baleen whales, mainly the balaenopterids, that involves rapid intake of large volumes of prey-laden sea water, with attendant throat distention and swallowing of prey once the water has been expelled through the baleen.

HABITAT The organisms and physical environment in a particular place; an organism's ecological support (maintenance) system.

HAREM A group of breeding females associated with a single male.

FLUKES *Humpback Whale*

HAREM *New Zealand Sea Lions*

HAUL OUT *(verb)* The process by which pinnipeds crawl or pull themselves out of the water onto land or ice.

HAULOUT *(noun)* The site at which pinnipeds come out of the water to rest or molt.

HERBIVOROUS Feeding mostly or entirely on plant matter; non-carnivorous. Sirenians are herbivorous.

HIGH ARCTIC Generally referring to islands within the Arctic Basin, such as Svalbard, northern Greenland, and the Canadian Arctic islands, where sea-ice coverage is heavy during much of the year.

HIGH-ENERGY Used to describe coastlines with large waves crashing against the beach; exposed or unprotected from intense wave action.

HINDFLIPPERS Fin-like rear limbs or appendages of a pinniped, used for propulsion by phocids and for steering by otariids. *See* Flippers.

HOME RANGE The area that an animal learns thoroughly and patrols regularly. The home range may or may not be defended; those portions that are defended constitute the territory.

HYBRID An offspring that arises from the interbreeding of two different taxa.

ICE FLOE A large mass of sea ice (pack ice) kept in motion by winds, currents, and wave action.

ICE SHEET A large expanse of glacial ice, sometimes of continental scale and reaching a thickness of more than 6,000 feet (1,800 m).

ICHTHYOLOGIST A student of, or expert on, fish.

INCIDENTAL Refers to the unintentional or accidental capture of nontarget animals.

INCISOR One of the chisel-shaped teeth in the front of the mouth of mammals.

INSHORE Nearshore; can also mean in embayments or other partially enclosed water bodies.

INTERGRADE Refers to intermediate forms; for example, populations with characteristics that are intermediate between those of populations adjacent on either side are said to be intergrading.

INTERNATIONAL UNION FOR THE CONSERVATION OF NATURE AND NATURAL RESOURCES (IUCN) (also known as The World Conservation Union) A conservation body established in 1948 and based in Switzerland that includes governments and nongovernmental organizations as members; best known for its Red Lists of threatened and endangered life-forms.

INTERNATIONAL WHALING COMMISSION (IWC) Body formed under the International Convention for the Regulation of Whaling, signed in Washington, D.C., in 1946; membership is open to any nation that formally adheres to the convention; responsible for regulation of the whaling industry and conservation of whales.

INTERTIDAL The zone between mean high-water and mean low-water levels; the littoral zone.

IN UTERO Within the uterus, or womb; development that occurs between conception and birth.

INVERTEBRATE An animal without a backbone (spinal column).

HAUL OUT *Weddell Seal*

HAULOUT *California Sea Lions*

JUG HANDLE *New Zealand Fur Seal*

KELP *South American Sea Lion*

IRRUPTION A sporadic or rare occurrence of a very great abundance of a species.

JUG HANDLE An expression adopted from sealers, used to describe the appearance of sea lions or fur seals resting at the surface with both hindflippers and one foreflipper exposed (to conserve heat). Animals in this position are said to be jugging.

JUVENILE Immature or preadult. In most marine mammals, the term is applied to weaned individuals that are not yet sexually mature.

KEEL A deepening or thickening of the body form, particularly in the caudal peduncle of some cetaceans. A keel can be dorsal, ventral, or both. A body that possesses a keel is said to be keeled.

KELP A type of algae that is anchored on rocky bottom near shore with its "stems" reaching to the surface; the longest of all marine plants.

KELP BELT The dense growth of kelp bordering a coastline, with sufficient mass to buffer wave action.

KELP FOREST A large expanse of kelp.

KELP LINE The margin of a kelp belt or forest.

KNOT A measurement of speed equal to one nautical mile per hour.

KRILL Small, shrimp-like marine crustaceans in the family Euphausiidae, which make up a large proportion of the zooplankton; especially abundant in the Antarctic, serving as the main food source of baleen whales and Crabeater Seals.

LACTATION The production of milk by female mammals to nurse young.

LAIR A resting place used by an animal, often for giving birth, nursing young, or hibernating; den.

LANUGO A downy or woolly covering of hair, especially on the fetus. In some seals it is retained for a short time after birth.

LATERAL At, pertaining to, or in the direction of the side; on either side of the vertical plane.

LATERALLY COMPRESSED Flattened in the vertical plane (from side to side).

LATITUDE Distance expressed in degrees and minutes, along a meridian north or south of the Equator. One degree of latitude is 60 nautical miles, thus one minute is one nautical mile. *High latitudes* are generally poleward of about 60 degrees north or

KRILL

LANUGO *Harp Seal*

LOBTAILING *Killer Whale*

LUNGE FEEDING *Fin Whales*

south of the Equator; *low latitudes,* within about 30 degrees north or south of the Equator.

LEAD A stretch, or lane, of open water that is left when two ice floes separate. Leads often serve as migratory corridors for marine mammals.

LEK An area, devoid of resources such as food, where males gather to display and females go to select mates.

LOBATE Describes a body part that has a rounded projection (lobe).

LOBTAILING When a whale or dolphin slaps the water surface with its flukes, sometimes repeatedly.

LONGLINE FISHERY A fishery that uses long lines with multiple baited hooks to catch fish such as halibut, tuna, sharks, and billfish.

LOW ARCTIC Generally refers to portions of the Arctic with less severe climatic conditions, particularly in relation to the extent and duration of sea ice coverage. Includes northern Quebec and Labrador, most of northern and western Alaska, and northernmost Scandinavia.

LUNGE FEEDING A feeding method used by some baleen whales in which they rush up to the water surface with their mouths agape.

MANDIBLE Bone of the lower jaw.

MANE Long thick hair on the neck of a mammal, including males of most species of eared seals.

MANGROVE A coastal marine forest in low latitudes, dominated by tree species in the families Rhizophoraceae and/or Avicenniaceae.

MARINE MAMMAL PROTECTION ACT (MMPA) A law passed by the United States Congress in 1972 to establish a comprehensive program to manage and conserve marine mammals. Responsibility for implementation and enforcement was given to either the Department of Commerce or the Department of the Interior, depending on the species, and an independent watchdog agency, the Marine Mammal Commission, was created.

MASS STRANDING Event in which several or many individual cetaceans come ashore.

MEDIAN NOTCH Cleavage between the paired flukes of a cetacean.

MELON The often bulging forehead of a toothed cetacean. The melon is lipid-rich and is thought to play an important role in echolocation.

MESOPELAGIC Living in the middle of the water column (midwater).

MANE *Antarctic Fur Seal*

MELON *Northern Bottlenose Whale*

MOLT *Southern Elephant Seal*

NEONATE *Hooded Seal*

MESOPLODONTS Beaked whales belonging to the genus *Mesoplodon.*

MIGRATION The process of moving from one habitat to another, often between breeding and feeding sites.

MOLAR A broad, flattened postcanine, or cheek, tooth adapted for grinding and crushing.

MOLLUSK An organism belonging to the phylum Mollusca, which includes bivalves (such as clams and mussels) and cephalopods (such as squid and octopus). Bivalves are usually sessile or burrowing; they possess a hinged shell and a hatchet-shaped foot.

MOLT The shedding of an outer layer of skin or fur, which is then replaced by new growth. *Catastrophic molt* refers to the abrupt (non-gradual) loss of hair clumps, typical of elephant and monk seals.

MONODONTID A member of the toothed whale family Monodontidae (Beluga and Narwhal).

MONOGAMY Breeding system in which the breeding unit consists of one male and one female; rare among mammals. *Serial monogamy* is a mating system of some pinnipeds, in which an adult male appears to remain with a female for a short period, mate with her, and then search the area for other females.

MORPH Any of the genetic forms, or individual variants, that account for the variety in shape and appearance (polymorphism) within a population or species.

MORPHOLOGY The form of an organism.

MORPHOMETRIC ANALYSIS Comparative study of variation, based on series of body measurements.

MOUTHLINE From the front tip of the jaws to the corner of the mouth.

MUDFLAT Level stretch of beach or intertidal zone with a muddy substrate.

MUSTELID Member of the carnivore family Mustelidae, including otters, weasels, skunks, wolverines, and badgers.

MUZZLE Forward-projecting part of the head of certain animals, including the nose and jaws; snout.

MYSID One of various elongate crustaceans that are usually transparent, or nearly so, and that are benthic, or deep-living.

MYSTICETE Whale species belonging to the suborder Mysticeti; baleen whale.

NATAL Of, pertaining to, or accompanying birth.

NATIONAL MARINE FISHERIES SERVICE (NMFS) Federal government agency in the United States Department of Commerce, a branch of the National Oceanic and Atmospheric Administration (NOAA), responsible for management and conservation of most marine mammals (but not sirenians, otters, Walruses, or Polar Bears).

NAUTICAL MILE Unit of measurement equal to 6,076 feet (1,852 m) or one minute of latitude.

NEONATE Newborn.

NERITIC Refers to the shallow water adjacent to a sea coast; coastal or near-shore.

NONSEASONAL Describes activities (for example, reproduction) that occur without a peak in frequency of occurrence.

OCEANIC Generally refers to the open-sea environment seaward of the continental shelf; blue-water.

ODONTOCETE Toothed cetacean; member of the suborder Odontoceti.

OFFSHORE Well away from the coast; pelagic.

OPPORTUNISTIC Describes animals that take advantage of whatever prey is available.

ORDER A taxon that is a subset of a class and that contains one or more families. *See* Taxonomy.

OTARIID Member of the carnivore family Otariidae, which consists of the fur seals and sea lions; eared seal.

PACK ICE *Crabeater Seals*

PACK ICE Sea ice, especially that which is unattached to land and usually moving and shifting to some extent; pack ice strongly influences polar ecosystems, obstructs navigation, and influences animal movements.

PAGOPHILIC Literally "ice-loving"; used to describe marine mammals that inhabit areas where there is considerable ice cover, at least seasonally.

PAIR BOND Close and long-lasting association between a male and female, usually involving cooperative rearing of young.

PANTROPICAL Occurring in all major tropical areas around the world.

PARASITE An organism that lives symbiotically with, and at the expense of, a host organism, usually without going so far as to cause the host's death.

PCBs Polychlorinated biphenyls (also known as chlorobiphenyls, or CBs); an array of man-made organochlorine compounds that are widespread, toxic, and persistent in the environment. They originate from use in electricity transformers, plastics, inks, lubricants, hydraulics, etc. These compounds accumulate in the tissues, especially the blubber, of marine mammals.

PEDUNCLE *See* Caudal peduncle.

PELAGE The fur or hair covering a mammal.

PELAGIC Of or relating to open offshore, as opposed to coastal, waters; also often used to refer to organisms that occur in the upper portions of the water column.

PERIPHERAL SEAS Those large water bodies that border, and are linked to varying degrees with, the various ocean basins.

PERMANENT ICE Generally, the stable masses of ice at and near the two poles (North and South) that does not melt or break up seasonally in all or most years.

PHOCID A species belonging to the carnivore family Phocidae; true, or earless, seal.

PHOCOENID A species belonging to the toothed whale family Phocoenidae; porpoise.

PHYSICAL MATURITY Stage in a mammal's life at which skeletal growth ceases.

PHYTOPLANKTON The plant forms of plankton.

PINNAE (*singular:* pinna) Ear flaps consisting of skin and cartilage; outer, or external, ears.

PINNIPED From the Latin for "wing-footed" or "finfooted," a term that encompasses three living families of the order Carnivora: Phocidae (true seals), Otariidae (fur seals and sea lions), and Odobenidae (the Walrus).

PLANKTON Passively drifting or weakly swimming organisms that occur in swarms near the surface of open waters.

POD A group of cetaceans, generally with the connotation that they are socially affiliated in some way.

POLAR In latitudes near one of the poles (North or South), typified by cold and ice-infested waters.

POLYGYNY The tendency of one male to mate with two or more females.

PINNAE *South African Fur Seal*

POD *Killer Whales*

PORPOISING *Short-beaked Common Dolphins*

RAFTING *Sea Otters*

POLYNYA A large area of water within the pack ice that remains open throughout the year, often providing refuge for air-breathing marine animals; maintained by winds, currents, or upwelling.

POPULATION The individuals of a given locality that potentially form a single interbreeding community.

PORPOISE Common name applied to species in the toothed whale family Phocoenidae. All porpoises are relatively small, have spade-shaped rather than conical teeth, and lack a distinct beak.

PORPOISING Refers to the way small cetaceans, and sometimes pinnipeds, make low, arcing leaps as they travel rapidly near the surface.

POSTCANINE Any of the teeth that occur behind the canine teeth; also called a cheek tooth.

POSTERIOR On or toward the back or hind part of the body.

PRECOCIAL Capable of a high degree of independent activity from birth.

PREMOLAR A bicuspid tooth located behind the canine but ahead of the first true molar.

PRESSURE RIDGE A pile of compressed rubble and ice fragments, often sizable, that extends above and below the surface of the ice; created by the opposing pressure of two ice floes.

PROBOSCIS A flexible, elongated snout of certain mammals.

PRODUCTIVE Referring, in the present context, to waters that provide large quantities of food for marine mammals.

PUP Newborn or young, unweaned pinniped.

RACE A geographical population that is not sufficiently differentiated to be formally designated as a subspecies.

RAFTING When marine mammals are not swimming but rather are more or less motionless at the surface, in a horizontal position.

RANGE The geographic area in which a species is usually found.

RECURVED Curved or bent backward.

RED TIDE A red or reddish-brown discoloration of surface waters caused by concentrations of certain microscopic organisms, particularly the often toxic dinoflagellates. Also called toxic water bloom.

REEF A formation of coral or rocks or a ridge of sand at

ROOKERY *South African Fur Seals*

ROOSTER TAIL *Dall's Porpoise*

ROSTRUM *Antarctic Minke Whale*

SADDLE *Killer Whales*

or near the surface. *Coral reefs* are complex ecological associations of bottom-living and attached calcareous, shelled marine invertebrates.

RESIDENT A nonmigratory individual or species living year-round in a specified area.

RIVER BASIN Network of streams and other water bodies that flow, or drain, into a particular river.

RIVER SYSTEM A river and its network of feeder streams, rivers, and lakes.

ROLL When a marine mammal surfaces for breath at a shallow angle, with a forward movement that exposes the nostrils followed by the arched back as the animal submerges again.

ROOKERY A site on land used by otariid pinnipeds for breeding and rearing young; breeding colony.

ROOSTER TAIL Spray of water created as a porpoise or dolphin surfaces at high speed; especially characteristic of Dall's Porpoise.

RORQUAL A species in the baleen whale family Balaenopteridae; balaenopterid.

ROSTRUM (*adjective:* rostral) Specifically the upper jaw of a cetacean; also sometimes used to refer to a beak that encompasses both the upper and lower jaws.

SAGITTAL CREST *California Sea Lion*

SADDLE A more or less saddle-shaped marking that straddles the dorsal midline and extends to a variable degree onto the sides; sometimes confused with cape.

SAGITTAL CREST Prominence on top of the cranium, causing a noticeably raised forehead on males of some otariid pinniped species.

SANDBAR A ridge of sand built up by currents; whales may strand if caught among sandbars when the tide goes out.

SANDFLAT Level, sandy stretch of beach or ocean bottom.

SCHOOL A large number of fish or other aquatic animals that swim, feed, or otherwise occur together with coordinated movements; sometimes applied to dolphins.

SEAL A term generically applied to pinnipeds, usually excluding the Walrus; "true" seals belong to the family Phocidae, while "fur" seals belong to the family Otariidae.

SEAMOUNT A submarine elevation that rises at least 2,900 feet (900 m) from the ocean bottom.

SEASONAL Used to describe certain activities (for example, reproduction) that occur during particular times of the year.

SEINE A large net that hangs vertically in the water. A *purse seine* is used to encircle large schools of fish; when the two ends are joined, the bottom is drawn in (pursed) with the fish trapped inside. A *beach seine* is operated from shore in shallow water, usually set by a boat and dragged shoreward, forming a barrier to the fish and other organisms inside the enclosed area.

SEMI-AQUATIC VEGETATION Plants that are rooted in the water or moist soil but that have most of their vegetative parts out of the water.

SESSILE Describes animals that are attached to a substrate.

SEXUAL DIMORPHISM When males and females of the same species differ in size (for example, the Sperm Whale), or when one sex has distinctive secondary sexual characteristics (such as a male Narwhal's tusk or a male Hooded Seal's inflatable nasal sac).

SEXUAL MATURITY The state in which an animal is physiologically capable of reproducing.

SHELF WATERS Waters over the continental shelf.

SHOALING Becoming very shallow (in reference to water depth); also, schooling (in reference to fish).

SIRENIAN Member of the mammalian order Sirenia, consisting of the manatees and the Dugong.

SKIM FEEDING Method of feeding by some species of baleen whales that involves slow and deliberate movement along or just below the water surface with the mouth open, continuously filtering the water as it streams through the baleen.

SLOPE WATERS Waters covering the continental slope.

SNOUT That part of an animal's head from the front margins of the eyes to the tip of the nose.

SOCIAL Referring to interactions between members of the same species; animals that frequently interact in a nonaggressive way are considered social.

SOLITARY Nonsocial; usually not found in groups.

SOUTHERN OCEAN Marine waters of the Southern Hemisphere south of the Subtropical Front at approximately 40°S. Also known as the Antarctic Ocean.

SPATULATE Shaped like a spoon or spatula; splayed or spread out.

SPECIES A group of interbreeding populations that are reproductively isolated from other such groups; taxon that is a subset of a genus and that may contain one or more subspecies. *See* Taxonomy.

SPECIFIC NAME Refers to the second of the two Latin names of a species, the first being the genus name.

SPERMACETI A liquid wax found in the head of the Sperm Whale and a few other species, with chemical properties that made it extremely valuable as a high-temperature lubricant and a base for cosmetics (such as cold cream).

SPERM COMPETITION Reproductive system in which a female mates with more than one male, and the males' sperm are presumed to "compete" for access to her egg(s). In such species, males tend to have larger testes and longer penises, and their goal is to produce large numbers of sperm per ejaculate and multiple ejaculates within a short time.

SPLASHGUARD Elevated area immediately in front of the blowholes of a baleen whale. The splashguard helps prevent water from entering the blowholes when they are open.

SEXUAL DIMORPHISM *Northern Elephant Seals*

SPLASH ZONE Area along a coast that is exposed to spray from waves as they crash against rocks.

SPYHOPPING When a cetacean raises its head vertically out of the water at least high enough for the eyes to be clear of the surface. The animal will often rotate slowly, apparently to scan the surrounding area visually.

STERN WAKE Waves and froth (turbulence) left behind as a vessel moves through the water.

STOCK A genetically separate population of a species (biological stock), or a discrete population subject to management (management stock).

STRANDING When a marine animal comes ashore or is cast ashore.

SUBADULT An individual older than an infant and younger than an adult; often used as a fourth category in age- or stage-class studies, preceded by infant and juvenile and followed by adult.

SUBANTARCTIC Transitional zone between Antarctic and temperate zone. *See* map on back endpaper.

SUBARCTIC Transitional zone between Arctic and temperate zone. *See* map on back endpaper.

SUBMERGENT Describes aquatic plants that are wholly submerged, as distinct from emergent or floating.

SUBSPECIES A population isolated geographically from other populations of the same species, and evolving in its own direction. *See* Taxonomy.

SUBSTRATE The bottom of a body of water, usually referring to its composition (sand, mud, rock, etc.); also, the surface on which a plant or animal attaches and grows.

SUBTROPICAL Refers to oceanic mid-latitude regions where the mean annual temperature ranges between 55 and 68°F (13–20°C). *See* map on back endpaper.

SPLASHGUARD *Bryde's Whale*

SPYHOPPING *Gray Whale*

SUBTROPICAL FRONT Approximately along 40°S latitude, where opposing currents cause a downward flow of surface waters. Also known as the Subtropical Convergence.

SUCTION FEEDING Capture of prey using suction, generally with the tongue employed as a piston to create vacuum pressure.

TAIGA Large stretches of boreal forest that cover much of Canada, Alaska, Siberia, and Scandinavia.

TAIL STOCK *See* Caudal peduncle.

TAXON (*plural:* taxa) A level of grouping, or category, in taxonomy, such as species, genus, or family.

TAXONOMY The theory and practice of classifying organisms. The highest-level taxon is the kingdom, which is typically broken down into subgroups called divisions (for plants) and phyla (for animals). Phyla are divided into classes (mammals are in the class Mammalia) and then into orders. Orders are subdivided into families. Within families, animals are grouped at other taxonomic levels, given here in sequence from larger groupings to smaller: subfamily, genus, species, and subspecies.

TELEMETRY Use of electronic instrumentation ("tagging") to obtain information on animal movements and other activities via radio signals; satellites are sometimes used to receive and forward the signals to a ground receiver.

TEMPERATE Between subpolar and subtropical regions, where the mean annual temperature ranges between 50 and 55°F (10–13°C). *See* map on back endpaper.

TERMINAL FLAPS The long fleshy tips on the ends of the digits on the hindflippers of eared seals.

TERRITORY An area occupied exclusively by one animal and defended by aggressive behavior or displays.

THERMOREGULATION The process of maintaining a constant internal body temperature despite changes in the outside environmental temperature.

THORACIC PATCH Color marking on the foreside of a dolphin.

THORAX The part of the body between the neck and the abdomen.

TIDAL BORE A very rapid rise of the tide in which the advancing water forms an abrupt front; usually occurs in certain shallow estuaries with large tidal ranges.

TIDAL FLAT A marshy or muddy area that is covered and uncovered by the rise and fall of the tide.

TIDE BAR A ridge of sand or gravel that is exposed at low tide and submerged at high tide.

TIDE POOL A depression within the intertidal zone that is alternately submerged and exposed with water remaining inside.

TRANSIENT Impermanent; used to describe nonresident Killer Whales.

TRAWL A bag or funnel-shaped net that catches fish and other organisms as it is towed through the water. A *bottom trawl* is dragged along the bottom to catch bottom-dwelling organisms (primarily fish). A *mid-water trawl* is towed in intermediate depths to catch invertebrates (such as squid) and fish (such as anchovy).

TRIBUTARY A river or stream that flows into a larger river or stream.

TROPICAL An area where the mean annual temperature is greater than 77°F (25°C) and no freezing occurs; generally refers to the region between the Tropics of Cancer (23°27'N) and Capricorn (23°27'S). *See* map on back endpaper.

TUBERCLE Small, knob-like prominence, such as the bumps on the edges of the flippers or dorsal fins of some cetaceans, or the knobs on the heads of Humpback Whales.

TUSK *Walrus*

TUSK *Narwhals*

TUNDRA Generally referring to the treeless vegetation that occurs in high latitudes, consisting mainly of lichens, mosses, and stunted shrubs. The subsoil of the tundra is permanently frozen (permafrost).

TUNICATE A globular, cylindrical, or sac-like marine animal that may be covered by a tough, flexible material and that may be sessile or pelagic.

TURBID Refers to water that has low transparency, normally due to large quantities of suspended or stirred-up particles or sediment; muddy.

TUSK An enlarged tooth that protrudes outside the mouth and is often used as a weapon. Walruses, Narwhals, Dugongs, and some beaked whales have tusks.

TUSSOCK GRASS Dominant vegetation of subantarctic shorelines, occurring up to 820 feet (250 m) above sea level; long-lived, up to 7 feet (2 m) high, with a core of very dense tissue and long narrow leaves emanating from the crown and cascading to ground level; plants closely packed together. (Also spelled "tussak" or "tussac.")

UMBILICUS The round, depressed scar on the median line of the abdomen where the fetal umbilical cord passed through; navel.

UNDERFUR Soft, fine, wavy hairs on a mammal's skin that grow in a clump immediately behind, and are covered and protected by, a single longer, stiffer guard hair.

UPWELLING Process by which nutrient-rich water rises from the bottom toward the surface, usually forced by currents, winds, or density gradients; normally results in increased production because nutrients are brought toward the surface where light is available to support photosynthesis.

URO-GENITAL AREA Portion of ventral surface around and near the excretory and genital orifices.

VAGRANT A wanderer, in the sense of an animal moving outside the usual limits of distribution for its species or population.

VENTRAL On or belonging to the lower surface of an animal.

TUSSOCK GRASS *Antarctic Fur Seal*

VIBRISSAE *California Sea Lion*

VENTRAL PLEATS Longitudinal furrows, or grooves, that extend backward from the chin of a balaenopterid (rorqual) whale, the purpose of which is to allow distention of the throat during feeding.

VERTEBRAE (*singular:* vertebra) The bones that make up the spine, or backbone, of an animal.

VESTIGIAL Describing an anatomical structure or organ that is an artifact, or vestige, and that was more developed and functional in an earlier form of the organism.

VIBRISSAE (*singular:* vibrissa) Stiff hairs that usually project from the face and function as sensory receptors; whiskers.

VOCALIZATION Sound produced by an organism via its vocal apparatus.

WAKE Waves and other turbulence left behind a moving vessel.

WAKE RIDING Swimming in the waves and other turbulence behind a moving vessel.

WATER COLUMN Anywhere between the surface and bottom of an ocean or other body of water.

WEST WIND DRIFT Principal circumpolar current in the Antarctic; eastward-flowing. Also called Antarctic Circumpolar Current.

WHALE A member of the mammalian order Cetacea; generally applied only to the larger species, including all of the baleen species and some of the toothed species.

WHALE LICE A family of amphipod crustaceans, Cyamidae, related to the skeleton shrimp that live on seaweeds, bryozoans, and hydroids. Whale lice feed on whale skin, and some species are known exclusively from their occurrence on the skin of whales. They tend to accumulate at sites where water flow is reduced, such as in folds of skin around the flippers, eyes, and blowholes, in ventral slits or grooves, lip margins, and wounds, and around barnacles on Gray Whales. Also called cyamids.

WHITECOAT A sealer's term usually reserved for the newborn Harp Seal, which is covered in white lanugo.

ZIPHIID A member of the toothed whale family Ziphiidae; beaked whale.

ZOOPLANKTON The animal forms of plankton.

Photo Credits

CREDITS ARE LISTED BY PAGE NUMBER (LEFT TO RIGHT, TOP TO BOTTOM). PHOTOGRAPHERS HOLD THE COPYRIGHT TO THEIR WORK.

9 James D. Watt/ Norbert Wu Productions
10A Doug Perrine/ Innerspace Visions*
10B Ingrid N. Visser/ Innerspace Visions
11A Brandon D. Cole/ www.brandoncole.com
11B Todd Pusser
11C Brandon D. Cole/ www.brandoncole.com
12A Norbert Wu/ Norbert Wu Productions
12B Dotte Larsen/ Bruce Coleman, Inc.
13A Leonard Lee Rue III/ Bruce Coleman, Inc.
13B Phillip Colla
13C Bob Cranston
13D François Gohier
14A Bill Curtsinger
14B Doug Perrine/ Innerspace Visions
15A Norbert Wu/ Norbert Wu Productions
15B Barbara Todd
16A George D. Lepp/ Photo Researchers, Inc.
16B Dan Guravich/ Photo Researchers, Inc.
17 Doug Perrine/Innerspace Visions
18 Sylvia Stevens
19A Hiroya Minakuchi/ Sphere Magazine
19B C. Allan Morgan
19C Frank S. Todd
20A Beth Davidow/WorldWild
20B Brandon D. Cole/ www.brandoncole.com
21A Mike Nolan/ Innerspace Visions
21B Adrian Dorst/ Ursus Photography
22A Bud Nielsen/Visuals Unlimited
22B Pieter Folkens
23 Franco Banfi/Innerspace Visions
24A James D. Watt/ Innerspace Visions
24B Joe McDonald/ Visuals Unlimited
25 Patrick J. Endres/ Visuals Unlimited
26 Hiroya Minakuchi/ Sphere Magazine
27A Todd Pusser/ Innerspace Visions
27B Norbert Wu/ Norbert Wu Productions
27C Norbert Wu/ Norbert Wu Productions
28 François Gohier
29 Flip Nicklin/Minden Pictures
30A Dave B. Fleetham/ Innerspace Visions
30B Paul Nicklen/ Ursus Photography
30C Richard Sears-MICS Photo
31 François Gohier/ Photo Researchers, Inc.
32 Bob Cranston
33A Richard Herrmann
33B Phillip Colla
37 Renee DeMartin
40 Brian & Cherry Alexander/ Innerspace Visions
41A Dan Guravich/ Photo Researchers, Inc.
41B Lynne Ledbetter/ Visuals Unlimited
44A Jeff Foott/Bruce Coleman, Inc.
44B Tom & Pat Leeson/ Photo Researchers, Inc.
45 Doc White/Innerspace Visions
48A Sylvia Stevens
48B Sylvia Stevens
50 Richard Herrmann
51 Ernest A. Janes/ Bruce Coleman, Inc.
53 Sylvia Stevens
60 Michael Fogden/ Bruce Coleman, Inc.
61A Nigel Dennis/ Photo Researchers, Inc.
61B Sylvia Stevens
64 Ingrid N. Visser/ Ursus Photography
65A Ingrid N. Visser/ Innerspace Visions
65B Art Wolfe/ Photo Researchers, Inc.
68A Sylvia Stevens
68B Scott A. Shaffer
70 François Gohier/ Photo Researchers, Inc.
71 Carolyn Gohier/ Photo Researchers, Inc.
74A Phillip Colla
74B Phillip Colla
75 Brent S. Stewart
78A Barbara Todd
78B Todd Pusser
78C Barbara Todd
80 François Gohier
81A Robin W. Baird/ Innerspace Visions
81B David Hosking/ Photo Researchers, Inc.
84A Tui De Roy/ Bruce Coleman, Inc.
84B Hiroya Minakuchi/ Sphere Magazine
85A Phillip Colla
85B Tui De Roy/ Bruce Coleman, Inc.
88A Yva Momatiuk & John Eastcott/Photo Researchers, Inc.
88B Phillip Colla
89A Yva Momatiuk & John Eastcott/ Photo Researchers, Inc.
89B Yva Momatiuk & John Eastcott/ Photo Researchers, Inc.
92A Dotte Larsen/ Bruce Coleman, Inc.
92B G. C. Kelley/ Photo Researchers, Inc.
93A John Sorensen
93B Richard Herrmann
96A Paul Nicklen/ Ursus Photography
96B Roy Tanami/ Ursus Photography
97 Brandon D. Cole/ www.brandoncole.com
100A Daniel Costa

100B Stuart Westmorland/ Photo Researchers, Inc.
101 Stuart Westmorland/ Photo Researchers, Inc.
104A Art Wolfe/ Photo Researchers, Inc.
104B Barbara Todd
105 Tui De Roy/ Bruce Coleman, Inc.
108A George Holton/ Photo Researchers, Inc.
108B C. Allan Morgan
109 Bill Curtsinger
112 Sylvia Stevens
113A Frank S. Todd
113B Bill Curtsinger
116A Brandon D. Cole/ www.brandoncole.com
116B E. R. Degginger/ Photo Researchers, Inc.
117 Roy Hamaguchi/ Ursus Photography
120A Art Wolfe/ Photo Researchers, Inc.
120B Bob Cranston
121 Mike Nolan/ Innerspace Visions
124A Kathy Frost
124B Kathy Frost
126 C. Ray/Photo Researchers, Inc.
127 Flip Nicklin/Minden Pictures
130A Carleton Ray/ Photo Researchers, Inc.
130B G. Carleton Ray/ Photo Researchers, Inc.
130C Kathy Frost
132 Sylvia Stevens
133 Sylvia Stevens
136–137 Bill Curtsinger
136A Charles McRae/ Visuals Unlimited
136B Pieter Folkens
137 Charles McRae/ Visuals Unlimited
140A Florian Graner/ Innerspace Visions
140B Robin W. Baird
141 Robin W. Baird/ Innerspace Visions
144 Bill Curtsinger
145A B. & C. Alexander/ Photo Researchers, Inc.
145B Pieter Folkens
148–149 Pieter Folkens
148A Pieter Folkens
148B Pieter Folkens
149 Roy Tanami/ Ursus Photography
152A Sylvia Stevens
152B Sylvia Stevens
153 J. Trotignon/Jacana/ Photo Researchers, Inc.
156 Dave B. Fleetham/ Innerspace Visions
157A Bill Curtsinger
157B Norbert Wu/ Norbert Wu Productions
160 Beth Davidow/WorldWild
161 Sylvia Stevens
164 François Gohier/ Photo Researchers, Inc.
165A Pieter Folkens
165B François Gohier
168 Ingrid N. Visser/ Innerspace Visions
169A Norbert Wu/ Norbert Wu Productions
169B Norbert Wu/ Norbert Wu Productions
172 Robin W. Baird
173 Sylvia Stevens
176A G. Carleton Ray/ Photo Researchers, Inc.
176B G. Carleton Ray/ Photo Researchers, Inc.
178 Robin W. Baird
179 C. Allan Morgan
181A Beth Davidow/WorldWild
181B Mike Nolan/ Innerspace Visions
181C Phillip Rosenberg
181D Pieter Folkens
181E Norbert Wu/ Norbert Wu Productions
182A François Gohier
182B François Gohier
182C Brandon D. Cole/ www.brandoncole.com
182D Ingrid N. Visser/ Ursus Photography
182E Pieter Folkens
182F Phillip Colla
182G Isabel Beasley
182H Mike Johnson/ Innerspace Visions
182I Robin W. Baird
183A Brandon D. Cole/ www.brandoncole.com
183B Brandon D. Cole/ Norbert Wu Productions
183C Masa Ushioda/ Innerspace Visions
183D John K. B. Ford/ Ursus Photography
183E C. Allan Morgan
183F François Gohier
185 François Gohier
186 Phillip Colla
187A Doug Perrine/ Innerspace Visions
187B Robert L. Pitman
187C Brandon D. Cole/ www.brandoncole.com
192A François Gohier/ Photo Researchers, Inc.
192B François Gohier
193 François Gohier/ Photo Researchers, Inc.
196A Brandon D. Cole/ www.brandoncole.com
196B Brandon D. Cole/ www.brandoncole.com
196C Brandon D. Cole/ www.brandoncole.com
197 Doug Perrine/ Innerspace Visions
200A Paul Nicklen/ Ursus Photography
200B Paul Nicklen/ Ursus Photography
201 Paul Nicklen/ Ursus Photography
206 Marilyn Kazmers/ Innerspace Visions
207A Dave B. Fleetham/ Visuals Unlimited
207B C. Allan Morgan
210 Brandon D. Cole/ www.brandoncole.com
211A Renee DeMartin
211B Renee DeMartin
211C Richard Sears-MICS Photo
211D Ingrid N. Visser/ Innerspace Visions
211E Brandon D. Cole/ Norbert Wu Productions
214A Robin W. Baird
214B Robert L. Pitman
215 Robin W. Baird
220 Robin W. Baird
221A Fujio Kasamatsu
221B Michael W. Newcomer
224 Doc White/Innerspace Visions
225A Beth Davidow/ Visuals Unlimited
225B Robert L. Pitman
228A Pieter Folkens
228B Pieter Folkens

228C Doug Perrine/ Innerspace Visions
229 Doug Perrine/ Innerspace Visions
232A Mike Nolan/ Innerspace Visions
232B François Gohier/ Photo Researchers, Inc.
233A François Gohier
233B Mark Ruth/ Innerspace Visions
236A François Gohier
236B Richard Sears-MICS Photo
237 Doc White/Innerspace Visions
239 François Gohier
242A C. Allan Morgan
242B Mike Nolan/ Innerspace Visions
242C Mike Nolan/ Innerspace Visions
243 Barbara Todd
246 Robert L. Pitman/ Innerspace Visions
247 Bernie Tershy
249 Sylvia Stevens
251A Fujio Kasamatsu
251B Robert L. Pitman/ Innerspace Visions
251C Mike Nolan/ Innerspace Visions
251D Amy Sierra Van Atten
251E Robert L. Pitman
251F Dylan Walker
256 Todd Pusser
256–257 Todd Pusser
257 Todd Pusser
262A John Sorensen
262B Robert L. Pitman
263 Todd Pusser
270A Sascha Hooker/ Innerspace Visions
270B Sascha K. Hooker
270C Robin W. Baird
271 Sascha Hooker/ Innerspace Visions
300 Todd Pusser
301 Hiroya Minakuchi/ Sphere Magazine
304A Brian D. Smith
304B Scott R. Benson
308A Fernando Trujillo/ Innerspace Visions
308B David K. Caldwell/ Steve Leatherwood Collection
309 Gregory Ochocki
312 John M. K. Wong
313A Mark Carwardine/ Innerspace Visions
313B Xianfeng Zhang
316A Glenn Williams/ Ursus Photography
316B John K. B. Ford/ Ursus Photography
317 François Gohier/ Photo Researchers, Inc.
320A Hiroya Minakuchi/ Sphere Magazine
320B John K. B. Ford/ Innerspace Visions
321 Don K. Ljungblad
324 John K. B. Ford/ Ursus Photography
325A Glenn Williams/ Ursus Photography
325B Glenn Williams/ Ursus Photography
327 Brandon D. Cole/ Norbert Wu Productions
328A Thomas Jefferson
328B Todd Pusser
328C Barbara Todd
329A Brandon D. Cole/ www.brandoncole.com
329B Doug Perrine/ Innerspace Visions
334A Todd Pusser/ Innerspace Visions
334B Beth Davidow/WorldWild
335 Frank S. Todd
338A Robert L. Pitman
338B Robert L. Pitman
340 Todd Pusser
341 Todd Pusser
344A Ingrid N. Visser/ Innerspace Visions
344B Brandon D. Cole/ Norbert Wu Productions
345 Brandon D. Cole/ www.brandoncole.com
348A Sascha K. Hooker
348B Robert L. Pitman/ Innerspace Visions
349 Robin W. Baird
352A Thomas Jefferson
352B Samuel K. Hung
352C Michael R. Heithaus
353 Samuel K. Hung
356A Paulo Flores/ Innerspace Visions
356B Fernando Trujillo/ Innerspace Visions
357 Paulo Flores/ Innerspace Visions
360 François Gohier
361 Norbert Wu/ Norbert Wu Productions
364A François Gohier/ Photo Researchers, Inc.
364B Michael R. Heithaus
365 Michael R. Heithaus
368 Doug Perrine/ Innerspace Visions
369A Robert L. Pitman/ Innerspace Visions
369B Brandon D. Cole/ www.brandoncole.com
372 James D. Watt/ Norbert Wu Productions
372–373 Brandon D. Cole/ www.brandoncole.com
373 Norbert Wu/ Norbert Wu Productions
376 Brandon D. Cole/ www.brandoncole.com
377 Mike Nolan/Innerspace Visions
382A Robert L. Pitman/ Innerspace Visions
382B Robin W. Baird
383 Doug Perrine/ Innerspace Visions
386 John Y. Wang/Naturart
386–387 Michael W. Newcomer
387 Robert L. Pitman/ Innerspace Visions
390 Todd Pusser
391 Hiroya Minakuchi/ Sphere Magazine
394A Marilyn Kazmers/ Innerspace Visions
394B Todd Pusser
396 Robin W. Baird
397A Robin W. Baird
397B Richard Sears-MICS Photo
400A Richard Sears-MICS Photo
400B Richard Sears-MICS Photo
401 Sascha Hooker/ Innerspace Visions
404–405 Brandon D. Cole/ www.brandoncole.com
404 Phillip Colla
405 John K. B. Ford/ Ursus Photography
408 Brandon D. Cole/ www.brandoncole.com
409A Todd Pusser
409B Todd Pusser
412A M. Iniguez

412B Todd Pusser/ Innerspace Visions
413 M. Iniguez
416 C. Allan Morgan
417 Mari Smultea
420–421 Todd Pusser
420A Todd Pusser/ Innerspace Visions
420B Robin W. Baird
424A Doug Perrine/ Innerspace Visions
424B Thomas Jefferson
425 Robert L. Pitman/ Innerspace Visions
428 Doug Perrine/ Innerspace Visions
429A Michael W. Newcomer
429B Michael W. Newcomer
432A Hiroya Minakuchi/ Sphere Magazine
432B Phillip Rosenberg
434 Phillip Colla
435 Mari Smultea
438–439 John K. B. Ford/ Ursus Photography
438 Brandon D. Cole/ Norbert Wu Productions
439 Brandon D. Cole/ www.brandoncole.com
442 Barbara Todd
443A Robin W. Baird
443B Robin W. Baird/ Innerspace Visions
443C Robin W. Baird
446A Pieter Folkens
446B C. Allan Morgan
446C Bill Curtsinger
447 Graeme Cresswell
450 Isabel Beasley
451 Isabel Beasley
453 Robert L. Pitman/ Innerspace Visions
454 John Y. Wang/Naturart
455A Flip Nicklin/ Minden Pictures
455B Robert L. Pitman
458A Samuel K. Hung
458B Samuel K. Hung
459 Frank S. Todd
462A John Y. Wang/Naturart
462B Armin Maywald/ Innerspace Visions
463A Pieter Folkens
463B Florian Graner/ Innerspace Visions
472 Todd Pusser
473A Robin W. Baird
473B Joe McDonald/ Innerspace Visions
475 Doug Perrine/ Innerspace Visions
476 Doug Perrine/ Innerspace Visions
480A Doug Perrine/ Innerspace Visions
480B Michael R. Heithaus
481 Doug Perrine/ Innerspace Visions
484–485 Douglas Faulkner/ Photo Researchers, Inc.
484 Phillip Colla
485 Douglas Faulkner/ Photo Researchers, Inc.
488 Doug Perrine/ Innerspace Visions
489A Doug Perrine/ Innerspace Visions
489B Jany Sauvanet/ Photo Researchers, Inc.
500A C. Allan Morgan
500B George Holton/ Photo Researchers, Inc.
501A Gary L. Friedrichsen
501B Herbert Clarke
501C François Gohier/ Photo Researchers, Inc.
501D Pieter Folkens/ Innerspace Visions
502A Sascha K. Hooker
502B Renee DeMartin
502C Pieter Folkens
502D Renee DeMartin
503A François Gohier
503B Todd Pusser/ Innerspace Visions
504A Phillip Colla
504B Barbara Todd
505 John Sorensen
506A Phillip Colla*
506B Yva Momatiuk & John Eastcott/ Photo Researchers, Inc.
506C Beth Davidow/WorldWild
507A Barbara Todd
507B Erwin C. Nielson/ Visuals Unlimited
507C Pieter Folkens
508A Todd Pusser
508B Tui De Roy/ Bruce Coleman, Inc.
508C Noel R. Kemp/ Photo Researchers, Inc.
508D Hans Reinhard/ Bruce Coleman, Inc.
509A Mike Nolan/ Innerspace Visions
509B Mike Nolan/ Innerspace Visions
509C Robert W. Hernandez/ Photo Researchers, Inc.
509D Sascha K. Hooker
510A John K. B. Ford/ Ursus Photography
510B Pieter Folkens
511A Robert W. Hernandez/ Photo Researchers, Inc.
511B Joe McDonald/ Visuals Unlimited
511C Brandon D. Cole/ www.brandoncole.com
512A C. Allan Morgan
512B Mark Newman/ Bruce Coleman, Inc.
512C Brian Rogers/ Visuals Unlimited
512D John K. B. Ford/ Ursus Photography
513A Sylvia Stevens
513B Brandon D. Cole/ www.brandoncole.com
513C Pieter Folkens
514 Pieter Folkens
515A C. Allan Morgan
515B François Gohier
516A Nada Pecnik/ Visuals Unlimited
516B John K. B. Ford/ Ursus Photography
517A Sylvia Stevens
517B Hiroya Minakuchi/ Sphere Magazine
526A E. L. Folkens
526B Genia Clapham
526C Randi Olsen
526D Brent S. Stewart
526E Maureen D. Powell

* Photos taken under scientific research permits issued by the National Marine Fisheries Service (**10A** Permit #633; **506A** Permit #882).

Index

NUMBERS IN BOLDFACE REFER TO THE SPECIES' MAIN DESCRIPTION IN THE SPECIES ACCOUNT SECTION.

Arctocephalus
- *australis,* 79
- *forsteri,* 76
- *galapagoensis,* 82
- *gazella,* 62
- *philippii,* 69
- *pusillus,* 58
- *townsendi,* 72
- *tropicalis,* 66

Baiji, 300, 310
Balaena mysticetus, 198
Balaenidae, 184, 190–201
Balaenoptera
- *acutorostrata,* 212
- *bonaerensis,* 218
- *borealis,* 226
- *edeni,* 222
- *musculus,* 234
- *physalus,* 230

Balaenopteridae, 184, 208–37
Baleen whales, 180–83, 184–237
Beaked Whale,
- Andrews', **286**
- Arnoux's, 249, 250, **258**
- Bahamonde's, 250
- Baird's, 250, **260**
- Blainville's, 251, **294**
- Cuvier's, 11, 249, **254**
- Gervais', **278**
- Ginkgo-toothed, **290**
- Gray's, 251, **282**
- Hector's, **274**
- Hubbs', **288**
- Layard's, 292
- Lesser, 284
- Longman's, 250, **266**
- North Atlantic, 280
- North Sea, 280
- Pygmy, 251, **284**
- Shepherd's, 249, **264**
- Sowerby's, 251, **280**
- Stejneger's, **296**
- True's, 251, **276**

Beaked whales, 180–83, 248–98
Bear, Polar, 13, 16, 36–37, **38**
Beluga, 11, 180–83, 316–17, **318**
Berardius
- *arnuxii,* 258
- *bairdii,* 260

Bhulan, 299, 302
Boto, 300, 306
Boto de dorsal espinhosa, 466
Bottlenose Dolphin,
- Common, 10, 182, 183, 346, **358**
- Indo-Pacific, 15, **362,** 501

Bottlenose Whale,
- Giant, 260
- Northern, 250, 251, **268,** 509
- Southern, 250, 251, **272**

Bufeo, 354

Callorhinus ursinus, 86
Caperea marginata, 202
Cephalorhynchus
- *commersonii,* 332
- *eutropia,* 336
- *heavisidii,* 339
- *hectori,* 342

Cetaceans, 180–473
Chancho marino, 466
Chinchmen, 46
Chungungo, 46
Common Dolphin,
- Long-beaked, **392**
- Short-beaked, **388,** 502, 512

Cystophora cristata, 146

Delphinapterus leucas, 318
Delphinidae, 326–451
Delphinus
- *capensis,* 392
- *delphis,* 388

Dolphin,
- Amazon River, 300, 301, **306**
- Atlantic Hump-backed, **350**
- Atlantic Spotted, 9, 15, 27, **370**
- Atlantic White-sided, **398**
- Black, 336
- Chilean, **336**
- Chinese White, 350
- Clymene, 328, **378**
- Commerson's, **332**
- Common Bottlenose, 10, 182, 183, 346, **358**
- Dusky, **406**
- Fraser's, **384**
- Ganges River, 182, 299, **302**
- Heaviside's, **339**
- Hector's, 329, **342**
- Hourglass, **414**
- Indian Hump-backed, 350
- Indo-Pacific Bottlenose, 15, **362,** 501
- Indo-Pacific Hump-backed, 328, **350**
- Indus River, 299, **302**
- Irrawaddy, 329, **448**
- La Plata, 314
- Long-beaked Common, **392**
- Northern Right Whale, **418**
- Pacific Hump-backed, 350
- Pacific White-sided, 327, **402,** 505
- Pantropical Spotted, **366,** 503
- Peale's, 16, **410**
- Pink, 300, 306
- Plumbeous, 350
- Risso's, 328, 329, **422,** 504
- Rough-toothed, 327, **346**
- Saddleback, 388
- Sarawak, 384
- Short-beaked Common, **388,** 502, 512
- Southern Right Whale, 328, **418**
- Spinner, 21, 24, 328, **374**
- Striped, 328, **380**
- White-beaked, **395**
- White Flag, 310
- Yangtze River, 300, **310**

Dolphins,
- Ocean, 180–83, 326–451
- River, 180–83, 299–315

Dugong, 30, 474–77, **478**

Dugong dugon, 478
Dugongidae, 474–81

Eared seals, 49–52, 54–55, 58–109
Elephant Seal,
- Northern, 22, **162,** 514
- Southern, 53, **158,** 510

Enhydra lutris, 42
Erignathus barbatus, 114
Eschrichtiidae, 184, 204–7
Eschrichtius robustus, 204
Eubalaena
- *australis,* 194
- *glacialis,* 190
- *japonica,* 190

Eumetopias jubatus, 94

Feresa attenuata, 430
Finback, 230
Fissipeds, marine, 36–48
Franciscana, **314**
Fur Seal,
- Amsterdam, 66
- Amsterdam Island, 66
- Antarctic, 53, **62,** 102, 178, 509, 517
- Australian, **58**
- Cape, 58
- Falkland, 79
- Galápagos, **82**
- Guadalupe, 19, 51, **72**
- Juan Fernández, 22, **69**
- Kerguelen, 62
- New Zealand, **76,** 508
- Northern, 19, 51, **86,** 506
- South African, 18, **58,** 511, 512
- South American, 12, **79**
- Southern, 79
- Subantarctic, **66,** 102
- Tasman, 58
- Tasmanian, 58

Fur seals, 36, 51–52, 54–55, 58–89

Globicephala
- *macrorhynchus,* 444
- *melas,* 440

Grampus griseus, 422

Halichoerus grypus, 138
Histriophoca fasciata, 128

Hump-backed Dolphin,
- Atlantic, **350**
- Indian, 350
- Indo-Pacific, 328, **350**
- Pacific, 350

Hydrodamalis gigas, 476
Hydrurga leptonyx, 177
Hyperoodon
- *ampullatus,* 268
- *planifrons,* 272

Indopacetus pacificus, 266
Inia geoffrensis, 306
Iniidae, 300–1, 306–9

Jiangzhu, 456

Killer Whale, 26, 180, 181, 182, 329, **436,** 473, 509, 511, 513
- False, 329, **433**
- Pygmy, 329, **430**

Kogia
- *breviceps,* 244
- *sima,* 244

Kogiidae, 238–39, 244–47

Lagenodelphis hosei, 384
Lagenorhynchus
- *acutus,* 398
- *albirostris,* 395
- *australis,* 410
- *cruciger,* 414
- *obliquidens,* 402
- *obscurus,* 406

Leptonychotes weddellii, 166
Lipotes vexillifer, 310
Lipotidae, 300, 310–13
Lissodelphis
- *borealis,* 418
- *peronii,* 418

Lobodon carcinophaga, 170
Lutra felina, 46

Manatee,
- Amazonian, 14, 476, 477, **486**
- Antillean, 482
- Florida, 476, 482
- West African, 477, **490**
- West Indian, 13, 477, **482**

Manati, 486
Marsopa de anteojos, 468
Marsopa espinosa, 466
Megaptera novaeangliae, 208
Mesoplodon
- *bidens,* 280
- *bowdoini,* 286
- *carlhubbsi,* 288
- *densirostris,* 294
- *europaeus,* 278
- *ginkgodens,* 290
- *grayi,* 282
- *hectori,* 274
- *layardii,* 292
- *mirus,* 276
- *peruvianus,* 284
- *stejnegeri,* 296

Minke Whale, 184, **212**
- Antarctic, 184, **218,** 513
- Dwarf, 187, **216**

Mirounga
- *angustirostris,* 162
- *leonina,* 158

Monachus
- *monachus,* 150
- *schauinslandi,* 154
- *tropicalis,* 52

Monk Seal,
- Caribbean, **52,** 146
- Hawaiian, 31, **154**
- Mediterranean, **150**

Monodon monoceros, 322
Monodontidae, 316–25
Mustelidae, 36–37, 42–48

Narwhal, 180–83, 316–17, **322,** 516
Neobalaenidae, 184, 202–3
Neophoca cinerea, 98
Neophocaena phocaenoides, 456
Nerpa, 134
Nordkaper, 190

Ocean dolphins, 180–83, 326–451
Odobenidae, 49–51, 52–53, 54, 110–13
Odobenus rosmarus, 110
Ommatophoca rossii, 174
Orca, 436
Orcaella brevirostris, 448
Orcinus orca, 436
Otaria flavescens, 106
Otariidae, 49–52, 54–55, 58–109

Otter,
Alaskan Sea, 42
California Sea, 42
Marine, 36, 37, **46**
Sea, 13, 27, 36, 37, **42,** 512
Southern Sea, 42
Otters, 36–37, 42–48

Pagophilus groenlandicus, 142
Peixe-boi, 486
Peponocephala electra, 426
Phoca
largha, 122
vitulina, 118
Phocarctos hookeri, 102
Phocidae, 49–51, 53, 56–57, 114–179
Phocoena
dioptrica, 468
phocoena, 460
sinus, 464
spinipinnis, 466
Phocoenidae, 452–73
Phocoenoides dalli, 470
Physeteridae, 238–43
Physeter macrocephalus, 240
Pilot Whale,
Long-finned, 17, 329, **440**
Short-finned, **444**
Pinnipeds, 36, 49–179
Platanista gangetica, 302
Platanistidae, 299, 302–5
Polar Bear, 13, 16, 36–37, **38**
Pontoporia blainvillei, 314
Pontoporiidae, 301, 314–15
Porpoise,
Burmeister's, 453, 455, **466**
Common, 460
Dall's, 181, 453, 455, **470,** 512
Finless, 452, **456**
Harbor, 29, 453, 454, **460,** 470
Spectacled, 454, **468**
White-bellied, 388
Porpoises, 180–83, 452–73
Pseudorca crassidens, 433
Pusa
caspica, 131
hispida, 125
sibirica, 134

Razorback, 230
Right Whale,
Biscayan, 190
Black, 190
Greenland, 198
North Atlantic, 187, **190,** 503
North Pacific, 187, **190**
Pygmy, 184, 185, 187, **202**
Southern, 20, 187, **194**
Right Whale Dolphin,
Northern, **418**
Southern, 328, **418**
River Dolphin,
Amazon, 300, 301, **306**
Ganges, 182, 299, **302**
Indus, 299, **302**
Yangtze, 300, **310**
River dolphins, 180–83, 299–315
Rorqual,
Common, 230
Lesser, 212
Rudolphi's, 226

Sea Lion,
Australian, **98**
California, 33, 75, **90,** 96, 507, 513, 517
Galápagos, 19, **90**
Hooker's, 102
Japanese, **90**
New Zealand, **102,** 507
Northern, 25, **94**
Patagonian, 106
South American, 26, **106,** 500, 508
Southern, 106
Steller, 94
Sea lions, 36, 51–52, 54–55, 90–109
Seal,
Amsterdam Fur, 66
Amsterdam Island Fur, 66
Antarctic Fur, 53, **62,** 102, 178, 509, 517
Australian Fur, **58**
Baikal, 14, **134**
Bearded, **114**
Bladder-nosed, 146
Cape Fur, 58
Caribbean Monk, **52,** 146
Caspian, **131**
Common, 118
Crabeater, **170,** 511
Falkland Fur, 79
Galápagos Fur, **82**
Gray, 51, **138**
Greenland, 142
Guadalupe Fur, 19, 51, **72**
Harbor, 33, 50, **118**
Harp, 24, **142,** 502, 508
Hawaiian Monk, 31, **154**
Hooded, **146,** 510
Juan Fernández Fur, 22, **69**
Kerguelen Fur, 62
Kuril, 118
Largha, **122**
Leopard, 36, **177**
Mediterranean Monk, **150**
New Zealand Fur, **76,** 508
Northern Elephant, 22, **162,** 514
Northern Fur, 19, 51, **86,** 506
Ribbon, **128**
Ringed, **125**
Ross, **174**
Saddleback, 142
South African Fur, 18, **58,** 511, 512
South American Fur, 12, **79**
Southern Elephant, 53, **158,** 510
Southern Fur, 79
Spotted, 122
Subantarctic Fur, **66,** 102
Tasman Fur, 58
Tasmanian Fur, 58
Weddell, 12, 50, **166,** 507
Seals,
Eared, 49–52, 54–55, 58–109
Fur, 36, 51–52, 54–55, 58–89
True, 36, 49–51, 53, 56–57, 114–179
Sirenians, 474–92
Sotalia fluviatilis, 354
Sousa
chinensis, 350
plumbea, 350
teuszii, 350
Sperm Whale, 11, 180, 181, 238–39, **240,** 504
Dwarf, 238, 239, **244**
Pygmy, 238, 239, **244**

Sperm whales, 180–83, 238–47
Spotted Dolphin,
  Atlantic, 9, 15, 27, **370**
  Pantropical, **366,** 503
Steller's Sea Cow, 474, **476**
Stenella
  attenuata, 366
  clymene, 378
  coeruleoalba, 380
  frontalis, 370
  longirostris, 374
Steno bredanensis, 346
Sunameri, 456
Susu, 299, 302

Tasmacetus shepherdi, 264
Tonina, 354, 466, 468
Tonina overa, 332
Toninha, 314
Trichechidae, 474–77, 482–92
Trichechus
  inunguis, 486
  manatus, 482
  senegalensis, 490
True seals, 49–51, 53, 56–57, 114–79
Tucuxi, 300, 327, **354**
Tursiops
  aduncus, 362
  truncatus, 358

Ursidae, 36–37, 38–41
Ursus maritimus, 38

Vaca marina, 486
Vaquita, 453, 455, **464**

Walrus, 13, 36, 49–51, 52–53, **110,** 516
Whale,
  Andrews' Beaked, **286**
  Antarctic Minke, 184, **218,** 513
  Arnoux's Beaked, 249, 250, **258**
  Bahamonde's Beaked, 250
  Baird's Beaked, 250, **260**
  Biscayan Right, 190
  Black Right, 190
  Blainville's Beaked, 251, **294**
  Blue, 30, 182, 184, 186, 187, **234**
  Bowhead, 30, 182, 183, 184, 185, 187, **198**
  Bryde's, 184, 187, **222,** 515
  Cuvier's Beaked, 11, 249, **254**
  Dense-beaked, 294
  Dwarf Minke, 187, **216**
  Dwarf Sperm, 238, 239, **244**
  False Killer, 329, **433**
  Fin, 183, 184, **230,** 502, 509
  Gervais' Beaked, **278**
  Giant Bottlenose, 260
  Ginkgo-toothed Beaked, **290**
  Goose-beaked, 249, 254
  Gray, 21, 27, 32, 182, 184, 185, 187, **204,** 501, 515
  Gray's Beaked, 251, **282**
  Greenland Right, 198
  Hector's Beaked, **274**
  Hubbs' Beaked, **288**
  Humpback, 10, 20, 23, 28, 180, 184, 185, 186, **208,** 237, 405, 432, 501, 502, 506
  Killer, 26, 180, 181, 182, 329, **436,** 473, 509, 511, 513
  Layard's Beaked, 292
  Lesser Beaked, 284
  Little Piked, 212
  Long-finned Pilot, 17, 329, **440**
  Longman's Beaked, 250, **266**
  Melon-headed, 329, **426**
  Minke, 184, **212**
  North Atlantic Beaked, 280
  North Atlantic Right, 187, **190,** 503
  Northern Bottlenose, 250, 251, **268,** 509
  North Pacific Right, 187, **190**
  North Sea Beaked, 280
  Pygmy Beaked, 251, **284**
  Pygmy Blue, 234
  Pygmy Killer, 329, **430**
  Pygmy Right, 184, 185, 187, **202**
  Pygmy Sperm, 238, 239, **244**
  Saber-toothed, 296
  Scamperdown, 282
  Sei, 184, 185, **226**
  Shepherd's Beaked, 249, **264**
  Short-finned Pilot, **444**
  Southern Bottlenose, 250, 251, **272**
  Southern Right, 20, 187, **194**
  Sowerby's Beaked, 251, **280**
  Spade-toothed, **250**
  Sperm, 11, 180, 181, 238–39, **240,** 504
  Stejneger's Beaked, **296**
  Strap-toothed, 251, **292**
  True's Beaked, 251, **276**
  White, 11, 318
Whales,
  Baleen, 180–83, 184–237
  Beaked, 180–83, 248–98
  Sperm, 180–83, 238–47
White-sided Dolphin,
  Atlantic, **398**
  Pacific, 327, **402,** 505

Zalophus californianus, 90
Ziphiidae, 248–98
Ziphius cavirostris, 254

Contributors

PIETER A. FOLKENS (Illustrator) is widely acknowledged as the finest contemporary illustrator of marine mammals. He has contributed to many books, including *The Sierra Club Handbook of Seals and Sirenians* and *The Encyclopedia of Marine Mammals* (Academic Press). He has also designed cetaceans for motion pictures and television, including the films *Star Trek IV, Flipper,* and *Free Willy.* He is a founding board member of the Alaska Whale Foundation, a non-profit marine mammal research and conservation organization, and spends his summers in Alaska studying the feeding ecology of Killer and Humpback Whales. Folkens lives in Benicia, California.

RANDALL R. REEVES *(Sperm Whales, Beaked Whales, River Dolphins, Beluga and Narwhal, Ocean Dolphins, Porpoises, Glossary)* has been involved in marine mammal work for over 25 years, ranging from field studies in the Arctic, the North Atlantic, and the Indus and Amazon Rivers, to archival research on the history of whaling. He coauthored *The Sierra Club Handbook of Whales and Dolphins* and *The Sierra Club Handbook of Seals and Sirenians* and edited *Conservation and Management of Marine Mammals* (Smithsonian). Reeves holds a Ph.D. from McGill University, Canada. Since 1997, he has served as chairman of the Cetacean Specialist Group of the World Conservation Union (IUCN). Reeves lives in Hudson, Quebec.

PHILLIP J. CLAPHAM *(Introduction, Baleen Whales)* is a leading expert on large whales. He has conducted research on a variety of whale species around the world and has written or contributed to several books, including *Humpback Whales* (Voyageur Press), *Whales of the World* (Voyageur Press), and *The Complete Book of North American Mammals* (Smithsonian). Clapham earned a Ph.D. in Biology from the University of Aberdeen in Scotland. He is a Research Associate at the Smithsonian Institution in Washington, D.C., and lives and works in Woods Hole, Massachusetts, where he directs a research program on large whales.

BRENT S. STEWART *(Polar Bear and Otters, Pinnipeds)* has been studying and writing about marine mammals since the late 1970s. He has published many articles on marine mammals and contributed to several books, including *The Smithsonian Book of North American Mammals* and *The Sierra Club Handbook of Seals and Sirenians.* He earned a Ph.D. from the University of California, Los Angeles, and a J.D. from the University of California, Berkeley. Stewart is a Senior Research Biologist at Hubbs-SeaWorld Research Institute in San Diego, California, and Marine Science and Foreign Affairs Officer with the State Department in Washington, D.C.

JAMES A. POWELL *(Sirenians)* is recognized as an international expert on manatees. For the past thirty years he has conducted field research on sirenians around the world from Florida to the West Indies and Belize, and has spent ten years in remote areas of western Africa studying the West African Manatee. He earned a Ph.D. in Zoology from Cambridge University, England. He is co-chair of the Sirenia Specialist Group of the World Conservation Union (IUCN). Powell is the Director for Aquatic Programs for the Wildlife Trust in Sarasota, Florida.

Acknowledgements

The authors and the illustrator collectively thank the many scientists, artists, and photographers we have worked with over the years and whose books and papers provided a wealth of information for this book.

Pieter Folkens thanks the numerous colleagues who assisted with research for and reviewed the illustrations: Robert Pitman, of the Southwest Fisheries Science Center, and Todd Pusser, of the National Marine Fisheries Service Observer Program, for their generous sharing of photographic references and honest critique of the illustrations; the late Stephen Leatherwood, whose contribution to this project would have been all the more obvious had his time not been cut short; Ken Norris for his early influence on everyone in the field; Fred Sharpe, of the Alaska Whale Foundation, and Thomas Johnson, of Albatross Guides, for ideas, observations, and comments; and Todd Telander for assistance on some of the pinniped and sirenian illustrations. In addition to the above, many other colleagues in the Society for Marine Mammalogy offered references, insights, opportunities, and ideas.

The accuracy and usefulness of the text is due in large part to the many experts in the field of marine mammal science who generously contributed their time and expertise.

Randall Reeves thanks Thomas Jefferson, of the Southwest Fisheries Science Center, and Merel Dalebout for their valued help with the cetacean text. Both provided constructive criticism of draft species accounts and some very helpful unpublished information.

Phillip Clapham thanks co-author Randall Reeves for carefully reviewing the general introduction and the baleen whale accounts, and Peter Best of the South African Museum in Cape Town for his assistance.

Brent Stewart thanks Gerald Kooyman, of the University of California, San Diego, for lending his expertise to the pinniped section. For their contributions of time, space, experiences, friendship, and knowledge, Stewart thanks Bud and Lana Antonelis, Jason Baker, John Bengtson, Marthan Bester, Peter Boveng, Fred Bruemmer, Dave Brillinger, Mike Cameron, Bob DeLong, Doug DeMaster, Bill Everett, Bill Fraser, Kathy Frost, Tom Gelatt, Mads-Peter Heide-Jorgensen, Steve Karl, Gala Klevezal, Steve Leatherwood, Niles Lehman, Tom Loughlin, Lloyd Lowry, Sharon Melin, Misha Mina, Pierre Pistorius, Tim Ragen, Johann Sigurjonsson, Don Siniff, Ian Stirling, Alexey Yablokov, Pamela Yochem, and his friends and colleagues at SeaWorld of San Diego and at the Bureau of Oceans at the U.S. Department of State. Thanks also for their endless moral support over the years to Frank Awbrey, George Bartholomew, Dave Caron, Chuck Cooper, Ron Dow, John Edwards, Bill Evans, Bob Hofman, Carl and Laura Hubbs, Don Kent, Mike Novacek, Frank Powell, Milt and Bill Shedd, Chuck Taylor, John Twiss, Pamela Yochem, and parents Frank and Frances Stewart.

James Powell thanks Daryl Domning, of Howard University, for reviewing the sirenian accounts and ensuring their accuracy. He also acknowledges John Reynolds and Sheri Barton, of the Mote Marine Laboratory, and Helen Marsh, of James Cook University, for their assistance, as well as the help of several colleagues from the Florida Marine Research Institute: Tom Pitchford, Katherine Frisch, Andy Garrott, Sentinel Rommel, and Megan Pitchford.

In addition, James Mead, of the Smithsonian Institution, provided guidance early on in the production of the guide.

We thank Andrew Stewart and the staff of Chanticleer Press for producing a book of such excellence. The success of the book is due largely to the skills and expertise of editor-in-chief George Scott, project editor Michelle Bredeson, and contributing editors Pamela Nelson, Amy K. Hughes, Anne O'Connor, and Patricia Fogarty. They were assisted in the editing process by editorial interns Mee-So Caponi, Annie Lok, and Flynne Wiley. Art director Drew Stevens and designers Ann Antoshak, Brian Boyce, Amy Klessen, Mauricio Rodriguez, and Bernadette Vibar took hundreds of images and tens of thousands of words of text and created a book that is both visually beautiful and eminently usable. Managing photo editor Ruth Jeyaveeran, assistant photo editor Laura Russo, and photo assistant Jennifer Braff sifted through thousands of photographs in their search for the stunning images that contribute so much to the beauty and usefulness of this guide. Alicia Mills, Arthur Riscen, and Katherine Thomason saw the book through the complicated production and printing processes. Production intern Alyssa Okun and office manager Sui Ping Cheung offered much support. Gary Antonetti of Ortelius Design produced the many detailed maps that appear throughout the book.

In addition, we thank all the photographers who gathered and submitted the gorgeous pictures that make this book a delight to view.

National Audubon Society Field Guides

- African Wildlife
- Birds (Eastern Region)
- Birds (Western Region)
- Butterflies
- Fishes
- Fossils
- Insects and Spiders
- Mammals
- Mushrooms
- Night Sky
- Reptiles and Amphibians
- Rocks and Minerals
- Seashells
- Seashore Creatures
- Trees (Eastern Region)
- Trees (Western Region)
- Tropical Marine Fishes
- Weather
- Wildflowers (Eastern Region)
- Wildflowers (Western Region)

Other National Audubon Society Illustrated Guides

The Sibley Guide to Birds

The Sibley Guide to Bird Life and Behavior

Prepared and produced by Chanticleer Press, Inc.

FOUNDING PUBLISHER: Paul Steiner
PUBLISHER: Andrew Stewart

STAFF FOR THIS BOOK
ASSOCIATE PUBLISHER: Alicia Mills
EDITOR-IN-CHIEF: George Scott
PROJECT EDITOR: Michelle Bredeson
EDITORS: Pamela Nelson, Amy K. Hughes, Anne O'Connor, Patricia Fogarty
EDITORIAL INTERNS: Mee-So Caponi, Annie Lok, Flynne Wiley

ART DIRECTOR: Drew Stevens
DESIGNERS: Ann Antoshak, Brian Boyce, Amy Klessen, Mauricio Rodriguez, Bernadette Vibar
MANAGING PHOTO EDITOR: Ruth Jeyaveeran
ASSISTANT PHOTO EDITOR: Laura Russo
PHOTO ASSISTANT: Jennifer Braff
PRODUCTION ASSOCIATE: Arthur Riscen
COLOR CORRECTION: Katherine Thomason
PRODUCTION INTERN: Alyssa Okun
OFFICE MANAGER: Sui Ping Cheung

MAPS: Gary Antonetti, Ortelius Design

All editorial inquiries should be addressed to:
Chanticleer Press
665 Broadway
Suite 1001
New York, NY 10012

To purchase this book or other National Audubon Society illustrated nature books, please contact:
Alfred A. Knopf
299 Park Avenue
New York, NY 10171
(800) 733-3000
www.randomhouse.com